Physical Chemistry in Depth

Johannes Karl Fink

Physical Chemistry in Depth

 Springer

Johannes Karl Fink
Montanuniversität Leoben
Institut für Chemie der Kunststoffe
Franz-Josef Str. 18
A-8700 Leoben
Austria
johannes.fink@mu-leoben.at

ISBN 978-3-642-42440-3   ISBN 978-3-642-01014-9 (eBook)
DOI 10.1007/978-3-642-01014-9
Springer Heidelberg Dordrecht London New York

Cover design: KünkelLopka GmbH, Heidelberg

Printed on acid-free paper

Springer is part of Springer Science+Business Media (www.springer.com)

Preface

Understanding natural sciences consists of introducing basically unallowed simplifications, but this technique is highly useful. Physical chemistry does not make an exception. Sometimes, the simplifications go either too far or in a direction so that the goal of more easy understanding will be obscured.

Some 30 years ago, I was introduced into physical chemistry. Meanwhile I studied a lot of new textbooks that appeared to find out whether they provide an easier understanding of the matter presented, to some extent with success. However, some of the issues still remained obscure to me. For this reason, I have revisited some of the topics presented in the textbooks and I try to explain with my own words, what caused and still causes problems for me to understand satisfactorily believing that the text presented here will be helpful also for at least some of the readers.

The material dealt with here is not treated strictly, but presented in a way that the path to a deeper understanding should be opened. There is always the danger that a simplification of the matter for the sake of didactic reasons is appraised as a wrong depiction. Actually, the borders are blurred.

Mostly, textbooks of physical chemistry exhibit a certain unique disposition, they start with thermodynamics, turn to kinetics, etc. This is also true for the material presented here. However, the text presented here is rather a supplement, not a standalone textbook. For this reason, some topics to which I was never pervasively engaged are not dealt with, or only marginally. On the other hand, I have included some topics that should be present in a textbook of physical chemistry.

I want to express my gratitude to the publisher and to the editorial staff who have been most supportive to this project, in particular to Dr. Marion Hertel, Birgit Münch, and Maria Arokiyaswamy.

Leoben
June 2009

Johannes Karl Fink

Contents

Chapter 1
Mathematics of Thermodynamics

There is a story about *Sommerfeld* and thermodynamics that is cited in several sources [1, p.43], including the Internet.

> When Sommerfeld was asked why he did not write a book on thermodynamics, because he had written books on nearly every topic of physics he replied:

> Thermodynamics is a funny subject. The first time you go through it, you don't understand it at all. The second time you go through it, you think you understand it, except for one or two points. The third time you go through it, you know you don't understand it, but by that time you are so used to that subject, it doesn't bother you anymore.

In fact, *Sommerfeld*[1] wrote a textbook on thermodynamics and statistics [2]. Sommerfeld preserved this volume as the last of totally five volumes on theoretical physics. He could not complete this volume. Statements like this of Sommerfeld are sometimes intended to console the newcomer, in case that he should not understand the subject fully at his first contact [1, p. 43]. On the other hand, such a statement may debauch to dismiss going into the depth of the topic, because it may not be important.

We will not do pure mathematics here, but we rather present sloppy mathematics with illustrations. Probably, thermodynamics is poorly or even not understood, because of incompetency in the mathematical background of the student. Try to understand this section, and try to understand more if you like.

Many of the relations in thermodynamics are based on the formal laws of the theory of functions of more than one variable. Therefore, a solid knowledge of the rules of partial differentiation and something more is fundamental to understand thermodynamics. Frequently, thermodynamics is not interested or able to deal with the respective function itself, but rather with the first total derivative. The mathematics of thermodynamics is described in some monographs of thermodynamics biased to mathematics [3–5].

---

[1] Arnold Johannes Wilhelm Sommerfeld, born Dec. 5, 1868, in Königsberg/East Prussia, died Apr. 26, 1951, in Munich by a traffic accident.

J.K. Fink, *Physical Chemistry in Depth*,
DOI 10.1007/978-3-642-01014-9_1, © Springer-Verlag Berlin Heidelberg 2009

1.1 Some Properties of Special Functions

A frequently emerging question is whether the Arrhenius factor should be written as

$$\exp\left(-\frac{E_A}{RT}\right) \quad \text{or} \quad \exp\left(-\frac{E_A}{kT}\right).$$

Here R is the gas constant and k is the Boltzmann constant. The answer is that it depends on the units in which the energy of activation is given. If the energy of activation is given in $\text{J mol}^{-1}\text{K}^{-1}$ then we use R. If the energy of activation is given in $\text{J particle}^{-1}\text{K}^{-1}$ then we use k, because $k = R/N_L$, where N_L is the Avogadro number. The Avogadro number has a physical dimension of particle mol^{-1}. Essentially the argument of the exponential function must be a dimensionless number.

On the other hand, we find sometimes

$$RT \ln p \quad \text{and sometimes} \quad RT \ln\left(\frac{p}{p_0}\right).$$

Under certain circumstances, both variants mean the same. If the standard pressure p_0 is one unit, say 1 at, then the *numerical value* for the reduced pressure is

$$N(p) = N(p)/N(p_0). \tag{1.1}$$

Equation (1.1) is valid if $N(p_0) = 1$. If we choose as standard pressure $760\,\text{mmHg}$, then Eq. (1.1) is no longer valid.

The procedure is not correct for the pressure itself and in general for a physical quantity, because a physical quantity consists of a number and the attached unit of the particular physical quantity. However, if we look at the integral definition of the logarithm

$$\ln x = \int_{t=1}^{t=x} \frac{dt}{t},$$

then we observe that the unit cancels automatically due to the definition of the logarithm. Thus, the logarithm purifies a physical quantity from its unit.

1.2 Functional Dependence

Throughout this text, we tacitly assume that the arguments of the functions and the functions themselves are real valued.

1.2.1 Concept of a Function

If we have the independent variables x, y, then $U(x, y)$ is a direct function of the variables x, y. If we have $z = f(x)$, then $U(z, y)$ is an indirect function with respect to x. In general, it is not clear whether $U(z, y)$ is really a function of x. For example, consider the case $z = x^0$. Further, consider the case $U(x, f''(y))$, with $f(y) = y^2$. To find out the actual set of independent variables, we would have to resolve the prescription of calculation.

Example 1.1. We consider the function

$$U(x, y) = \frac{x - y^2}{x + y}.$$

We build now the functions $f(x, y) = (x + y)^2$ and $g(x, y) = x + y$. Then

$$U(f(x, y), g(x, y)) = 0.$$

□

When we make the statement

$$U(x, y) = x + y, \tag{1.2}$$

then we define the function. We establish a prescription how to calculate the function value using variables. For instance, we can calculate $U(2, 3)$ to be 5. Very often in mathematics, and consequently in the mathematics of thermodynamics, we work in fact with functions that have not been defined in detail before. For example, it is a common task to get the definition of a function by solving a differential equation.

With the statement

$$U(x, y) = U_0, \tag{1.3}$$

where U_0 is a constant, we fix the result of the function. Equation (1.3) is a constraint equation. Together with Eq. (1.2) we have $U_0 = x + y$. In the function definition, Eq. (1.2), we are free to choose any pair of x and y to calculate $U(x, y)$, whereas the constraint equation allows only to choose either x or y freely. If one of the variables is fixed, the other variable is also fixed.

We can form the total differential of first order of $U(x, y)$, which appears as

$$dU(x, y) = \frac{\partial U(x, y)}{\partial x} dx + \frac{\partial U(x, y)}{\partial y} dy. \tag{1.4}$$

If we form from Eq. (1.3) the total differential of first order we get $dU(x, y) = 0$. Together with Eq. (1.4) again a constraint emerges, as x and y can no longer be varied independently. However, we lose information using the differential form

$dU(x, y) = 0$, since the differential form fixes the total energy to some arbitrary integration constant, while Eq. (1.3) fixes the function $U(x, y)$ exactly to U_0.

1.2.2 Separation of Variables

Let $f(x)$ be a function of x and $g(y)$ a function of y. If the relation

$$f(x) = g(y) \tag{1.5}$$

holds, then there must be some relation $f(x) = C = g(y)$, where C is a constant. If C is not a constant, then there must be some functional dependence of x and y, i.e., $y = y(x)$.

1.2.3 Power Laws

Power laws are somewhat sneaky. A typical power law is

$$y = x^\alpha. \tag{1.6}$$

When α is an integer, we may consider the power law as a special form of a polynomial. Forming the differentials of any order will not cause problems, since we will end eventually with zero. However, if α is not an integer, the situation is different. For instance, $y = x^{4/3}$ will result in $y' = 4/3x^{1/3}$, whereas $y = x^{2/3}$ will result in $y' = 2/3x^{-1/3}$. These derivatives show a completely different behavior, if the limit to zero, $x \to 0$, is inspected.

In general, if $\alpha > 0$ the first derivative of the power law approaches zero, but if $\alpha < 0$, the first derivative of the power law diverges to infinity. Moreover, it is not possible to expand a power law in a Taylor series with an arbitrary number of terms. An example for a power law is given in Sect. 3.4.3.1.

1.3 Change of the Arguments

Example 1.2. We exemplify a change of the argument for $f(x) = x^2$. Then

$$\frac{d f(x)}{dx} = 2x \quad \text{or} \quad d f(x) = 2x \, dx.$$

We introduce $y = 1/x$. Thus, $dy/dx = -1/x^2 = -y^2$ and $dy = -y^2 dx$. Substituting for x we have

$$f(1/y) = (1/y)^2$$

and

$$\frac{\mathrm{d}f(1/y)}{\mathrm{d}y} = -\frac{2}{y^3} = -\frac{2x}{y^2} = -\frac{1}{y^2}\frac{\mathrm{d}f(x)}{\mathrm{d}x} = \frac{\mathrm{d}x}{\mathrm{d}y}\frac{\mathrm{d}f(x)}{\mathrm{d}x} = \frac{\mathrm{d}f(x)}{\mathrm{d}y}. \tag{1.7}$$

Thus, a change in the argument will not change the differential of the function, i.e., $\mathrm{d}f(1/y) = \mathrm{d}f(x)$. In fact, we are exemplifying the chain rule. The chain rule is explained in Sect. 1.5. □

The notation $\mathrm{d}f(1/y)$ suggests to replace every x in the actual definition of the function by $1/y$. On the other hand, there is no need to write in the argument list $1/y$, but rather y. However, then we are handling with a function that is defined in a different way. We have to express this more unambiguously as we give to the function a different name. Therefore, we could write more correctly

$$f(x) = F(y)$$
$$f(x) = x^2$$
$$y = x^{-1}$$
$$F(y) = y^{-2} \tag{1.8}$$
$$\frac{\mathrm{d}f(x)}{\mathrm{d}y} = \frac{\mathrm{d}F(y)}{\mathrm{d}y}$$
$$\frac{\mathrm{d}x}{\mathrm{d}y}\frac{\mathrm{d}f(x)}{\mathrm{d}x} = \frac{\mathrm{d}F(y)}{\mathrm{d}y}$$

Of course, this is also true in general. Let $y = g(x)$ and the inverse function $x = G(y)$. Then Eq. (1.8) can be rewritten as

$$f(x) = F(y)$$
$$y = g(x)$$
$$F(y) = f(G(y))$$
$$\frac{\mathrm{d}f(G(y))}{\mathrm{d}y} = \frac{\mathrm{d}F(y)}{\mathrm{d}y}$$
$$\frac{\mathrm{d}G(y)}{\mathrm{d}y}\frac{\mathrm{d}f(x)}{\mathrm{d}x} = \frac{\mathrm{d}F(y)}{\mathrm{d}y}$$

Be careful with a function of two or more arguments, i.e.,

$$\mathrm{d}U(S, V, n) = T\mathrm{d}S - p\mathrm{d}V + \mu\mathrm{d}n. \tag{1.9}$$

Moving from S as independent variable into T results in

$$\mathrm{d}U(T, V, n) = C_v\mathrm{d}T + \frac{\partial U(T, V, n)}{\partial V}\mathrm{d}V + \frac{\partial U(T, V, n)}{\partial n}\mathrm{d}n.$$

We emphasize that the negative pressure $-p = \frac{\partial U(S,V,n)}{\partial V}$ arises from the differential of the energy with respect to volume with the entropy and the volume as arguments: $U(S, V, n)$. This is not equal to $\frac{\partial U(T,V,n)}{\partial V}$, where the energy is considered as a function of temperature, volume, and mol number, $U(T, V, n)$.

$$\frac{\partial U(S, V, n)}{\partial V} \neq \frac{\partial U(T, V, n)}{\partial V}.$$

On the other hand, for an isochoric change $dV = 0$. This means that $T dS = C_v dT$. It does not matter whether the heat capacity is a function of entropy and volume $C_v = C_v(S, V, n)$ or a function of temperature and volume $C_v = C_v(T, V, n)$.

By the use of the chain rule, we can get

$$\frac{\partial C_v(T(S, V), V)}{\partial S} = \frac{\partial C_v}{\partial T}\frac{\partial T}{\partial S} = \frac{\partial \ln C_v}{\partial \ln T}.$$

In contrast, since the temperature is a function of both S and V, $T(S, V)$, the derivative with respect to the volume is more complicated, namely

$$\frac{\partial C_v(T(S, V), V)}{\partial V} = \frac{\partial C_v}{\partial T}\frac{\partial T}{\partial V} + \frac{\partial C_v}{\partial V}.$$

1.4 Total Differential

A differential of the form

$$df = P(x, y)dx + Q(x, y)dy$$

is called a total differential of an exact differential, if the integral $\int df$ is not dependent on the integration path. The necessary condition for exactness is that

$$P(x, y) = \frac{\partial f(x, y)}{\partial x} = f_x,$$

$$Q(x, y) = \frac{\partial f(x, y)}{\partial y} = f_y.$$

Observe the shorthand notation f_x and f_y. We can also build further derivatives of higher order, $\partial P/\partial y = f_{xy}$ and $\partial Q/\partial x = f_{yx}$. Since $f_{xy} = f_{yx}$,

$$\frac{\partial P}{\partial y} = \frac{\partial Q}{\partial x}.$$

This is the law of *Schwarz*, which will be discussed in detail in a separate section. Thermodynamics makes extensive use of the law of *Schwarz*.

In Eq. (1.9), $dU(S, V, n)$ is the total differential of first order. The temperature, the pressure, and the chemical potential are to be treated as the partial derivatives of the energy

$$T(S, V, n) = \frac{\partial U(S, V, n)}{\partial S},$$

$$-p(S, V, n) = \frac{\partial U(S, V, n)}{\partial V},$$

$$\mu(S, V, n) = \frac{\partial U(S, V, n)}{\partial n}.$$

The arguments of the energy must be S, V, n, because of the form of the right-hand side of Eq. (1.9). There are also total differentials of higher order. We denominate a second-order total differential by d^2. For example, at constant mol number, we can suppress n as argument and the second-order total differential of the energy is

$$d^2U(S, V) = \frac{\partial^2 U(S, V)}{\partial S^2} dS^2 + 2\frac{\partial^2 U(S, V)}{\partial SV} dV dS + \frac{\partial^2 U(S, V)}{\partial V^2} dV^2. \quad (1.10)$$

1.5 Chain Rule

We define a function $U(S, V, n)$. Then, using the chain rule for a function with multiple variables, the differential with respect to a general variable X is

$$\frac{\partial U}{\partial X} = \frac{\partial U}{\partial S}\frac{\partial S}{\partial X} + \frac{\partial U}{\partial V}\frac{\partial V}{\partial X} + \frac{\partial U}{\partial n}\frac{\partial n}{\partial X}.$$

In the common meaning of U as the energy, we get

$$\frac{\partial U}{\partial X} = T\frac{\partial S}{\partial X} - p\frac{\partial V}{\partial X} + \mu\frac{\partial n}{\partial X}.$$

We examine now simple examples of functional dependencies of the set of variables (S, V, n). If each of the variables in the set is essentially independent of X, as is suggested by the notation, then

$$\frac{\partial U}{\partial X} = 0.$$

We discuss now a special case, $X = S$. Then

$$\frac{\partial S}{\partial X} = 1; \quad \frac{\partial U}{\partial X} = T.$$

Analogous relations hold for $X = V$ and $X = n$. We proceed now to treat $S = S(T, V, n)$ and form the differential of the energy with respect to the volume

$$\frac{\partial U(S(T, V, n), V, n)}{\partial V} = T\frac{\partial S(T, V, n)}{\partial V} - p. \tag{1.11}$$

Resolving the calculus of the function S in the variable list, we have

$$U(S(T, V, n), V, n) = U(T, V, n). \tag{1.12}$$

Actually the function on the right-hand side of Eq. (1.12) should be renamed, because the set of variables has a different meaning. For example, the area of a circle is $A_r(r) = r^2\pi$, when r is the radius of the circle. Likewise, we can express the area of the circle as a function of its diameter d, when $A_d(d) = d^2\pi/4$. Thus $A_r(r) = A_d(d)$ is correct. However, $A_r(r) \neq A_r(d)$ if $d = 2r$. Thus, the notation of Eq. (1.12) is misleading. A computer program would give perfectly wrong results if the functions are literarily used in this way.

We do not stop at this stage, but recall that the total differential of the Helmholtz energy is

$$dF(T, V, n) = -SdT - pdV + \mu dn.$$

Accordingly, Maxwell's relations yield

$$\frac{\partial S(T, V, n)}{\partial V} = \frac{\partial p(T, V, n)}{\partial T}. \tag{1.13}$$

Inserting Eq. (1.13) into Eq. (1.11) and using Eq. (1.12) results in

$$\frac{\partial U(T, V, n)}{\partial V} = T\frac{\partial p(T, V, n)}{\partial T} - p. \tag{1.14}$$

Equation (1.14) is famous, because the energy as a function of temperature and pressure is zero for an ideal gas. It is often derived in a lengthier way.

We proceed to a more general case and treat now the differential

$$\frac{\partial U(S(T, p, n), V(T, p, n), n)}{\partial p} = T\frac{\partial S(T, p, n)}{\partial p} - p\frac{\partial V(T, p, n)}{\partial p}.$$

Using Maxwell's relation in the free enthalpy total differential,

$$\frac{\partial U(T, p, n)}{\partial p} = -T\frac{\partial V(T, p, n)}{\partial T} - p\frac{\partial V(T, p, n)}{\partial p}. \tag{1.15}$$

As Eq. (1.14), Eq. (1.15) is zero for an ideal gas.

Example 1.3. In thermodynamics, we will frequently encounter the expression

$$\frac{\partial \ln x_1}{\partial n_1} dn_1 \tag{1.16}$$

with

$$x_1 = \frac{n_1}{n_1 + n_2}.$$

Then we have the relations

$$\frac{\partial \ln x_1}{\partial n_1} = \frac{n_2}{n_1(n_1 + n_2)} = \frac{(x_1 - 1)^2}{n_2 x_1},$$

$$\frac{\partial n_1}{\partial x_1} = \frac{n_2}{(x_1 - 1)^2}.$$

Therefore, Eq. (1.16) can be rewritten as

$$\frac{\partial \ln x_1}{\partial n_1} dn_1 = \frac{dx_1}{x_1} = d \ln x_1. \qquad (1.17)$$

In addition, for $x_1 = 1 - x_2$ and for small x_2

$$d \ln x_1 = d \ln(1 - x_2) \approx -dx_2.$$

□

In certain cases, extremes with constraints can be obtained by application of the chain rule.

Example 1.4. We assume that

$$U'(S', V', n') \quad \text{and} \quad U''(S'', V'', n'')$$

are some functions with

$$\frac{\partial U'(S', V', n')}{\partial S'} = T'; \qquad \frac{\partial U''(S'', V'', n'')}{\partial S''} = T''$$

$$\frac{\partial U'(S', V', n')}{\partial V'} = -p'; \qquad \frac{\partial U''(S'', V'', n'')}{\partial V''} = -p''.$$

$$\frac{\partial U'(S', V', n')}{\partial n'} = \mu'; \qquad \frac{\partial U''(S'', V'', n'')}{\partial n''} = \mu''$$

Further,

$$S' + S'' = S_{\text{tot}},$$
$$V' + V'' = V_{\text{tot}},$$
$$n' + n'' = n_{\text{tot}},$$

where the quantities with index tot are constants. Then

$$\frac{\partial U'(S', V', n')}{\partial S'} + \frac{\partial U''(S'', V'', n'')}{\partial S'} = T' + \frac{\partial U''(S'', V'', n'')}{\partial S''}\frac{dS''}{dS'} = T' - T''.$$

Similarly,

$$\frac{\partial U'(S', V', n')}{\partial V'} + \frac{\partial U''(S'', V'', n'')}{\partial V'} = -p' + \frac{\partial U''(S'', V'', n'')}{\partial V''}\frac{dV''}{dV'}$$

$$= -p' + p''$$

$$\frac{\partial U'(S', V', n')}{\partial n'} + \frac{\partial U''(S'', V'', n'')}{\partial n'} = -\mu' + \frac{\partial U''(S'', V'', n'')}{\partial n''}\frac{dn''}{dn'}$$

$$= \mu' - \mu''$$

So, by inserting the constraints in the energy equation, the conditions of equilibrium can be obtained via the chain rule. However, the method of undetermined coefficients is more versatile. □

Example 1.5. The energy of an ideal gas is

$$U = n(C_v T + C_2).$$

We treat C_v and C_2 as constants. To get the chemical potential μ it is suggestive to build

$$\frac{\partial U}{\partial n} = C_v T + C_2.$$

This procedure is wrong, because in this special case the energy is a function of the temperature, which is itself, among others, a function of the mol number: $T = T(S, V, n)$. For this reason, the correct expression for the chemical potential is

$$\frac{\partial U}{\partial n} = C_v T(S, V, n) + C_2 + nC_v\frac{\partial T(S, V, n)}{\partial n}. \qquad (1.18)$$

This example shows the importance of clearly indicating which are the independent variables of the function under consideration. □

1.5.1 Euler's Chain Relation

Let the volume V be a function of pressure p and temperature T, $V = V(T, p)$. The total differential of the volume is

$$dV(T, p) = \frac{\partial V(T, p)}{\partial T}dT + \frac{\partial V(T, p)}{\partial p}dp.$$

At constant volume, $dV = 0$, and the differential, e.g., dp, can be regarded now as a function of temperature and volume, $dp(T, V)$. We get immediately

$$0 = \frac{\partial V(T, p)}{\partial T} + \frac{\partial V(T, p)}{\partial p} \frac{\partial p(T, V)}{\partial T}.$$

Further rearrangement results in

$$\frac{\partial V(T, p)}{\partial p} \frac{\partial p(T, V)}{\partial T} \frac{\partial T(V, p)}{\partial V} = -1. \tag{1.19}$$

This is an example for Euler's chain relation. Euler's chain relationship can be derived more formally with Jacobian determinants, as shown in Example 1.23. Rewriting Eq. (1.19) without arguments results in

$$\frac{\partial V}{\partial p} \frac{\partial p}{\partial T} \frac{\partial T}{\partial V} = -1.$$

It is suggestive to treat the differential as an ordinary fraction, i.e.,

$$\frac{\Delta V}{\Delta p} \frac{\Delta p}{\Delta T} \frac{\Delta T}{\Delta V},$$

where the terms would cancel to $+1$. However, using the ideal gas equation, it can be easily verified that Eq. (1.19) is correct.

Example 1.6. In an electron tube, a triode, the anode current I_a is a function of the anode voltage U_a and the grid voltage U_g. The change of the anode current is

$$dI_A(U_a, U_g) = \frac{\partial I_a(U_a, U_g)}{\partial U_a} dU_a + \frac{\partial I_a(U_a, U_g)}{\partial U_g} dU_g = \frac{1}{R} dU_a + S dU_g.$$

R is the tube resistance and S is the mutual conductance. At constant I_g, the inverse amplification factor D is

$$D = -\frac{\partial U_g(U_a, I_g)}{\partial U_a}.$$

Notice the minus sign in the definition. Thus, by putting $dI_g = 0$, $DSR = 1$. This is the *Barkhausen*[2] equation, or tube equation.   □

---

[2] Heinrich Georg Barkhausen, born Dec. 2, 1881, in Bremen, Germany, died Feb. 20, 1956, in Dresden, Germany.

1.6 Substantial Derivative

In transport phenomena, there appears important types of derivations called the substantial derivative, material derivative, hydrodynamic derivative, or derivative following the motion. Consider a function $c(x, y, z, t)$. The symbol c suggests that we are thinking of a concentration, but the function can be more general. Then the derivative with respect to time is simply

$$\dot{c}(x, y, z, t) = \frac{\partial c(x, y, z, t)}{\partial t}.$$

We are watching the concentration from a fixed position (x, y, z). However, if the coordinates are moving with some velocity

$$x = x_0 + v_x t; \quad y = y_0 + v_y t; \quad z = z_0 + v_z t,$$

then the situation changes. We are watching the concentration now from a moving coordinate system. The derivative with respect to time is now

$$\frac{Dc}{Dt} = \frac{\partial c}{\partial t} + v_x \frac{\partial c}{\partial x} + v_y \frac{\partial c}{\partial y} + v_z \frac{\partial c}{\partial z} = \frac{\partial c}{\partial t} + \mathbf{v} \cdot \nabla c. \tag{1.20}$$

Equation 1.20 is the substantial derivative of the function $c(x, y, z, t)$. It can be naturally obtained by an application of the chain rule. Usually there is a flow in the system under consideration, and the velocity is just the flow of matter, energy, or the like with respect to the fixed system. However, the calculation is valid for any movement of the coordinate system.

1.7 Conversion of Differentials

We summarize some simple formulas for the calculus in two variables, some of them presented already with special examples. Let $f(x, y)$ be some function. Assume that we could resolve x as a function of f and y, i.e., $x = x(f, y)$ and the same should be true for y, i.e., $y = y(f, x)$. According to the chain rule,

$$\frac{\partial x}{\partial f} = \frac{\partial x}{\partial y} \frac{\partial y}{\partial f} \tag{1.21}$$

and rearrangement of Eq. (1.21) gives

$$\frac{\partial x}{\partial y} = \frac{\frac{\partial x}{\partial f}}{\frac{\partial y}{\partial f}}.$$

Again, let $f(x, y)$ be a function:

$$df(x, y) = df = \frac{\partial f}{\partial x}dx + \frac{\partial f}{\partial y}dy.$$

We resolve to dx and get

$$dx(f, y) = \frac{df}{\frac{\partial f}{\partial x}} - \frac{\frac{\partial f}{\partial y}}{\frac{\partial f}{\partial x}}dy. \tag{1.22}$$

Thus, for constant f, its differential $df = 0$ and Eq. (1.22) becomes

$$\frac{\partial x(f, y)}{\partial y} = -\frac{\frac{\partial f}{\partial y}}{\frac{\partial f}{\partial x}}.$$

The method presented here can be found in some textbooks. However, arranging and rearranging differentials into quotients and back has been considered as a possible pitfall and more strictly as an illegal procedure in modern mathematics. See, for example, the discussion after Eq. (1.19).

A powerful method for the conversion of differentials uses Jacobian determinants, introduced in Sect. 1.14.6. We anticipate the conversion starting from Eq. (1.22) in terms of Jacobian determinants:

$$\frac{\partial(x, f)}{\partial(y, f)} = \frac{\partial(x, f)/\partial(x, y)}{\partial(y, f)/\partial(x, y)} = -\frac{\frac{\partial(x,f)}{\partial(x,y)}}{\frac{\partial(y,f)}{\partial(y,x)}}.$$

Further, an indispensable tool for the conversion of differentials is the law of *Schwarz*, dealt with in Sect. 1.9.

1.8 Legendre Transformation

Sometimes, the Legendre transformation is also addressed as Legendre transform. However, we want to emphasize here that the term *transform* is a shorthand notation for integral transforms [6].

We take now $dU(S, V) = TdS - pdV$ and subtract $d(TS)$.

$$dU(S, V) - d(TS) = TdS - pdV - SdT - TdS$$
$$= -SdT - pdV.$$

The left-hand side now obviously should be a function of T and V, let it be $dF(T, V) = -SdT - pdV$. The physical relevance of the *Legendre*[3] transformation

---

[3] Adrien Marie Legendre, born Sep. 18, 1752, in Paris, France, died Jan. 10, 1833, in Paris.

will be shown later. It changes the independent variable into the dependent variable.

1.8.1 One-Dimensional Case

Actually, we mean that the derivative of function goes in place of the argument and the argument goes in place of the function. If we have a function $f(x)$, then in the graph the abscissa will be x and the ordinate will be $f(x)$. Thus, we plot the pair $(x, f(x))$. Here, x is the independent variable. In the Legendre transformation, we will use the pair $(f'(x), f(x) - f'(x) * x)$ for plotting. Thus, f' is the independent variable and x will be eventually resolved as a function of f'. Note that in pure mathematics, the Legendre transformation is defined with opposite sign. In addition, we will eliminate the formerly independent variable x in favor of f' as independent variable.

Example 1.7. Consider the function $y(x) = x^2$. Then $y'(x) = 2x$ and the Legendre transformation is $L(y') = y - xy' = x^2 - 2x^2 = -x^2$. In terms of y', $L(y') = -y'^2/4$. On the other hand, the Legendre transformation of the function $g(p) = -p^2/4$ is $p^2/4$. Here $g'(p) = -p/2$, which turns out to be $L(g') = g'^2$. Actually, the Legendre transformation is involutory in nature. □

1.8.2 Bijective Mapping

In order to have a single-valued transformation, in the range of interest, the slope $f'(x)$ should always be different in the range of interest. In the case that $f'(x_1) = f'(x_2)$, there would be two points in the abscissa of the transformed function with possible different ordinate values. Thus, $f'(x)$ must be a strictly monotone function. In general, a function $f(x)$ is monotonically increasing if $f(x + \Delta x) \geq f(x)$ for a positive Δx. Still more restrictive, a function is strictly monotonically increasing, if $f(x + \Delta x) > f(x)$.

In Fig. 1.1, a graphical example of a Legendre transformation is shown. At the right top graphics a function is plotted. Below the right top, the derivative is plotted. Left from the derivative, the Legendre transformation is plotted, however, rotated. Thus the user can move from a top right point of the original function straight down to the derivative and from there horizontal left to get the Legendre transformation. In regions where the derivative is not single valued, also the Legendre transformation is so. By the way, in Fig. 1.1 we have used the function

$$f(x) = (x - 3) + 0.2(x - 3)^2 + 0.4 \exp(-2(x - 3)^2)$$

to illustrate the procedure.

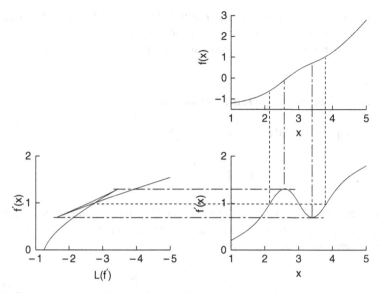

Fig. 1.1 Graphical example of a Legendre transformation

1.8.3 Taylor Expansion

We present still another view of the Legendre transformation. A function $f(x)$ can be expanded around a point x_0 as

$$f(x_0 + \Delta x) = f(x_0) + f'(x_0)\Delta x + \ldots.$$

Using $\Delta x = x - x_0$, $y = f(x)$, and neglecting higher order terms we arrive at

$$y = f(x) = f(x_0) - f'(x_0)x_0 + f'(x_0)x. \tag{1.23}$$

Equation (1.23) is just the equation of a tangent through the point x_0 of the function $y = kx + d$. Here $k = f'(x_0)$ and $d = f(x_0) - f'(x_0)x_0$. Observe that the intercept of this line at $x = 0$ is the Legendre transformation at $x = x_0$. To complete the procedure of the transformation, we have to eliminate x_0 in favor of $f'(x_0)$ as independent variable.

We can think that both the intercept d and the slope k are functions of the abscissa, i.e., $d = d(x_0)$ and $k = k(x_0)$. If the graph of a function $f(x)$ exhibits a common tangent, say at the points x_0 and x_1, then we have

$$k(x_0) = k(x_1) \quad \text{and} \quad d(x_0) = d(x_1)$$

for these points. Thus, a common tangent can be localized by forming the Legendre transformation and eliminating the originally independent variable in favor of the slope. In this representation the graphs of the transformations will cut at the points of common tangents. This situation is illustrated in Fig. 1.1 by the dashed lines. In contrast, the dash-dotted lines represent the inflexion points.

In general, the procedure is also valid for two different functions, $f(x)$ and $g(x)$.

1.8.4 Two-Dimensional Transformation

Take as example $dU(S, V) = T dS - p dV$. We resolve to

$$dS(U, V) = \frac{1}{T} dU + \frac{p}{T} dV.$$

By this rearrangement, S is the function and U is one argument. Note that the entropy $S(U, V)$ is a function of the energy and the volume. Next, we perform the Legendre transformation with respect to U:

$$dL\left(\frac{1}{T}, V\right) = -U d\frac{1}{T} + \frac{p}{T} dV. \qquad (1.24)$$

We annotate that in Eq. (1.24) for all the functions involved, the argument list containing the independent variables is $\left(\frac{1}{T}, V\right)$, i.e., $U\left(\frac{1}{T}, V\right)$ and $p\left(\frac{1}{T}, V\right)$.

The use of Legendre transformations in chemical thermodynamics has been extensively exemplified [7, 8]. Besides, in thermodynamics, the Legendre transformation is used in Hamilton – Lagrange mechanics.

Example 1.8. The Legendre transformation can be used to get functions with changed arguments. Observe that

$$\frac{\partial F(T, V, n)}{\partial n} = \mu(T, V, n) \quad \text{and} \quad \frac{\partial U(S, V, n)}{\partial n} = \mu(S, V, n).$$

Watch the difference in arguments in both expressions for the chemical potential μ. In the chemical potential obtained from the energy U, we can resolve the entropy from $T = \partial U(S, V, n)/\partial n$ and substitute $S = f(T, V, n)$ to get the chemical potential as a function of temperature, volume, and mol number, $\mu = \mu(T, V, n)$. This expression for the chemical potential is the same as obtained from the free energy $F(T, V, n)$.

In contrast, it is not allowed to substitute $S = f(T, V, n)$ directly into the equation for the energy to get $U(T, V, n)$, because

$$\partial U(T, V, n)/\partial n \neq \mu.$$

□

1.9 Law of Schwarz

With all the first total differentials in thermodynamics the law of *Schwarz* holds. We write it here as $df(x, y) = f_x dx + f_y dy$. This is of fundamental importance in thermodynamics.

Note that $f_{x,y} = f_{y,x}$, which is actually the law of *Schwarz*. The order of differentiation is immaterial. In this way, a lot of useful and less useful thermodynamic relations can be derived.

Example 1.9. We try here the law of *Schwarz* with Eq. (1.24). We substitute now $1/T = x$. The partial derivatives are

$$\frac{\partial U(x, V)}{\partial V} = -\frac{\partial (xp(x, V))}{\partial x} = -x\frac{\partial p(x, V)}{\partial x} - p(x, V). \tag{1.25}$$

Now we notice that $dx = -x^2 dT$ and substitution into Eq. (1.25) yields

$$\frac{\partial U(T, V)}{\partial V} = -\frac{\partial (xp(x, V))}{\partial x} = T\frac{\partial p(T, V)}{\partial T} - p(T, V). \tag{1.26}$$

Here we made use of the change of the arguments as explained in Eq. (1.7). Equation (1.26) is an important thermodynamic relation, which we have derived in a rather unusual way. □

On the other hand, we could have changed $1/T$ into T directly in Eq. (1.24).

$$dL(T, V) = U(T, V)T^{-2}dT + p(T, V)T^{-1}dV. \tag{1.27}$$

It is not allowed to multiply with a common factor, like T^2, as is suggested from the special form of Eq. (1.27). More precisely, the change of the internal structure of the function by changing the variables must be taken into account.

Forming the derivatives of the coefficients of Eq. (1.27), we obtain

$$\frac{1}{T^2}\frac{\partial U(T, V)}{\partial V} = \frac{1}{T}\frac{\partial p(T, V)}{\partial T} - \frac{p}{T^2}.$$

Observe that T and V are here the independent arguments (or variables). So T is not a function of V and vice versa. This also causes the relation $\partial T/\partial V = 0$ and $\partial V/\partial T = 0$ (not ∞!).

Using Jacobian determinants, as explained in more detail in Sect. 1.14.6, for the energy U in natural variables S, V, n, the law of Schwarz can be expressed as [9]

$$\frac{\partial(T, S, n)}{\partial(V, S, n)} = \frac{\partial(-p, V, n)}{\partial(S, V, n)}.$$

From this relation it follows by rearrangement that

$$\frac{\partial(T, S, n)}{\partial(-p, V, n)} = -1. \tag{1.28}$$

In three variables, there are two additional analogous equations. Similar relations are valid also for other thermodynamic functions. The relation Eq. (1.28) is claimed to be the formulation of the first law of thermodynamics in the form of Jacobian determinants [9].

1.10 Pfaffian Forms

A Pfaffian form is also known as a total differential equation [10, pp. 326–330]. This type of differential equation plays an important role in thermodynamics. Consider the vector function $\mathbf{f}(\mathbf{x})$ of the vector argument $\mathbf{x}$. The scalar product $\mathbf{f}(\mathbf{x}) \cdot d\mathbf{x}$ is

$$\mathbf{f}(\mathbf{x}) \cdot d\mathbf{x} = \sum_{i=1}^{n} F_i(x_1, x_2, \ldots, x_n) dx_i. \tag{1.29}$$

For $n = 3$, the necessary and sufficient condition that the Pfaffian form Eq. (1.29) is integrable is

$$\mathbf{f}(\mathbf{x}) \cdot (\nabla \times \mathbf{f}(\mathbf{x})) = 0.$$

For higher n, the necessary and sufficient condition that the Pfaffian form Eq. (1.29) is integrable is

$$F_p \left[\frac{\partial F_r}{\partial x_q} - \frac{\partial F_q}{\partial x_r} \right] + F_q \left[\frac{\partial F_p}{\partial x_r} - \frac{\partial F_r}{\partial x_p} \right] + F_r \left[\frac{\partial F_q}{\partial x_p} - \frac{\partial F_p}{\partial x_q} \right] = 0. \tag{1.30}$$

The integers p, q, r are then any in the range of $1, 2, \ldots, n$. Observe the similarity to the law of *Schwarz* in square brackets in Eq. (1.30). A Pfaffian form can be integrated, if an integrating factor λ exists, so that

$$d\phi = \sum_{i=1}^{n} \lambda F_i dx_i.$$

The integrating factor satisfies any of the equations for $i = 1, 2, \ldots, n$.

$$-\frac{d\lambda}{\lambda} = \sum_{i=j}^{n} \frac{1}{F_i} \left[\frac{\partial F_i}{\partial x_j} - \frac{\partial F_j}{\partial x_i} \right] dx_j.$$

If two integrating factors λ_1, λ_2 are found, then a solution is given by $\lambda_1/\lambda_2 = C$.

1.11 Differentials of Functions

1.11.1 The First Differential – Extremes of Functions

We start with a function of one independent variable, $f(x)$. This function is assumed to have an extremum at $\frac{df(x)}{dx} = 0$ at a certain value x_0. In the close environment of the extremum $f(x_0)$ will not change when we change x_0. Therefore, it is also stated that $f(x_0)$ is stationary in the respective region.

We refine the statement concerning the extremum. In the extremum, the function could be a minimum of a maximum. $f(x_0)$ is a minimum, when $\frac{d^2 f(x)}{dx^2} > 0$ at $x = x_0$. On the other hand, $f(x_0)$ is a maximum, when $\frac{d^2 f(x)}{dx^2} < 0$ at $x = x_0$.

We extend the arguments to functions with more than one argument. We consider now a function $f(x, y)$. The first total derivative is

$$d f(x, y) = f_x \, dx + f_y \, dy \tag{1.31}$$

An extremum will be at $f_x(x, y) = 0$ and $f_y(x, y) = 0$ at some $x = x_0$ and $y = y_0$.

1.11.2 The Second Differential – Type of Extremes of Functions

It is common from elementary school that the second derivative test is used to decide whether a stationary point of a function $f(x)$ is a maximum, a minimum, or an inflection point. Actually, the second differential is the change in slope of a curve. It provides information as to whether a function runs convex or concave with respect to the x-axis.

Example 1.10. Consider the motion of a ball under gravity on a plate. The gravity acts normal to the surface. The situation is illustrated in Fig. 1.2.

If the plate is completely plain, the ball can be moved laterally without energy input or energy output. If the plate is not plain but inclined, energy is needed to achieve the motion in both lateral and vertical positions. Since the second derivative gives information about the curvature, the second derivative can be used to state something about the ease of displacement with respect to the energy needed to achieve this process. □

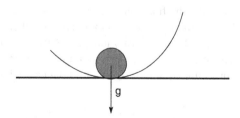

Fig. 1.2 Motion of a ball
under gravity under constraint

Consider the function $U \propto S^{4/3}$. Then the first derivative is $U' \propto S^{1/3}$ and the second derivative is $U'' \propto S^{-2/3}$. Observe that the first derivative runs with infinite slope into the origin. For this reason, the second derivative approaches infinity in the limit of the origin. In general, the function $f(x) = x^{1+\alpha}$ becomes anomalous immediately when the exponent deviates from 1.

1.11.2.1 Two Variables

To find out if a stationary function is a minimum or a maximum is more complicated for more than one variable. There are special cases, e.g., two-dimensional inflexion points, saddle points. The second total derivative for $f(x, y)$ is $d^2 f(x, y)$. It reads as

$$d^2 f(x, y) = f_{xx} dx^2 + 2 f_{xy} dx dy + f_{yy} dy^2. \tag{1.32}$$

In general, $f_{xx} f_{yy} - f_{xy}^2 > 0$ at $x = x_0$ and $y = y_0$ in order that the function could be either a minimum or a maximum. If $f_{xx} > 0$ (or $f_{yy} > 0$) at $x = x_0$ and $y = y_0$, then the function is a minimum. If $f_{xx} < 0$ (or $f_{yy} < 0$) at $x = x_0$ and $y = y_0$, then the function is a maximum.

A minimum of a function $f(x, y, \ldots)$, such as energy, occurs when the function is stationary, i.e., $df(x, y, \ldots) = 0$ and when the second derivative is positive, i.e., $d^2 f(x, y, \ldots) > 0$.

However, it may happen that the second derivative $d^2 f(x, y, \ldots)$ is zero at $df(x, y, \ldots) = 0$. In this case, we must seek for a higher order derivative. If the total differential of second order is zero, then the third-order differential $d^3 f$ has terms like Δx^3 in the expansion. These terms do not establish a minimal curve. However, the differential of fourth order again is symmetric with respect to changes that increase and decrease the argument. In other words, when the even term of second order is zero, the odd term of third order must also be zero to guarantee stability. This procedure is called the higher order derivative test.

$$\begin{aligned} d^{2n} f(x, y, \ldots) > 0 &\rightarrow \text{Minimum} \\ d^{2n} f(x, y, \ldots) < 0 &\rightarrow \text{Maximum} \end{aligned} \qquad n = 1, 2, \ldots.$$

1.11.2.2 Extremes with Constraints

Consider the function $f(x, y) = x^2 + y^2$. Clearly, at $x = 0$ and $y = 0$ the function has a minimum. We want to find out whether there is a minimum under the condition that x and y do not vary independently, but are connected by the relation $x + y = 1$. This case is called an extremum under constraints. The latter equation is called the constraint equation. We can find out easily by substituting

$$f(x, y) = f(x) = x^2 + (1 - x)^2 = 1 - 2x + 2x^2.$$

We find the extremum at $f_x = 0 = -2 + 4x$, or $x = 1/2$. Since $f_{xx} = 4$, we conclude that we have a minimum. Further, we can find the associated $y = 1/2$ from the constraint equation.

On the other hand, we can use the method of undetermined multipliers to find out the extremum. First, we normalize the constraint equation into $x + y - 1 = 0$. In general the constraint equation would then have the form $g(x, y) = 0$. Then we add the constraint equation with a factor to the equation where to find out the extremum. Thus, we form

$$h(x, y) = f(x, y) + \lambda g(x, y). \tag{1.33}$$

We now have three unknowns λ, x, y and three equations to evaluate.

$$\begin{aligned} h_x &= 0 \\ h_y &= 0 \\ g(x, y) &= 0. \end{aligned} \tag{1.34}$$

Example 1.11. We show the method of undetermined multipliers with the function $f(x, y) = x^2 + y^2$ and with the constraint equation $x + y - 1 = 0$. We form an analogy to Eq. (1.33)

$$h(x, y) = x^2 + y^2 + \lambda(x + y - 1).$$

Next, we form the derivatives according to Eq. (1.34)

$$\begin{aligned} h_x &= 2x + \lambda &= 0 \\ h_y &= 2y + \lambda &= 0 \\ g(x, y) &= x + y - 1 &= 0 \end{aligned}$$

and we evaluate x and y. Using Eq. (1.32) for $f(x, y)$ we can verify that the extremum is a minimum. □

The method of undetermined multipliers has the advantage that we do not need the constraint equation itself, but rather the derivatives are needed, because the procedure needs $\lambda d g(x, y)$.

Example 1.12. Consider a dissociation reaction

$$A \rightleftharpoons 2B.$$

Therefore, for the change of the mol numbers we have

$$2dn_A + dn_B = 0 = dC(n_A, n_B). \tag{1.35}$$

This is the constraint equation in differential form. The free enthalpy at constant temperature and pressure ($dT = 0$, $dp = 0$) is

$$dG(n_A, n_B) = \mu_A dn_A + \mu_B dn_B. \tag{1.36}$$

Adding the constraint equation Eq. (1.35) to Eq. (1.36) we obtain

$$dG(n_A, n_B) + \lambda dC(n_A, n_B) = \mu_A dn_A + \mu_B dn_B + 2\lambda dn_A + \lambda dn_B,$$

from which follows

$$2\mu_A + \lambda = 0$$
$$\mu_B + \lambda = 0.$$

Therefore, $\mu_A = \mu_B/2$ is the condition to make the free enthalpy to an extremum.
□

The constraint equations have to be used with care. When constraints are merged, the physical meaning of a constraint may become different as originally intended.

Example 1.13. Consider the constraints

$$dV' = 0 \text{ and } dV'' = 0. \tag{1.37}$$

We can add these two constraints to obtain

$$dV' + dV'' = 0. \tag{1.38}$$

The statement of Eq. (1.38) is not the same as what Eq. (1.37) states. Namely, Eq. (1.37) states that both the volume V' and the volume V'' are constant, whereas Eq. (1.38) states that merely the sum of both volumes are constant. Instead, the individual volumes may vary. When we use Eq. (1.39)

$$dV' + dV'' = 0 \text{ and } dV' = 0, \tag{1.39}$$

then the meaning is the same as that of Eq. (1.37). We learn from this story that we may not simplify constraints in some unallowed way.   □

1.11.2.3 Comparison Between Direct Substitution and Undetermined Multipliers

If we have a function of variable such as $f_k(x_1, x_2, \ldots, x_k)$ then we have totally k variables. The notation f_k serves to indicate that f is a function with k variables. Without constraints we get k equations to find the extremum, namely

$$\frac{\partial f_k(x_1, x_2, \ldots, x_k)}{\partial x_1} = 0$$
$$\frac{\partial f_k(x_1, x_2, \ldots, x_k)}{\partial x_2} = 0$$

$$\cdots \cdots \cdots \cdots \cdots$$

$$\frac{\partial f_k(x_1, x_2, \ldots, x_k)}{\partial x_k} = 0.$$

(1.40)

So there is a sufficient number of equations to solve the extremal problem. If there is a constraint relation between $x_1, x_2, \ldots, x_k$, say $c_k(x_1, x_2, \ldots, x_k) = 0$, then we can resolve this relation, say into x_k, and substitute into Eq. (1.40) to get a set of equations like

$$\frac{\partial f_{k-1}(x_1, \ldots, x_{k-1})}{\partial x_1} = 0$$

$$\cdots \cdots \cdots \cdots \cdots$$

$$\frac{\partial f_{k-1}(x_2, \ldots, x_{k-1})}{\partial x_{k-1}} = 0.$$

Here f_{k-1} is another function compared to f_k. Further, constraint relations c_l may contain the full set of variables $x_1, \ldots, x_k$ or an appropriate subset of the variables. With each additional constraint relation

$$c_{k-1}(x_1, \ldots, x_k) = 0$$
$$c_{k-2}(x_1, \ldots, x_k) = 0$$
$$\ldots = 0$$

we can successively reduce the number of variables in the set of the partial differentials, Eq. (1.40). So, we arrive at one differential,

$$\frac{\partial f_1(x_1)}{\partial x_1} = 0,$$

when we have $k - 1$ relations. If we would have k constraint relations $c_1, \ldots, c_k$ then basically all the variables $x_1, \ldots, x_k$ are fixed and the differentials no longer make sense. In other words, in such a case, we can solve the set of variables by means of the constraint relations.

The constraint relations may still contain other variables that do not occur as arguments in the function f. In this case, accordingly, more constraint relations are needed to obtain a fully determined solution of the extremal problem. A common example is a system that is coupled with some environment. In the energy equation for the system only the entropy of the system appears $U_{sys}(S_{sys}, \ldots)$, but in a constraint equation the total entropy may be held constant, i.e.,

$$S_{sys} + S_{env} - S_{tot} = 0.$$

If we are using the method of undetermined multipliers that is attributed to *Lagrange*[4], we increase for each constraint equation c_l also the number of variables by one (λ_l). However, we do not reduce the number of partial differentials. Therefore, the situation is essentially the same as by direct substitution. We show the situation in Table 1.1.

Table 1.1 Extremals in functions with more than one variable obtained by direct substitution or undetermined multipliers

Type	Variables	equations
Conventional substitution		
Function	k	
Constraints	$0 \ldots k - 1$	
Differential	$k \ldots 1$	
Lagrange multipliers		
Function	k	
Constraints	$0 \ldots k - 1$	$0 \ldots k - 1$
Differential	k	

1.11.2.4 Changing Variables

We consider a differential form

$$0 = \xi\, dX + \eta\, dY + \zeta\, dZ. \tag{1.41}$$

We can resolve Eq. (1.41) in three ways:

$$dX(Y, Z) = -\frac{\eta}{\xi} dY - \frac{\zeta}{\xi} dZ$$

$$dY(X, Z) = -\frac{\xi}{\eta} dX - \frac{\zeta}{\eta} dZ$$

$$dZ(X, Y) = -\frac{\xi}{\zeta} dX - \frac{\eta}{\zeta} dY.$$

In making one function of these three functions stationary, we have information of about two differentials, namely, just those that are not the coefficient at the respective function.

1.11.3 The Third Derivative – How the Function Runs Away

The third derivative rarely emerges in physics: it is treated shabbily. We explain now the physical meaning of the third derivative. For illustration we expand

---

[4] Joseph Louis Lagrange, born Jan. 25, 1736, in Torino, Italy, died Apr. 10, 1813, in Paris, France.

$h(x) = f(x + \Delta x) - f(x + \Delta x).$

$$f(x + \Delta x) = f(x) + f'(x)\Delta x + \frac{1}{2}f''(x)\Delta x^2 + \frac{1}{6}f'''(x)\Delta x^3 + \dots$$
$$-f(x - \Delta x) = -f(x) + f'(x)\Delta x - \frac{1}{2}f''(x)\Delta x^2 + \frac{1}{6}f'''(x)\Delta x^3 + \dots .$$

At a stationary point, i.e., when $f'(x) = 0$

$$h(x) = \frac{1}{3}f'''(x)\Delta x^3 + \dots .$$

Suppose, we are in a minimum. Then $h < 0$ means that the function increases less steep at $x + \Delta x$, than at $x - \Delta x$.

As an example we take $f(x) = x + 1/x$. We show the graphical representation of this curve in Fig. 1.3. The curve has a minimum at $x = 1$. It increases obviously less steep to the right, i.e., at increasing abscissa behind the minimum. The third derivation is $f''' = -6x^{-4}$. So at the minimum the third derivative is negative. We have summarized the situation in Table 1.2.

We extend the concept now to a function of two variables, $f(x, y)$. The Taylor expansion of $f(x + \Delta x, y + \Delta y)$ is

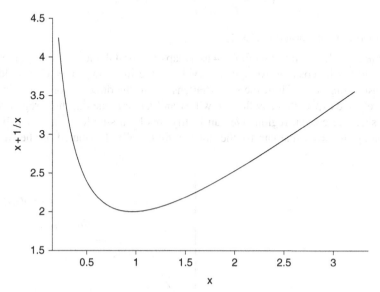

Fig. 1.3 Graph of the function $f(x) = x + \frac{1}{x}$

Table 1.2 Behavior of higher derivatives at a stationary point

Type	Second derivative dx	Third derivative [a]
Minimum $d^2 f(x) > 0$	$dx > 0$ $d^3 f(x) < 0$	
Maximum $d^2 f(x) < 0$	$dx > 0$ $d^3 f(x) > 0$	

[a] In order that the function is more flat at increasing x

$$f(x + \Delta x, y + \Delta y) =$$

$$\frac{1}{0!} f(x, y) + \frac{1}{1!}[f_x \Delta x + f_y \Delta y] + \frac{1}{2!}[f_{xx}\Delta x^2 + 2f_{xy}\Delta x \Delta y + f_{yy}\Delta y^2]+$$

$$\frac{1}{3!}[f_{xxx}\Delta x^3 + 3f_{xxy}\Delta x^2 \Delta y + 3 f_{xyy}\Delta x \Delta y^2 + f_{yyy}\Delta y^3] + \ldots$$

Here f_x, etc., stands for the partial derivative with respect to x at point (x, y). In short we can express the series in terms of total differentials of higher order, i.e.,

$$f(x+dx, y+dy) = \frac{1}{0!} f(x, y) + \frac{1}{1!}d^2 f(x, y) + \frac{1}{2!}d^2 f(x, y) + \frac{1}{3!}d^3 f(x, y) + \ldots.$$

Now we form the difference $f(x+\Delta x, y+\Delta y) - f(x - \Delta x, y - \Delta y)$, however, with the constraint $f_x \Delta x + f_y \Delta y = 0$. This means that we are moving along a tangent line along some point x, y, or in other words, in the limiting case of infinitesimal $(\Delta x, \Delta y)$, the total differential of first order, $d f(x, y) = 0$. Note that the equation of a tangent at point (x_0, y_0) of some function $f(x, y) = C$ is

$$y = -\frac{f_x(x_0, y_0)}{f_y(x_0, y_0)}(x - x_0) + y_0.$$

The situation is illustrated in Fig. 1.4.

Example 1.14. Let us interpret Fig. 1.4 as a map of a coastal line. The line $f(x, y) = C_1$ should be the coast at average sea level and the line $f(x, y) = C_2$ should be the coast at high tide. Thus, the coast flattens out in the direction of increasing x. Intuitively, we expect that any flotsam will strand rather at the flat shore region than at the shelving coastal region. We can justify this by a simple assumption. If the sites along the coast that can fix the flotsam are equally distributed on the strand,

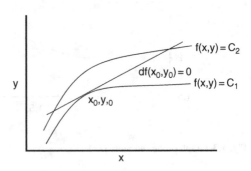

Fig. 1.4 Tangent path of a function

then in the flat coastal region the area flooded by the tide is more in the flat region, thus the probability of a body to be stranded is increased in the flat region. Now, obviously the shape of the curve $f(x, y)$ along the tangent tells us in which direction the coast flattens out. Actually, the coast line is the isodynamic path with respect to gravitational energy. □

In Example 1.14, the position of the flotsam along the shore is not fixed. This situation arises, because idealized, the sea level is at the same height along the shore.

Commonly in thermodynamics, the situation is different. In thermodynamics, the isodynamic path is fixed by the first total differential $dU = 0$ or some other thermodynamic function. The stability is guaranteed by the total differential of second order, d^2U. If $d^2U > 0$ then the state is stable and if $d^2U < 0$, the state is not stable. In these cases for the thermodynamic analysis, the total differentials of first order and second order are sufficient. Obviously, the limiting case of $d^2U = 0$ arises when the system passes from stable to unstable. This situation can be addressed as a critical state. Here, for the thermodynamic analysis, the third derivative enters the game. As pictorially explained in Example 1.14, fluctuations are likely in critical phenomena, e.g., the critical opalescence.

1.11.4 Reciprocal Extremal Problems

In a rectangle with fixed perimeter p, the area A becomes a maximum if the rectangle is a square. In contrast, in a rectangle with fixed area, the perimeter becomes a minimum if the rectangle is a square. This fact may be easily verified by plotting the functions, after substituting the area into the function of the perimeter and vice versa,

$$A = xy; \quad p = 2(x + y).$$

Here x and y are the lengths of the edges of the rectangle. Problems of this type frequently appear in mathematical physics, in particular in the more advanced calculus of variation. We will address this set of somehow related problems as reciprocal extremal problems.

Particularly related to thermodynamics is the second law. The second law states that at constant energy the entropy tends to reach a maximum in equilibrium. The reciprocal statement is that at constant entropy the energy tends to reach a minimum in equilibrium. Often in thermodynamics the principle of minimum energy is deduced from the principle of maximum entropy via a thermodynamic process that is not conclusive, in that as the initial assumption of maximum entropy is violated. These methods originate to Gibbs. Other proofs run via graphical illustrations.

Before we access the problem, we state presuming the result that the correspondence of a maximum to a minimum in a reciprocal extremal problem is not necessarily true in general. Consider a function $z(x, y)$. Its total differential is

$$dz(x, y) = \xi(x, y)dx + \eta(x, y)dy. \tag{1.42}$$

At $dx = 0$, an extremum will be at $\eta(x, y) = 0$. The extremum will be a maximum if $\partial\eta(x, y)/\partial y < 0$. Thus, the conditions for a maximum of z will be

$$dx = 0; \quad \eta(x, y) = 0; \quad \frac{\partial\eta(x, y)}{\partial y} < 0. \tag{1.43}$$

We exchange now the function with the independent variable x and get by simple rearranging

$$dx(y, z) = \frac{1}{\xi(x, y)}dz - \frac{\eta(x, y)}{\xi(x, y)}dy. \tag{1.44}$$

Now, x is no longer an independent variable, but turns into a function of y and z, i.e., we have to substitute $x \to x(y, z)$ even on the right-hand side.

$$dx(y, z) = \frac{1}{\xi(x(y, z), y)}dz - \frac{\eta(x(y, z), y)}{\xi(x(y, z), y)}dy. \tag{1.45}$$

According to the chain rule we have

$$\frac{\partial\eta(x(y, z), y)}{\partial y} = \frac{\partial\eta(x, y)}{\partial y} + \frac{\partial\eta(x, y)}{\partial x}\frac{\partial x(y, z)}{\partial y}. \tag{1.46}$$

The first term on the right-hand side of Eq. (1.46) refers to the derivation at constant x and the second term evaluates the functional dependence $x = x(y, z)$. From Eq. (1.42), at constant z, we may verify that the last term in Eq. (1.46) is

$$\frac{\partial x(y, z)}{\partial y} = -\frac{\eta}{\xi} = \zeta(y, z). \tag{1.47}$$

Thus, for $\eta = 0$, it disappears. In order to get information whether a minimum or a maximum is present, we need the derivative

$$\frac{\partial\zeta(y, z)}{\partial y} = -\eta_y\xi^{-1} + \eta\xi^{-2}\xi_y = -\eta_y\xi^{-1}. \tag{1.48}$$

For brevity we have used the subscript notation $(_y)$ in Eq. (1.48) to indicate the partial differentials. The last equality in Eq. (1.48) emerges from the extremum conditions $\eta = 0$. The second derivative we are searching for turns out to be

$$\frac{\partial^2 x(y, z)}{\partial y^2} = -\frac{1}{\xi}\frac{\partial\eta}{\partial y} \tag{1.49}$$

and finally we summarize the conditions that the reciprocal quantity to z, namely x, is a minimum, opposed to Eq. (1.43) as

$$dz = 0; \quad \eta(x, y) = 0; \quad \xi(x, y) \neq 0; \quad -\frac{1}{\xi(x, y)} \frac{\partial \eta(x, y)}{\partial y} > 0. \quad (1.50)$$

The conditions of minimum and maximum, Eqs (1.43) and (1.50), differ only in the term $\xi(x, y)$. It depends on the sign of this term, whether we find the minimum – maximum reciprocity or not.

1.12 Euler's Theorem

The theorem of *Euler* is a similarity relation. We illustrate the theorem with a simple example.

Example 1.15. We calculate the area A of a rectangle using the function

$$A = f(a, b) = ab,$$

where a is the length of the rectangle and b is the width of the rectangle. When we scale both length and width with the same factor λ, then the area of the rectangle is

$$f(\lambda a, \lambda b) = \lambda a \lambda b = \lambda^2 f(a, b).$$

Because the factor λ occurs with a power of 2, we say that the area is a homogeneous function of the length and the width of order 2. □

Another example for a homogeneous function is a homogeneous polynomial. For instance, $(x + y)^k$ is a homogeneous polynomial of kth order. In addition, various functions in mechanics and thermodynamics are homogeneous functions of first order.

1.12.1 Homogeneous Variables

Homogeneity is a property that a system does not change, besides it becomes smaller if we cut it into pieces. Homogeneity refers to a property that fills the three-dimensional space, i.e., the volume with the same property. Each partial volume can be placed into another part of the total system without observing a change.

Examples for homogeneous variables are typically expressed originally in densities, thus normalized to the unit of volume, e.g.,

- The mass density
- The molar density
- The electrical charge density in an electrolyte

Another way to get homogeneous variables is to form molar quantities from extensive variables. Note that the equation of continuity and other derived equations consider typically a small unit of volume $dV = dx\,dy\,dz$ which is the starting point for the derivation of the respective law.

1.12.2 General Form

We present the general form of Euler's homogeneity theorem. A function is homogeneous of order k, when

$$\lambda^k f(x, y, \ldots) = f(\lambda x, \lambda y, \ldots). \tag{1.51}$$

We investigate some properties of such functions. Forming the differential with respect to λ of Eq. (1.51) we get

$$\frac{df}{d\lambda} = \lambda^{k-1} f(x, y, \ldots) = \frac{\partial f}{\partial \lambda x}\frac{d\lambda x}{d\lambda} + \frac{\partial f}{\partial \lambda y}\frac{d\lambda y}{d\lambda} + \ldots \tag{1.52}$$

Therefore, the right-hand equation in Eq. (1.52) can be further simplified into

$$\lambda^{k-1} f(x, y, \ldots) = \frac{\partial f}{\partial \lambda x}x + \frac{\partial f}{\partial \lambda y}y + \ldots \tag{1.53}$$

Moreover, Eq. (1.53) is valid for all λ. Thus, also for $\lambda = 1$ and for this special case we obtain

$$f(x, y, \ldots) = \frac{\partial f}{\partial x}x + \frac{\partial f}{\partial y}y + \ldots$$

In thermodynamics, the case $k = 1$ is of importance. Therefore, we have

$$\lambda U(S, V, n, \ldots) = U(\lambda S, \lambda V, \lambda n, \ldots). \tag{1.54}$$

In analogy to the homogeneity of mathematics, Eq. (1.54) defines in thermodynamics a homogeneous system. Conversely, the relation is valid only for the thermodynamic properties that are valid in a homogeneous system, such as entropy, volume, number of moles, electric charge. These variables are proportional to the size of the homogeneous system. Doubling the size will double the energy under consideration. In fact, we may consider this property as very characteristic for a phase. Instead of Eulerian homogeneity we may address this special property as thermodynamic similarity.

In a heterogeneous system, this can still be true for the individual phases, if certain parameters such as the surface do not play a role. If the volume is increased by a certain factor λ, the surface would grow by a factor of $\lambda^{2/3}$. The surface problem and other corresponding problems are still discussed controversially [11–15].

Of practical importance are electrochemical systems with electrodes. Electrical and gravitational systems may be of a pseudo first order with respect to charge and mass additions, when only the contribution of the external field to the energy is considered. However, this does not mean that these electrical and gravitational contributions should be included into the Gibbs – Duhem equation [16]. Of course, it is a matter of definition of the thermodynamic variables, whether the thermodynamic similarity exists or not.

Example 1.16. The well-known kinetic energy is $mv^2/2$. Likewise, using the momentum $\mathbf{P} = m\mathbf{v}$ and introducing molar quantities instead of mass $n = m/M$, we can write the kinetic energy as a function of momentum $\mathbf{P}$ and mol number n as

$$U(\mathbf{P}, n) = \frac{\mathbf{P}^2}{2Mn}. \tag{1.55}$$

We drop now the vector notation and think of absolute quantities of the momentum. Otherwise we could not readily form the differential. Using the differentials

$$\frac{\partial U(P, n)}{\partial P} = \frac{P}{Mn} = v, \qquad \frac{\partial U(P, n)}{\partial n} = -\frac{P^2}{2Mn^2} = \mu,$$

we can find the Eulerian homogeneity is fulfilled. Namely, we can write the energy as $U(P, n) = vP + \mu n$. Note that the chemical potential is negative. This means, when mass is added, the energy will drop. This is the situation of the inelastic touch. However, the conservation of both energy and momentum must be fulfilled. □

For homogeneous systems, a thermodynamic function can also be represented in the form

$$U(S, V, n) = TS - pV + \mu n. \tag{1.56}$$

The form of Eq. (1.56) suggests that the total energy can be split into the individual energy forms belonging to the corresponding variables, i.e., the thermal energy is located in TS, the energy of compression or expansion is located in $-pV$, and the chemical energy is located in μn. However, this view is not correct. Starting from a state with a certain pressure p and a certain volume V we cannot exhaust the total energy of compression or expansion in the amount of pV. For example, consider the adiabatic expansion. In this case, the temperature and the pressure will also change in the course of the process.

Rearranging Eq. (1.56),

$$1 = \left[\frac{TS}{U}\right] - \left[\frac{pV}{U}\right] + \left[\frac{\mu n}{U}\right]$$

is obtained. For an ideal gas,

$$1 = \left[\frac{\tilde{S}}{\tilde{C}_v}\right] - \left[\frac{R}{\tilde{C}_v}\right] + \left[1 - \frac{\tilde{S}}{\tilde{C}_v} + \frac{R}{\tilde{C}_v}\right].$$

Question 1.1. Coming back to Eq. (1.56) again, we recall that it is often stated that the energy cannot be determined absolutely, but only relative with an unknown additive constant. Where in Eq. (1.56) is the additive constant?

Example 1.17. We show that we can modify Eq. (1.56) by substituting

$$S = S_0 \exp(x); \quad V = V_0 \exp(y); \quad n = n_0 \exp(z)$$

into $U(S, V, n) \rightarrow U(S_0 \exp(x), V_0 \exp(y), n_0 \exp(z))$, as well as in the declaration of U. Now the derivatives with respect to x, y, z are

$$\frac{\partial U}{\partial x} = TS; \quad \frac{\partial U}{\partial y} = -pV; \quad \frac{\partial U}{\partial z} = \mu n,$$

using the chain rule and substituting back. □

Since the independent variables are the set (S, V, n), the dependent variables T, p, μ are in general functions of the independent variables, i.e.,

$$T = T(S, V, n); \quad p = p(S, V, n); \quad \mu = \mu(S, V, n).$$

On the other hand, we wrote for the energy $dU(S, V, n) = T dS - p dV + \mu dn$. Therefore, the complete *Legendre* transformation must be zero:

$$dU - d(TS) + d(pV) - d(n\mu) = -S dT + V dp - n d\mu = 0.$$

This is an example of the *Gibbs – Duhem*[5] relations. We cannot omit the chemical potential in this case, because if we are changing right after the similarity the mol numbers are also increasing, together with entropy and volume. In fact, $U = U(S, V, n)$, even when we are often only writing $U = U(S, V)$. The latter is allowed for processes in which the mol numbers are not changing, $dn = 0$, e.g., for the isothermal expansion of a gas. But, nevertheless, the energy is always a function of the mol numbers of the phase or the phases.

1.12.3 Molar Energy

If we allow $\lambda = 1/n$ in Eq. (1.54), then we obtain

$$U(S, V, n, \ldots) = nU\left(\frac{S}{n}, \frac{V}{n}, 1, \ldots\right). \tag{1.57}$$

---

[5] Pierre Maurice Marie Duhem, born Jun. 10, 1861, in Paris, died Sep. 14, 1916, in Cabrespine.

In the right-hand side of Eq. (1.57) we have to replace in the declaration of the energy, the entropy, the volume, etc., by the respective molar quantities. Further, we would have to replace in the function declaration every explicate occurrence of n by 1. Therefore, the energy is referring to 1 mol of substance. n no longer appears as a variable in the variable list and can be removed from the list of variables. We can rewrite the molar energy as

$$U\left(\frac{S}{n}, \frac{V}{n}, 1, \ldots\right) = \tilde{U}(\tilde{S}, \tilde{V}, \ldots).$$ (1.58)

In this way, Eq. (1.57) can be rewritten by the use of Eq. (1.58) as

$$U(S, V, n, \ldots) = n\tilde{U}(\tilde{S}, \tilde{V}, \ldots).$$

If we have two subsystems 1 and 2 of the same kind, then the functional dependence must also be the same, i.e., $\tilde{U}_1 = \tilde{U}_2 = \tilde{U}$. If both systems also exhibit equal molar variables, $\tilde{S}_1 = \tilde{S}_2 = \tilde{S}$, $\tilde{V}_1 = \tilde{V}_2 = \tilde{V}$, then

$$U(S, V, n_1 + n_2, \ldots) = (n_1 + n_2)\tilde{U}(\tilde{S}, \tilde{V}, \ldots),$$ (1.59)

otherwise the linear relation of Eq. (1.59) is not valid. This reflects once more a property of a phase. In each part of the phase, the molar variables are the same and we cannot decide from the variables in which part of the phase we are. Strictly speaking, a phase under the influence of gravity is not a phase, even at homogeneous gravitational force. The energy is dependent on the height. Only in a thin horizontal shell, all the parts of the phase would be at the same height.

1.12.4 Complete Variable Sets

The complete Legendre transformation of Eq. (1.56) is zero, because naturally

$$U(S, V, n) - TS + pV - \mu n = 0.$$

We can use this fact to check whether the variable set and the functional dependence of the variable set obey the demand that the complete Legendre transformation is zero. Only then, we have a function of the energy that obeys thermodynamic similarity.

Example 1.18. Verify that for the function

$$U(S, V, n) = n\left(C_1 \exp\left(\frac{S}{nC_v}\right)\left(\frac{V}{n}\right)^{-R/C_v} + C_2\right),$$

$U(S, V, n) - TS + pV - \mu n = 0$ holds. Now introduce molar quantities to get

$$\tilde{U}(\tilde{S}, \tilde{V}) = \left(C_1 \exp\left(\frac{\tilde{S}}{C_v}\right) \tilde{V}^{-R/C_v} + C_2 \right)$$

and verify that $\tilde{U}(\tilde{S}, \tilde{V}) - T\tilde{S} + p\tilde{V} \neq 0.$ □

1.12.5 Thermodynamic Similarity of Transformed Functions

There is a potential pitfall in the use of the Legendre transformation of Eulerian homogeneous equations. We start from the representation of the energy U of a single phase in natural variables as a total differential. We use purposely a sloppy notation, as sometimes common,

$$dU = T\,dS - p\,dV + \mu\,dn. \tag{1.60}$$

From the differentials in Eq. (1.60) it is obvious that the set of variables of the energy U is (S, V, n). Because in a single phase all the variables in the variable set are mutually proportional, i.e., $S \propto n$, $V \propto n$, we expect homogeneity. Equation (1.60) can be represented in integrated form as

$$U(S, V, n) = TS - pV + \mu n.$$

Another way to express homogeneity is

$$\frac{\partial U}{\partial S} S + \frac{\partial U}{\partial V} V + \frac{\partial U}{\partial n} n - U = 0.$$

Observe that the variables $T, -p, \mu$ are not directly proportional to the mol number n. Therefore, if all the variables of the system are scaled by a common factor, the energy is scaled by the same factor.

We inspect now the Legendre transformation of the energy with respect to volume, the free energy, or Helmholtz energy F:

$$dF = -S\,dT - p\,dV + \mu\,dn. \tag{1.61}$$

Now the set of variables of the free energy F is (T, V, n). Analogously, for the energy, U, its integral could be $F(T, V, n) = -TS - pV + \mu n$. However, this view is wrong and this is the pitfall. Actually,

$$F(T, V, n) \neq -TS - pV + \mu n. \tag{1.62}$$

Obviously, this function behaves on scaling of entropy, volume, and mol number in the same way like the energy. If these variables of the system are scaled by a common factor, the free energy is scaled by the same factor. However, this is only true if we have for F the independent variable list (S, V, n). But in Eq. (1.62) we have explicitly $F(T, V, n)$. In particular, we have $S(T, V, n)$, $p(T, V, n)$, and $\mu(T, V, n)$. In contrast, for the energy U, in Eq. (1.60) we have $T(S, V, n)$, $p(S, V, n)$, and $\mu(S, V, n)$.

Thus we have with $F(S, V, n)$

$$\frac{\partial F}{\partial S}S + \frac{\partial F}{\partial V}V + \frac{\partial F}{\partial n}n - F = 0,$$

but with $F(T, V, n)$

$$\frac{\partial F}{\partial T}T + \frac{\partial F}{\partial V}V + \frac{\partial F}{\partial n}n - F \neq 0.$$

If we omit the first term, the derivative with respect to T, the world is in order again.

We illustrate the situation once more in a practical example. From Eq. (1.56) we form the Legendre transformation with respect to the entropy S, $F = U - TS$ to get the free energy or Helmholtz energy

$$F(T, V, n) = -p(T, V, n)V + \mu(T, V, n)n. \tag{1.63}$$

In contrast to the energy representation, the dependent variables are now functions of the set (T, V, n) which are the independent variables. The entropy term, which appears in the energy representation, has been swallowed by the transformation. Due to Maxwell's relations

$$\frac{\partial \mu(T, V, n)}{\partial T} = -\frac{\partial S(T, V, n)}{\partial n}.$$

We can form the differential with respect to the temperature T in Eq. (1.63) to get

$$\frac{\partial F(T, V, n)}{\partial T} = -\frac{\partial p(T, V, n)}{\partial T}V + \frac{\partial \mu(T, V, n)}{\partial T}n = -S(T, V, n).$$

Thus, for an ideal gas we have

$$-nR - \frac{\partial S(T, V, n)}{\partial n}n = -S(T, V, n). \tag{1.64}$$

In the same way, for an ideal gas,

$$\frac{\partial F(T, V, n)}{\partial V} = 0 + \frac{\partial \mu(T, V, n)}{\partial V}n = -p(T, V, n) = -\frac{nRT}{V}.$$

Similarly, from

$$\frac{\partial F(T, V, n)}{\partial n} = -\frac{\partial p(T, V, n)V}{\partial n} + \frac{\partial \mu(T, V, n)}{\partial n}n + \mu(T, V, n) = \mu(T, V, n)$$

we obtain for an ideal gas

$$-RT + \frac{\partial \mu(T, V, n)}{\partial n}n = 0. \tag{1.65}$$

Example 1.19. Use the fundamental form for the energy of an ideal gas, Eq. (4.8), or a simplified equation, e.g.,

$$U(S, V, n) = C \times n \exp\left(\frac{S}{n\tilde{C}_v}\right)\left(\frac{V}{n}\right)^{-R/\tilde{C}_v}$$

to verify the relations Eq. (1.64) and Eq. (1.65).   □

1.13 Series

1.13.1 Taylor Series

The *Taylor*[6] expansion of a function $f(x)$ is

$$f(x + \Delta x) = f(x) + f'(x)\Delta x + \frac{1}{2!}f''(x)\Delta x^2 + \ldots \tag{1.66}$$

We can rewrite Eq. (1.66) in an operator notation as

$$f(x + \Delta x) = \exp\left(\Delta x \frac{d}{dx}\right)[f(x)].$$

Here we use the convention that $d^0 f(x)/dx^0 = f(x)$. For a function in two variables, the Taylor expansion is

$$f(x + \Delta x, y + \Delta y) = f(x, y)$$
$$+ f_x(x, y)\Delta x + f_y(y)\Delta y$$
$$+ \frac{1}{2!}\left[f_{xx}(x, y)\Delta x^2 + 2f_{xy}(x, y)\Delta x\Delta y + f_{yy}(x, y)\Delta y^2\right] + \ldots \tag{1.67}$$

---

[6] Brook Taylor, born Aug. 18, 1685, in Edmonton, Middlesex, England, died Dec. 29, 1731, in Somerset House, London, England.

In operator notation Eq. (1.67) looks like

$$f(x + \Delta x, y + \Delta y) = \exp\left(\Delta x \frac{\partial}{\partial x} + \Delta y \frac{\partial}{\partial y}\right)[f(x, y)]. \qquad (1.68)$$

Likewise, we can rewrite Eq. (1.68) as

$$f(x + \Delta x, y + \Delta y) = \exp\left(\Delta x \frac{\partial}{\partial x}\right) \exp\left(\Delta y \frac{\partial}{\partial y}\right)[f(x, y)]. \qquad (1.69)$$

1.13.2 Series Reversion

Consider a series

$$y = a_1 x + a_2 x^2 + a_3 x^3 + \dots. \qquad (1.70)$$

We emphasize that $a_0 = 0$. We want to find the coefficients of the series

$$x = A_1 y + A_2 y^2 + A_3 y^3 + \dots. \qquad (1.71)$$

Inserting Eq. (1.71) in Eq. (1.70) results in [17]

$$y = a_1 A_1 y + \left(a_2 A_1^2 + a_1 A_2\right) y^2 + \left(a_3 A_1^3 + 2a_2 A_1 A_2 + a_1 A_3\right) y^3 + \dots. \qquad (1.72)$$

To find the coefficients A_i, we have to equate all the coefficients in Eq. (1.72) to zero. This results in

$$A_1 = \frac{1}{a_1}; \quad A_2 = -\frac{a_2}{a_1^3}; \quad A_3 = \frac{\left(2a_2^2 - a_1 a_3\right)}{a_1^5}; \quad \dots.$$

If the series does not start with a first power term, but with a higher power term, y with broken powers will appear.

A function is invertible if it is bijective. This means that to every x there is a unique y and vice versa. For example, $y = x^2$ is not bijective, because for a given $y > 0$, $x = \pm\sqrt{y}$, there are two solutions for x.

If a function is strict monotone, it has an inverse function. Strict monotones are functions that will be increasing or decreasing, with increasing x besides one unique single point. For example, $y = x^3$ is strict monotone. At $x = 0$, $y' = 0$, but only here. Otherwise, $y' > 0$. The inverse function of a monotone increasing function is again a monotone increasing function.

If a function has no unique solution for each y, then its inverse function is ambiguous. Strictly, the inverse function is not a function at all. The graph of the inverse function is the mirror by the $y = x$ line.

Monotone functions are of some interest in thermodynamics. For example, the entropy should be a monotone function of the temperature. Otherwise, different values of the entropy could have the same temperature.

1.14 Analysis

The modern notation of analysis goes back to J. W. Gibbs [18]. Analysis includes the handling of vectors and matrices. We need a few terms emerging in analysis and will discuss them subsequently.

1.14.1 Vectors

1.14.1.1 Scalar Product

The scalar product of two vectors

$$\mathbf{a} = \begin{pmatrix} a_x \\ a_y \\ a_z \end{pmatrix} = \mathbf{i}a_x + \mathbf{j}a_y + \mathbf{k}a_z; \quad \mathbf{b} = \begin{pmatrix} b_x \\ b_y \\ b_z \end{pmatrix} = \mathbf{i}b_x + \mathbf{j}b_y + \mathbf{k}b_z$$

is a scalar. Here, $\mathbf{i}, \mathbf{j}, \mathbf{k}$ are the unit vectors along the x, y, z axis.

We write $\mathbf{a} \cdot \mathbf{b} = a_x b_x + a_y b_y + a_z b_z$. Often the dot is omitted. In particular, $\mathbf{a} \cdot \mathbf{a} = a_x^2 + a_y^2 + a_z^2$. We address $\mathbf{a} \cdot \mathbf{a}$ also as $\mathbf{a}^2$ and $(\mathbf{a} \cdot \mathbf{a})^{1/2} = \text{abs}(\mathbf{a}) = |\mathbf{a}|$. This is the length of a vector. Often we write the length of a vector simply by omitting the arrow $|\mathbf{a}| = a$.

1.14.1.2 Vector Product

The vector product or cross product of two vectors $\mathbf{a}$ and $\mathbf{b}$ is again a vector:

$$\mathbf{a} \times \mathbf{b} = \begin{pmatrix} a_y b_z - a_z b_y \\ a_z b_x - a_x b_z \\ a_x b_y - a_y b_x \end{pmatrix}.$$

1.14.1.3 Gradient

Basically, the gradient applies to a scalar function in three variables, say $f(x, y, z)$. The ∇ operator in cartesian coordinates is a differential vector operator, like

$$\nabla = \begin{pmatrix} \frac{\partial}{\partial x} \\ \frac{\partial}{\partial y} \\ \frac{\partial}{\partial z} \end{pmatrix}.$$

The gradient ∇f is a vector

$$\nabla f(x, y, z) = \begin{pmatrix} \frac{\partial f(x,y,z)}{\partial x} \\ \frac{\partial f(x,y,z)}{\partial y} \\ \frac{\partial f(x,y,z)}{\partial z} \end{pmatrix}.$$

1.14.1.4 Divergence

The divergence is a special differential form of the scalar product. In cartesian coordinates the divergence is

$$\nabla \cdot f(x, y, z) = \frac{\partial f(x, y, z)}{\partial x} + \frac{\partial f(x, y, z)}{\partial y} + \frac{\partial f(x, y, z)}{\partial z}.$$

1.14.1.5 Curl

The curl of a vector is a special differential form of the vector product. The curl of a vector $\mathbf{r}$ with its components r_x, r_y, r_z is

$$\nabla \times \mathbf{r} = \begin{pmatrix} \frac{\partial r_z}{\partial y} - \frac{\partial r_y}{\partial z} \\ \frac{\partial r_x}{\partial z} - \frac{\partial r_z}{\partial x} \\ \frac{\partial r_y}{\partial x} - \frac{\partial r_x}{\partial y} \end{pmatrix}.$$

Inspect the components of the curl and compare them with the law of *Schwarz*.

Example 1.20. We can use the vector notation in thermodynamics. Show that for a thermodynamic fundamental form $U(S, V, n)$ the gradient with respect to the components of a vector containing the extensive variables $\mathbf{e} = \begin{pmatrix} S \\ V \\ n \end{pmatrix}$ is a vector with the components $T, -p, \mu$,

$$\nabla U(\mathbf{e}) = \begin{pmatrix} T \\ -p \\ \mu \end{pmatrix}.$$

Show that for a potential function, the curl is a zero vector. A DERIVE® session is given in Table 1.3. Further, we can express homogeneity property of the energy $U(S, V, n)$ as a vector function with the set of extensive variables as the scalar product of the vector $\mathbf{e}$ with the gradient of $U(\mathbf{e})$

$$U(\mathbf{e}) = \mathbf{e} \cdot \nabla U(\mathbf{e}).$$

□

Table 1.3 Thermodynamics with DERIVE®

```
InputMode:= Word
CaseMode:= Sensitive
Branch:= Any
U(S,V,n):=
[S,V,n]
GRAD(U(S,V,n),[S,V,n])
[DIF(U(S,V,n),S),DIF(U(S,V,n),V),DIF(U(S,V,n),n)]
CURL(GRAD(U(S,V,n),[S,V,n]),[S,V,n])
    *GRAD(U(S,V,n),[S,V,n])
0
CURL(GRAD(U(S,V,n),[S,V,n]),[S,V,n])
[0,0,0]
```

1.14.1.6 Surface Area

We characterize a small region dA of a surface area around a certain point lying on the surface by a normal vector to the surface $d\mathbf{A}$ starting at this point with a length dA. The surface surrounds a closed volume. The normal vector to the surface $d\mathbf{A}$ should point outward.

1.14.1.7 Divergence Theorem

Let $\mathbf{F}$ be a continuous vector in space. The divergence theorem, or *Gauß – Ostrogradsky* theorem states that

$$\int_V (\nabla \cdot \mathbf{F}) dV = \int_S \mathbf{F} \cdot d\mathbf{S}.$$

Here V is the volume and $\mathbf{S}$ is the normal vector of a surface. The left integral is taken over the whole volume V and the right-hand side integral is taken over the surface S that encloses the particular volume V.

The divergence theorem finds use in the derivation of the virial. Consider the motion of a collection of particles. To the ith particle a force $\mathbf{F}_i$ is exerted at some position $\mathbf{r}_i$. The sum over each particle of the scalar product is the virial $\sum_i \mathbf{r}_i \cdot \mathbf{F}_i$ [19, p. 129]. The time-averaged virial is equal to the average kinetic energy

$$-\frac{1}{2}\overline{\sum_i \mathbf{r}_i \cdot \mathbf{F}_i} = \frac{1}{2}\overline{\sum_i \dot{\mathbf{r}}_i \cdot \dot{\mathbf{r}}_i} = \overline{K}.$$

Assuming that forces appear only at the border of the system, the virial is in integral form:

$$\frac{1}{2}\int_S \mathbf{r} \cdot d(S\mathbf{F}) = \frac{p}{2}\int_S \mathbf{r} \cdot d\mathbf{S}. \tag{1.73}$$

Using the divergence theorem for Eq. (1.73) and taking care that $\nabla \mathbf{r} = 3$, we find

$$\overline{K} = \frac{3}{2}pV. \tag{1.74}$$

1.14.2 Matrix

A matrix A is an array of elements $\{a_{ij}\}$ as

$$A = \begin{pmatrix} a_{11} & a_{12} & \cdots \\ a_{21} & a_{22} & \cdots \\ a_{l1} & \cdots & a_{lm} \end{pmatrix}.$$

If $l = m$ then we have a square matrix. The identity matrix I is a square matrix, with all diagonal elements $a_{ii} = 1$, all other elements being zero.

$$A = \begin{pmatrix} 1 & 0 & \cdots \\ 0 & 1 & \cdots \\ 0 & \cdots & 1 \end{pmatrix}.$$

Using the Kronecker delta we can write $a_{ij} = \delta_{ij}$ or the identity matrix. A diagonal matrix is like an identity matrix, but the diagonal elements are not 1, but possibly different from 1. All other elements are zero.

From the view of matrices, vectors are special cases of matrices. The column vector is a matrix with only one column $\{a_{i1}\}$ and the row vector is a matrix with one row $\{a_{1j}\}$.

In the transpose matrix A^T to A the rows and columns are interchanged, i.e.,

$$A^T = \begin{pmatrix} a_{11} & a_{21} & \cdots \\ a_{12} & a_{22} & \cdots \\ a_{1m} & \cdots & a_{ml} \end{pmatrix}.$$

Two matrices A, B of equal size are added, $C = A + B$, by adding the elements $c_{ij} = a_{ij} + b_{ij}$. The product of a matrix $C = AB$ is defined as

$$c_{ik} = \sum_j a_{ij} b_{jk}.$$

Example 1.21. For the matrix $A = \begin{pmatrix} a_{11} & a_{21} \\ a_{12} & a_{22} \end{pmatrix}$ and the vector $B = \begin{pmatrix} b_{11} \\ b_{21} \end{pmatrix}$ we obtain the product $C = AB$ as

$$C = \begin{pmatrix} a_{11}b_1 + a_{12}b_2 & a_{21}b_1 + a_{22}b_2 \end{pmatrix}.$$

$C = \begin{pmatrix} c_{11} & c_{12} \end{pmatrix}$ turns out as a row vector, whereas $B = \mathbf{b}$ is in fact a column vector.
$\square$

The product of two transposes satisfies $(AB)^T = A^T B^T$.

1.14.3 Tensor

A tensor $\mathbf{T}$ is an array of scalars or numbers of scalar functions

$$\mathbf{T} = \begin{pmatrix} T_{xx} & T_{xy} & T_{xz} \\ T_{yx} & T_{yy} & T_{yz} \\ T_{zx} & T_{zy} & T_{zz} \end{pmatrix}.$$

So a tensor is a special case of a matrix. The product of a tensor $\mathbf{T}$ with a column vector $\mathbf{A} = \begin{pmatrix} A_x \\ A_y \\ A_z \end{pmatrix}$ is again a row vector. The order of multiplication is important.

$$\mathbf{TA} = \begin{pmatrix} T_{xx} A_x + T_{xy} A_y + T_{xz} A_z \\ T_{yx} A_x + T_{yy} A_y + T_{yz} A_z \\ T_{zx} A_x + T_{zy} A_y + T_{zz} A_z \end{pmatrix}^T. \tag{1.75}$$

Note that we have added the superscript T to indicate a column vector, written as transpose. Equation (1.75) is a special case of matrix multiplication. A tensor is symmetric, when $T_{xy} = T_{yx}$, $T_{xz} = T_{zx}$, and $T_{yz} = T_{zy}$. The product of multiplication of a tensor with a vector reduces to a multiplication of the vector with a scalar, if the diagonal elements are equal, i.e., $T_{xx} = T_{yy} = T_{zz}$ and all other elements of the tensor are zero. Moreover, if $T_{xx} = T_{yy} = T_{zz} = 1$ and all other elements are zero, we have the unit tensor.

If we have

$$\mathbf{T} = \begin{pmatrix} p & 0 & 0 \\ 0 & p & 0 \\ 0 & 0 & p \end{pmatrix},$$

then

$$\mathbf{TA} = \begin{pmatrix} p A_x & p A_y & p A_z \end{pmatrix}. \tag{1.76}$$

1.14.4 Determinants

The determinant of a matrix A is denoted as $det\, A$. For a 2×2 matrix

$$A = \begin{pmatrix} a_{11} & a_{12} \\ a_{21} & a_{22} \end{pmatrix},$$

its determinant $det\,A$ is

$$det\,A = a_{11}a_{22} - a_{12}a_{21}.$$

The determinant of a quadratic matrix A can be expanded into minors. A minor determinant M_{ij} is obtained by omitting the row i and the column j from the determinant. The determinant of the matrix A is then

$$det\,A = \sum_i a_{ij}(-1)^{i+j}M_{ij}.$$

Here j remains fixed. This expansion works even for the determinant of a 2×2 matrix. Some important properties of determinants are

1. Interchanging two rows or columns changes the sign of the determinant.
2. Multiples of rows and columns can be added together without changing the value of the determinant.
3. Multiplication of a row by a constant c multiplies the determinant by this constant.
4. A determinant with a row or a column containing exclusively elements that are zero has a value of zero.
5. A determinant with two equal rows or columns has a value of zero.

1.14.5 Systems of Linear Equations

Consider a set of linear equations

$$\begin{aligned}
a_{11}x_1 + a_{12}x_2 + \ldots &= y_1 \\
a_{21}x_1 + a_{22}x_2 + \ldots &= y_2 \\
\ldots\ldots\ldots\ldots\ldots\ldots
\end{aligned} \tag{1.77}$$

The known coefficient system of equations can be represented by the matrix A and the vector $\mathbf{y}$, the unknowns by the vector $\mathbf{x}$

$$A = \begin{pmatrix} a_{11} & a_{12} & \ldots \\ a_{21} & a_{22} & \ldots \\ \ldots & \ldots & \ldots \end{pmatrix}; \quad \mathbf{y} = \begin{pmatrix} y_1 \\ y_2 \\ \ldots \end{pmatrix}; \quad \mathbf{x} = \begin{pmatrix} x_1 \\ x_2 \\ \ldots \end{pmatrix}.$$

We can express the equation system Eq. (1.77) in matrix form:

$$A\mathbf{x} = \mathbf{y}^T.$$

Then the unknown variables x_i can be obtained by the rule of *Cramer*[7] (sometimes also addressed as Kramer's rule). The ith component of the vector $\mathbf{x}$, x_i can be obtained as

$$x_i = \frac{det(A)_k}{det(A)}.$$

Here $det(A)_k$ is the determinant, when in the matrix A the k^k column is replaced by $\mathbf{y}$. We do not detail here the conditions needed to obtain a valid solution, but obviously $det(A) \neq 0$.

1.14.6 Jacobian Determinant

The Jacobian determinant is the determinant of the Jacobian matrix. Both are sometimes simply addressed as Jacobian. If we have a set of functions

$$
\begin{aligned}
y_1 &= y_1(x_1, x_2, \ldots, x_n) \\
y_2 &= y_2(x_1, x_2, \ldots, x_n) \\
&\cdots\cdots\cdots \\
y_n &= y_n(x_1, x_2, \ldots, x_n)
\end{aligned}
,
$$

then the Jacobian matrix is

$$
J = \begin{pmatrix}
\dfrac{\partial y_1}{\partial x_1} & \cdots & \dfrac{\partial y_1}{\partial x_n} \\
\dfrac{\partial y_2}{\partial x_1} & \cdots & \dfrac{\partial y_2}{\partial x_n} \\
& \cdots\cdots & \\
\dfrac{\partial y_n}{\partial x_1} & \cdots & \dfrac{\partial y_n}{\partial x_n}
\end{pmatrix}
$$

and the Jacobian determinant $det(J)$ is the determinant of this matrix J. For reasons becoming obvious immediately, the Jacobian determinant is written as

$$det(J) = \frac{\partial(y_1, y_2, \ldots, y_n)}{\partial(x_1, x_2, \ldots, x_n)}. \tag{1.78}$$

The Jacobian determinant has the following properties [20, pp.128–130]:

$$\frac{\partial(y_1)}{\partial(x_1)} = \frac{\partial(y_1, x_2, \ldots, x_n)}{\partial(x_1, x_2, \ldots, x_n)} \tag{1.79}$$

---

[7] Gabriel Cramer, born Jul. 31, 1704, in Geneva, Switzerland, died Jan. 4, 1752, in Bagnols-sur-Cèze, France.

The left-hand side of Eq. (1.79) emerges as the ordinary partial differential. Exchange of one element changes the sign:

$$\frac{\partial(y_1, y_2, \ldots, y_n)}{\partial(x_1, x_2, \ldots, x_n)} = -\frac{\partial(y_2, y_1, \ldots, y_n)}{\partial(x_1, x_2, \ldots, x_n)}.$$

The Jacobian determinant can be handled like an ordinary fraction:

$$\frac{\partial(y_1, y_2, \ldots, y_n)}{\partial(x_1, x_2, \ldots, x_n)} = \frac{\partial(y_1, y_2, \ldots, y_n)}{\partial(z_1, z_2, \ldots, z_k)} \frac{\partial(z_1, z_2, \ldots, z_k)}{\partial(x_1, x_2, \ldots, x_n)}$$

$$\frac{\partial(y_1, y_2, \ldots, y_n)}{\partial(x_1, x_2, \ldots, x_n)} = \left(\frac{\partial(x_1, x_2, \ldots, x_n)}{\partial(y_1, y_2, \ldots, y_n)}\right)^{-1}.$$

The properties of Jacobian determinants allow a formalism to convert differentials occurring in thermodynamics [21].

1.14.6.1 Maxwell's Relations

We set up total differential of a function $\Phi(X, Y)$ as

$$d\Phi(X, Y) = \frac{\partial\Phi(X, Y)}{\partial X}dX + \frac{\partial\Phi(X, Y)}{\partial Y}dY = \xi(X, Y)dX + \eta(X, Y)dY.$$

By the law of Schwarz, the relation

$$\frac{\partial\xi(X, Y)}{\partial Y} = \frac{\partial\eta(X, Y)}{\partial X}$$

holds. In Jacobian notation, the law of Schwarz reads as

$$\frac{\partial(\xi, X)}{\partial(Y, X)} = \frac{\partial(\eta, Y)}{\partial(X, Y)}.$$

Exchange of the elements in the denominator of the left-hand side results in

$$-\frac{\partial(\xi, X)}{\partial(X, Y)} = \frac{\partial(\eta, Y)}{\partial(X, Y)}.$$

Thus, the numerator can be formally exchanged by $\partial(\xi, X) = -\partial(\eta, Y)$, or else

$$\frac{\partial(\xi, X)}{\partial(\eta, Y)} = -1.$$

Example 1.22. We want to get an expression for

$$\frac{\partial S(T, V)}{\partial V}.$$

As a Jacobian determinant, we rewrite the expression as

$$\frac{\partial S(T, V)}{\partial V} = \frac{\partial(S, T)}{\partial(V, T)}.$$

Recalling that in the denominator the set of variables belonging to the Helmholtz energy appears, we substitute the numerator by $\partial(S, T) = -\partial(p, V) = \partial(V, p)$. Thus,

$$\frac{\partial S(T, V)}{\partial V} = \frac{\partial(S, T)}{\partial(V, T)} = \frac{\partial(V, p)}{\partial(V, T)} = \frac{\partial(p, V)}{\partial(T, V)} = \frac{\partial p(T, V)}{\partial T}.$$

We emphasize that the numerator must be of the form $\partial(\xi, X)$ and the denominator must correspond to the form $\partial(X, Y)$. Otherwise the method fails, e.g., if the numerator is of the general form $\partial(\xi, Y)$. This procedure could be useful in the formal derivation of Eq. (1.13). □

Example 1.23. In the Jacobian determinant

$$\frac{\partial(x, y)}{\partial(z, y)},$$

by consecutive cyclic permutation of the variables $x \to y, y \to z, z \to x$, we get the Jacobian determinants

$$\frac{\partial(x, y)}{\partial(z, y)}; \quad \frac{\partial(y, z)}{\partial(x, z)}; \quad \frac{\partial(z, x)}{\partial(y, x)}.$$

The product of these determinants is

$$\frac{\partial(x, y)}{\partial(z, y)} \frac{\partial(y, z)}{\partial(x, z)} \frac{\partial(z, x)}{\partial(y, x)} = \frac{\partial(x, y)}{\partial(y, x)} \frac{\partial(y, z)}{\partial(z, y)} \frac{\partial(z, x)}{\partial(x, z)} = (-1)(-1)(-1) = -1.$$

Thus,

$$\frac{\partial(x, y)}{\partial(z, y)} \frac{\partial(y, z)}{\partial(x, z)} \frac{\partial(z, x)}{\partial(y, x)} = -1. \tag{1.80}$$

Equation (1.80) is Euler's chain relationship. The relationship can be resolved into

$$\frac{\partial(x, y)}{\partial(z, y)} = -\frac{\partial(x, z)}{\partial(y, z)} \frac{\partial(y, x)}{\partial(z, x)}.$$

This should not be confused with an indirect function of one variable, e.g., $x(z)$, where $z = z(y)$. Here,

$$\frac{\partial x}{\partial y} = +\frac{\partial x}{\partial z}\frac{\partial y}{\partial z}.$$

□

1.14.7 Matrix Notation of State Functions

The matrix notation in this section is similar to the formalism in the theory of elasticity. For brevity, we denote

$$\frac{\partial^2 U(S, V)}{\partial S \partial V} = U_{SV}, \text{ etc.}$$

If we expand $T(S, V)$ and $-p(S, V)$ in a total differential of first order, we get the set of equations [20, p. 361]

$$\begin{aligned}
dT(S, V) &= T_S dS + T_V dV &= U_{SS} dS + U_{SV} dV \\
-dp(S, V) &= -p_S dS - p_V dV &= U_{VS} dS + U_{VV} dV
\end{aligned} \quad (1.81)$$

We can express Eq. (1.81) in a matrix form, c.f. Sect. 1.14.5:

$$\begin{pmatrix} dT \\ -dp \end{pmatrix}^T = \begin{pmatrix} U_{SS} & U_{SV} \\ U_{VS} & U_{VV} \end{pmatrix} \begin{pmatrix} dS \\ dV \end{pmatrix}. \quad (1.82)$$

The superscript T refers to the transposed vector. By simple algebraic manipulation, we can invert Eq. (1.82) into

$$\begin{pmatrix} -dS \\ dV \end{pmatrix}^T = \begin{pmatrix} G_{TT} & G_{Tp} \\ G_{pT} & G_{pp} \end{pmatrix} \begin{pmatrix} dT \\ dp \end{pmatrix}. \quad (1.83)$$

Obviously, in Eq. (1.83) the partial derivatives of the free enthalpy appear. Moreover, Eqs. (1.82) and (1.83) are somehow related. In fact, by resolving Eq. (1.82), the system

$$\begin{aligned}
-dS(T, p) &= \frac{U_{VV} dT + U_{SV} dp}{U_{SV} U_{VS} - U_{SS} U_{VV}} \\
dV(T, p) &= \frac{U_{SS} dT + U_{VS} dp}{U_{SV} U_{VS} - U_{SS} U_{VV}}
\end{aligned} \quad (1.84)$$

is obtained. Comparing Eqs. (1.84) and (1.83) gives the derivatives of G in terms of U. Observe that $U_{SS} U_{VV} - U_{SV} U_{VS}$ is the determinant from the 2×2 matrix in the derivatives of U occurring in Eq. (1.82).

1.14.8 Shaw Tables

The use of Shaw tables [22] will be explained by an example [23]. Find

$$\frac{\partial G(T, V)}{\partial T} = \frac{\partial(G, V)}{\partial(T, V)} = \frac{Sa + Vb}{-a} = -S - V(b/a) \tag{1.85}$$

$$b/a = \frac{\partial(p, V)}{\partial(V, T)} = -\frac{\partial(p, V)}{\partial(T, V)} = -\frac{\partial p(T, V)}{\partial T}.$$

In the first step, develop the partial derivative into a Jacobian determinant. Then look up in Table 1.4 both numerator and denominator and insert the abbreviation. In the final step, from the table footnotes again expand the abbreviation into a Jacobian determinant.

It has been argued that the use of Shaw tables is not necessary [24], so we try the calculation without the use of Shaw tables once more.

Table 1.4 Shaw tables [22]

–	P	V	T
P	0	b	1
V	−b	0	a
T	−1	−a	0
S	−c	−n	−b
U	−Tc+Pb	−Tn	−Tb−Pa
H	−Tc	−Tn+Vb	−Tb+Vl
A	Sl+Pb	Sa	−Pa
G	Sl	Sa+Vb	Vl

–	S	U	H
P	c	Tc−Pb	Tc
V	n	Tn	Tn−Vb
T	b	Tb+Pa	Tb−Vl
S	0	Pn	−Vc
U	−Pn	0	−TPn−V(Tc−Pb)
H	Vc	TPn+V(Tc−Pb)	0
A	−Sb−Pn	−S(Tb+Pa)−TPn	−S(Tb−Vl)−P(Tn−Vb)
G	−Sb+Vc	−S(Tb+Pa)+V(Tc−Pb)	−S(Tb−Vl)+TVc

–	A	G
P	−Sl−Pb	−Sl
V	−Sa	−Sa−Vb
T	Pa	−Vl
S	Sb+Pn	Sb−Vc
U	S(Tb+Pa)+TPn	S(Tb+Pa)−(Tc−Pn)
H	S(Tb−Vl)+P(Tn−Vb)	S(Tb−Vl)−TVc
A	0	PSa+V(Sl+Pb)
G	−PSa−V(Sl+Pb)	0

a = ∂(V,T) = −∂(T,V); b = ∂(P,V) = −∂(V,P) = ∂(T,S) = −∂(S,T); c = ∂(P,S) = −∂(S,P); 1 = ∂(P,T) = −∂(T,P): n = ∂(V,S) = −∂(S,V).

$$\frac{\partial G(T, V)}{\partial T} = \frac{\partial(G, V)}{\partial(T, V)} = \frac{\partial(G, V)}{\partial(T, p)} \frac{\partial(T, p)}{\partial(T, V)}. \tag{1.86}$$

Here we have converted the variable set for the Gibbs free energy into its natural variables by multiplying the numerator and denominator by $\partial(T, p)$ with subsequent rearranging. Expanding the Jacobian determinant of Eq. (1.86), but keeping the Jacobian notation except to the final equality results in

$$\frac{\partial(G, V)}{\partial(T, p)} \frac{\partial(T, p)}{\partial(T, V)} =$$

$$\left[\frac{\partial(G, p)}{\partial(T, p)} \frac{\partial(V, T)}{\partial(p, T)} - \frac{\partial(G, T)}{\partial(p, T)} \frac{\partial(V, p)}{\partial(T, p)} \right] \frac{\partial(T, p)}{\partial(T, V)} =$$

$$\frac{\partial(G, p)}{\partial(T, p)} + \frac{\partial(G, T)}{\partial(p, T)} \frac{\partial(p, V)}{\partial(T, V)} = -S(p, T) + V(p, T) \frac{\partial p(T, V)}{\partial T}.$$

When the partial derivatives are directly inserted in the expansion of the Jacobian determinant, it may be more difficult to verify which terms will cancel. The calculation without the use of the Shaw tables is more lengthy. The experienced user may not need the Shaw tables. However, these tables give valuable hints on how to substitute properly to arrive at the desired conversion.

1.15 Coordinate Systems

If we are plotting a pair of data, we can use a two-dimensional cartesian coordinate system. This is the most common application of visualization of data.

If we want to plot a couple of data containing three parameters instead of a pair, we can visualize the data in a perspective plot or in a contour plot. Contour plots are common in cartography and in meteorology. In cartography, the iso-hypses are connected as lines, whereas in meteorology often isotherms and isobars are plotted as contours.

Besides the well-known cartesian system there are a lot of other systems in use in science. These include polar cylindrical coordinates, spherical polar coordinates, ellipsoidal coordinates, parabolic cylindrical coordinates, and cylindrical bipolar coordinates.

Particularly favorable systems are orthogonal coordinate systems. In orthogonal coordinate systems the coordinates are mutually at right angles for each point in space and the scalar product of their unit vectors follow the orthogonality relation, i.e., $\mathbf{u}_i \cdot \mathbf{u}_j = \delta_{ij}$.

If the origin of a cartesian coordinate system is moved by x_0, y_0, z_0, the coordinates, denoted by primes, are transformed into $x' \to x - x_0$, $y' \to y - y_0$, $z' \to z - z_0$. When a cartesian coordinate system is rotated by an angle of ϕ in the x, y-plane, i.e., around the z-coordinate, and the point (x, y, z) remains fixed, the

z-coordinate is unchanged and the x and y coordinates are transformed by the matrix operation

$$\begin{pmatrix} x' \\ y' \\ z' \end{pmatrix} = \begin{pmatrix} \cos\phi & \sin\phi & 0 \\ -\sin\phi & \cos\phi & 0 \\ 0 & 0 & 1 \end{pmatrix} \cdot \begin{pmatrix} x \\ y \\ z \end{pmatrix}. \tag{1.87}$$

The matrix appearing in Eq. (1.87) is the rotation matrix. On the other hand, if in a fixed coordinate system, a certain vector is rotated, the coordinate system being fixed, the rotation matrix is the transposed matrix of that appearing in Eq. (1.87). The elements in the rotation matrix are the direction cosines of the angles between the old and new vectors.

The rotation matrix has some interesting properties. The diagonal matrix with $a_{ij} = \delta_{ij}$ yields an identity. The matrix

$$\begin{pmatrix} z \\ y \\ x \end{pmatrix} = \begin{pmatrix} 0 & 0 & 1 \\ 0 & 1 & 0 \\ 1 & 0 & 0 \end{pmatrix} \cdot \begin{pmatrix} x \\ y \\ z \end{pmatrix}$$

mirrors the coordinates.

Orthogonal coordinates in two dimensions can be easily set up using complex variables. Consider a complex variable $z = x + \imath y$. For each function $f(z)$, the following holds:

$$\frac{f(z)}{\partial x} = \frac{f(z)}{\partial z}\frac{\partial z}{\partial x} = \frac{f(z)}{\partial z}; \quad \frac{f(z)}{\partial y} = \frac{f(z)}{\partial z}\frac{\partial z}{\partial y} = \imath \frac{f(z)}{\partial z}.$$

Forming the second derivatives reveals that the Laplace equation for two variables holds for any complex function $f(z)$:

$$\Delta f(z) = \frac{f^2(z)}{\partial x^2} + \frac{f^2(z)}{\partial y^2} = 0.$$

The result of the function is again a complex number, say

$$w = u(x, y) + \imath v(x, y) = a + \imath b.$$

In the last equality we have inserted deliberately the constants. The functions thus reflect isolines at constant a and $\imath b$, respectively. If the Laplace equation holds for the function $f(z)$, it must also hold for both real and imaginary parts. In Fig. 1.5, we show the real and imaginary parts of the function $f(z) = z^2$ as contour plots for a series of fixed a and b.

In some areas arising in physical chemistry, a transformation of the cartesian coordinates into another system is useful. For instance, in the derivation of the law of Hagen and Poiseuille, the introduction of polar coordinates is helpful. Another

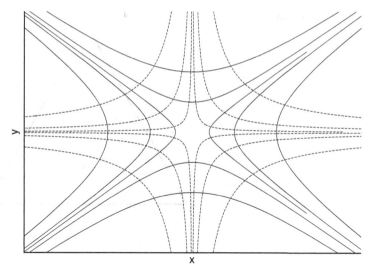

Fig. 1.5 Real and imaginary parts of the quadrupole potential $f(z) = z^2$, $a = x^2 - y^2$, and $b = 2xy$. contour plot for a series of fixed a and b

example is the quantum mechanical treatment of the hydrogen atom, where spherical coordinates are introduced.

1.15.1 Triangle Coordinates

In chemistry, we use often the Gibbs triangle coordinates or tetrahedron coordinates. We emphasize here that the three-dimensional analogue of the triangle coordinates is the tetrahedra coordinates. Triangle coordinates can plot only the mole fractions. However, with the transformation

$$x_2 = \frac{2x_1}{1 - (2x_1)^2}$$

we can scale the range from $-\infty$ to $+\infty$, when x_1 varies from -0.5 to $+0.5$. This procedure suggests centering the zero at the middle of the edges. Transforming back, the positive root gives the solution

$$x_1 = \frac{\sqrt{(4x_2^2) - 1}}{4x_2}.$$

However, we cannot map a three-dimensional event into two dimensions. This comes because within the triangle coordinates we cannot change all the three coordinates independently. In Fig. 1.6, it can be clearly seen that the distances a, b, c are mutually dependent on each other.

Fig. 1.6 Triangle coordinates

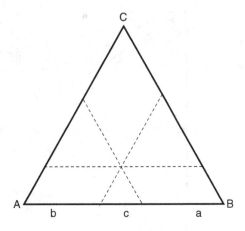

References

1. Laidler, K.J., Meiser, J.H.: Physical Chemistry. The Benjamin Cummings Publishing Company, Menlo Park, CA (1982)
2. Sommerfeld, A.: Vorlesungen über theoretische Physik. In: F. Bopp, J. Meixner (eds.) Thermodynamik und Statistik, vol. 5, 2nd edn. Harri Deutsch, Frankfurt (2002 [reprinted])
3. Giles, R.: Mathematical Foundations of Thermodynamics, *International Series of Monographs on Pure and Applied Mathematics*, vol. 53. Pergamon Press, Oxford (1964)
4. Landsberg, P.T.: Thermodynamics and Statistical Mechanics. Dover Publications, New York (1990)
5. Owen, D.R.: A First Course in the Mathematical Foundations of Thermodynamics. Undergraduate Texts in Mathematics. Springer-Verlag, New York (1984)
6. Weisstein, E.W.: Transform. From MathWorld – A Wolfram Web Resource: [electronic] http://mathworld.wolfram.com/Transform.html (2009)
7. Alberty, R.A.: Legendre transforms in chemical thermodynamics. J. Chem. Thermodyn. **29**(5), 501–516 (1997)
8. Alberty, R.A.: Use of Legendre transforms in chemical thermodynamics: International union of pure and applied chemistry, physical chemistry division, commission on thermodynamics. J. Chem. Thermodyn. **34**(11), 1787–1823 (2002)
9. Degenhart, C.: Thermodynamics. A summary. [electronic] christof-degenhart.de/TeX/td.ps.gz (1995)
10. Zwillinger, D.: Fokker-Planck equation. In: Handbook of Differential Equations, 2nd edn., pp. 254–258. Academic Press Inc., Harcourt Brace Jovanovich, Publishers, Boston, MA (1992)
11. Couchman, P.R., Jesser, W.A.: On the thermodynamics of surfaces. Surf. Sci. **34**(2), 212–224 (1973)
12. Eriksson, J.C.: Thermodynamics of surface phase systems: V. Contribution to the thermodynamics of the solid-gas interface. Surf. Sci. **14**(1), 221–246 (1969)
13. Guidelli, R.: Superficial work and surface stress at solid electrodes: A thermodynamic assessment. J. Electroanal. Chem. **453**(1–2), 69–77 (1998)
14. Guidelli, R.: A reply to the question of whether the internal energy of solid interfaces can be of a non-homogeneous nature. J. Electroanal. Chem. **472**(2), 174–177 (1999)
15. Láng, G., Heusler, K.E.: Can the internal energy function of solid interfaces be of a nonhomogeneous nature? J. Electroanal. Chem. **472**(2), 168–173 (1999)
16. Sørensen, T.S., Compañ, V.: On the Gibbs-Duhem equation for thermodynamic systems of mixed Euler order with special reference to gravitational and nonelectroneutral systems. Electrochim. Acta **42**(4), 639–649 (1997)

17. Weisstein, E.W.: Series reversion. From MathWorld – AWolframWeb Resource: [electronic] http://mathworld.wolfram.com/SeriesReversion.html (2004)

18. Gibbs, J.W.: Dynamics, vector analysis and multiple algebra, electromagnetic theory of light, etc. In: The Collected Works, vol. 2. Yale University Press, New Haven, CT (1948 [reprinted])

19. Lindsay, R.B.: Concepts and Methods of Theoretical Physics. Dover Publications, Inc., New York (1969)

20. Callen, H.B.: Thermodynamics. John Wiley, New York (1961)

21. Crawford, F.H.: On Jacobian methods in thermodynamics. Am. J. Phys. **17**, 397 (1949)

22. Shaw, A.N.: The derivation of thermodynamic relations for a simple system. Trans. R. Soc. (London) **A234**, 299–328 (1935)

23. Wood, S.: Thermodynamics of simple systems. [electronic] http://www.sci.uidaho.edu/geol555/topic_2.htm (2006). Course GEOL 555

24. Pinkerton, R.C.: A Jacobian method for the rapid evaluation of thermodynamic derivatives, without the use of tables. J. Phys. Chem. **56**, 799–800 (1952)

Chapter 2
Foundations of Thermodynamics

In Chapter 1 mathematical facts are dealt with exclusively. Now we are going to the physics of it. We may mention several important preliminaries, usually treated shabbily. We start to talk with the term energy and use the term system in this chapter without having it defined before. However, the term system is discussed soon afterward. Thus, we do not talk about the individual terms here in an axiomatic way. An axiomatic introduction into thermodynamics is, if it is possible at all, quite more difficult to understand. It is unnecessary to say that there are many impressive textbooks in thermodynamics [1–6].

2.1 Idealization

A few comments on idealization are in order here. We obtain physical laws by abstracting reality. Often, only the concept of simplification of the real world by abstraction allows a mathematical description of a physical phenomenon. On the other hand, problems and inconsistencies arise by the simplification procedure. Here we present some examples:

An ideal gas is abstracted as a couple of points that are moving freely in a confined space. The points have a zero extension in space. Further, they do not exert mutual forces. It is highly understandable that points with a zero extension would not be a source of forces. However, even when the points have zero extension, they carry mass. Therefore, applying common sense, we conclude that each point must have an infinite density. Moreover, these points show a heat capacity.

On first glance, this fact is also not understandable. However, if we are accustomed that the points have mass, then we could accept that the heat capacity is caused by the kinetic energy. This is in order again as we restrict ourselves to a translational energy.

We present still another example, Newton's law of motion. This law states that under the action of force a particle will undergo acceleration. The acceleration is inversely proportional to the mass. In the treatment, often the particles are considered as point masses, as in an ideal gas.

J.K. Fink, *Physical Chemistry in Depth*,
DOI 10.1007/978-3-642-01014-9_2, © Springer-Verlag Berlin Heidelberg 2009

These innocently looking assumptions run into problems, if we extend the concept of force to the concept of energy. If we observe a particle at a short moment we do not know how much energy the particles are possessing. However, if we observe the path of a particle, then we can state how much energy it has gained or lost when moving along the path. The energy is the path integral, in one dimension, we find

$$U = - \int F(x) \mathrm{d}x. \tag{2.1}$$

On the other hand, we can trace back the energy to the force:

$$F = - \frac{\mathrm{d}U(x)}{\mathrm{d}x}. \tag{2.2}$$

If we are approaching the concept of force from the energy view $U(x)$, we can state about the energy without knowing the force at all. If we are treating the energy as the basic quantity, the question arises as to how a point-shaped particle experiences that in a small distance; apart from the particle there exists a smaller energy and it could move to the position where the smaller energy is.

In other words, a point-shaped particle does not know about the gradient of the energy. We could introduce the concept that the point-shaped particle does not stand in its position, and the motion is fluctuating around a certain position. In this way, the particle could find out that besides its instantaneous position is a place where it would possess less energy. On the other hand, an extended particle in space could experience a gradient of the energy in space.

In common sense, we are accustomed to such gradients. If a drinking glass is positioned near the border of the table plate, a lot of people intuitively feel to place the glass more in the center of the table, to prevent it from falling down. This mental dictate emerges because it is believed that somebody could push the glass unintentionally and so cause the glass to reach the border and fall down eventually. If there would be nobody who could touch the glass unintentionally, then the habit to center a glass on the table is completely unnecessary. The glass would not fall down, no matter how high the table is.

2.2 Energy

Energy has the physical dimensions of $\mathrm{kg\,m^2s^{-2}}$. From its nature, energy is a nonnegative quantity. From a thermodynamic system, we can withdraw all the energy available, but we cannot borrow energy that is not there. For this reason, the energy has an absolute zero. If we deal with the energy available, then we mean the energy that is accessible to us from a system. A system may still contain energy that is not accessible to us by the tools we have.

For example, if we have a gas in a container with a piston, we can access the energy of expansion. If the container is adiabatically insulated, we cannot access the thermal energy directly. However, the gas is cooling down on adiabatic expansion. Therefore, we can access the thermal energy indirectly via the energy of expansion. After utmost expansion, the gas may still contain energy that could be set free by a chemical reaction or by a nuclear reaction. However, if we cannot initiate these reactions, we cannot access the energy. Therefore, we may set the energy, when all the energy of expansion is gathered as zero.

2.2.1 Extensive and Intensive Quantities

Energy can always be disassembled into two terms, one being extensive and the other being intensive. If we scale the system, then the extensive term will be scaled by the scale factor, but the intensive term does not change. Thus, the extensive term may contain the mass, the volume, the electric charge, and the like. We can distinguish several types of extensive variables. An extensive variable does not necessarily contain the mass in its physical dimension, for example, the volume of a system. On the other hand, the pressure is energy conjugated to the volume and contains the mass as basic unit.

We recall that in mechanics the energy is the negative integral of a force by the path. This concept holds ideally for masses that are concentrated in a mathematical point, so-called point masses:

$$U = -\int \mathbf{F} \, d\mathbf{s} + C, \tag{2.3}$$

where C is some constant of integration. If the mass is constant we can rewrite Eq. (2.3) as

$$U = -\int \frac{\mathbf{F}}{m} d(m\mathbf{s}) + C = -\int \mathbf{F}_u d(m\mathbf{s}) + C. \tag{2.4}$$

In Eq. (2.4) we have replaced the force $\mathbf{F}$ by the force acting on the unit mass $\mathbf{F}_u$ and the total mass we have pushed in the differential. The force and the path are vectors. If the point mass body is moving in the direction of the force then it is losing energy. If the body is moving against the force, it is accumulating energy. On the other hand, when the body loses mass at the same position, the energy also changes. We can address the intensive variable just as a specific force $\mathbf{F}_u$. The extensive variable is then a generalized mass path m s. This notation conforms more to the kinetic energy that is $v\,dP$, i.e., the energy-conjugated pair of velocity and momentum. The energy forms are close to a convenient description of the quantities in which a system may take up or release energy.

Since a macroscopic system has an extension in space, not all parts of the system can be at the same height. Therefore, the system is not homogeneous. In order to deal with a homogeneous system with gravitational energy, we must consider an infinitely thin shell where all parts of the system can be at the same height.

The way to create extensive and intensive variables, as exemplified in Eq. (2.4), does not work in general. For example, the volume is an extensive variable, but the mass is captured in the corresponding intensive variable, i.e., the pressure. Therefore, dividing the pressure by mass is not what we would expect. In the sense of thermodynamic homogeneity, scaling in the gravitational field is restricted to a direction in space where no field is actually acting.

Despite these problems, homogeneity is rather characterized by a certain portion of space where the properties are not changing. Thus, scaling the volume of a region in space where properties such as pressure, charge density, entropy density are constant seems to be a more fundamental process in order to identify extensive and intensive variables.

The volume is rather a basic concept of a region, as it appears in the derivation of all kinds of equations of continuity. For this reason, we introduce quantities that refer to the unit of volume. We will use the superscript $\check{X}$ in order to indicate that this is X per volume. Remember that we are using $\tilde{C}_v$ for the molar heat capacity (at constant volume), $\hat{C}_v$ for the specific heat capacity, $\bar{C}_v$ for a reduced heat capacity, i.e., divided by a reference value, and now $\check{C}_v$ for the heat capacity density. The quantities $\check{X}$ are basically X densities, for example, the mass density or the molar density:

$$\check{m} = \frac{\partial m}{\partial V} = \rho; \quad \check{n} = \frac{\partial n}{\partial V} = c. \tag{2.5}$$

For the mass density and the molar density, we have common symbols available. Usually the molar density is referred to as concentration. However, there are other densities, such as the density of entropy, density of electrical charge, or density of momentum, which are

$$\check{S} = \frac{\partial S}{\partial V} = \rho_S; \quad \check{Q}_{el} = \frac{\partial Q_{el}}{\partial V} = \rho_{el}; \quad \check{P} = \frac{\partial P}{\partial V} = \rho_P. \tag{2.6}$$

For these quantities we have to introduce special symbols. The differentials in the definition indicate that we go to infinite small volume to get the property $\check{X}$. If $\check{X}$ is not dependent on position, we have homogeneity in the system or region under consideration. In this case, we can extend the volume arbitrarily and the system will still remain homogeneous. As explained in Sect. 1.12, the energy of a homogeneous system can be written as

$$U = TS - pV + \mu n. \tag{2.7}$$

We are now coming into some detail to explain an intensive variable in the sense of thermodynamics. If we are forming the density of a system, the intensive variable will not be involved in the differential, since it is a constant. For this reason, in a homogeneous system described by Eq. (2.7) the energy density will emerge as

$$\check{U} = T\check{S} - p\check{V} + \mu\check{n} = T\check{S} - p + \mu c.$$

Note that $\check{V} = 1$. If the intensive variable varies in space, this is an indication that the system is not in equilibrium. When such a system equilibrates, it may but need not become a homogeneous system.

The concept of a homogeneous system will still hold in a system under gravity, or other fields, when we are changing the volume of the system only in a direction normal to the direction of gravity. Otherwise, in a more general case we must deal with an ensemble of infinitesimal small subsystems that make up the macroscopic system.

Intensive variables are independent on proper thermodynamic scaling, i.e., when all relevant extensive variables are multiplied by the same common factor, i.e., $S \rightarrow \lambda S, V \rightarrow \lambda V, \mu \rightarrow \lambda \mu$.

Extensive quantities cannot be essentially negative. This view is in opposition to the phlogiston theory that stated that weight increase by the oxidation reaction is caused by an escape of phlogiston that has a negative weight. Naturally, in modern science here we could think about the electric charge. We emphasize that the sign of the electric charge has been associated to the particle that it is bearing, initially completely arbitrary. We can deal with one charged particle, with two charged particles, and so on, but as such, we cannot deal with a vanished and thus minus one charged particle. A deficit in charge arises only by the interaction with other particles.

In fact, charges of opposite sign are neutralized, as they come close together only. If we are far away from the charges, we cannot identify them, but they are there. If the charge changes to opposite sign then this means that there is an excess of particles bearing the opposite charge.

It is a matter of the philosophical view, whether we treat a neutralization reaction like a chemical reaction and think that a new species has been formed or whether the particles are simply close together. Going one step further we may treat the reaction of an electron with a proton forming a neutron as a new species being formed or also that both species are close together only. However, from the properties exerted, it is more straightforward to deal with a new species.

Summarizing, we recall a common and simple definition of intensive and extensive variables. If two identical systems are gathered, the extensive term will double, whereas the intensive term will remain the same. Examples are the pairs (T, S), $(-p, V)$, (μ, n), (φ, q); i.e., (temperature, entropy), (pressure, volume), (chemical potential, mol number), (electrical potential, electrical charge). Often the extensive quantity may flow into or out of the system.

2.2.2 Change of Energy

We discuss again the topic of intensive and extensive variables in a formal way, namely in terms of exchange of energy. A change in energy form can be written in differential form, such as

$$dU = \xi \, dX. \tag{2.8}$$

In Eq. (2.8) we have two terms, one being before the differential, the other in the differential. The former ξ is addressed as the intensive variable, the latter X is the extensive variable. Thus the intensive variable ξ is, in fact, a function of the extensive variable, i.e., $\xi = \xi(X)$, and in general of other extensive variables that govern the system as shown in Eq. (2.9). The intensive variable is referred to as a thermodynamic potential, but only in certain cases:

$$dU(X_1, X_2, \ldots) = \xi_1(X_1, X_2, \ldots)dX_1 + \xi_2(X_1, X_2, \ldots)dX_2 + \ldots \qquad (2.9)$$

From methodology, is seems to be advantageous to include all derivatives shown in a general way in Eq. (2.9) to the group of thermodynamic potentials.

2.2.3 Range of Energy

The energy of a system under consideration can be smaller in comparison to another system. Despite this fact, as such, as a standalone quantity, the absolute energy can never be negative. On the other hand, a finite system, i.e., a system composed of a finite number of particles n, cannot bear an infinite energy. In mathematical symbols we have,

$$0 \leq n < \infty \Rightarrow 0 \leq U < \infty,$$

i.e., the range of the energy is between zero and infinity in a finite system.

For this reason, if the intensive quantity bears a negative sign, such as the pressure, then

$$0 \leq U(\infty) - U(V_0) = - \int_{V=V_0}^{V \to \infty} p \, dV < \infty,$$

even when the volume approaches ∞ in the limit.

Conversely, at the other end of the scale the energy approaches zero if the number of particles in the system approaches zero. Obviously, the *nothing* bears no energy, thus

$$n \to 0 \Rightarrow U \to 0.$$

2.3 Energy Forms

We summarize in Table 2.1 the physical dimensions of various energy forms. We emphasize that the list is not complete. In square brackets the physical dimension of the extensive variable has been given. The intensive variable has the energy-conjugated physical dimension. In the following sections, we discuss briefly some issues of the energy forms detailed in Table 2.1.

Table 2.1 Physical dimensions of various energy forms

Energy name	Energy form	Physical dimension of intensive variable
Energy of compression (pressure × volume)	$-p\,dV$	$\mathrm{J[m^3]^{-1}}$
Thermal energy (temperature × entropy)	$T\,dS$	$\mathrm{J[JK^{-1}]^{-1}}$
Surface energy (surface tension × area)	$\sigma\,dA$	$\mathrm{J[m^2]^{-1}}$
Gravitational energy (gravitational acceleration × mass height)	$g\,d\,mh$	$\mathrm{J[kg\,m]^{-1}}$
Kinetic energy (velocity × momentum)	$v\,dP$	$\mathrm{J[kg\,m^2s^{-2}]^{-1}}$
Material energy[a] (chemical potential × mol number)	$\mu\,dn$	$\mathrm{J[mol]^{-1}}$
Electrical energy (electric potential × charge)	$U_e\,dQ$	$\mathrm{J[A\,s]^{-1}}$
Electric dipole energy (electrical field × polarization)	$\mathbf{E}_e\,d\mathbf{P}$	$\mathrm{J[m\,A\,s]^{-1}}$
Magnetic energy (magnetization × auxiliary magnetic field)	$\mathbf{M}\cdot\mathbf{H}$	$\mathrm{J[kg\,A^{-1}s^{-2}]^{-1}}$

Note: Acceleration, height, velocity, and momentum are vectors. Usually the acceleration is pointing down and height is pointing up. Therefore, the correct description would be $-\mathbf{g}\,d(\overrightarrow{mh})$. The gravitational energy is related to the kinetic energy by $\frac{d(mh)}{dt} = dp$ and $\frac{dv}{dt} = g$.
[a] The term chemical energy is misleading.

2.3.1 Energy of Compression

2.3.1.1 Energy

This energy form uses the volume V as extensive variable and the pressure p as intensive variable. It is associated with a negative sign. Elementary thermodynamics most often makes use of the energy of compression to demonstrate instructive examples. The energy of compression is often termed volume energy that includes both the energy of compression and the energy of expansion. These two types of energies are essentially the same. Compression means that $dV < 0$ whereas expansion means $dV > 0$.

We should not confuse the energy of volume with the energy of deformation. The energy of deformation plays a role for solids. It can change even when the volume does not change. The energy of volume changes only with the volume.

The following is valid for liquids and gases. We can interpret the pressure as a vector, pointing outward and normal to the surface border of the system. When the surface border of size A is moving by $d\mathbf{s}$ then the increase of volume is $d\mathbf{V} = A\,d\mathbf{s}$. Recall Eq. (2.3). If we modify this equation appropriately, we find immediately the explanation to why the volume energy has the negative sign:

$$U = -\int \mathbf{F}\,d\mathbf{s} + \mathrm{C} = -\int \frac{\mathbf{F}}{A}\,d(A\mathbf{s}) + \mathrm{C} = -\int p\,d\mathbf{V} + \mathrm{C}.$$

The pressure is not a force normalized by the mass, but normalized by some area. A closer look shows that in liquids or gases the pressure should not be a vector, because it is independent of the direction. The pressure acts not only at the border of a liquid or gaseous system but also inside the system in any direction. However, we could argue that we can measure the pressure only by inserting a pressure-sensitive device. In this way, we have to create a surface. However, we observe that in a pressure gradient the motion of the material will occur. Therefore, the pressure gradient and also the pressure itself are basically inside the body.

From the view of energy, we can imagine the pressure as an energy density, namely the pressure has the physical unit of $J m^{-3}$.

We will show now that the pressure is in general a tensor. We consider a small surface area A and we establish the force in this region as $\mathbf{F} = \mathbf{T}A$. A is a vector normal to the surface, but the force could point in another direction, because $\mathbf{T}$ is a tensor. The physical dimension of $\mathbf{T}$ is just a pressure.

We can insert the expression into the basic equation of energy, but we have to follow a somehow modified procedure than in Eq. (2.3), because now the area is a vector that is mapped in the tensor. We can come back to the special case of Eq. (2.3) when the tensor $\mathbf{T}$ has the same form as used in Eq. (1.76) and if the motion ds is normal to the area, i.e., the normal vector of the area $\mathbf{A}$ and the motion ds are mutually parallel.

2.3.1.2 Volume as Thermodynamic Variable

Usually thermodynamic variables are as such that the energy has its low value, when the variable is approaching zero. For instance, the kinetic energy has the velocity v as intensive variable and the momentum $P = mv$ as extensive variable. As the momentum approaches zero, the velocity approaches zero, and thus the kinetic energy approaches zero. The same is true for the entropy, where the temperature is the intensive variable T and the entropy S is the extensive variable.

For the energy of compression the situation is different. If the volume for an ideal gas approaches infinity, then the energy approaches a minimum. We can see this also by the fact that in the differential the minus sign appears, i.e.,

$$dU = T dS - p dV + \mu dn.$$

However, we can reverse the situation. Similar to crystallography, we can move into the reciprocal space. We assign Υ as the reciprocal volume with the property $\Upsilon = 1/V$. We find immediately

$$\frac{d\Upsilon}{dV} = -V^{-2} = -\Upsilon^2. \tag{2.10}$$

Therefore, inserting for dV, the minus sign in the energy differential will cancel:

$$dU = T dS + p\Upsilon^{-2} d\Upsilon + \mu dn. \tag{2.11}$$

The form of Eq. (2.11) indicates that now the independent variables are Υ, S, and n, i.e., $U = U(S, \Upsilon, n)$, $T = T(S, \Upsilon, n)$, $p = p(S, \Upsilon, n)$.

We illustrate a simple case. For the isentropic expansion of an ideal gas we have $pV^\kappa = C(S, n)$, or in the new variables $p\Upsilon^{-\kappa} = C(S, n)$. We calculate the energy for the isentropic expansion to infinite volume as

$$U = C(S) \int_{x=\Upsilon}^{x=0} x^{\kappa-2} dx = \frac{C(S, n)}{\kappa - 1} \left(\Upsilon^{\kappa-1} - 0^{\kappa-1} \right) + C. \tag{2.12}$$

The expression for the energy will only be finite, if we have $\kappa - 1 > 0$. Observe that the Eulerian homogeneity relation still holds for an ideal gas, if V/n is replaced by n/Υ.

2.3.2 Thermal Energy

2.3.2.1 Energy

The inclusion of thermal energy into the theory of mechanics caused the success of thermodynamics. Originally and continuously until now thermal energy is noted as Q and the differential dQ. Then the increase of energy is subdivided into mechanical work dW and into thermal energy dQ:

$$dU = dQ + dW. \tag{2.13}$$

Equation (2.13) opposes the thermal energy to all other work in the system somewhat arbitrary and as such completely unjustified. The symbol d is used to show that dQ is not a total differential, but rather $dQ = TdS$ will be the appropriate notation.

We do not make use of these considerations, but we start directly with a statement valid in general for every energy form: The thermal energy has an extensive and an intensive part. The extensive variable of the thermal energy is the entropy S and the intensive variable is the thermodynamic temperature or absolute temperature T.

In common sense, the input of thermal energy into a system increases the temperature. In terms of entropy, the input of thermal energy means that entropy enters the system. A flow of entropy is basically a flow of heat consisting of a flow of entropy. When the temperature rises, then in the older literature this heat is also addressed as perceptible heat (in German: fühlbare Wärme). This results from the fact that we can detect it by our senses.

An increase of entropy in a system is not always accompanied with an increase of temperature. Similarly, an increase of temperature does not mean that the entropy increases.

Example 2.1. In the first step of the Carnot process, thermal energy will flow into a gas. This inflow is accompanied with an expansion of the gas, in order to run the process isothermal. In the last step of the Carnot process, the gas is compressed adiabatically. Strictly, we mean that the compression occurs at constant entropy, $dS = 0$. Here an increase of the temperature occurs, even when the entropy remains constant. On the other hand, since

$$\frac{\partial T(S, V, n, \ldots)}{\partial S} > 0, \tag{2.14}$$

we can safely expect that in a process, where S will increase, but all other variables $V, n, \ldots$ will remain constant, also the temperature T will increase. Equation (2.14) arises from the stability condition.   □

2.3.2.2 Entropy

The entropy is a mass-like quantity and cannot take negative values. It is often stressed that entropy does not follow a conservation law. Dealing with compression energy, it can be shown that it is possible to construct systems where the total volume is not conserved. Therefore, it should not be surprising that also entropy would not be conserved.

In fact, according to *Clausius*, the entropy tends to become a maximum $S \rightarrow$ max. When the entropy has reached the maximum we have $dS = 0$ or $S = C$. Therefore, all processes running in thermodynamic equilibrium, at $dS = 0$, are running with constant total entropy, even when the entropy in the subsystems may change.

Sometimes it is stated that entropy is not defined for a nonequilibrium process. We want to relativize this statement. Consider a pure mechanical process, e.g., a moving pendulum. Setting up the energy equation, we can clearly subdivide into kinetic and potential energy for the ideal pendulum. The ideal pendulum will never stop with motion. If we add a friction term, we can identify the total energy lost by the system to a gain of thermal energy and thus to the entropy gain. Thus, for the nonideal pendulum the entropy content is clearly defined even in the nonequilibrium state. However, in other thermodynamic systems it may cause problems to identify the kinetic energy clearly. In such situations, in addition to the content of kinetic energy the content of entropy of the system is not clearly defined.

2.3.2.3 Temperature

The thermodynamic temperature has its *zero*, when an input of entropy into a system will not cause an increase of energy, i.e., when $dU = T dS$, and

$$dS > 0; \quad T = 0 \quad \Rightarrow \quad dU = 0. \tag{2.15}$$

This definition is sound, but is not strictly applicable in practice. Note that in this case, even $dS < 0 \Rightarrow dU = 0$ holds. This behavior is in connection with the monotony of entropy with temperature. We will explain this issue elsewhere.

Nevertheless, even when the *zero* of T is fixed by Eq. (2.15), the scale is not fixed, because there is no freedom to insert a nonzero constant factor:

$$dU = (\beta T)d(S\beta^{-1}).$$

The absolute temperature is defined in that way that at the triple point of water $T = 273.16$ K. If the physical unit of the temperature is K, then this means the unit of the entropy is $J\,K^{-1}$. Any other temperature can be accessed using the ideal gas thermometer.

Due to the relation $T = C\,pV$, where C is a unknown calibration constant, this calibration constant can be measured by bringing the ideal gas in contact with a body at triple temperature of water. Then, by measurement of pressure and volume of the ideal gas which is in contact with a body of unknown temperature, this temperature can be measured. Again, this prescription of measurement of the absolute temperature is highly impracticable and suffers from the fact that an ideal gas exists, as the name indicates only ideally.

Another possibility to access the ratio of two absolute temperatures is to measure the efficiency η of a *Carnot* process running between two temperatures T_h and T_l, a higher temperature and a lower temperature. If the process is running with one temperature at the triple point then via the efficiency the other temperature is known. The efficiency is independent of the working medium and is the ratio of work energy W withdrawn to the heat throughput Q into the working medium at the higher temperature T_h:

$$\eta = \frac{W}{Q} = \frac{T_h - T_l}{T_h}.$$

In this way, the temperatures are completely defined by a measurement of work and heat. Here the knowledge of the mechanical heat equivalent is necessary. Further, the practical implementation of the method faces problems with nonideal conditions, e.g., friction losses.

2.3.3 Surface Energy

The extensive variable of the surface energy is the area and the intensive variable is the surface tension. The surface energy is a typical example for an energy form, where the extensive variable can be increased and decreased nearly arbitrarily.

We first deal with a liquid in contact with vacuum. The surface tension σ is a vector that points into the inner region of the liquid. In static equilibrium, the vector points normal to the surface, whereas the surface vector $d\mathbf{A}$ which is also a normal vector points outward from the liquid. Since the two vectors are opposite in direction, the negative sign in the energy equation cancels.

If the liquid is not in contact with vacuum, but with a vapor or a liquid, then the above considerations hold. However, we must see the situation from both phases. The surface tension will be modified by the neighborhood of another phase, even in a different way for the individual phases. The particular change of the surface tension also governs the miscibility. Since both phases are coupled in the case by a common interface, the areas of both will increase or decrease simultaneously. In the case of miscibility there is no longer a separating surface area.

2.3.4 Gravitational Energy

With the law of gravity, Newton was able to explain why Kepler's laws described the planetary motion. The law of gravity is an example of a central force. The force is directed along the center of mass of two bodies. The mathematical formulation is given as follows:

$$\mathbf{F} = G\frac{m_1 m_2}{r^2}\frac{\mathbf{r}}{r}. \tag{2.16}$$

Here $\mathbf{F}$ is the force between two masses m_1 and m_2 separated by a distance $\mathbf{r}$. The last term $\mathbf{r}/r$ builds the unit vector along the distance of the two masses. G is the gravitational constant. The gravitational constant is a constant number. A unit mass that is located on the surface of the earth is subject to an acceleration, i.e., force on the unit mass of $g = 9.8$ m s^{-2}. We obtain this by inserting the numerical values of $G = 6.6726 \times 10^{-11}$ m^{3}kg^{-1}s^{-2}, mass of earth $m_2 = 5.98 \times 10^{24}$ kg, and radius of earth $r = 6.37 \times 10^6$ m.

Therefore, we obtain for the force of a mass m near the surface of the earth

$$\mathbf{F} = m\mathbf{g}. \tag{2.17}$$

This equation is commonly used to calculate the change in gravitational energy $\Delta U = -m\mathbf{g}\Delta\mathbf{h}$, if a mass is lifted by a distance of $\Delta\mathbf{h}$ against the earth acceleration $\mathbf{g}$. We emphasize that usually $\mathbf{g}$ is pointing downward, say in z-direction, and $\mathbf{h}$ is pointing upward. If Eq. (2.17) is written in scalar form, then the negative sign of the earth acceleration is pulled out and the minus sign cancels. Equation (2.17) is highly satisfactory under laboratory conditions. However, $\mathbf{g}$ is slightly dependent on the altitude, as can be clearly seen by inspecting Eq. (2.16).

In thermodynamics, gravitational work has also been set up with gh as the gravitational potential and dm as the differential form of the extensive variable [7]. Here the preferred form is g as the gravitational potential and $d(mh)$ as the differential form of the extensive variable.

2.3.5 Kinetic Energy

The extensive variable for the kinetic energy is the momentum P and the intensive variable is the velocity $\mathbf{v}$. Since $\mathbf{P} = m\mathbf{v}$, the intensive and the extensive variables are highly coupled. In fact, both variables are vectors. Since the kinetic energy integrates to $U = m\mathbf{v}^2/2 + C$, from the consideration of the range of energy the integration constant C should be zero to allow a zero energy and when not below zero, the mass m should be positive.

Example 2.2. Consider the free falling rigid body. The total energy U is

$$U = \frac{mv^2}{2} + mgh.$$

In differential form we obtain using $P = mv$ and $q = mh$

$$dU = v\,dP + g\,dq.$$

Now apply the law of Schwarz to get

$$\frac{\partial v}{\partial q} = \frac{\partial g}{\partial P}.$$

Note that we have dismissed the chemical potential and we deal with P and q as independent variables. The reader should discuss issues arising here.   □

2.3.6 Material Energy

We discuss now various forms of energy related to matter. We will see that this from of energy arises very naturally.

2.3.6.1 Chemical Potential

Charles *Kittel* states in his book *Introduction to Solid State Physics* [8] about the chemical potential:

> A vague discomfort at the thought of the chemical potential is still characteristic of a physics education. This intellectual gap is due to the obscurity of the writings of *Gibbs*[1] who discovered and understood the matter 100 years ago.

Studying the scientific papers (*writings*) of J.W. Gibbs, the reader may find that Gibbs is a highly honest *writer*, in contrast to other contemporary *writers*.

The extensive variable for material energy is the number of moles n and the intensive variable is the chemical potential μ. We avoid the term chemical energy. The material energy is associated with chemical reactions, but it is more general. The material energy simply refers to an increase of the energy when a certain amount of material is put into the system, when all the other thermodynamic variables remain constant. There is no need that a chemical reaction occurs.

---

[1] Josiah Willard Gibbs, born Feb. 11, 1839, in New Haven, CN., USA, died Apr. 28, 1903, in New Haven, USA.

2.3.6.2 Electrochemical Potential

The electrochemical potential was introduced by *Guggenheim*.[2] The electrochemical potential μ_e is the sum of the chemical potential μ and the electric potential:

$$\mu_e = \mu + \tilde{Q}\varphi = \mu + zF\varphi, \tag{2.18}$$

where $\tilde{Q}$ is the molar electrical charge, z is the elementary charge number, F is the *Faraday* constant, and φ is the electric potential. The Faraday constant refers to the electric charge of 1 mol of elementary charges.

The electrochemical potential is used to describe the equilibrium of charged particles. It arises naturally from the fact that when a charged particle enters or leaves the system, it takes its charge with it. Thus the molar charge $\tilde{Q}$ in the system remains constant, and we have $U(S, V, Q, n)$ as the energy function, where $Q = n\tilde{Q}$.

According to the rules for forming partial differentials in functions with multiple arguments, $\partial U/\partial n$ yields Eq. (2.18). Actually, there is no need to *define* the electrochemical potential, as this potential is introduced in some textbooks. As pointed out above, the electrochemical potential emerges from the fact that the electric charge and thus the electric energy are coupled to matter in an unique manner. For related cases, refer to Example 5.2 on p. 186.

2.4 Energy Representation

We deal with a simple case of a homogeneous system of one component. The energy should be a function of entropy, volume, and mol number, $U = U(S, V, n)$. Recall that it is necessary to include the mol number as parameter for sake of homogeneity. In a homogeneous system, we can represent the energy as the sum of the terms

$$U(S, V, n) = TS - pV + \mu n. \tag{2.19}$$

We deal with the situation of constant energy, thus $U(S, V, n) = $ U. Next , we substitute $TS \rightarrow y$, $-pV \rightarrow x$, and $\mu n \rightarrow z$. With these substitutions, Eq. (2.19) turns into

$$\frac{x}{U} + \frac{y}{U} + \frac{z}{U} = 1.$$

Recall that in cartesian coordinates [9]

$$\frac{x}{a} + \frac{y}{b} + \frac{z}{c} = 1,$$

---

[2] Edward Armand Guggenheim born 1901 at Manchester died 1970.

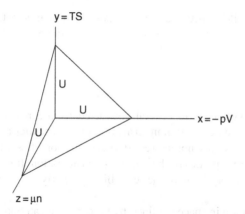

Fig. 2.1 Isodynamic path of a homogeneous one-component system

represents a plane that cuts the x-axis at a, the y-axis at b, and the z-axis at c. For this reason, the isodynamic path in an one component homogeneous system may be represented as the motion in a plane, with the coordinates TS, $-pV$, and μn. The situation is illustrated in Fig. 2.1. On the other hand, since the enthalpy H is $U + pV$, we get for isenthalpic processes

$$\frac{y}{H} + \frac{z}{H} = 1.$$

Next, we inspect the isentropic path for an ideal gas in terms of TS, $-pV$, and μn. Using the ideal gas equation, we find

$$\frac{pV}{TS} = \frac{nR}{S} = -\frac{x}{y}.$$

From the *Gibbs* form of the energy of an ideal gas, we get

$$\frac{z}{y} = -1 + n\frac{\tilde{C}_v + R}{S}.$$

Therefore, for an ideal gas, at constant entropy and mol number, the variables x, y, z are linearly dependent, as the motion is in a plane passing the origin.

There is still another way to substitute for the natural variables. By substitution of $S \to \exp(y)$, $V \to \exp(x)$, and $n \to \exp(z)$, we get

$$\frac{\partial U}{\partial y} = \frac{\partial U}{\partial S}\frac{\partial S}{\partial y} = TS.$$

Similar expressions hold for x and z. In a plot of x, y, z of constant energy, the distance of the isodynamic lines in the x, y, z directions represents the respective energy forms. Thus, if TS becomes larger at some isodynamic path, the distance of certain adjacent isodynamic curves in the direction of the ordinate (y) shrinks.

2.5 System

A thermodynamic system is in general abstracted from a real object of interest. Issues not needed for a description in terms of thermodynamics are omitted. More-over, the scientist is literally not interested in certain properties, often in the partic-ular shape and other properties of the object of consideration. It is left to the skill of the scientist to map every issue of the real object properly into the thermodynamic system.

A system is a region in space confined by borders. We can obtain information of the properties of a system via its thermodynamic variables. To get the information we have to probe the variables by the exchange of certain energy forms. Thus, the system itself is a black box that outs itself by exchange processes. We can never be sure of what is really inside a system.

Example 2.3. Consider a system that is a container with a piston. But suppose that the container is closed and we can see only the plunger. By exchange of force onto the plunger, we can find out that the force F depends on the position of the plunger l according to

$$Fl^\alpha = C.$$

Now in the container there may be a gas adiabatically enclosed or a sophisticated gear mechanism connected to a spring that simulates this behavior. That is why the system is a black box. □

In the thermodynamic sense a system is described by its extensive quantities. From these quantities, the energy can be calculated somehow. Basically, all energy forms must be taken into consideration. But most can be neglected. Those that never change can be safely neglected. In a laboratory, this may be the gravitational energy. Those that contribute very less to the energy may be neglected. In most chemical problems, surface energy does not play any role.

Note that the phase rule is based on the assumption that the description is gov-erned by S, V, n_i for each phase. The thermodynamic description of a system is confined to quantities that belong, in fact, to the system. We can use this as definition of a system. If we are focusing on a reaction vessel in the hood, then the bottles in the refrigerator beside are not part of the description of the system and will never enter.

The energy of a system may be completely characterized by its extensive vari-ables, such as S, V, n. If we would choose to replace some extensive variables by the corresponding intensive variables, then information is lost. On the other hand, certain functions of the energy, i.e., the *Legendre* transformations utilize inten-sive variables, but there is no loss of information, because the functions can be transformed back into the original form. We have to clearly differentiate the variables of the system itself and the variables that exchange energy forms with the system.

Example 2.4. We allow a system to accept compression work $-p\,dV$, $dV < 0$. To make it more clear, we write $-p_{sys}\,dV_{sys}$. We mean that the pressure of the system is p_{sys} and that the volume shrinks by dV_{sys}. This does not necessarily mean that another system, say the environment, will lose just this energy. In particular, if the pressure of the environment is higher than the pressure of the system $p_{env} > p_{sys}$, $dV_{env} = -dV_{sys}$, then the environment will lose $-p_{env}\,dV_{env}$, $dV_{env} > 0$, which is more than the system gains. Of course, no energy is lost according to the first law of thermodynamics. The energy will be recovered most probably in the form of kinetic energy and eventually as thermal energy. □

A system may exchange with the environment (in fact also a system) energy. It will never exchange any intensive quantity. This is basically a nonsense. However, the intensive quantity may change as a result of an exchange of any extensive quantity. The energy change of the system is exclusively governed by the parameters of the system itself.

Example 2.5. To a big gasometer a small piston is attached. When the piston is moving a small distance out to increase the volume by dV, the gasometer loses its energy, i.e., $-p\,dV$. It does not matter what is behind the piston.

There may be an engine that raises a weight, there may be a spring that takes up the energy, there may be gas that takes up the energy released. The energy that is taken up behind the piston can be zero in the worst case, if there is vacuum and $+p\,dV$ in the best case if the outer pressure is the same as in the gasometer itself. If the outer pressure is greater than the pressure in the gasometer, then the gas of the gasometer will not expand. The intended process is not possible at all. If the gas expands into vacuum, the interesting question arises of what will happen with the energy that is lost at the first glance. □

We may not conclude, from the energy take up in any environment of the system, on how much energy the system has lost in fact. We can only conclude what is the minimum energy lost. Only in the case of an exchange of energy in an equilibrium state we can state that no particular energy form will be lost on exchange.

When a certain energy form is delivered into a system, we cannot state that the system bears this particular energy form. We can merely state that in the system some have changed by the input of an energy form.

Example 2.6. We can allow some entropy ΔS to flow in a gas at constant volume. In this case, the entropy of the system increases by ΔS. If the entropy before the process is S, then the new entropy is $S + \Delta S$. Now we can withdraw this entropy by thermal work, but we can also allow the energy that was introduced to be pulled out by mechanical work $-p\,dV$. □

In this way, a system is a transformer of energy forms. This is not true in general. A necessary condition to transform energy forms is that the intensive variables are functions of the extensive variables. In order to transform thermal energy into energy of expansion we must have

$$p = p(S, V) \quad \text{and} \quad T = T(S, V).$$

If we would have

$$p = p(V) \quad \text{and} \quad T = T(S), \tag{2.20}$$

then the input of thermal energy could not cause an increase in pressure. However, such an action is necessary to transform the particular energy form. In the case of Eq. (2.20) we can withdraw thermal energy that has been put into the system only as thermal energy as such and the same is true for the volume energy.

2.5.1 Types of Systems

Several ways of classification of systems are possible. One classification refers to the way how energy forms can be exchanged and another more recent classification refers to the type of entropy.

2.5.1.1 Exchange of Energy Forms

Historically we differentiate between

1. Isolated systems.
2. Closed systems.
3. Open systems.

- The isolated system has no interaction with an environment. It cannot exchange any energy form with its environment.
- The closed system cannot exchange matter, but it can exchange other energy forms with its environment.
- The open system not only can exchange matter, but also can exchange other energy forms with its environment.

The reason for this classification arises from the various *Legendre* transformations of the energy. However, these transformations allow for a more subtle classification.

For example, when free exchange of entropy of the system at constant temperature $(dT = 0)$ may occur, but no exchange of volume energy is allowed $(dV = 0)$, we have for $dn = 0$, $d(U - TS) = -S dT - p dV + \mu dn = 0$. Thus, the Helmholtz energy is zero for this type of system. Other cases can be easily constructed in the same manner.

2.5.1.2 Entropy Behavior

We can define general properties of the entropy [10, 11]:

1. Super additivity: $S(x + y) \geq S(x) + S(y)$,
2. Homogeneity: $S(\lambda x) = \lambda S(x)$,
3. Concavity: $S(\lambda x) + (1 - \lambda)y) \geq \lambda S(x) + (1 - \lambda)S(y)$.

Whether these three properties may, or may not, be independently fulfilled or not we can distinguish 2^3 different cases. These cases are shown in Table 2.2.

For example, the entropy of an ideal gas is

$$S = k \ln \left[\frac{a U V^g}{N^h} \right].$$

Here k, a, g, and h are positive constants. V is the volume, U is the internal energy, and N is the number of particles in the gas. Three classes of thermodynamics can be formed:

$$\text{Type 1:} h = g + 1$$
$$\text{Type 4:} h < g + 1$$
$$\text{Type 6:} h > g + 1.$$

Type 1 refers to ordinary thermodynamics, Type 4 is the classical analogue of the thermodynamics of the *Schwarzschild*[3] black hole, it violates the property of concavity, however, it keeps the super additivity.

Table 2.2 Classes of thermodynamic systems

Number	Super additivity	Homogeneity	Concavity
1	+	+	+
2	+	−	+
3	+	+	−
4	+	−	−
5	−	+	+
6	−	−	+
7	−	+	−
8	−	−	−

+: fulfilled
−: not fulfilled

---

[3] Karl Schwarzschild, born Oct. 9, 1873, in Frankfurt am Main, died May 11, 1916, in Potsdam.

2.6 Conservation Laws

There are several natural conservation laws. We have a natural feeling for some of these conservation laws. Distances in a distance table should follow certain conservation laws. Traveling from A to B directly should not show a longer distance than traveling from A to B via C. But in practice, some of these distance tables show this unexpected property – due to errors.

2.6.1 Heat of Reaction

In contrast to this is the law of *Heß*.[4] The law of Heß states that the heat of reaction is independent of the reaction path. The law of Heß can be regarded as a precursor of the first law of thermodynamics.

2.6.2 Mass

The mass is constant. However, this is only true in a nonrelativistic situation. The equation of continuity expresses the conservation of mass. In relativity theory the mass m increases with its velocity v by

$$m = \frac{m_0}{\sqrt{1 - v^2/c^2}},$$

where m_0 is the mass at rest and c is the velocity of light.

2.6.3 Charge

The electric charge is constant. It is claimed that no exception has been observed. On the other hand, in the course of neutralization reactions, obviously charges disappear. A closer look into modern physics disguises the disappeared charges as being merely close together. So it is a matter of philosophy, the belief in modern physics, or the belief in illusion whether the charges after neutralization are there, not there, or merely disguised.

---

[4] Heß, Germain Henri (Герман Иванович Гесс) born Aug. 7, 1802, in Geneva died Dec. 12, 1850, in St. Petersburg.

2.6.4 Momentum

The conservation of momentum and angular momentum is a fundamental law in mechanics. *Kepler*'s second law can be easily justified by the conservation of angular momentum.

The momentum of a system remains constant if there are no external forces acting on the system. This is actually Newton's first law, the law of inertia. The conservation of momentum is valid in classical mechanics. However, *Heisenberg*'s[5] principle of uncertainty states that

$$\Delta x \Delta P \geq \frac{\hbar}{2}.$$

So, considering very small distances, a possible fluctuation of the momentum could occur that violates the conservation of momentum.

Note the laws of angular momentum and the laws of momentum are related by the operator $\nabla \times$ applied to the respective equations of momentum.

2.6.5 Volume

In science, the total volume of space is constant. The space is treated to have a constant extension. However, cosmology tells us that the space is expanding. Of course, we may ask what is behind the Schwarzschild radius and include a suspected space into consideration. The Schwarzschild radius r refers to the size of a black hole with a mass m. In a simplified treatment, it can be derived by setting the escape velocity v of a black hole equal to the velocity of light c. The velocity of light is $c = 2.99792458 \times 10^8 \, \mathrm{m\,s^{-1}}$. The escape velocity v is

$$v = c = \sqrt{\frac{2Gm}{r}},$$

where G is the gravitational constant which is $G = 6.672 \times 10^{-11} \, \mathrm{N\,m^2 kg^{-2}}$.

On the other hand, even in a much simpler case the volume of a system may not be constant. Consider a system composed of two cylinders of different diameters with two pistons connected by a plunger. As the plunger moves the total volume of this system changes. Of course, the surrounding environment would compensate these changes.

---

[5] Werner Karl Heisenberg, born Dec. 5, 1901, in Würzburg died Feb. 1, 1976, in Munich.

2.6.6 Surface Area

In contrast to volume, the surface area is definitely not constant. Surface area can be generated in an arbitrary way. Consider the formation of a foam.

2.6.7 Baryon Number

In a closed system, the baryon number remains constant. If a proton can decay, this conservation law is no longer valid. In particle physics there are also other conservation laws, e.g., the conservation of parity, the conservation of color, and others.

2.6.8 Energy

The first law of thermodynamics can be formulated as the conservation of energy.

2.6.9 Time Reversal

All time-dependent mechanical processes, such as a motion of a particle through the space, are still valid on time reversal. This means if we change the sign of time in a physical law, still a possible process will emerge. *Maxwell* observed that the second law of thermodynamics is not consistent on time reversal [12]. Time reversal plays a role in Loschmidt's paradox.

2.7 Process

A thermodynamic process is anything that changes the thermodynamic variables in the system or environment. A thermodynamic process can be effected in various ways. Note the difference between a thermodynamic process and the way to effect the process in practice, i.e., the realization of the thermodynamic process.

Example 2.7. We consider the isothermal expansion of a gas. Purposely we do not deal with an ideal gas, but with a real gas. At fixed mol number, the energy of the ideal gas is governed by two relevant variables, i.e., (S, V). Before the expansion, at start, we mark the variables by the subscript s. At start, the energy is $U_s = U(S_s, V_s)$. After the isothermal expansion, the relevant variables have been changed by ΔS and ΔV. Therefore, at the end of expansion, which we indicate by the subscript e we have $S_e = S_s + \Delta S$ and $V_e = V_s + \Delta V$. In order to have an isothermal expansion we need the condition

$$\frac{\partial U(S, V)}{\partial S}\bigg|_{S_s, V_s} = T = \frac{\partial U(S, V)}{\partial S}\bigg|_{S_e, V_e}.$$

For the process as such, it is not important at which path we have reached $S_s \to S_e$ and $V_s \to V_e$.

On the other hand, for the realization of the particular process, it is important how we could reach this state. For example, the expansion could be achieved by allowing to expand against a certain pressure or to allow it to expand into vacuum. If the gas is doing work, then energy is lost, and thermal energy in the form of entropy has to be fed in a certain amount. Expanding into vacuum may change the internal energy and also some thermal energy is needed to maintain the temperature. □

2.8 Transformers

We consider now a nonstandard exchange of an extensive variable. Two pistons may be connected with a gear or they may have simply different area so that $dV_1 \neq -dV_2$. But there may be a factor, not necessarily a constant, so that $dV_1 + \lambda dV_2 = 0$. So

$$dU(V_1, V_2) = -p_1 dV_1 - p_2 dV_2 = (-p_1 + p_2/\lambda)dV_1 \qquad (2.21)$$

and to have the energy stationary, $dU(V_1, V_2) = 0$. The engine that effects $dV_1 + \lambda dV_2 = 0$ we will call a transformer for the respective variable. The factor λ is dimensionless. The experimental setup is very simple, it consists of two pistons with different areas connected by a common plunger as shown in Fig. 2.2. The factor λ is the ratio of the areas. If the piston associated to the volume V_1 has an area of A_1 and the piston associated to the volume V_2 has an area of A_2, then $\lambda = A_1/A_2$.

The energy will become stationary when $-p_1 + p_2/\lambda = 0$. This means that when the energy becomes stationary, the pressures in the two subsystems need not be equal. To calculate the stationary value, we must have some information about the functional dependence of the pressure with the volume, i.e.,

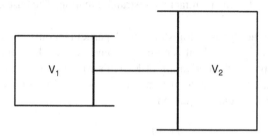

Fig. 2.2 Two pistons of different area connected with a common plunger

$$p_1 = p_1(V_1); \quad p_2 = p_2(V_2).$$

Another example for a transformer is a chemical equilibrium with dissociation, e.g.,

$$J_2 \rightleftharpoons 2J. \tag{2.22}$$

Here $dn_J = -2dn_{J_2}$. On the other hand, a simple sublimation $J_{2,s} \to J_{2,g}$ is a standard exchange with respect to the mol numbers.

Example 2.8. We can replace formally the variables p and V in Eq. (2.21) by T and S. In this way, we obtain the corresponding thermal problem to the volume energy as

$$dU(S_1, S_2) = -T_1 dS_1 - T_2 dS_2 = (-T_1 + T_2/\lambda)dS_1. \tag{2.23}$$

Equation (2.23) is completely analogous to Eq. (2.21), but there is some important difference. It is possible to conduct a thermodynamic process in which $dU = 0$ and $T_1 \neq T_2$. Setting $dS_1 + \lambda dS_2 = 0$, we get from the first equality in Eq. (2.23) $T_1 dS_1 + T_2 dS_2 = 0, \lambda = T_2/T_1$. Inserting this into the second equality of Eq. (2.23), we get the identity $T_1 = T_1$. Therefore, it is necessary to obtain the factor λ from an independent relation, if we want to calculate some specific values of S_1 and S_2 where the energy would be stationary. However, we can state that for stationary energy $dU = 0$, $T_1 \to T_2$ implies $dS_1 \to -dS_2$, which means the conservation of entropy. $\square$

Two systems (') and (''), not necessarily reservoirs, but big systems, can be thermally coupled by a Carnot cycle. This is in practice a heat pump, or if it runs in the reverse direction, a heat engine. In this case, we have $S' + S'' = 0$, thus if the systems are big, for one revolution of the Carnot cycle $dS' + dS'' = 0$. This constraint does not imply that $T' = T''$. Namely, the Carnot engine is an active device that releases or donates energy. In the Carnot engine, the change of entropy S' is not directly connected with the change of entropy S'', even when the cross balance $dS' + dS'' = 0$ over a full turn holds.

In general, a transformer for an extensive variable X_1 is characterized by $X'_1 + \lambda X''_1 = C$ with $\lambda \neq 1$. This is in fact a constraint equation. The case of $\lambda = 1$ is the standard exchange.

There are still other types of transformers. A particular constraint equation could look like $dX'_1 + dX''_2 = 0$. In this case, the conversion of different energy forms occurs. For example, in the derivation of Kelvin's equation of surface tension such a constraint appears. The volume of a sphere $V = 4r^3\pi/3$ is connected to the area $A = 4r^2\pi$ via the differential equation

$$\frac{dV(r)}{dr} - A = 0.$$

2.9 Reservoir

A reservoir is a system that does not change its intensive variable when the corresponding energy-conjugated variable is changed. For example, for any man-made system that may change its volume, the atmosphere is a reservoir with respect to pressure. Here we idealize that the change of volume of this man-made system will be fast in comparison to the natural changes of pressure due to weather. The ocean is a reservoir with respect to the chemical potential. If we add a little bit of salt, then the chemical potentials of the components in the ocean will remain essentially constant.

It appears that a system is close to a reservoir if it is sufficiently big. Of course, there is a more subtle possibility to establish a reservoir if there is an engine that keeps the intensive variable under consideration somehow constant. We may think of a thermostat as a reservoir for temperature. In this case, the system is not standing alone, however, because the engine adds or removes entropy to keep the temperature of the system constant.

We may address the big systems as reservoirs of first order and the engine-controlled reservoirs as reservoirs of second order. However, dealing with the concept of a reservoir it is immaterial how the properties of a reservoir are achieved.

Example 2.9. A thermal reservoir is a system with the property

$$\frac{\partial T(S,\ldots)}{\partial S} \to 0.$$

Thus, on input or output of entropy the temperature will remain utmost constant.
□

More general, a reservoir with respect to a thermodynamic variable X has the property of

$$\frac{\partial U^2(X,\ldots)}{\partial^2 X} = 0.$$

The general method on how to couple a system of interest with a reservoir is shown in the literature [13].

2.10 Constraint Makers

A constraint maker is a system that is doing something in a rather mysterious way. Usually such a system is essential for the proper work of other thermodynamic systems, but only marginal attention is focussed to explain the properties of such a system. Thereby the obscure system is idealized in some way. Examples for constraint makers are the

- Rigid wall
- Diathermic wall in heat conduction
- Porous plug in the Joule – Thomson experiment
- Semipermeable membrane in osmometry
- Electrodes
- Interphase borders
- Filter membranes in general

2.10.1 Rigid Wall

The rigid wall is stiff. When pressure is applied, it does not change its shape and size. If a thermodynamic system is enclosed in a rigid wall, its volume remains constant, i.e., it ensures that $dV = 0$ under any change of the thermodynamic variables of the system.

2.10.2 Diathermic Border

The diathermic border has no volume. So it cannot take up volume energy. The diathermic border faces a temperature T_h from one surface and a temperature T_l from the other surface. Here $T_h > T_l$. When entropy $dS(T_h)$ passes into the border across a surface, the diathermic border ejects $-dS(T_l)$ at the opposite surface following the relation

$$T_h dS(T_h) + T_l dS(T_l) = 0.$$

In this way, the diathermic border is a generator for entropy, even when it cannot store entropy. Moreover, the diathermic border acts as a valve for entropy. If $T_h < T_l$ then it does not allow an inflow of entropy, $dS(T_h)$.

2.10.3 Porous Plug

Also, the porous plug in the *Joule – Thomson* experiment is an idealized device. It has certain properties, such as lack of volume, common to the diathermic border, but it does not allow a flow of entropy without flow of matter. When matter flows though the porous plug, the pressure and the temperature are reduced. This is associated with an increase of entropy.

2.10.4 Semipermeable Membrane

The semipermeable membrane in osmometry is a valve for the flow of certain kind of matter. It must be sufficiently stiff to withstand external forces resulting from a pressure difference on both sides.

2.10.5 Electrode

An ion may be imagined as a neutral matter that carries a charge. Of course, the charge has also some matter. So we will be more precise in the next sentence. An electrode serves as a separator for ions and the charge. It is permeable to the charge, but is not permeable to the neutral part of the ion. When an ion moves to the electrode, the charge is stripped off from the ion and the ion is discharged. Think about the similarity to the semipermeable membrane.

2.10.6 Interphase Border

When matter passes through an interphase border, e.g., through a liquid – gas interphase border, the matter evaporates and takes up thermal energy.

2.10.7 Filter Membrane

In the same way as for electrodes, we invent a filter membrane for entropy. When matter moves to this filter, the filter strips the entropy that the matter is carrying with it and leaves the entropy on the entrance side and the matter may move on the other side of the membrane. This device is an ideal cooler. To be in accordance with the second law of thermodynamics, a filter membrane for entropy must be an active device.

2.10.8 Examples of Constraints

In thermodynamics, often the extremum of a function of energy or the energy itself has to be found out to get an equilibrium state. This extremum is searched with certain additional conditions. These conditions are addressed as constraint equations. For example, the energy of a system $U(V', V'')$, consisting of two subsystems ($'$ and $''$), will become a minimum under the condition of constant volume $V_{tot} = V' + V''$.

It is important to verify that the constraint is in fact physically taking place at the process under consideration. If the total volume should be constant, both volumes can be coupled by a plunger. The plunger ensures that if one volume expands by dV' the other volume will shrink by $dV'' = -dV'$.

Constraints are reflecting to some extent the process under consideration. The set $dV' = 0$ and $dV'' = 0$ implies $dV' + dV'' = 0$, but the first two constraints are stronger. Further, the constraint $dS' - \tilde{S}'dn' = 0$, which refers to a process at constant molar entropy, acts both on the entropy and on the mol number, when used in context with Lagrange multipliers.

In other words, the form of the constraint equations governs the intensive variables in equilibrium. If the volume of an individual system (') is kept constant, then the pressure in a neighborhood coupled system (") is not necessarily the same. However, if the wall is porous, then a gas will move through the porous wall to establish equal pressure. This process does not occur by an exchange of volume energy, but by an exchange of matter. However, if the wall is only semipermeable, again a pressure difference may be there in equilibrium in a multi-component system.

To illustrate the improper use of constraints, we deal with the Carnot cycle with an ideal gas as working system. In the isothermal expansion, the gas takes up thermal energy from the reservoir and does work of expansion to a further external system. Thus, the energy change of the working gas is

$$dU(T, V, n) = C_v dT + \frac{\partial U}{\partial V} dV + \frac{\partial U}{\partial n} dn = 0. \qquad (2.24)$$

On the right-hand side of Eq. (2.24), the first and the third terms cancel because $dT = 0$ and $dn = 0$, and the second term cancels, because for an ideal gas, $\frac{\partial U(T,V,n)}{dV} = 0$. The statement $dU = 0$ remains still valid, if we change the variables from $(T, V, n) \to (S, V, n)$. With $dn = 0$ we have

$$dU(S, V, n) = T dS - p dV = 0. \qquad (2.25)$$

In the Carnot cycle, the process indicated in Eq. (2.25) occurs twice. First, the gas is expanded isothermally in contact with the reservoir at high temperature (h) and after the adiabatic step, the gas is compressed isothermally in contact with the reservoir at lower temperature (l). We place now formally the indices to indicate the two steps and get

$$\begin{aligned} dU_h(S_h, V_h, n) &= T_h dS_h - p_h dV_h &= 0 \\ dU_l(S_l, V_l, n) &= T_l dS_l - p_l dV_l &= 0 \end{aligned} \qquad (2.26)$$

We have now the constraint for a reversible process $dS_h + dS_l = 0$. However, we may not add up both equations (2.26) and conclude that $T_l = T_h$, etc. We would use the constraint in a wrong way.

In fact, even when the constraint $dS_h + dS_l = 0$ exists, this constraint does not appear actually in the process, because the entropy of the reservoir with lower temperature does not immediately increase when the entropy of the reservoir with higher temperature decreases. The process is conducted in a different way. Rather the constraints appearing in Eq. (2.26) are valid as such. The increase of entropy under isothermal conditions in the working gas causes directly an expansion of the volume of the working gas, thus doing work to an external system and vice versa.

2.11 Fundamental Equations

The term *fundamental equation* as introduced by Gibbs is now also known as a Gibbs fundamental form [14, pp. 55–353]. It should not be confused with the Gibbs function, which is also addressed to as the Gibbs free energy or free enthalpy, but the Gibbs function. Gibbs himself had introduced what we here call Gibbs function in a different way.

First, we follow the ideas of Gibbs. A fundamental form is a relation of the energy, entropy, volume, and the masses or mol numbers, such as

$$\Phi(U, S, V, n_1, \ldots, n_k) = 0. \tag{2.27}$$

There are $k + 3$ variables. From the property of homogeneity, the number of these variables can be reduced by one, by dividing by the volume. In this way, the densities of energy, entropy, and masses will emerge. If only one component is there, also the division by n_1 is useful. The function Φ relates now $k + 2$ variables, or totally there are $k + 1$ independent variables. Preferably Φ will be resolved to U, with the set of variables $S, V, n_1, n_2, \ldots, n_k$ of independent variables. If the energy is a known function of these variables, then also the differential may be formed as

$$dU = T dS - p dV + \mu_1 dn_1 + \mu_2 dn_2 + \ldots + \mu_k dn_k. \tag{2.28}$$

$T, p, \mu_1, \mu_2, \ldots, \mu_k$ are variables, but at the same time known functions of the independent variables $S, V, n_1, n_2, \ldots, n_k$. We can obtain them by forming the partial derivatives

$$\frac{\partial U}{\partial S} = T$$
$$\frac{\partial U}{\partial V} = -p$$
$$\frac{\partial U}{\partial n_1} = \mu_1 \tag{2.29}$$
$$\ldots = \ldots$$
$$\frac{\partial U}{\partial n_k} = \mu_k.$$

So we have totally $2k + 5$ variables

$$U, S, V, n_1, n_2, \ldots, n_k, T, p, \mu_1, \mu_2, \ldots, \mu_k,$$

but of them only $k + 2$ are really independent. On the other hand, Eqs. (2.28) and (2.29) provide $k + 3$ relations.

A single equation from all these relations can be deduced called Gibbs fundamental equation. Not only the energy equation, but all Legendre transformations of the energy equation, such as enthalpy H, free energy F, free enthalpy G, the Massieu function, are fundamental equations. We summarize some types of fundamental equations in Table 2.3. As variables we use just those common in chemical thermodynamics. The fundamental equations shown in Table 2.3 are completely equivalent because they can be commuted one into another by the Legendre transformation without any loss of information. The variables given for the particular functions are addressed as the natural variables. For example, the energy U has the set of $S, V, n_i, \ldots$ as natural variables. If we want to use the temperature T instead of the entropy, we lose information, because the temperature is related to the entropy as the differential of the energy with entropy. Namely, the solution of the entropy from the temperature involves one unknown constant of integration.

The functions with the dimension of energy are sometimes addressed as thermodynamic potentials. This should not be confused with other thermodynamic functions that are also addressed as potential, such as the chemical potential or the electric potential.

There are basically two types of Legendre transformations, one type based on the energy and the other type based on the entropy. There are various other types of Massieu functions [15]. The Gibbs – Duhem equation is a very special type of Legendre transformation. It is a full transformation and is a relation of the intensive variables only.

To illustrate what is not a fundamental equation we compare a relation Eq. (2.27) with that given in Eq. (2.30).

$$\Psi(U, T, V, n_1, \ldots, n_k) = 0. \tag{2.30}$$

Both relations differ only in that in Eq. (2.30), the temperature T appears instead of the entropy. From Eq. (2.27) we can deduce the temperature by $\partial U \partial S = T$. However, in Eq. (2.30) we are losing just this relation. So we cannot get back the

Table 2.3 Types of fundamental equations

Type			Differential
Energy	$dU(S, V, n_i, \ldots)$	$=$	$+T\,dS - p\,dV + \sum_{i=1}^{i=K} \mu_i\,dn_i$
Entropy	$dS(U, V, n_i, \ldots)$	$=$	$+T^{-1}dU + pT^{-1}dV - \sum_{i=1}^{i=K} \mu_i T^{-1}dn_i$
Enthalpy	$dH(S, p, n_i, \ldots)$	$=$	$+T\,dS + V\,dp + \sum_{i=1}^{i=K} \mu_i\,dn_i$
Free Energy	$dF(T, V, n_i, \ldots)$	$=$	$-S\,dT - p\,dV + \sum_{i=1}^{i=K} \mu_i\,dn_i$
Free Enthalpy	$dG(S, p, n_i, \ldots)$	$=$	$-S\,dT + V\,dp + \sum_{i=1}^{i=K} \mu_i\,dn_i$
Massieu	$d\Phi(T^{-1}, V, n_i, \ldots)$	$=$	$-U\,dT^{-1} + pT^{-1}dV - \sum_{i=1}^{i=K} \mu_i T^{-1}dn_i$
Gibbs-Duhem Equation	0	$=$	$-S\,dT + V\,dp + \sum_{i=1}^{i=K} n_i\,d\mu_i$

entropy S. Note that the partial derivatives for $U(T, V, n_1, \ldots, n_k)$, in general, are not the same as for $U(S, V, n_1, \ldots, n_k)$ given in Eq. (2.29).

2.12 Partial Quantities

Here we exemplify the concept of partial quantities in terms of the volume in a system composed of two components. However, the concept is more general. We assume that the volume is a function of the mol numbers, $V = V(n_1, n_2)$. Then the total differential is

$$dV(n_1, n_2) = \frac{\partial V}{\partial n_1} dn_1 + \frac{\partial V}{\partial n_2} dn_2. \tag{2.31}$$

Because of the thermodynamic similarity, i.e., the volume is a homogeneous function of first order

$$V = V_1 n_1 + V_2 n_2.$$

By division by $n_1 + n_2$, thus introducing molar quantities, we obtain

$$\tilde{V} = V_1 x_1 + V_2 x_2 = V_1(1 - x_2) + V_2 x_2. \tag{2.32}$$

The partial volumes V_1 and V_2 do not need to be constant quantities, in particular in nonideal systems.

Obviously, in Eq. (2.32) the functional dependence changes now to x_1 or x_2 as independent variables, since the constraint $x_1 + x_2 = 1$ holds.

In Fig. 2.3, a plot of the total molar volume against the mole fraction is shown. From Eq. (2.32), immediately the relation

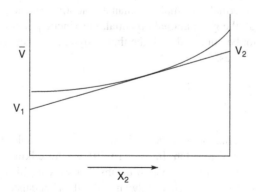

Fig. 2.3 Plot of molar volume against mole fraction

$$\frac{\partial \tilde{V}}{\partial x_2} = -V_1 + V_2 \qquad (2.33)$$

is obtained. The tangent cuts the ordinate at $x_2 = 0$ at

$$\tilde{V} - \frac{\partial \tilde{V}}{\partial x_2} x_2,$$

where the molar volume and the slope have to be taken at the particular point x_2, where the tangent touches the curve. From Eq. (1.23) follows that the tangent is some kind of the Legendre transformation.

Using Eqs. (2.32) and (2.32), we obtain finally the relation.

$$\tilde{V} - \frac{\partial \tilde{V}}{\partial x_2} x_2 = V_1.$$

2.13 Thermodynamic Derivatives

2.13.1 Potentials

The term potential originates from the Latin word *potentia* which means power. It is used in the sense of possibility in contrast to the term actual. There are several classes of thermodynamic derivatives of interest. Starting from the energy in natural variables, $U(S, V, n)$, the derivatives

$$\frac{\partial U(S, V, n)}{\partial S} = T; \quad \frac{\partial U(S, V, n)}{\partial V} = -p; \quad \frac{\partial U(S, V, n)}{\partial n} = \mu$$

are the common intensive variables, which are in general functions of the set S, V, n. The intensive variables are energy conjugated to their respective extensive variables. The quantity μ is addressed as chemical potential. Consequently, the other derivatives should be the thermal potential, the manal potential. Further, there should be the electric potential, the gravitational potential, the kinetic potential, etc. All these quantities should not be confused with the thermodynamic potentials that are basically energies themselves.

2.13.2 Capacities

The term capacity originates from the Latin words *capax* as able to hold much, or from *capere* as to take or to catch. In ordinary life we are talking, for example, of a vat that has the capacity of 10 l, which means that we can fill in a maximum of 10 l. The best-known capacity in thermodynamics is the heat capacity, either at constant volume or at constant pressure. Other terms concerning heat capacity appear in the magnetocaloric effect, more commonly addressed as adiabatic demagnetization. However, there may also be a capacity with respect to compression energy. The heat capacity is defined as

$$C_v = \frac{\partial U(T, V, n)}{\partial T}; \quad C_p = \frac{\partial H(T, p, n)}{\partial T}.$$

Using Jacobian determinants, we can write

$$C_v = \frac{\partial (U, V, n)}{\partial (T, V, n)} = \frac{\partial (U, V, n)}{\partial (S, V, n)} \frac{\partial (S, V, n)}{\partial (T, V, n)} = T(S, V, n) \frac{\partial S(T, V, n)}{\partial T},$$

and in an analogous way for C_p

$$C_p = \frac{\partial (H, p, n)}{\partial (T, p, n)} = \frac{\partial (H, p, n)}{\partial (S, p, n)} \frac{\partial (S, p, n)}{\partial (T, p, n)} = T(S, p, n) \frac{\partial S(T, p, n)}{\partial T}.$$

Obviously, the general procedure to get a capacity is to formulate a thermodynamic function initially in its natural variables, say $U(S, V, n)$. Then a variable is replaced by its energy-conjugated variable, such as $U(T, V, n)$, and the partial differential with respect to this variable is formed to get the capacity, namely $\partial U(T, V, n)/\partial T$.

We try this procedure with the replacement of the volume by the negative pressure in the energy and in the Helmholtz energy

$$C_S = \frac{\partial U(S, -p, n)}{\partial (-p)}; \quad C_T = \frac{\partial F(T, -p, n)}{\partial (-p)}.$$

Thus, we are developing a capacity with respect to the energy of compression, rather than to heat energy. We arrive at

$$C_S = -p(S, V, n) \frac{\partial V(S, -p, n)}{\partial (-p)}; \quad C_T = -p(T, V, n) \frac{\partial V(T, -p, n)}{\partial (-p)}.$$

For example, for an ideal gas,

$$C_T = -p(T, V, n) \frac{\partial V(T, -p, n)}{\partial (-p)} = -\frac{nRT}{p} = -V.$$

In other words, if we would compress an ideal gas to zero volume, it can no longer store energy or free energy. In general, observe the following functional dependence of the capacities:

$$C_v(T, V, n); \quad C_p(T, p, n); \quad C_S(S, -p, n), \quad C_T(T, -p, n).$$

2.13.3 Susceptibilities

The term susceptibility originates from the Latin word *susceptus*, the past participle of *suscipere* in the sense of sustain or support. So we have not presented a real translation into basic English words.

Most common, the term susceptibility is used as electric susceptibility and magnetic susceptibility. In this connection, the susceptibility is the amount of dipole moment generated in a field and this per unit of volume. Dipole moment and field apply both to the electric and magnetic case. In physics, both electric and magnetic susceptibility are defined in a way so that they are dimensionless quantities.

2.13.3.1 Electric Susceptibility

The electric susceptibility χ_e is defined as

$$\chi_e = \frac{P}{\varepsilon_0 E}.$$

Here ε_0 is electric permittivity of vacuum, P is the dielectric polarization density, and E is the local electric field strength.

The dielectric polarization density has the units of $m^{-2} \, s \, A$ (dipole moment per volume) and the electric field strength has the units of $V \, m^{-1} = m \, kg \, s^{-3} \, A^{-1}$.

The electric permittivity is related to the permeability of vacuum μ_0 by

$$\varepsilon_0 \mu_0 = c^2.$$

In other words, the product of the electric permittivity of vacuum and the permeability of vacuum is the square of the velocity of light. The permeability is related to magnetism and has the units of $m \, kg \, s^{-2} \, A^{-2}$. It is magnetic induction per length. Thus, the electric permittivity has the units of $m^{-3} kg^{-1} s^4 A^2$. It is an electrical capacity per length. From these considerations, the electric susceptibility turns out as a dimensionless quantity.

2.13.3.2 Magnetic Susceptibility

The magnetic susceptibility χ_m is defined as

$$\chi_m = \frac{M}{\mathcal{H}}.$$

Here M is the magnetization and $\mathcal{H}$ is the local magnetic field. Magnetization has the unit of $A \, m^{-1}$ and the magnetic field has the same unit of $A \, m^{-1}$. Therefore, the magnetic susceptibility is a dimensionless quantity.

2.13.3.3 Generalized Susceptibilities

From the foregoing it is appealing to relate generalized susceptibilities χ_ξ as the differential of an arbitrary extensive variable X with respect to its energy-conjugated intensive variable ξ,

$$\frac{\partial X}{\partial \xi}.$$

Examples are

$$\frac{C_v}{T} = \frac{\partial S(T, V)}{\partial T}; \quad \beta_T = -\frac{1}{V}\frac{\partial V(T, p)}{\partial p}; \quad \beta_S = -\frac{1}{V}\frac{\partial V(S, p)}{\partial p}.$$

Obviously, there is a relationship between the heat capacity and something like a susceptibility. β_T and β_S are the isothermal compressibility and the isentropic compressibility. The common definition of compressibility and electric and magnetic susceptibility suggests that susceptibility should be defined rather as

$$\chi_\xi = \frac{1}{V}\frac{\partial X}{\partial \xi} \quad \text{than} \quad \chi_\xi = \frac{1}{X}\frac{\partial X}{\partial \xi}.$$

It becomes now clear that the susceptibilities describe the response of the extensive variable on a change of the associated intensive variable. At first glance, it seems to be sufficient to place only one index, because the nature of the extensive variable can be found out. The generalized susceptibilities cannot be derived readily from the energy equation, but are accessible from the Legendre transformations of the energy. For instance, for

$$dG(T, p, n) = -S(T, p, n)dT + V(T, p, n)dp + \mu(T, p, n)dn,$$

a susceptibility can be derived as

$$\frac{\partial S(T, p, n)}{\partial T} = -\frac{\partial^2 G(T, p, n)}{\partial T^2}, \tag{2.34}$$

etc. Thus, the susceptibility appears to be a second-order derivative in terms of the natural variables of the energy-derived functions, the thermodynamic potentials.

Closer inspection of Eq. (2.34) indicates that it is possible to form also derivatives of the same kind, e.g., from the Helmholtz energy, so it is advisable to add the set of variables to the respective susceptibility.

Susceptibilities play a role in the theory of thermodynamic stability. Moreover, they can provide information useful for thermodynamic state functions.

2.13.4 Mixed Derivatives

There is a further class of important derivatives that can be traced neither back to a capacity nor a susceptibility. The most well known of these is the thermal expansion coefficient. Here the differential of an extensive variable with respect to a nonenergy-conjugated variable appears, such as

$$\alpha = \frac{1}{V} \frac{\partial V(T, p)}{\partial T}.$$

This type of differential can be converted into other expressions by using the Maxwell relations. The thermal expansion coefficient α originates naturally from the Gibbs free energy, since for the Gibbs free energy the set (T, p, n) are the natural variables.

2.14 Thermodynamic Similarity

Imagine a gas consisting of a single component in a big vessel. There should not be any external forces acting on the gas, such as gravity. It should be sufficient to describe the gas in the thermodynamic sense completely by three variables, i.e., the entropy, the volume, and the number of particles.

To find out the state of the gas, we probe a small, but still macroscopic, region within the vessel with the volume V. We will find a certain mol number of particles n and we will find a certain content of entropy S in this particular probed volume. If we probe in another region a volume of the same size, we will find the same values of entropy and mol number. Thus, from the thermodynamic data afterward we cannot find out from which region we had probed.

We address the two regions from where we had probed as two subsystems of the gas in the vessel. If we combine the two subsystems into a single one, it is intuitively obvious that in the combined subsystem the volume, the number of particles, and further the entropy will be additive. Also, the energy will be additive. This is the condition of a homogeneous system, which is also addressed as a phase.

If the vessel contains both liquid and gas, then the additivity will no longer hold in general, but is, if at all, restricted to carefully selected regions. In particular, it must be probed along an interphase border, in a certain position to catch equal amounts of the phases and surface.

The mathematical formulation of homogeneity of a phase occurs by the theorem of *Euler* on homogeneous functions. The theorem of *Euler* and the postulate that the energy is a homogeneous function of first order implies a property that resembles similarity. However, this is not a geometrical similarity, as we may see, and the energy function involving the volume V and a surface A, e.g., $U(S, V, A, n)$. The homogeneity demands that both volume and surface should grow by the same factor λ to result in

$$\lambda U(S, V, A, n) = U(\lambda S, \lambda V, \lambda A, \lambda n). \tag{2.35}$$

Note that on choosing $\lambda = 1/n$ will make n to disappear as independent variable in the argument list. The right-hand side of Eq. (2.35) turns into

$$U\left(\frac{S}{n}, \frac{V}{n}, \frac{A}{n}, 1\right) = U(\tilde{S}, \tilde{V}, \tilde{A}),$$

in that the respective molar quantities emerge. Therefore, also the energy will refer to molar quantities. The left-hand side of Eq. (2.35) tells us that the energy of an arbitrary-sized system can be converted into the molar energy by dividing the energy by the mol number n.

On the other hand, geometric similarity wants a growth like

$$
\begin{aligned}
V &\propto & \lambda & & \propto l^3 \\
S &\propto & \lambda & & \propto l^3 \\
n &\propto & \lambda & & \propto l^3 \\
A &\propto & \lambda^{2/3} & & \propto l^2
\end{aligned}
\tag{2.36}
$$

In Eq. (2.36), l is a length. The situation becomes still more puzzling, when gravitation is in effect. In the case of gravitational energy we could think that the extensive variable is $gd(mh)$, where m is the mass and h is the height (position) in the gravitational field. An increase of the system by a factor of λ would cause the mass to increase by a factor of λ^1, since it runs like the mol number. If we treat the gravitational energy as $(mg)dh$ as usual then only the height has to be increased.

We will address the homogeneity according to Eq. (2.35) as thermodynamic similarity. This similarity outs itself that an enlargement of the whole system is effective, when two systems of the same kind are put into one system. For instance, with regard to kinetic energy, if we consider two particles with the same velocity, we may integrate these two particles into one system to double the momentum, but having the same velocity.

2.15 Gibbs – Duhem Equations

If we use $dU(S, V) = T\,dS - p\,dV$, the Gibbs – Duhem equations demand $-S\,dT + V\,dp = 0$. However, this application of the Gibbs – Duhem equations is wrong. In fact, a similarity of the system is relying, i.e.,

$$U(\alpha S, \alpha V, \alpha n) = \alpha U(S, V, n). \tag{2.37}$$

Forming the differential of Eq. (2.37) with respect to α and evaluating at $\alpha = 1$ results in

$$TS - pV + \mu n = U(S, V, n). \tag{2.38}$$

Forming the total differential from Eq. (2.38) results in

$$T\,dS + S\,dT - p\,dV - V\,dp + \mu\,dn + n\,d\mu = dU(S, V, n)$$

and from the form of dU

$$S\,dT - V\,dp + n\,d\mu = 0. \tag{2.39}$$

We must include all variables that contribute to similarity. For instance, if in a system the surface energy is relevant, the surface does not increase linearly with the volume of a similar system and a Gibbs – Duhem equation does not hold longer.

Example 2.10. We illustrate the Gibbs – Duhem equation for a two-component system. Equation (2.39) turns now into

$$S\,dT - V\,dp + n_1\,d\mu_1 + n_2\,d\mu_2 = 0. \tag{2.40}$$

We expand the chemical potential in a Taylor series:

$$
\begin{aligned}
d\mu_1 &= \frac{\partial \mu_1}{\partial T}\,dT + \frac{\partial \mu_1}{\partial p}\,dp + \frac{\partial \mu_1}{\partial n_1}\,dn_1 + \frac{\partial \mu_1}{\partial n_2}\,dn_2 \\
d\mu_2 &= \frac{\partial \mu_2}{\partial T}\,dT + \frac{\partial \mu_2}{\partial p}\,dp + \frac{\partial \mu_2}{\partial n_1}\,dn_1 + \frac{\partial \mu_2}{\partial n_2}\,dn_2.
\end{aligned}
\tag{2.41}
$$

Observe that by the law of *Schwarz*:

$$
\begin{aligned}
\frac{\partial \mu_1}{\partial T} &= -\frac{\partial S}{\partial n_1} = -\tilde{S}_1, \\
\frac{\partial \mu_1}{\partial p} &= +\frac{\partial V}{\partial n_1} = +\tilde{V}_1.
\end{aligned}
$$

Inserting these relations into Eq. (2.41) results in

$$\left(n_1\frac{\partial \mu_1}{\partial n_1} + n_2\frac{\partial \mu_2}{\partial n_1}\right) dn_1 + \left(n_1\frac{\partial \mu_1}{\partial n_2} + n_2\frac{\partial \mu_2}{\partial n_2}\right) dn_2 = 0 \tag{2.42}$$

since both entropy and volume will cancel. For example, the total entropy of the system is $S = n_1\tilde{S}_1 + n_2\tilde{S}_2$ and similarly holds for the volume. Further, dn_1 and dn_2 can be varied independently so that both terms in brackets in Eq. (2.42) will equate to zero. □

Often it is advantageous to divide Eq. (2.39) by the sum of the mol numbers, here $\sum_i n_i = n$. In this case, the molar quantities appear

$$\tilde{S}dT - \tilde{V}dp + d\mu = 0. \tag{2.43}$$

On dividing Eq. (2.40) by $\sum_i n_i = n_1 + n_2$, the mole fractions and the average molar quantities per total moles of mixture, $S \to S/(n_1 + n_2) = \tilde{S}$ and $V \to V/(n_1 + n_2) = \tilde{V}$, will emerge:

$$\tilde{S}dT - \tilde{V}dp + x_1 d\mu_1 + x_2 d\mu_2 = 0. \tag{2.44}$$

Equations (2.43) and (2.44) no longer contain terms that are dependent on the size of the system. A similar procedure as above results in

$$
\begin{aligned}
x_1 \frac{\partial \mu_1}{\partial x_1} + x_2 \frac{\partial \mu_2}{\partial x_1} &= 0 \\
x_1 \frac{\partial \mu_1}{\partial x_2} + x_2 \frac{\partial \mu_2}{\partial x_2} &= 0.
\end{aligned}
\tag{2.45}
$$

2.16 The Principle of Finite Energy

We consider the isothermal expansion of an ideal gas. Expanding the gas loses the energy:

$$-\int_{V_{Start}}^{V_{End}} p\, dV = -\int_{V_{Start}}^{V_{End}} \frac{RT}{V} dV. \tag{2.46}$$

If we expand to infinite volume, we can extract infinite energy. On the other hand, the amount of gas is not infinite. We presume that a finite amount of substance may contain only a finite amount of energy. Therefore, the ideal gas law cannot be valid in this case. We should have a relation like $p = V^{-\kappa}$ with $\kappa \geq 1$.

Actually, the ideal gas law in Eq. (2.46) is sound when the gas is in contact with a thermal reservoir. The thermal reservoir has by definition an infinite capacity. Here, we want to focus that interest to the case of finite entropy and the consequences for heat capacity, when the absolute temperature approaches zero, as important in statements on the third law of thermodynamics.

For historical reasons we emphasize the following:

- In the derivation of the speed of sound, originally the ideal gas law was inserted, before it was recognized that the process is adiabatic.
- Although it should be clear, we may emphasize that an isothermal expansion may occur only if heat is delivered from a thermal reservoir.

Next, we consider another example for the principle of finite energy. According to Newton's law of gravitation, Eq. (2.16), the potential gravitational energy is $U_g = -Gm_1m_2/r$. G is the gravitational constant, m_1 and m_2 are the masses involved, and r is the mutual distance of the masses. Now, the potential gravitational energy turns to infinity, if the distance approaches zero. In order to avoid infinite energy, we must not allow a zero distance. Thus, the principle of finite energy suggests that the masses have a finite extension so that they cannot approach zero distance. In other words, the principle of finite energy rejects the concept of point masses.

2.17 The Minimum Energy Principle

The principle of conservation of energy says that in the universe or more abstract in a closed system the energy is constant. This statement is different that the energy is tending to a minimum. We experience this principle in everyday mechanics.

For example, if we raise a body and allow it to fall down, we would feel that due to the action of gravity the body will search a state where its energy is a minimum. However, here we are considering only the potential energy of the body. We will not take into account that the potential energy is meanwhile transformed into kinetic energy and that the floor where the body is coming at rest would warm up in the course of the crash.

Further, we have withdrawn gravitational energy from the system that provides the gravitational energy. Therefore, in summary we have balanced the energy only for a part of the actors that are important to a body that is falling down.

We deal now with the statement of minimum of energy in detail and somewhat heuristically. Since the energy of a closed system is a real constant, we cannot speak strictly of a minimum in the sense that a neighboring state would have more energy than the state under consideration. However, since the energy is constant for such a system, every process that changes the thermodynamic variables $X, Y, \ldots$ of the energy $U(X, Y, \ldots)$ as function of them must proceed that the energy would not change, i.e., $dU(X, Y, \ldots) = 0$. This does not mean that the energy is a minimum in the sense that

$$U(X + \Delta X, Y + \Delta Y, \ldots + \Delta \ldots) > U(X, Y, \ldots). \tag{2.47}$$

Instead, $U(X + \Delta X, Y + \Delta Y, \ldots + \Delta \ldots) = U(X, Y, \ldots) = C_1$. We divide now the system under consideration into two subsystems U_1 and U_2. Since the original system was chosen arbitrarily, we expect the same property of attaining a minimum energy also for the subsystems. However, the subsystems now can exchange energy among them.

The statement that the energy would tend to a minimum means that the system would give or release energy. In contrary, if we postulate that the energy should tend to a maximum this would mean that the system would like to take or to catch energy. So, if both subsystems have the property to release energy as far as possible then we intuitively feel that there would be a state of minimum energy in that way that the energy is somewhat equally distributed among the two subsystems.

On the other hand, if there should be a maximum of energy, than the tendency is reverse. Each subsystem would like to catch energy at the cost of the other subsystem. So, the particular subsystem that is stronger would accumulate energy, as long as it can suck energy from the other system. In the limiting case in one system a zero energy would remain. For these reasons we could feel that in general the postulate of minimum energy is more sound than the postulate of maximum energy.

Nevertheless, the principle of minimum energy is not always reasonable. Astronomers have found black holes. These systems are accumulating energy from other systems in their neighborhood and would never release energy. Also, the atom is a system related to this type. Here we use the term atom rather in the original sense than in the nowadays chemical or physical sense, i.e., we mean a system that may not further be subdivided. Therefore, it can release neither energy nor mass. We annotate that mass could also be interpreted as energy. The situation is still more complicated. A black hole should emit radiation. We will discuss this issue in detail in Sect. 10.7.

2.18 Energy of a Quadratic Form

2.18.1 Energy

We will point out here the difference of constant energy and the minimum of energy. We exemplify the consideration by a spring, which is a quadratic form in terms of the variables. However, there are a lot of other examples for that the energy is a quadratic form. For example, the kinetic energy is quadratic in terms of the momentum. Further, the electric energy of a plate condenser is a quadratic form in terms of the electric charge.

2.18.2 Energy of a Spring

The energy of a spring is

$$U = \frac{k}{2} (x - x_0)^2 . \tag{2.48}$$

Here $x - x_0$ is the extension of the spring apart from it equilibrium position and k is the force constant. This means that when the spring extends to its equilibrium position, its energy U equals zero. We consider this spring as an ideal spring. We can see easily, if we form the derivative with respect to the position, Hooke's law will emerge:

$$-\frac{dU}{dx} = F = -k (x - x_0) .$$

We transform Eq. (2.48) now to dimensionless variables by putting $\xi = (x/x_0 - 1)$ and $\Upsilon = 2U / (k x_0^2)$ and we arrive at

$$\Upsilon = \xi^2 .$$

If we have more than one identical spring with equal force constants k and equal equilibrium position x_0, we insert an index

$$\Upsilon_{tot} = \Upsilon_1 + \Upsilon_2 + \ldots = \xi_1^2 + \xi_2^2 + \ldots .$$

The total length of the springs when switched in series is $x = x_1 + x_2 + \ldots$. In dimensionless variables we find

$$\xi_{tot} = \frac{x_1 - x_0 + x_2 - x_0 + \ldots}{x_0} = \xi_1 + \xi_2 + \ldots$$

2.18.3 Constant Energy and Length

We demand now that the dimensionless total energy Υ_{tot} and the dimensionless total length ξ_{tot} should be constant. Further, we deal with two springs only. In this case, we have now the following two equations:

$$\begin{aligned} \Upsilon_{tot} &= \xi_1^2 + \xi_2^2 \\ \xi_{tot} &= \xi_1 + \xi_2 \end{aligned} . \tag{2.49}$$

We have only two variables ξ_1 and ξ_2 in these two equations. Therefore, we could have one unique solution in a linear set of equations and since one of the equations is quadratic and the equations are completely symmetric with respect to ξ_1, ξ_2 we obtain a set of solutions:

$$\xi_1 = \frac{\xi_{tot} \pm \sqrt{2\Upsilon - \xi_{tot}^2}}{2}$$

$$\xi_2 = \frac{\xi_{tot} \mp \sqrt{2\Upsilon - \xi_{tot}^2}}{2} .$$

We show the situation in Fig. 2.4. The isodynamic curves are forming circles around the origin, whereas the curves of equal dimensionless length are straight lines with a slope of -1.

When we extend Eq. (2.49) for three springs, we get

$$\begin{aligned} \Upsilon_{tot} &= \xi_1^2 + \xi_2^2 + \xi_3^2, \\ \xi_{tot} &= \xi_1 + \xi_2 + \xi_3, \end{aligned} \tag{2.50}$$

and we easily see that we have now one variable, say ξ_3, for free adjustment. In other words, we have under favorite selection of the energy and the total length a continuous path along which we can change one of the variables $\xi_i, i = 1, 2, 3$ and can still fulfill Eq. (2.50).

The graphical representation of Fig. 2.4 can be easily extended in three dimensions. In three dimensions, the dimensionless isodynamic curves are now spheres

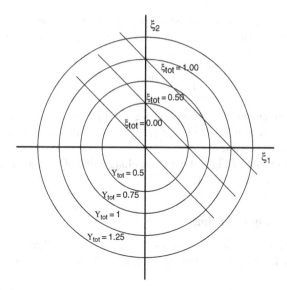

Fig. 2.4 Graphical representation of the dimensionless energy and the dimensional length

around the origin, whereas the curves of equal dimensionless length are straight planes. The planes are cutting the spheres as a spherical segment. The intersecting curve is then a cycle, which is the graphical solution.

2.18.4 Minimum of Energy

The situation becomes completely different when we ask for the energy to be an extremum at a given length.

2.18.4.1 Two Coupled Springs

For two coupled springs $\Upsilon_{tot}(\xi_1, \xi_2)$ as a function of the energy of the springs under consideration, Υ is no longer a constant. If the other condition regarding the length remains still the same, the total length will be a constant. Therefore, we have the equations:

$$\Upsilon_{tot}(\xi_1, \xi_2) = \xi_1^2 + \xi_2^2$$
$$\xi_{tot} = \xi_1 + \xi_2. \tag{2.51}$$

For the first equation we want to find a minimum. The second equation in Eq. (2.51) is a constraint equation. We can easily insert the second equation of Eq. (2.51) in the first equation to obtain

$$\Upsilon_{tot}(\xi_1, \xi_2) = \xi_1^2 + (\xi_{tot} - \xi_1)^2.$$

By substitution, we have now reduced the problem to a function of one variable, ξ_1, find the extremum under the constraint, form the derivative as usual and obtain

$$\frac{d\Upsilon_{tot}}{d\xi_1} = 4\xi_1 - 2\xi_{tot}.$$

Therefore, the dimensionless elongation is $\xi_1 = \xi_2 = \xi_{tot}/2$. Eventually we from the second derivative which is $d^2\Upsilon_{tot}/d\xi_1^2 = 4$. This indicates that basically only a minimum energy is possible.

In the representation of Fig. 2.4 we move along a line of constant length and look for the curve of minimum energy.

2.18.4.2 Three Coupled Springs

The dimensionless total energy becomes $\Upsilon_{tot}(\xi_1, \xi_2, \xi_3)$ for three coupled springs. Further, the total length should again be a constant. We set up the constraint equation now as

$$\xi_{tot} - \xi_1 - \xi_2 - \xi_3 = \Phi(\xi_1, \xi_2, \xi_3) \equiv 0. \tag{2.52}$$

In the argument set Φ, only ξ_1, ξ_2, ξ_3 appear, indicating that ξ_{tot} is not a variable. We introduce now the more convenient method of undetermined multipliers by Lagrange. In the differential form this equation reads as

$$-d\xi_1 - d\xi_2 - d\xi_3 = 0.$$

To find the extremum under the constraint, we will multiply Eq. (2.52) with an undetermined multiplier λ and add a zero valued function $\Phi(\xi_1, \xi_2, \xi_3)$ to the expression for $\Upsilon(\xi_1, \xi_2, \xi_3)$. In detail, we form a function

$$\Psi(\xi_1, \xi_2, \xi_3) = \Upsilon(\xi_1, \xi_2, \xi_3) + \lambda\Phi(\xi_1, \xi_2, \xi_3).$$

Forming now the differential of Ψ we get

$$d\Psi(\xi_1, \xi_2, \xi_3) = (2\xi_1 - \lambda)\,d\xi_1 + (2\xi_2 - \lambda)\,d\xi_2 + (2\xi_3 - \lambda)\,d\xi_3. \tag{2.53}$$

Therefore, we have to equate the set of equations

$$2\xi_1 - \lambda = 0; \quad 2\xi_2 - \lambda = 0; \quad 2\xi_3 - \lambda = 0$$

to zero which means that $\xi_1 = \xi_2 = \xi_3$. This means that also the energy of each individual spring is the same. We observe here that the energy distributes itself in same amounts in a system of equal subsystems, or in other words, the energy follows an equipartition law.

We carefully point out the difference between the meaning of Eq. (2.50) where we are dealing with a constant energy and Eq. (2.53) where we are searching for a minimum or more generally for an extremum of the energy. We now make use of Eq. (2.52) and find

$$\xi_1 = \xi_2 = \xi_3 = \frac{\xi}{3}. \tag{2.54}$$

Depending on whether the springs are pulled or pushed, or left freely corresponding to $\xi > 0$, $\xi < 0$, or $\xi = 0$ also the individual elongations ξ_1, ξ_2, ξ_3 reflect the state.

Finally, we should always inspect the second derivative of the energy $d^2\Upsilon$ to obtain information about the type of stationarity. The second total differential is simply

$$d^2\Upsilon(\xi_1, \xi_2, \xi_3) = 2\left(d\xi_1^2 + d\xi_2^2 + d\xi_3^2\right),$$

since the mixed terms cancel.

We return now again to the condition of constant energy. However, we set now the total length as a variable. We have now

$$\xi_{tot}(\xi_1, \xi_2, \xi_3) = \xi_1 + \xi_2 + \xi_3$$

and $\Upsilon_{tot} - \xi_1^2 + \xi_2^2 + \xi_3^2 = 0$. To find an extremum of the dimensionless length at constant energy we proceed in the same way as we found the minimum of the energy in Eq. (2.53). We form the function $\Phi(\xi_1, \xi_2, \xi_3) = \Upsilon - \xi_1^2 - \xi_2^2 - \xi_3^2$ and further $\Psi(\xi_1, \xi_2, \xi_3) = \lambda\Phi(\xi_1, \xi_2, \xi_3) + \xi_{tot}(\xi_1, \xi_2, \xi_3)$. Then the equation in question will turn out as

$$d\Psi(\xi_1, \xi_2, \xi_3) =$$
$$(-2\xi_1\lambda + 1)\, d\xi_1 + (-2\xi_2\lambda + 1)\, d\xi_2 + (-2\xi_3\lambda + 1)\, d\xi_3. \tag{2.55}$$

Obviously, Eq. (2.55) gives the same result as the treatment where we searched for a minimum energy. Again, we find the result of Eq. (2.54).

2.18.5 Spring Coupled with Two Gases

Consider the system shown in Fig. 2.5. Here, two gases (1) and (2) are separated by a freely movable piston. Further the gas (2) is coupled by a movable piston with a spring. The spring is fixed on the other end. The total volume V_{tot} of both gases is the sum of their individual volumes V_1 and V_2

$$V_{tot} = V_1 + V_2. \tag{2.56}$$

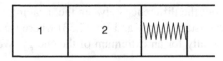

Fig. 2.5 Two ideal gases (*1*) and (*2*) coupled by a spring

The energy of the spring is related exclusively to the total volume in some way, $U_s = U_s(V_{tot})$. We could calculate the dependence of the energy of the spring on the length of the spring and relate the energy of the spring to the total volume, but we will not do this in detail.

The energy of the gas is a function of entropy S, volume V, and mol number n. Since we have two gases we will provide the indices (1) and (2). Thus

$$U_1 = U_1(S_1, V_1, n_1); \quad U_2 = U_2(S_2, V_2, n_2).$$

The total energy of the system is

$$U_{tot} = U_1 + U_2 + U_s.$$

We assume now that the walls of the systems are impermeable to entropy and matter. Further we consider a state of equilibrium. This assumption implies the constraints $dS_1 = 0$, $dS_2 = 0$, $dn_1 = 0$, $dn_2 = 0$. For this reason we may express the total energy merely as a function of the volumes, i.e.,

$$U_{tot}(V_{tot}, V_1, V_2) = U_1(V_1) + U_2(V_2) + U_s(V_{tot}). \tag{2.57}$$

Using the constraint equation (2.56) together with the method of undetermined multipliers,

$$(-V_{tot} + V_1 + V_2)\lambda = 0,$$

we get immediately

$$dU_{tot} = \left(\frac{\partial U_s}{\partial V_{tot}} - \lambda\right)dV_{tot} - (p_1 - \lambda)dV_1 - (p_2 - \lambda)dV_2. \tag{2.58}$$

Thus, without knowing the special form of the energy of the spring, we come to the conclusion that in equilibrium the pressures of the gases must be equal, i.e., $p_1 = p_2$.

The constraints with respect to entropy and mol number allow different temperatures and different chemical potentials for the gases (1) and (2). The spring, of course, ideally does not have entropy, mol number, and volume. The dependence of the energy of the spring on the total volume arises, because the length of the spring is coupled with the total volume of both gases.

We may point out a special case in that the energy of the spring does not change with length at all. Then $\lambda = 0$ from the first term on the right-hand side of Eq. (2.58) and the gases will expand to zero pressure.

2.19 Maximum of Entropy

We examine now the entropy of two subsystems under the constraint of constant volume and constant energy. We have two systems with the energies $U_1(S_1, V_1)$ and $U_2(S_2, V_2)$. These energies are functions of their entropies denoted as S_1, S_2 and their volumes V_1, V_2. The total energy and the total volume should be constant:

$$\begin{aligned} U_{tot} - U_1(S_1, V_1) - U_2(S_2, V_2) &= 0 \\ V_{tot} - V_1 - V_2 &= 0 \end{aligned} \qquad (2.59)$$

We form now the total entropy and add the constraint equations (2.59) multiplied with undetermined coefficients λ_U, λ_V:

$$\begin{aligned} \Psi_S(S_1, S_2, V_1, V_2) = \\ S_1 + S_2 + \\ \lambda_U(U_{tot} - U_1(S_1, V_1) - U_2(S_2, V_2)) + \\ \lambda_V(V_{tot} - V_1 - V_2). \end{aligned} \qquad (2.60)$$

To find the extremum we form in Eq. (2.60) the partial differentials and we substitute also $\partial U(S, V)/\partial S = T(S, V)$ and $\partial U(S, V)/\partial V = -p(S, V)$ introducing the appropriate index 1 or 2:

$$\begin{aligned} d\Psi_S(S_1, S_2, V_1, V_2) = \\ (1 - \lambda_U T_1(S_1, V_1)) \, dS_1 + \\ (1 - \lambda_U T_2(S_2, V_2)) \, dS_2 + \\ (\lambda_U p_1(S_1, V_1) - \lambda_V) \, dV_1 + \\ (\lambda_U p_2(S_2, V_2) - \lambda_V) \, dV_2. \end{aligned} \qquad (2.61)$$

Equating all differentials in Eq. (2.61) to zero and eliminating λ_U, λ_V results in

$$\begin{aligned} T_1(S_1, V_1) &= T_2(S_2, V_2) \\ p_1(S_1, V_1) &= p_2(S_2, V_2) \end{aligned} \qquad (2.62)$$

In this procedure we do not search for an extremum for the total energy. Instead, we assume the total energy to be constant. We do not use a condition of constant entropy but we search an extremum of the entropy under the constraint of constant energy and constant volume. Nevertheless, it turns out that under these constraints not only the pressures of the two systems are equal but also the temperatures.

However, if we search for an extremum of the energy under the constraint of constant total volume $V_{tot} - V_1 - V_2 = 0$ and of constant total entropy $S_{tot} - S_1 - S_2 = 0$, the same result would be obtained. In fact, we would have to evaluate an expression like

$$\Psi_U(S_1, S_2, V_1, V_2) =$$
$$U(S_1, S_2, V_1, V_2) +$$
$$\lambda_S(S_{tot} - S_1 - S_2) +$$
$$\lambda_V(V_{tot} - V_1 - V_2)$$

in a similar manner like Eq. (2.61).

2.19.1 Entropy Gain in an Ideal Gas

It could be that Eq. (2.62) is not satisfactory by inspection and common sense. We consider a system as shown in Fig. 2.6. Here the total volume V_{tot} is constant and is the sum of the individual volumes V_1 of gas (1) and V_2 of gas (2), i.e., $V_{tot} - V_1 - V_2 = 0$. When we start with equal amounts of gas with the volume at the left side V_1 smaller than the volume at the right side V_2, it is suggestive that the pressure on the left side is higher than the pressure on the right side. In general, this may not be true, because the pressure is dependent also on the temperature of the respective gas. However, we assume $p_1 > p_2$.

When we allow the plunger to move freely the volume at the left side will increase and the volume at the right side will decrease correspondingly. Therefore, we naturally feel that the process will continue until both pressures become equal. This means that the condition $p_1(S_1, V_1) = p_2(S_2, V_2)$ from Eq. (2.62) is fulfilled, but not necessarily the condition of equal temperature.

In fact, the calculation of maximum entropy finds the most possible maximum entropy. We have not given a constraint for the entropy in Eqs. (2.60) and (2.61). However, in a nonequilibrium process the entropy should increase. Thermodynamics does not provide an answer as to which way the entropy would increase. We

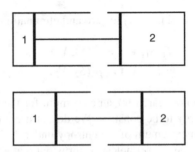

Fig. 2.6 Two ideal gases (1) and (2) coupled by a piston

could obtain an answer, if we are dealing with one gas only in the way as the experiment of *Gay-Lussac*[6] runs. This means that a single gas is expanding into vacuum. In the case of two gases, the situation is already more complicated and we must state additional constraints to find the way in which the gases would equilibrate.

2.20 Coupled Energy Forms

We inspect now an arbitrary system together with an environment system. The arbitrary system should have a certain energy dependent on the entropy, the volume, and the mol number, as usual $U_{sys}(S_{sys}, V_{sys}, n_{sys})$. The environment should have an energy dependent on certain variable that we will specify later as $U_{env}(X_1, X_2, \ldots)$. The energy of both systems should be constant, i.e.,

$$U_{tot} = U_{sys}(S_{sys}, V_{sys}, n_{sys}) + U_{env}(X_1, X_2, \ldots). \tag{2.63}$$

For illustration we may think that the system consists of an ideal gas. The system and its environment is presented schematically in Fig. 2.7.

The energy of the systems as can be deduced easily from the mathematical form of $U(S, V, n)$ is solely dependent on the entropy S, the volume V, and the mol number n of the system. In fact, when we know the entropy and the volume we can find out the energy. We recall that for an ideal gas the energy is something like

$$U_{sys}(S_{sys}, V_{sys}, n_{sys}) = n \left(C_1 \exp \left(\frac{S_{sys}}{\tilde{C}_v n_{sys}} \right) \left(\frac{V_{sys}}{n_{sys}} \right)^{-R/\tilde{C}_v} + C_2 \right). \tag{2.64}$$

The adiabatic coefficient κ is defined as $\kappa = \tilde{C}_p/\tilde{C}_v$. Since $R = \tilde{C}_p - \tilde{C}_v$, $R/\tilde{C}_v = \kappa - 1$. For a monoatomic gas $\kappa = 5/3$ and for a diatomic gas $\kappa = 7/3$. This can be concluded from the equipartition theorem.

Further, we could express the entropy in a dimensionless quantity, in units of C_v, i.e., $\check{S}_{sys} = S_{sys}/C_v$. We can adjust a dimensionless energy as $\check{U}_{sys}(\check{S}_{sys}, \check{V}_{sys}) = \exp(\check{S}_{sys})\check{V}_{sys}^{-2/3}$. We will discuss now certain possibilities of changing the energy of the system.

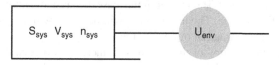

Fig. 2.7 An ideal gas as system and its environment

2.20.1 Environment Takes Up Volume Energy

We construct the environment as an engine that can take up volume energy, $U_{env} = U_{env}(V_{sys})$. In particular, the environment knows about the volume of the system, thus

$$U_{env} = f(V_{sys}). \tag{2.65}$$

Without elaborating the nature of engine of the environment in detail we can imagine that the energy storage of the environment can be a spring coupled with a gear that has a variable gear transmission ratio depending on the volume of the system. In this way, a variable pressure can be transduced to the system depending on the volume, as shown schematically in Fig. 2.7.

We now make a closer assumption on the nature of $f(V_{sys})$ given in Eq. 2.65. We assume that

$$U_{env} = f(V_{sys}) = U_{tot} - C_3 V_{sys}^{-R/C_v} V_{sys}^{\beta}.$$

Inserting this expression into Eqs. (2.63) and (2.64) we find:

$$n \left(C_1 \exp \left(\frac{S_{sys}}{\tilde{C}_v n_{sys}} \right) \left(\frac{V_{sys}}{n_{sys}} \right)^{-R/\tilde{C}_v} + C_2 \right) - C_3 V_{sys}^{-R/C_v} V_{sys}^{\beta} = 0.$$

We adjust $C_2 = 0$ and find, introducing a new (positive) constant C that gathers all the former constants,

$$S_{sys} = C\beta \ln V_{sys}.$$

We can discuss now the following cases:

1. If $\beta = 0$, then the entropy of the system will not change provided that the volume of the system changes. This is a reversible change of the state. This does not mean that the temperature will be constant in this case, because the temperature is also a function of the volume. The temperature will decrease with expanding volume according to energy equation of the ideal gas.
2. If $\beta > 0$, then the entropy of the system will increase provided that the volume of the system increases. This is an irreversible change of the state.
3. If $\beta < 0$, then the entropy of the system will decrease provided that the volume of the system increases. This case is an impossible change of the state, since the second law of thermodynamics does not allow a spontaneous process where the overall entropy decreases. We emphasize that the environment according to the model does not bear any entropy.

2.21 Dimensionless Relative Variables

We inspect now the thermodynamic variables and functions of the various energy forms by making the respective extensive variable dimensionless.

We illustrate the principle first with the kinetic energy. The kinetic energy U of a mass point with mass m expressed in terms of the momentum P is $U(P) = P^2/(2m)$, which is well known in classical mechanics. We form the differential, thinking that the mass is constant and obtain

$$dU(P) = \frac{P}{m} dP. \tag{2.66}$$

We can rearrange Eq. (2.66) to obtain

$$dU(P) = \frac{P^2}{m} \frac{dP}{P} = \frac{P^2}{m} d \ln P. \tag{2.67}$$

Here dP/P is a dimensionless increase of the momentum, the relative increase of the momentum. Consequently, the first term P^2/m on the right-hand side of Eq. (2.67) is an energy. Formally P^2/m is the increase of the kinetic energy when the relative increase of the momentum dP/P equals 1.

We could further substitute $\ln P = z$ and further $P = \exp z$ into Eq. (2.67) but we will leave it as is. However, we can state that the differential is of the order P^2.

We may repeat the same procedure for other energy forms. We collect some energy forms in Table 2.4. For the volume energy in Table 2.4 we have converted the volume V into its reciprocal $1/V$ to eliminate the negative sign in the differential. Further, in Table 2.4 functions of the extensive variables appear. This is the pressure p now as a function of the logarithm of the reciprocal volume $1/V$ or the temperature as a function of the logarithm of the entropy for the thermal energy. We can resolve these functions under certain conditions, i.e., the pressure as a function of volume at constant entropy or the temperature as a function of entropy at constant volume, etc. For the isentropic expansion of an ideal gas we have $pV^\kappa = C_1(S)$ and for the temperature change at constant volume we have $T = \exp(S/C_v)$ when C_v is a constant. We can insert these dependencies for the special cases as done in the third column of Table 2.4.

Inspecting Table 2.4 we find that the kinetic energy runs with the order p^2/m, the gravitational energy runs with the order mgh, the volume energy runs with the order $C(S)/V^{\kappa-1}$, and finally the thermal energy runs with the order $\exp(S/C_v)S$.

Table 2.4 Energy forms with relative dimensionless argument

Energy form	Differential	Special case
Kinetic energy	$P^2/m \ d \ln(P)$	
Gravitational energy	$mgh \ d \ln(mh)$	
Volume energy	$pV \ d \ln(1/V)$	$C(S)/V^{\kappa-1} \ d \ln(1/V)$
Thermal energy	$TS \ d \ln(S)$	$\exp(S/C_v)S \ d \ln(S)$

This means among the energy forms considered here, the thermal energy runs in the highest order.

We make a little bit more precise of what and how we understand the order discussed here. In the case of a polynomial, we deal with the exponents. For example, the polynomial $x^2 + x + 1$ has the highest exponent x^2 and thus the order of 2. Here we mean which function is growing faster with increasing argument. If we have two functions $f(x)$ and $g(x)$ both being real positive then $f(x)$ has a higher order if $f(x) - g(x) > 0$ for some range in x.

2.22 Stability

In this section, we anticipate the problem of stability of a thermodynamic system briefly, which needs understanding of equilibrium and the second law. We may start from the equilibrium instead from the inequalities for entropy, which are discussed in Sect. 3.3 and vice versa. Say, we have free exchange of a certain extensive variable of state and we ask why it will not happen that an exchange will not occur in a certain direction.

Example 2.11. There are two systems $U_1(S_1)$ and $U_2(S_2)$ that can freely exchange S. Suppose $T_1 = T_2$. Why it does not happen that entropy will flow from one to the other on a small disturbance so that one system will release entropy and the other will catch it. Note that in the case of two soap bubbles of equal radi in fact this will occur. The soap bubble system is not stable. □

Imagine a system composed of two subsystems with fixed volume, etc. Only entropy can be varied. Therefore, we suppress variables other than entropy in the variable list. We emphasize that the consideration can be extended to more than one variable. The energy is a function of the entropies of the individual subsystems. The total energy is additive with respect to the energies of the subsystems

$$U(S_1, S_2) = U_1(S_1) + U_2(S_2).$$

Think about a change of energy in such a way that the total entropy remains constant. We will exchange a small amount of entropy ΔS. Initially the energy is $U(S_1, S_2)$. Moreover, we start from an equilibrium position which means that $T_1 = T_2 = T$. After varying the energy we have $U(S_1 + \Delta S, S_2 - \Delta S)$. Inspect the first total differential:

$$\begin{aligned} dU_1(S_1 + \Delta S) &= +\frac{\partial U(S_1)}{\partial S_1} \Delta S = T \Delta S \\ dU_1(S_2 - \Delta S) &= -\frac{\partial U(S_2)}{\partial S_2} \Delta S = -T \Delta S. \end{aligned} \tag{2.68}$$

Thus, the first derivative $dU = dU_1 + dU_2$ equals zero. Now inspect the second total differential:

$$d^2U_1(S_1 + \Delta S) = \frac{1}{2}\frac{\partial^2 U(S_1)}{\partial S_1^2}\Delta S^2$$

$$d^2U_2(S_2 - \Delta S) = \frac{1}{2}\frac{\partial^2 U(S_1)}{\partial S_1^2}\Delta S^2.$$
(2.69)

The second total differential d^2U will be positive if $dT/dS > 0$. In this case, the total energy will be a minimum with respect to an exchange of entropy. This is the condition of stability. A process that varies the entropy will not take place, if

$$\frac{\partial T}{\partial S} > 0.$$
(2.70)

Otherwise, the world would collapse or disproportionate to an infinitely cold state and to an infinitely hot state. We emphasize, if the material condition

$$\frac{\partial S}{\partial T} > 0$$
(2.71)

is not valid we would not exist at all, nor our world could exist. This is more fundamental than the statement of *Clausius* on the second law.

We generalize now stability to more variables.

Example 2.12. Consider the function

$$f(x, y) = ax^2 + bxy + cy^2.$$

The function has an extremum at $x = 0$, $y = 0$. We show a plot of the function with $a = 1, c = 1, b = 3$ in Fig. 2.8. If the discriminant $D \equiv 4ac - b^2 > 0$, the contour plot will show ellipses, with a definite maximum, otherwise a saddle point emerges. In fact, we are dealing with a binary quadratic form in two real variables. The binary quadratic form is positive definite if its discriminant is positive.   □

Now, we assume that the internal energy is a function of the extensive variables $X_1, X_2, \ldots, X_n$, i.e., $U = U(X_1, X_2, \ldots, X_n)$. We normalize the energy that in the minimum, indexed by m with $X_1 = X_{1,m}, X_2 = X_{2,m}, \ldots, X_n = X_{n,m}$ it reaches zero. We can now expand the energy around the minimum into a Taylor series. The result is

$$\begin{aligned}
U(X_1, X_2, \ldots, X_n) &= U(X_{1,m}, X_{2,m}, \ldots, X_{n,m}) \\
&+ \frac{\partial U}{\partial X_1}(X_1 - X_{1,m}) + \frac{\partial U}{\partial X_2}(X_2 - X_{2,m}) + \ldots + \frac{\partial U}{\partial X_n}(X_n - X_{n,m}) \\
&+ \frac{\partial^2 U}{\partial X_1^2}(X_1 - X_{1,m})^2 + \frac{\partial^2 U}{\partial X_1 X_2}(X_1 - X_{1,m})(X_2 - X_{2,m}) + \ldots \\
&+ \frac{\partial^2 U}{\partial X_n^2}(X_n - X_{n,m})^2.
\end{aligned}$$
(2.72)

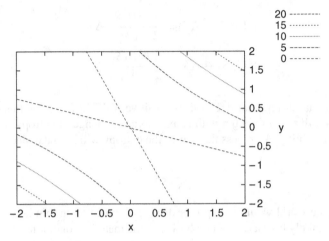

Fig. 2.8 Plot of $f(x, y) = x^2 + 3xy + y^2$

The first line on the right-hand side of Eq. (2.72) is zero. The second line is the total differential of first order, which is zero when the function is stationary. The further terms are the total differential of second order. This differential must be positive in order that the function provides a minimum.

From the theory of positive definite bilinear forms it has been shown that all the coefficients must be positive, for the condition of a minimum [2, Appendix G, pp. 361–366]. From the view of mathematics, we are dealing with definite positive quadratic forms.

We point out still another formal issue of Eq. (2.70) and (2.70). The natural variables of the energy are the set $(S, V, n, \dots)$. These variables are extensive in nature. Now Eq. (2.71) deals with the reciprocal of Eq. (2.70). Forming the reciprocal, the independent variables change. In particular, writing the independent variables explicitly, when we use $\partial T(S, V, \dots)/\partial S$, the reciprocal turns into $\partial S(T, V, \dots)/\partial T = C_v(T, V, \dots)/T$. This is not very exciting.

However, observe that the corresponding pressure changes from $\partial p(S, V, \dots)/\partial V$ into $\partial V(S, p, \dots)/\partial p$. Thus, keep in mind that the entropy is still the dependent variable. As an exercise, we change the variables using Jacobian determinants. For simplicity, we are using two independent variables in the list of variables:

$$\frac{\partial p(S, V)}{\partial V} = \frac{\partial(p, S)}{\partial(V, S)} = \frac{\partial(p, S)/\partial(V, T)}{\partial(V, S)/\partial(V, T)} =$$

$$\frac{\dfrac{\partial p(T, V)}{\partial V} \dfrac{\partial S(T, V)}{\partial T} - \dfrac{\partial p(T, V)}{\partial T} \dfrac{\partial S(T, V)}{\partial V}}{\dfrac{\partial S(T, V)}{\partial T}} =$$

$$\frac{\partial p(T, V)}{\partial V} - \frac{T}{C_v(T, V)} \left(\frac{\partial p(T, V)}{\partial T} \right)^2 . \tag{2.73}$$

Observe the square in the last term of Eq. (2.73). It effects that the squared term is always positive.

2.23 Phases

Commonly, in a system with multiple phases, the phase is defined as a region; when moving into this region an abrupt change in the physical properties is observed. Thus, the phase is rather characterized by its sharp border.

As well known, in equilibrium of a PVT system, the intensive variables such as temperature T, pressure p, and the chemical potentials of the individual components μ_i in each phase are equal. This is the starting point of the derivation of the phase rule.

One important prerequisite for the derivation of the phase rule is that the phases are distinguishable. Obviously, the phases cannot be distinguished by their intensive parameters, except composition. Therefore, other properties must hold in order to characterize a phase. In order to detect a phase, at least one property, either a direct thermodynamic property such as the extensive properties or a property that is dependent on these properties, such as diffraction, must differ in each phase. Here we annotate that there is agreement that a thermodynamic potential contains the complete information in order to describe a thermodynamic system unambiguously. Among the extensive variables of a phase, the mol numbers just tell about the size of the phase. In contrast, the mole fractions give information about the constitution of the components in the phase.

If all direct and indirect properties in each region of a system are equal, then it does not make sense to talk about multiple phases. In particular, the properties in a liquid – gas system approach one another at the critical point. At the critical point the two phases disappear.

It is somehow suggestive to postulate that if the direct extensive properties are equal, also the derived properties are equal. For example, in a two-component system, if the molar entropy, the molar volume, and the mole fraction are the same, also other properties that are functions of these parameters should be the same. We are talking about properties that are beyond the equations of state. For example, the refractive index should be the same for each phase and different phases will no longer be detectable.

Likewise, we may state that a proper handling of the phase rule requires a complete set of thermodynamic variables. Properties derived from this complete set allow unambiguously to state whether there are two or more phases present or not. This postulate does not appear directly in the thermodynamic laws.

References

1. Adkins, C.J.: Equilibrium Thermodynamics, 3rd edn. Cambridge University Press, Cambridge, UK (1983)
2. Callen, H.B.: Thermodynamics. John Wiley, New York (1961)

3. Callen, H.B.: Thermodynamics and an Introduction to Thermostatics. John Wiley, New York (1985)
4. Haddad, W.M., Chellaboina, V.S., Nersesov, S.G.: Thermodynamics: A Dynamical Systems Approach. *Princeton Series in Applied Mathematics*. Princeton University Press, Princeton (2005)
5. Landsberg, P.T.: Thermodynamics and Statistical Mechanics. Dover Publications, New York (1990)
6. Owen, D.R.: A First Course in the Mathematical Foundations of Thermodynamics. Undergraduate Texts in Mathematics. Springer-Verlag, New York (1984)
7. Alberty, R.A.: Use of Legendre transforms in chemical thermodynamics: International union of pure and applied chemistry, physical chemistry division, commission on thermodynamics. J. Chem. Thermodyn. **34**(11), 1787–1823 (2002)
8. Kittel, C.: Introduction to Solid State Physics, 8th edn. Wiley, New York (2004)
9. Weisstein, E.W.: Plane. From MathWorld – A Wolfram Web Resource: [electronic] http://mathworld.wolfram.com/Plane.html (2004)
10. Landsberg, P.T.: Entropies galore! Braz. J. Phys. **29**(1), 46–49 (1999)
11. Landsberg, P.T.: Fragmentations, mergings and order: Aspects of entropy. Phys. Stat. Mech. Appl. **305**(1–2), 32–40 (2002)
12. Maxwell, J.C.: Theory of Heat, 9th edn. Dover Publications, Inc., Mineola, NY (2001). First published in 1871
13. de Medeiros, J.L., Cardoso, M.J.E.d.M.: On the derivation of thermodynamic equilibrium criteria. Fluid Phase Equilib. **136**(1–2), 1–13 (1997)
14. Gibbs, J.W.: On the equilibria of heterogeneous substances. In: The Collected Works, vol. 1. Yale University Press, New Haven, CT (1948 [reprinted])
15. Planes, A., Vives, E.: Entropic formulation of statistical mechanics. J. Stat. Phys. **106**(3–4), 827–850 (2002)

Chapter 3
The Laws of Thermodynamics

The British scientist and author Charles Percy *Snow*[1] described the three laws of thermodynamics as following:

1. You cannot win (that is, you cannot get something for nothing, because matter and energy are conserved).
2. You cannot break even (you cannot return to the same energy state, because there is always an increase in disorder; entropy always increases).
3. You cannot get out of the game (because absolute zero is unattainable).

3.1 Zeroth Law of Thermodynamics

The need to formulate the zeroth law was felt after the first law and the second law were established. The zeroth law was regarded to be more fundamental than the other laws. So it was pushed at the start of the thermodynamic laws and thus numbered by zero. There is still some discussion about its status in relation to the other three laws.

The zeroth law of thermodynamics states that when a system ($'$) is in thermal equilibrium with a system ($''$), and the system ($''$) is in thermal equilibrium with a system ($'''$), then the system ($'$) is also in thermal equilibrium with the system ($'''$). Therefore, the equilibrium is transitive. Frequently it is added that the zeroth law is the base for building a thermometer.

We emphasize here that the statements of the zeroth law are referred to as a special case of thermal equilibrium. More general statements of the same kind apply also to other kinds of equilibria. For instance, the existence of a manal (pressure) equilibrium allows us to build a pressure gauge, or the equilibrium of the electric potential allows us to build a *Wheatstone* bridge.

We consider two systems ($'$) and ($''$) with their energies U' and U'' dependent on the thermodynamic variables $U' = U'(X_1', X_2', \ldots)$ and $U'' = U''(X_1'', X_2'', \ldots)$.

---

[1] Baron Charles Percy Snow of Leicester, born Oct. 15, 1905, in Leicester, Leicestershire, England, died Jul. 1, 1980, in London.

J.K. Fink, *Physical Chemistry in Depth*,
DOI 10.1007/978-3-642-01014-9_3, © Springer-Verlag Berlin Heidelberg 2009

Here, $X'_1, X'_2, \ldots, X''_1, X''_2, \ldots$ are the extensive variables of the systems. We do not specify these variables, but they are $S, V, \ldots$ in practice. The two systems should be coupled together by the relations

$$
\begin{aligned}
dX'_1 + dX'_1 &= 0 \\
dX'_2 + dX''_2 &= 0 \quad .
\end{aligned}
\tag{3.1}
$$

$$
\cdots\cdots\cdots
$$

Then the change of the total energy is

$$
dU' + dU'' =
$$

$$
\xi'_1 dX'_1 + \xi'_2 dX'_2 + \ldots +
$$

$$
\xi''_1 dX''_1 + \xi''_2 dX''_2 + \ldots .
$$

We allow X'_1 and X''_1 run freely under the constraint of Eq. (3.1), first line, but we fix all other extensive variables. Then a variation in a way that the total energy remains stationary must be such that the intensive variables $\xi'_1 = \xi''_1$.

We state the total system is in equilibrium with respect to the energy form $\xi_1 dX_1$. We have omitted now the primes. If $\xi_1 = T$ and consequently $X_1 = S$, we are talking about thermal energy. The zeroth law of thermodynamics states that in thermal equilibrium the temperatures of coupled systems are the same. As pointed out in detail, the zeroth law of thermodynamics is a special case of equilibrium, namely thermal equilibrium.

We can see the special character if we generalize. We allow all extensive variables to change according to the constraint of Eq. (3.1). Then in equilibrium $\xi'_n = \xi''_n$.

3.2 First Law of Thermodynamics

The first law of thermodynamics deals with energy conservation. We start with energy conservation in pure mechanical systems and extend this principle to thermodynamics.

3.2.1 Conservation of Energy in Mechanical Systems

In a system of mass points $1 \ldots k$ on an individual mass point i the force should act according to Eq. (3.2):

$$
X_i = -\frac{\partial \Phi}{\partial x_i}; \quad Y_i = -\frac{\partial \Phi}{\partial y_i}; \quad Z_i = -\frac{\partial \Phi}{\partial z_i}.
\tag{3.2}
$$

We use here a Cartesian coordinate system. X_i, Y_i, and Z_i are the forces in x-direction, y-direction, and z-direction, respectively, that are acting on the ith particle.

The right-hand side of Eq. (3.2) is a partial derivative of a function $\Phi(x, y, z)$. Note that this function is only dependent on the position of the particle i. The function Φ is called the potential energy of the particle.

According to *d'Alembert*[2] we introduce now the forces of inertness. A particle that is under stress will try to lower the stresses and will follow the stresses. The forces of inertness are

$$-m_i \frac{\partial^2 x_i}{\partial t^2}, \quad -m_i \frac{\partial^2 y_i}{\partial t^2}, \quad -m_i \frac{\partial^2 z_i}{\partial t^2}.$$

The forces of inertness and the forces arising from the potential energy sum to zero:

$$m_i \frac{\partial^2 x_i}{\partial t^2} + \frac{\partial \Phi}{\partial x_i} = 0$$

$$m_i \frac{\partial^2 y_i}{\partial t^2} + \frac{\partial \Phi}{\partial y_i} = 0 \qquad (3.3)$$

$$m_i \frac{\partial^2 z_i}{\partial t^2} + \frac{\partial \Phi}{\partial z_i} = 0.$$

We multiply the three equations of (3.3) successively by dx_i/dt, dy_i/dt, and dz_i/dt, respectively, a sum over all particles $1 \ldots k$ to obtain

$$\sum_{i=1}^{i=k} m_i \left(\ddot{x}_i \dot{x}_i + \ddot{y}_i \dot{y}_i + \ddot{z}_i \dot{z}_i \right) + \sum_{i=1}^{i=k} \left(\frac{\partial \Phi}{\partial x_i} \dot{x}_i + \frac{\partial \Phi}{\partial y_i} \dot{y}_i + \frac{\partial \Phi}{\partial z_i} \dot{z}_i \right) = 0. \qquad (3.4)$$

Using the symbols introduced by *Gibbs* we can write more compactly in terms of vector analysis with $\mathbf{r}_i = i x_i + j y_i + k z_i$

$$\sum_{i=1}^{i=k} m_i \dot{\mathbf{r}}_i \ddot{\mathbf{r}}_i + \frac{\sum_{i=1}^{i=k} \nabla \Phi d \mathbf{r}_i}{dt} = 0. \qquad (3.5)$$

Using the identity $\dot{\mathbf{r}}_i \ddot{\mathbf{r}}_i = \frac{1}{2} d(\dot{\mathbf{r}}_i \dot{\mathbf{r}}_i)/dt$ and identifying the total differential of first order $d\Phi = \nabla \Phi d \mathbf{r}$ we can identify the first term in Eq. (3.4) as the total kinetic energy and the second term as the total potential energy. Consequently, we can integrate Eq. (3.5) with respect to time and rewrite as

$$K + \Phi = C_1. \qquad (3.6)$$

This means that the sum of kinetic energy and potential energy is a constant C_1. Equation (3.6) expresses the energy principle in an isolated mechanical system. Namely, the sum of kinetic energy and potential energy is constant. This type of

---

[2] Jean Le Rond d'Alembert, born Nov. 17, 1717, in Paris, died Oct. 29, 1783, in Paris.

mechanical system is also addressed to as a conservative system. The potential is a function only of the positions $\mathbf{r}_i$ of the individual particles $1 \ldots k$ and not of other variables such as time. On the other hand, the kinetic energy is a function of only the velocity of the individual particles $\dot{\mathbf{r}}_i$ and not of position.

Equation (3.6) is widely used in the treatment of idealized mechanical systems. Actually, friction forces are not included in the treatment. If we include friction forces then the Eqs. (3.3) change into

$$m_i \frac{\partial^2 x_i}{\partial t^2} + \frac{\partial \Phi}{\partial x_i} = - k \frac{\partial x_i}{\partial t}$$
$$m_i \frac{\partial^2 y_i}{\partial t^2} + \frac{\partial \Phi}{\partial y_i} = - k \frac{\partial y_i}{\partial t} \quad .$$
$$m_i \frac{\partial^2 z_i}{\partial t^2} + \frac{\partial \Phi}{\partial z_i} = - k \frac{\partial z_i}{\partial t}$$

On the right-hand side, a negative term appears that acts against the forces of inertia. We can interpret the right-hand side as the start of a series expansion, with the zeroth-order differential with respect to time being zero. A similar treatment as before results in

$$\frac{d(K + \Phi)}{dt} = -k \sum_{i=1}^{i=k} \dot{\mathbf{r}}_i^2. \qquad (3.7)$$

Since the right-hand term of Eq. (3.7) is always negative, in a system subject to friction the total energy can only decrease with time. In fact, eventually the system will come to rest. The right-hand term in Eq. (3.7) is a dissipation term and has the form similar to a kinetic energy. The left-hand side term is no longer a so-called conservative term, i.e., time independent.

3.2.2 Destructive Interference

Sometimes, confusion arises in the phenomenon of destructive interference of waves. If two waves are allowed to interfere with a phase angle of π, then the waves annihilate. Thus the question arises where the energy is left. Some people think that there is a violation of the energy principle. However, even when waves annihilate, the energy is locally recovered before interference occurs along the path of light. In other words, destructive interference is accompanied by reflection [1, 2].

3.2.3 Pauli's Solution to Preserve Conservation of Energy

The conservation of energy was deeply accepted by all scientists at the start of the twentieth century. It was found by mass spectrometry that the atomic weights are not just the sum of the individual constituting elementary particles. According to the theory of relativity, the mass defect was explained as a gain in the bond energy of the elementary particles. Further, the discovery of radioactivity demonstrated clearly the amount of energy that is dormant in the atoms itself. A very striking example is the story of Pauli. A biography of *Pauli*[3] has been written by Enz [3].

It was observed that in the case of the β-decay the first law, i.e., the conservation of energy, is violated. Pauli was so much convinced in the correctness of the conservation of energy that he postulated the existence of a hitherto unknown particle. This particle should take away the missing energy. An English translation of Pauli's famous letter, given subsequently, can be found elsewhere [4].

Physikalisches Institut der Eidg. Technischen Hochschule Zürich
   Gloriastr.   Zürich, den 4.12.1930
   Offener Brief an die Gruppe der Radioaktiven bei der Gau-Vereinigung zu Tübingen
   Liebe Radioaktive Damen und Herren
   wie der Überbringer dieser Zeilen, den ich huldvollst anzuhören bitte, Ihnen des näheren auseinandersetzen wird, bin ich angesichts der 'falschen' Statistik der N- und Li 6-Kerne, sowie des kontinuierlichen β-Spektrums auf einen verzweifelten Ausweg verfallen, um den 'Wechselsatz' der Statistik und den Energiesatz zu retten. Nämlich die Möglichkeit, es könnten elektrisch neutrale Teilchen, die ich Neutronen nenne will, in den Kernen existieren, welche den Spin 1/2 tragen und das Ausschließungsprinzip befolgen und sich von Lichtquanten außerdem noch dadurch unterscheiden, daß sie nicht mit Lichtgeschwindigkeit laufen. Die Masse der Neutronen müßte von derselben Größenordnung wie die Elektronenmasse sein und jedenfalls nicht größer als 0,01 Protonenmasse.
   Das kontinuierliche β-Spektrum wäre dann verständlich unter der Annahme, daß beim β-Zerfall mit dem Elektron jeweils noch ein Neutron emittiert wird, derart, daß die Summe der Energien von Neutron und Elektron konstant ist ...
   Ich traue mich vorläufig aber nicht, etwas über diese Idee zu publizieren, und wende mich erst vertrauensvoll an Euch, liebe Radioaktive, mit der Frage, wie es um den experimentellen Nachweis eines solchen Neutrons stände, wenn dieses ebensolches oder etwas 10mal größeres Durchdringungsvermögen besitzen würde wie ein γ-Strahl.
   Ich gebe zu, daß mein Ausweg vielleicht von vornherein wenig wahrscheinlich erscheinen mag die Neutronen, wenn sie existieren, wohl längst gesehen hätte. Aber nur wer wagt gewinnt, und der Ernst der Situation beim kontinuierlichen β-Spektrum wird durch einen Ausspruch meines verehrten Vorgängers im Amte, Herrn Debye, beleuchtet, der mir kürzlich in Brüssel gesagt hat: 'O, daran soll man am besten gar nicht denken, so wie an die neuen Steuern.' Darum soll man jeden Weg zur Rettung ernstlich diskutieren.
   Also, liebe Radioaktive, prüfet und richtet.
   Leider kann ich nicht persönlich in Tübingen erscheinen, da ich infolge eines in der Nacht vom 6. zum 7. Dez. in Zürich stattfindenden Balles hier unabkömmlich bin.
   Mit vielen Grüßen an Euch,
   Euer untertänigster Diener,
   W. Pauli

---

[3] Wolfgang Pauli, born Apr. 25, 1900, in Vienna, Austria, died Dec. 15, 1958, in Zürich, Switzerland.

What *Pauli* is addressing as neutron in the letter quoted above was experimentally observed in 1956 by *Cowan*[4] and *Reines*.[5] This particle was called *neutrino*, already before the experimental visualization, by *Fermi*.[6]

Another example is the physics of *Aristotle*.[7] The conservation of energy was not postulated at all, but it was clear that particles that are in movement would come to rest after some time. It was Newton who introduced a conservation principle of energy, and later the heat was also included in the forms of energy.

However, at the latest at the discovery of black holes we may be in doubt if we have knowledge on how much energy is really in the world. We do not have access to the energy of a black hole.

3.2.4 Formulation of the First Law

The first law can be stated that the energy is not dependent on the path, in that we have reached certain parameters that characterize the energy. These parameters are in general the thermodynamic variables. This law can be stated also mathematically that the energy is a sole function of the thermodynamic variables themselves. In other words, we do not need other variables to ensure that the energy has just this value. Therefore, we may write the energy in the usual form:

$$U = U(S, V, \ldots) \tag{3.8}$$

but not

$$U = U\left(S, V, \ldots, \int_S f(U)\mathrm{d}S\right),$$

or similar. In thermodynamics, Eq. (3.8) is called as state function and this function allows formulating $\mathrm{d}U$ as an exact differential.[8] We can easily see that this is also the conservation of energy, because if we are going back, on any path to the original values of the arguments, the energy change is completely reversed. However, we did not make statements on the conservation of the arguments $S, V, \ldots$ here.

This view is somehow an idealization, in contrast to common sense. For example, to bring a weight from sea level to a certain mountain, either with a cable car or on the road, the first law says that the same amount of energy is needed. This is against

---

[4] Clyde Lorrain Cowan Jr., born Dec. 6, 1919, in Detroit, MI, died May 24, 1974, in Bethesda, MD. USA.

[5] Frederick Reines, born Mar. 16, 1918, in Paterson, NJ, died Aug. 26, 1998, in Orange, CA. USA.

[6] Enrico Fermi, born Sep. 29, 1901, in Rome, Italy, died Nov. 28, 1954, in Chicago, IL. USA.

[7] Aristotle, born 384 BC in Chalcidice, Greece, died Mar. 7, 322 BC in Chalcis, Euboea.

[8] Sometimes the intensive variables as functions of the extensive variables will be addressed as state functions. Often, the intensive variables are homogeneous with an order of zero.

the natural experience. Every one knows that a car will consume as more fuel, as longer the path. So, if you surround the mountain or take the direct road it makes a difference. One of the basic features in thermodynamics is the idealization. We may neglect the friction forces and we may try to regain the energy, if the car is moving temporarily downhill, etc., and then the concept will work.

We emphasize that the formulation according to Eq. (3.8) does not guarantee that the function U is not independent of its path. Conversely, from the value of the state function, we cannot conclude on the values of its arguments.

Example 3.1. Academic teachers often stress the term state function. But most of them cannot exemplify any process that is not a state function. Here we show one example, the ideal gas

$$\tilde{V}(T, p) = \frac{RT}{p}.$$

We go first from room temperature $T > 0$ and ordinary pressure $p > 0$ to zero pressure $p \to 0$ and then to zero temperature $T \to 0$

$$\lim_{T \to 0} \lim_{p \to 0} \tilde{V}(T, p) = \infty.$$

Then we go from room temperature and ordinary pressure to zero temperature and then to zero pressure.

$$\lim_{p \to 0} \lim_{T \to 0} \tilde{V}(T, p) = 0.$$

Observe the difference. For an ideal gas, the zero temperature and the zero pressure are a singular point.   □

3.3 Second Law of Thermodynamics

Several famous scientists have contributed to certain aspects of the second law of thermodynamics, among them are *Carnot, Joule, Kelvin, Clausius, Planck,* and *Boltzmann.* Various formulations of the second law have been created. This process still continues. Subsequently we will mainly focus on the formulation of *Clausius.*

The principle of maximum entropy is widespread in other disciplines besides thermodynamics, for example, in information theory, economics. In fact, it implies to maximize the probability of a state of a system under certain constraints [5].

3.3.1 Formulation of Clausius

The formulation of the second law according to *Clausius*[9] in the version from 1865 consists essentially of two statements.

---

[9] Rudolf Julius Emanuel Clausius, born Jan. 2, 1822, in Kölin, died Aug. 24, 1888, in Bonn.

Die Energie der Welt ist konstant. Die Entropie der Welt strebt einem Maximum zu. The energy of the world is constant. The entropy of the world tends to a maximum.

Remarkably enough, the second statement on the entropy does not address an equilibrium state. Nevertheless, it is excessively used for the foundations of equilibrium thermodynamics.

We annotate that Clausius states about the total entropy of the world. If the world consists of thermodynamic subsystems, each containing some entropy, then his statement concerns the total entropy, which is the sum of the entropies of the subsystems. In order to get the total entropy to a maximum, it is postulated that the subsystems cannot be made thermally insulated. In other words, the formulation of Clausius denies the existence of thermal nonconductive materials.

Let us say first that the second law of thermodynamics is not really needed in equilibrium thermodynamics. We could state that in an equilibrium process the total entropy will remain constant. This is just characteristic for an equilibrium process.

There are only a few nonequilibrium processes frequently dealt with in the textbooks, namely

- The *Joule – Thomson* effect
- The *Gay-Lussac* experiment

From the original statement of Clausius other formulations can be derived that are equivalent and more applicable to other practical issues.

3.3.2 Other Formulations

Other formulations of the second law have appeared in science. Some of these originate even earlier than the formulation of Clausius.

Still another more complicated formulation, which is related to Clausius, states that

a process in which the sum of the change in the total entropy of the system and the change in the total entropy of the surroundings is negative is impossible.

In simple words, this means that the total entropy cannot decrease by any process. This formulation is a somewhat more rigorous formulation of the second statement of Clausius concerning that the entropy tends to become a maximum. If this formulation states that a process associated with a decrease of entropy is not possible, it allows a process where the entropy does not increase, i.e., a process where the entropy remains constant. Such a process is an idealized process. We should emphasize that in a real process in which the thermodynamic parameters are changing always an increase of entropy occurs, even when the process can be directed close to the ideal process where the entropy remains constant. Actually, a process where the entropy does not change plays an important role in theoretical consideration.

3.3.2.1 Formulation of Thomson

This statement is often attributed to Lord *Kelvin*[10] and Max *Planck*.[11] It was formulated by Kelvin around 1851. The formulation of Kelvin and Planck deals with the conversion of heat to work:

> It is impossible to build a heat engine that produces a net work output by exchanging heat with a single fixed-temperature region.

This statement was later refined by Planck. On first glance, the formulation of the second law of Clausius and the formulation of Kelvin and Planck do not seem to have much in common. However, it is shown in elementary texts that both formulations are equivalent.

3.3.2.2 Carathéodory's Axiom

Carathéodory[12] made significant contributions to the calculus of variations, the theory of point set measure, and the theory of functions of a real variable. In physics, he established an axiomatic formulation of thermodynamics.

To start, we repeat a simple statement: If a state A contains more entropy than the state B, then there is an adiabatic process to go from B to A, but not vice versa. Carathéodory [6] reformulated the second law by stating that in the neighborhood of any state there exists states, which are adiabatically inaccessible from it. It is important to derive Carathéodory's axiom without the assumption that $\partial U/\partial T \neq 0$ [7].

3.3.2.3 Recent Formulations

There are still more recent other formulations of the second law. Kelvin's formulation of the second law is not universally true. It fails for negative absolute temperature states. Of course, most thermodynamic systems do not have such states. However, there are some systems. Therefore, the second law has been reformulated: It is impossible to transfer an arbitrarily large amount of heat from a standard heat source with processes terminating at a fixed state of the working system [8].

It has been shown that concavity of the entropy is a consequence of the second part of the second principle and extensivity. Then the energy results to be convex, on the assumption that it be a monotonic function of the entropy [9, 10]. A relation of the concavity of the entropy to thermodynamic stability has been established [11].

---

[10] William Thomson was raised to the peerage as Lord Kelvin in Jan. 1892. Born Jun. 26, 1824, in Belfast, Ireland, died Dec. 17, 1907, in Nethergall, Scotland).

[11] Max Karl Ernst Ludwig Planck, born Apr. 23, 1858, in Kiel, Germany, died Oct. 4, 1947, in Göttingen, Germany.

[12] Constantin Carathéodory born Sep. 13, 1873, in Berlin, Germany, died Feb. 2, 1950, in Munich.

3.3.2.4 The Second Law in Statistics

The second law of thermodynamics is a macroscopic law concerning the phenomenon of nonreversibility. *Boltzmann,*[13] using the methods of statistical mechanics and probability theory, also derived the second law of thermodynamics. He argued that a system will preferably cross over from a state of low probability into a state of higher probability. The reverse step is possible. Whereas in systems containing a small number of particles the reverse step can occur with a probability clearly above zero, in the case of large particle systems, this probability becomes so small that the reverse step is practically impossible.

Fluctuation Theorem

The theory of fluctuation deals with the problem that a system far from equilibrium could move away from the direction of equilibrium for a short time, which is in contrast to the second law [12, 13]. The fluctuation theorem is also connected to the phenomenological formalism of reaction kinetics, in particular, to unimolecular reactions [14]. The fluctuation theorem predicts appreciable and measurable violations of the second law of thermodynamics for small systems over short time scales.

The fluctuation theorem was experimentally demonstrated, by following the trajectory of a colloidal particle captured in an optical trap that is translated relative to surrounding water molecules. From the particle trajectories, the entropy production and consumption along the trajectory were calculated. The fraction of second law-defying trajectories that are in contrary to the second law could be sampled. The results of these experiments show that entropy consumption can occur over colloidal length and time scales [15].

Thus, the second law of thermodynamics has been shown not to hold for microscopic systems. Therefore, a fundamental limit concerning miniaturization is pointed out by these experiments, because it suggests that the devices envisaged by nanotechnology would not behave like simple scaled-down versions of their larger counterparts. In fact, they could sometimes run backward.

3.3.3 Clausius' Statements in Detail

3.3.3.1 Direction of Heat Flow

We now start discussing in detail with another, still earlier formulation due to *Clausius*:

> Wärme kann nie von einer niedrigeren Temperatur zu einer höheren Temperatur fließen.
> Heat will not flow deliberately from the colder body to the hotter body.

---

[13] Ludwig Eduard Boltzmann, born Feb. 20, 1844, in Vienna, Austria, died Sep. 5, 1906, in Duino, Italy.

In our model, there is nothing else in the world than a hotter body and a colder body and an engine that can switch on or off the heat flow, without consuming energy thereby. We presume that the hotter body has the higher temperature $T_h > T_l$. So the colder body has the lower temperature, T_l. The outflow from heat, as in general for energy, we associate with a negative quantity. Consequently, an energy loss for the colder body would be $dQ_c = T_l dS_l < 0$, but does not occur, according to the above statement. We express the previous wordy statements in a logical equation:

$$(T_h > T_l \quad AND \quad T_l dS_l < 0) = FALSE. \tag{3.9}$$

Equation (3.9) can be transformed into

$$(T_h > T_l \quad AND \quad T_l dS_l \geq 0) = TRUE. \tag{3.10}$$

So, in addition to $T_h > T_l$ we have $T_h > T_l \geq 0$ to establish $dS_l \geq 0$. This means at the same time that the entropy of the body with the higher temperature will decrease, $dS_h < 0$.

3.3.3.2 Change of Entropy in Isodynamic Processes

In an isodynamic process we have $dU = 0$. If only heat is exchanged in two subsystems, we have for this special system

$$dU = 0 = T_l dS_l + T_h dS_h.$$

Substituting $dS = dS_l + dS_h$, we arrive at

$$dS = dS_l \frac{T_h - T_l}{T_h}.$$

The expression $(T_h - T_l)/T_h$ is positive for $T_l/T_h < 1$. For this reason, the total entropy dS increases for this condition and for $dS_l > 0$.

We discuss the change of entropy in detail, if two systems (') and ('') are in contact. The energy of these systems should be in the form $U(S, V, n)$, as is usual for many systems dealt with in thermodynamics. We want to keep the volume V and the mol number n of these systems constant. Therefore, the energy is a function of the variable entropy alone. We presume that the entropy can be resolved in terms of the energy, i.e., a function of the form $S = S(U)$ exists.

The entropy of an individual system is dependent on parameters belonging just to this individual system, no mind whether it is isolated or whether it is coupled to another system. Thus we have for the entropy of the systems (') and ('')

$$S' = S'(U'), \quad S'' = S''(U'').$$

The total entropy is the sum of the individual entropies and a function of the individual entropies:

$$S(U', U'') = S'(U') + S''(U'').$$ (3.11)

We want to have the coupled system isolated, which means that the sum of the individual energies is constant, i.e., $U' + U'' = U$. This means in differential form that $dU' + dU'' = 0$.

We form now the differential of Eq. (3.11) with respect to say U' and observe that U'' may be interpreted as a function of U', e.g., $dU''/dU' = -1$. Further we introduce the general notation $dU/dS = T$. Using the chain rule we obtain

$$\frac{dS(U', U'')}{dU'} = \frac{1}{T'(U')} - \frac{1}{T''(U'')}.$$ (3.12)

Thus the total entropy becomes stationary if the temperatures of both systems are equal. We emphasize that in more complicated systems we can handle the constraint of constant energy more conveniently by the method of undetermined multipliers. In our derivation we have used the chain rule for derivation.

We form now the second derivative which indicates whether the curvature of the function is concave of convex, i.e., whether a maximum or a minimum is prevalent. A second differentiation of Eq. (3.12) results in

$$\frac{d^2 S(U', U'')}{dU'^2} = -\frac{1}{T'(U')^2} \frac{dT'(U')}{dU'} - \frac{1}{T''(U'')^2} \frac{dT''(U'')}{dU''}.$$ (3.13)

The minus sign in the second term of Eq. (3.12) is kept because the application of the chain rule demands $dU''/dU' = -1$. Now

$$\frac{dU(T, V, n)}{dT} = C_v \quad \rightarrow \quad \frac{dT(U, V, n)}{dT} = \frac{1}{C_v}.$$

Note the correct arguments of the functions at a change of the dependent variable $U \rightarrow T$. We can simplify Eq. (3.13) into

$$\frac{d^2 S(U', U'')}{dU'^2} = -\frac{1}{T'^2 C'_v} - \frac{1}{T''^2 C''_v}.$$ (3.14)

We discuss now the terms in Eq. (3.14) in detail. In order to have a maximum for the entropy we must have the right-hand side of Eq. (3.14) negative. Since the temperature occurs as a square term, the heat capacity must be positive. We can split the term $T^2 C_v$ still in a different way, namely $T \times C_v T$. We can identify the term $C_v T$ as a thermal energy. Basically, any energy cannot drop below zero, thus $C_v T \geq 0$ and the temperature in the first factor of $T \times C_v T$ must be positive in a system that maximizes its entropy, at least in the simple example we are discussing here. Actually, here we do not deal with the limiting case of zero energy and temperature.

3.3.3.3 The Second Law for Other Energy Forms

Now, we have accepted that heat will not flow deliberately from the cold body to the hot body. Heat is an energy form with the temperature T as the intensive variable and the entropy S as the extensive variable. We re-examine the problem with another energy form. Suppose two bodies with different pressures are in contact via a movable piston. Here the energy is the volume energy $-p\,dV$. The negative pressure $-p$ is the intensive variable and the volume V is the extensive variable. We know very well that the body with the higher pressure will expand and the body with the lower pressure will compress, if the volume can be freely changed. We start with the same argumentation as above and conclude

$$p_h > p_l$$
$$-dV_h < 0. \tag{3.15}$$

Here we have attached the minus sign to the volume. Eliminating the minus sign in the second equation of Eq. (3.15) we arrive at

$$dV_h > 0.$$

In this way, we have a very similar situation as in the case of deliberate entropy flow. In a system consisting of two bodies with different pressures, the body with the higher pressure will deliberatively expand.

We can still emphasize quite a lot of other examples. For instance, if two charged electrical condensers are connected via a conductive wire, then the electrical charge will flow deliberately only from the condenser with the higher voltage to the condenser with the lower voltage.

Therefore, we could conclude that the second law of thermodynamics is a special law of a still more general principle. This means that if two bodies can freely exchange a certain energy form, then the extensive variable of the body with the higher intensive variable will decrease, but not deliberately increase. We formulate this also in terms of formulas. The energy form should consist of an intensive variable ξ and an extensive variable X. We presume that ξ is a positive quantity, i.e., $\xi > 0$. Then one body has a ξ_h and the second body has a ξ_l. The extensive variable X_h can deliberately only decrease:

$$\xi_h > \xi_l, \quad dX_h < 0.$$

In spite of these examples, the second law of thermodynamics has an extraordinary position in thermodynamic sciences, as will be explained in detail.

3.3.3.4 Foundations of the Second Law

In terms of the methodology of scientific concepts, the first law of thermodynamics and the second law of thermodynamics are basic statements that cannot be proved further. These laws play a similar fundamental role as axioms in mathematics. Unlike axioms in mathematics the basic laws of thermodynamics like other basic laws in natural sciences, e.g., *Maxwell*'s equations, are based on observations. These observations are generalized. The result of these observations is extended to a basic law, also addressed as a postulate.

If only a single observation does not follow such a postulate, you may throw away the complete theory based on the respective postulates. This statement should not be taken as strict and literal. There are possibilities to escape from a dilemma.

We discuss once more the statements of *Clausius*. The energy of the world is constant. The entropy of the world tends to a maximum. We reformulate now the term *world*. For a laboratory experiment we think that can simulate a small world, which is a completely isolated thermodynamic system. We do not know what is at the border of the world; however, for our laboratory world we think there are no constraints at the border. This idea corresponds to a real world that should be embedded in an empty space.

Actually, the first law of thermodynamics states that for a completely isolated system the energy is constant. So we are confident that the laboratory world has some similarity with the real world in the thermodynamic sense. We will address thermodynamic processes where the energy of the system under consideration remains constant as isodynamic processes.

The second part of the statement of Clausius suggests that the entropy of the world tends to become a maximum. This statement is often accepted as such, as a postulate. However, we may ask still more into depth: if the entropy tends to a maximum, then a maximum should exist at all.

Types of Maxima

We discuss briefly, what is a maximum, before we go into more depth. Some types of maxima are shown in Fig. 3.1.

Ordinary Maximum

In the ordinary maximum the maximum is reached at finite values of X. Approaching from small values of X we can pass through the maximum. After reaching the maximum, the function decreases again. At this type of maximum $dS = 0$ and $d^2S < 0$. We can approach this maximum from both sides of the abscissa, i.e., either increasing X or decreasing X, if we are left either from the maximum or right from the maximum. Near the maximum, the second derivative is negative.

Degenerated Maximum

In the degenerated maximum the maximum is reached at infinite values of X. Approaching from small values of X we cannot pass really through the maximum.

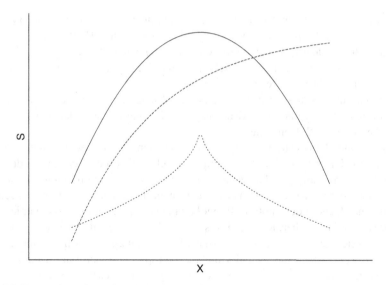

Fig. 3.1 Types of maxima: *Top*: Ordinary maximum, *Mid*: Degenerated maximum, it will be reached at infinity, *Bottom*: A needle-like maximum

At this type of maximum the entropy tends to zero $dS \to 0$ as $X \to \infty$. Approaching the maximum the second derivative is smaller than zero, $d^2 S < 0$, but it tends to zero as we are approaching the maximum.

Needle-Like Maximum

In the case of the needle-like maximum, the function is really discontinuous at the maximum. Approaching this type of maximum, the slope is positive, i.e., $dS > 0$, but also $d^2 S > 0$. We will not consider or allow this type of maximum. At the maximum $\partial S/\partial X \to \pm\infty$.

The statement of Clausius that the entropy tends to approach a maximum is somewhat ambiguous in daily language. We can interpret this statement that the entropy.

1. tends to approach and eventually reaches an ultimate maximum, or that
2. entropy tries to approach a maximum as far as possible.

These variants result in different consequences. Suppose that the world consists of two subsystems, separated by a wall via which no energy form can be exchanged. In each subsystem processes can run, which maximize the entropies of the individual subsystems. In this case, the second variant of the statement above is valid, because, if the wall would not be there, possibly an even higher value of the entropy could be reached.

If we think rather of the first variant, then we imply that we cannot fabricate a wall that is essentially impermeable with respect to transfer of any energy form. It is in common sense that a material has basically a finite heat conductivity, otherwise a Dewar vessel that keeps warm for ever could be constructed. Further, any material exhibits a finite electric conductivity. On the other hand, the theory of diffusion demands that a membrane is completely impermeable to particles that are greater in diameter than the larger pore size.

Summarizing, the first variant of the extended statement postulates certain properties of any existing material. Also this particular interpretation leads to the phenomenon of the death from heat.

The second, less restrictive variant of the statement is used for certain processes in good approximation. For example, an adiabatic process can be conducted, because the surrounding wall does not conduct heat into the environment, at least as long the experiment is in action. In practice, it seems to be that more often the second variant is used, and it does not locally violate the principles of thermodynamics.

The second law of thermodynamics is believed to be a fact that cannot boil down to more basic statements. It is somewhat like a postulate, or in mathematics, an axiom. However, from the formal view, the entropy can reach only a maximum, if a maximum in fact exists. Otherwise, entropy cannot reach a maximum. Here we confine ourselves to nonpathological functions, as encountered in physics. From elementary mathematics, there are two conditions necessary for a maximum, namely $dS = 0$ and $d^2S < 0$. If we are somewhat away from the maximum, we have the condition $dS > 0$. In this case, the sign of the independent variable tells the direction where to find the maximum.

We exemplify with a gas, thus we have the variables S, V, n. Since we keep the particle number n as constant, we will not deal with this variable further. By simple arrangement, we obtain for the entropy

$$dS(U, V) = \frac{1}{T(U, V)}dU + \frac{p(U, V)}{T(U, V)}dV. \tag{3.16}$$

To make it more clear, we have added the variable set (U, V) to each function. The first part of the statement says that the energy is constant, therefore $dU = 0$. In order to have the entropy stationary, we have

$$dS = \frac{p(U, V)}{T(U, V)} = 0. \tag{3.17}$$

As pointed out already in general, if we are away from the maximum, the condition

$$dS = \frac{p(U, V)}{T(U, V)}dV > 0 \tag{3.18}$$

holds. So we have to choose dV either $dV > 0$ or $dV < 0$ to satisfy Eq. (3.18). Next, we have to show that the curvature is concave with respect to the abscissa, in order that we are dealing really with a maximum. Thus,

$$\frac{\partial^2 S(U, V))}{\partial V^2} = -\frac{p(U, V)}{T^2(U, V)} \frac{\partial T(U, V)}{\partial V} + \frac{1}{T(U, V)} \frac{\partial p(U, V)}{\partial V} \leq 0. \qquad (3.19)$$

At the maximum, the first term approaches zero because $p/T \to 0$, if the other terms remain finite. So we have to inspect the second term. We emphasize once more that we are dealing at constant energy. For an ideal gas, it is known that at the Gay – Lussac experiment, the temperature remains constant at isodynamic expansion. Therefore, we could directly obtain an expression for the second term.

However, we will treat the matter more generally and we recalculate the differential for temperature and volume as the independent variables. We use functional determinants, also addressed as Jacobian determinants. In terms of functional determinants, we can rewrite

$$\frac{\partial p(U, V)}{\partial V} =$$

$$= \frac{\partial(p, U)}{\partial(V, U)} = \frac{\partial(p, U)}{\partial(V, T)} \frac{\partial(V, T)}{\partial(V, U)} = \frac{\partial(p, U)}{\partial(V, T)} \left(\frac{\partial(V, U)}{\partial(V, T)}\right)^{-1} =$$

$$= \left(\frac{\partial p(T, V)}{\partial V} C_v(T, V) - \frac{\partial p(T, V)}{\partial T} \frac{\partial U(T, V)}{\partial V}\right) \frac{1}{C_v(T, V)} =$$

$$= \frac{\partial p(T, V)}{\partial V} - \frac{\partial p(T, V)}{\partial T} \frac{\partial U(T, V)}{\partial V} \frac{1}{C_v(T, V)}. \qquad (3.20)$$

Inspection of the last line in Eq. (3.20) and intuitive evaluation reveals that $\partial p(T, V)/\partial V < 0$, $\partial p(T, V)/\partial T > 0$, and $C_v > 0$.

Probably for the term $\partial U(T, V)/\partial V < 0$, we have no intuition. However, recall the well-known relation

$$\frac{\partial U(T, V)}{\partial V} = -p + T \frac{\partial p(T, V)}{\partial T},$$

which is zero for an ideal gas and a/V^2 for a *van der Waals* gas. So we can exemplify that the terms are meeting the requirement that the entropy would approach a maximum in fact for at least an ideal gas and a van der Waals gas.

Observe that we can trace back the last line in Eq. (3.20) essentially to terms that appear in the analysis of stability of thermodynamic systems. For example, in the analysis of stability emerges that $C_v > 0$ for a stable system. In general, the generalized susceptibilities must be positive.

We emphasize that Eq. (3.16) is a very special case. In fact, we could repeat the same considerations with a general variable like

$$dU(S, X) = T(S, X)dS + \xi(S, X)dX.$$

We presume now that the temperature T is independent of the variable X, thus

$$\frac{\partial T(S, X)}{\partial X} = 0 = \frac{\partial \xi(S, X)}{\partial S}. \tag{3.21}$$

This means that at constant entropy we cannot increase the temperature by forcing a change of the variable X. In other words, the temperature is a function of the entropy alone, i.e., $T = T(S)$. The second equality in Eq. (3.21) arises from the inequality of Schwarz. Thus $\xi = \xi(X)$.

Examples for real systems like this are the plate condenser with X as the charge or the kinetic energy with X as the momentum. In the former case, we may think of the discharge of a condenser and concomitant increase of the temperature at constant energy. In the latter case, we may think that a projectile is lowering its kinetic energy in a viscous medium at constant total energy. Thus the system consists of the moving projectile and the viscous medium at rest.

Rearranging and changing the independent variables, we get

$$dS(U, X) = \frac{1}{T(U, X)} dU - \frac{\xi(U, X)}{T(U, X)} dX. \tag{3.22}$$

Now the entropy becomes a function of the energy U and the formal variable X. We form now the differential of the entropy at constant energy as

$$\frac{\partial S(U, X)}{\partial X} = -\frac{\xi(U, X)}{T(U, X)}.$$

The second derivative becomes

$$\frac{\partial^2 S(U, X)}{\partial X^2} = -\frac{\partial \xi(U, X)}{\partial X}\frac{1}{T(U, X)} + \frac{\xi}{T^2(U, X)}\frac{\partial T(U, X)}{\partial X}. \tag{3.23}$$

We evaluate now the partial derivatives in Eq. (3.23) in terms of the independent variables S and X:

$$\frac{\partial \xi(U, X)}{\partial X} =$$

$$\frac{(\partial \xi, U)}{\partial(X, U)}\frac{(\partial X, S)}{\partial(X, S)} = \frac{(\partial \xi, U)}{\partial(X, S)}\frac{(\partial X, S)}{\partial(X, U)} =$$

$$\left(\frac{\partial \xi(S, X)}{\partial X}T(S, X) - \frac{\partial \xi(S, X)}{\partial S}\xi(S, X)\right)\frac{1}{T(S, X)} =$$

$$\frac{\partial \xi(S, X)}{\partial X} \tag{3.24}$$

and

$$\frac{\partial T(U, X)}{\partial X} =$$

$$\frac{(\partial T, U)}{\partial (X, U)} \frac{(\partial S, X)}{\partial (S, X)} = \frac{(\partial T, U)}{\partial (S, X)} \frac{(\partial S, X)}{\partial (X, U)} =$$

$$\left(\frac{\partial T(S, X)}{\partial S} \frac{\partial U(S, X)}{\partial X} - \frac{\partial T(S, X)}{\partial X} \frac{\partial U(S, X)}{\partial S} \right) \left(-\frac{1}{T(S, X)} \right) =$$

$$-\frac{\xi(S, X)}{C_X}. \quad (3.25)$$

In Eqs. (3.24) and (3.25) we have used the final lines of Eq. (3.21). Further, C_X is the heat capacity at constant X. Going back to Eq. (3.23), we get

$$\frac{\partial^2 S(U, X)}{\partial X^2} = -\frac{\partial \xi(S, X)}{\partial X} \frac{1}{T(U, X)} - \frac{\xi^2(S, X)}{T^2(U, X)} \frac{1}{C_X}. \quad (3.26)$$

Summarizing, we could append to the famous statement of Clausius "The energy of a closed system is constant, the entropy approaches a maximum," the additional remark: *It must be in this way. Otherwise, we would violate the stability conditions of the system.* It would not be stable at all. This addendum would contribute to a more basic intuitive understanding why the entropy should approach a maximum.

We emphasize that we did not discover a new or more basically justified axiom of thermodynamics. Namely, the minimum principle of energy is derived from the maximum principle of entropy, and the stability theorems are derived from the minimum principles of energy enthalpy, free energy, and free enthalpy. Thus, we cannot construct an axiomatical approach to thermodynamics by simply adding the stability theorem to the axioms of thermodynamics.

Instead, we have traced back the maximum principle of entropy to the properties of matter. It would be possible to imagine a world in that the heat capacity is negative, as well as other material properties such as compressibility have a reverse sign to that we are accustomed. However, such a world would not be intrinsically stable, if we assume that energy has a giving property. Sloppily spoken, the energy of a system tries to give another system with lower energy its energy. If a system tries to accumulate energy, we do not find stability. This situation is comparable to unfair business rivalry in economics.

Systems that do not tend to approach a maximum of entropy might be basically of interest in cosmology, if the big bang stops to bang. However, we should annotate that at the time when Clausius formulated the second law, nothing was known about the big bang.

Example 3.2. We deal now again with the specialized case of

$$dU = T(S, V)dS - p(S, V)dV$$

using an ideal gas. Rearrangement gives

$$dS(U, V) = \frac{1}{T(U, V)}dU + \frac{p(U, V)}{T(U, V)}dV. \tag{3.27}$$

The condition for $S \to max$ at constant energy U is

$$\frac{\partial(p(U, V)/T(U, V))}{\partial V} < 0.$$

In contrast to Eq. (3.22), the negative sign cancels, because the pressure p introduced the negative sign. Further, observe that the partial differential has to be evaluated at constant energy, thus we have the set of independent variable of (U, V). This is different from the definition equation of the energy, where we have the set (S, V) as independent variables. We want to express Eq. (3.27) in terms of the independent variables (T, V). Using the method of Jacobian determinants, as explained in Sect. 1.14.6, we get

$$\frac{\partial(p(U, V)/T(U, V))}{\partial V} \equiv \frac{\partial p/T}{\partial V}\bigg|_U = \frac{\partial(U, p/T)}{\partial(U, V)} = \frac{\partial(U, p/T)/\partial(T, V)}{\partial(U, V)/\partial(T, V)}.$$

A Mathematica® session to get the expansion of the Jacobian determinant is shown in Table 3.1.

Table 3.1 Conversion of independent variables by Jacobian determinants with Mathematica®

Det[D[{p[U,V]/T[U,V]},{{V}}]]
$\dfrac{p^{(0,1)}[U, V]}{T[U, V]} - \dfrac{p[U, V]T^{(0,1)}[U, V]}{T[U, V]^2}$
Det[D[{U,p[U,V]/T[U,V]},{{U,V}}]]
$\dfrac{p^{(0,1)}[U, V]}{T[U, V]} - \dfrac{p[U, V]T^{(0,1)}[U, V]}{T[U, V]^2}$
Det[D[{U[T,V],p[T,V]/T},{{T,V}}]]
$\dfrac{p[T, V]U^{(0,1)}[T, V]}{T^2} - \dfrac{U^{(0,1)}[T, V]p^{(1,0)}[T, V]}{T} + \dfrac{p^{(0,1)}[T, V]U^{(1,0)}[T, V]}{T}$
Det[D[{U[T,V],V},{{T,V}}]]
$U^{(1,0)}[T, V]$
U$^{(1,0)}$[T, V] := Cv
$U^{(0,1)}[T, V]:=$**T** $* $**p$^{(1,0)}$[T, V] $-$ p[T, V]**
Det[D[{U[T,V],p[T,V]/T},{{T,V}}]]/ Det[D[{U[T,V],V},{{T,V}}]]
$-\dfrac{p[T,V]^2}{T^2} + \dfrac{Cvp^{(0,1)}[T,V]}{T} + \dfrac{2p[T,V]p^{(1,0)}[T,V]}{T} - p^{(1,0)}[T, V]^2$
$\overline{\qquad\qquad Cv \qquad\qquad}$
FullSimplify
[Det[D[{U[T,V],p[T,V]/T},{{T,V}}]]/Det[D[{U[T,V],V},{{T,V}}]]]
$\dfrac{CvTp^{(0,1)}[T, V] - (p[T, V] - Tp^{(1,0)}[T, V])^2}{CvT^2}$

Thus, the conversion into the variables yields

$$\frac{\partial(p(U, V)/T(U, V))}{\partial V} =$$

$$\frac{1}{T}\frac{\partial p(T, V)}{\partial V} - \frac{1}{C_v T^2}\left(p(T, V) - \frac{\partial p(T, V)}{\partial T}\right)^2. \quad (3.28)$$

In order to have the first term of Eq. (3.28) negative, the pressure should decrease with increasing volume. The second term in brackets does not play any role in the statement of the second law. Because of the negative sign and the square, it is always negative. This means that the statement of the second law in this form explicitly allows any sign of the coefficient of tension and related coefficients such as the ratio of thermal expansion and compressibility coefficient. Recall the relation

$$dV(T, p) = \frac{\partial V(T, p)}{\partial T}dT + \frac{\partial V(T, p)}{\partial p}dp$$

that relates the coefficient of thermal expansion, compressibility, and the coefficient of tension. However, note that the compressibility as such is restricted by the first term. Therefore, only the coefficient thermal expansion can have any sign. For this reason, a density maximum as it occurs in water is thermodynamically allowed. □

3.3.4 Maximum Entropy – Minimum Energy

It is often stated that the maximum entropy principle implies at the same time a minimum energy. This is basically correct; however, the arguments are often not clearly presented.

Question 3.1. If we have $dU = TdS + \ldots$, how can the energy U tend to approach a minimum, when the entropy S wants to approach a maximum?

We start with a formal derivation of the equivalency of the minimum principle of energy and maximum principle of entropy and illustrate subsequently the principle with a few examples in back-breaking work. A more detailed treatment based on the concavity of entropy, dealing with thermodynamic potentials, can be found in the literature [16].

3.3.4.1 Formal Derivation of the Principle

Recall the condition of equilibrium and maximum entropy

$$\frac{\partial S(U, X)}{\partial X} = 0; \quad \frac{\partial S^2(U, X)}{\partial X^2} < 0. \quad (3.29)$$

Here X is a general extensive variable. From

$$dS(U, X) = \frac{\partial S(U, X)}{\partial U} dU + \frac{\partial S(U, X)}{\partial X} dX$$

we get

$$\frac{\partial U(S, X)}{\partial X} = -\left(\frac{\partial S(U, X)}{\partial X}\right)\left(\frac{\partial S(U, X)}{\partial U}\right)^{-1}$$

$$= -T(U, X)\frac{\partial S(U, X)}{\partial X} = 0. \quad (3.30)$$

The last equality in Eq. (3.30) follows from Eq. (3.29).

Forming the second derivative in Eq. (3.30) results in

$$\frac{\partial^2 U(S, X)}{\partial X^2} = -T(U, X)\frac{\partial S^2(U, X)}{\partial X^2} > 0. \quad (3.31)$$

The mathematical tools to get these equations properly are detailed in Sect. 1.11.4. Assuming a positive temperature $T(U, X)$, it follows that the energy is a minimum. A comparison of the condition of maximum entropy and minimum energy reveals that both conditions are different only by the factor $-T(U, X)$.

3.4 Third Law of Thermodynamics

The third law of thermodynamics is claimed to be related to, but not dependent on, the second law of thermodynamics. It is also known as the Nernst heat theorem. The history of the third law has been reviewed by Hiebert and Kox [17, 18]. The third law was published by Nernst in 1906 [19]. Subsequently we cite from *Nernst*'s original paper:

Hiernach gilt für die maximale Arbeit A und die Änderung der gesamten Energie U folgende Beziehung

$$A - U = T\frac{dA}{dT},$$

worin T die absolute Temperatur bedeutet. Nur wenn A unabhängig von der Temperatur wäre, so würde bei allen Temperaturen

$$A = U$$

sein und es würde zugleich auch U von der Temperatur unabhängig sein müssen. Die letztere Forderung würde für die spezifischen Wärmen der an einer Reaktion sich beteiligenden Substanzen gewisse Beziehungen mit sich bringen, die in der Regel nicht erfüllt sind.

...

Obwohl, wie bereits oben erwähnt, die Größen A und U im Allgemeinen nicht einander gleich sind, so ist es doch sehr auffällig, daß, wenigstens bei nicht zu hohen Temperaturen, in der Regel der Unterschied beider Größen innerhalb mäßiger Grenzen bleibt. Freilich sind bei diesem Vergleich Gase und verdünnte Lösungen auszuschalten, weil bei diesem bekanntlich Q, nicht aber A von der Konzentration unabhängig ist. Schon lange war mir nun in dieser Hinsicht aufgefallen, daß bei galvanischen Kombinationen, bei welchen in der Gleichung des stromliefernden chemischen Prozesses nur feste Körper und sehr konzentrierte Lösungen vorkommen, die Unterschiede zwischen A und U auffällig klein sind; ferner sei auch an das Verhalten der sogenannten idealen konzentrierten Lösungen erinnert. So drängte sich die Annahme auf, daß in solchen Fällen in der nächsten Nähe des absoluten Nullpunktes ein völliges Zusammenfallen beider Größen stattfindet, und es würde als Grenzgesetz

$$\lim_{T \to 0} \frac{dA}{dT} = \lim_{T \to 0} \frac{dQ}{dT}$$

sich ergeben.

English translation:

After this, for the maximum work A and the change of the total energy U the following relation holds

$$A - U = T \frac{dA}{dT},$$

Where T is the absolute temperature. Only, if A would be independent of the temperature, at every temperatures would be

$$A = U$$

and in addition U would be independent from temperature. The latter postulate would establish certain relations for the specific heats of the substances involved in a chemical reaction, which are usually not fulfilled.

$$\cdots$$

Even though, as mentioned above, the quantities A and U are not equal in general, it is conspicuous, at least at not to high temperatures, that the difference remains in moderate size.

Admittedly, gases and diluted solutions have to be excluded, because as generally known, Q, but not A is dependent on the concentration.

Already long before in this aspect I realized that in galvanic elements in which in the equation of the current delivering process only solid bodies and highly concentrated solutions are prevalent, the differences between A and U are remarkably small; further the behavior of so-called ideally concentrated solutions should be revisited.

The the assumption is suggested that in such cases closely to the absolute zero point a complete coincidence of both quantities could occur and the limiting law of

$$\lim_{T \to 0} \frac{dA}{dT} = \lim_{T \to 0} \frac{dQ}{dT}$$

would result.

In another paper [20], Nernst states about the series development of the heat in terms of temperature.

Assuming that Q may be expanded in powers of T as

$$Q = Q_0 + \alpha T + \beta T^2 + \gamma T^3 + \dots,$$

the integration of the expression for the second law, $A - Q = T\, dA/dT$, gives

$$A = Q_0 + aT - \alpha T \ln T - \beta T^2 - \frac{\gamma}{2} T^3.$$

Ŏn the further assumption that the limit of $dA/dT =$ limit of dQ/dT for $T = 0$, it is proved that $a = 0$ and $\alpha = 0$.

$$\lim_{T \to 0} dA/dT = \lim_{T \to 0} dQ/dT$$

Therefore, $Q = Q_0 + \beta T^2 + \gamma T^3$ and $A = Q_0 - \beta T^2 - \frac{\gamma}{2}T^3$. In general, only the first two terms of these series are needed.

Historically, the third law of thermodynamics emerges from the heat theorem, by *Nernst*,[14] which states: A chemical reaction between pure crystalline phases that occurs at absolute zero produces no entropy change. This means that adiabatic and isothermal processes approach each other at very low temperatures. The importance of Nernst's theorem is that it gives a solid base for the calculation of thermodynamic equilibria.

We discuss now the statement of Nernst and the problems behind in our contemporary notation. The molar Helmholtz energy of a reaction $\Delta \tilde{F}(T, \tilde{V})$ and likewise the molar Gibbs energy of a reaction $\Delta \tilde{G}(T, p)$ can be expressed as

$$\begin{aligned}
\Delta \tilde{F}(T, \tilde{V}) &= \Delta \tilde{U} - T \Delta \tilde{S} = \Delta \tilde{U} + T \frac{\partial \Delta \tilde{F}}{\partial T} \\
\Delta \tilde{G}(T, p) &= \Delta \tilde{H} - T \Delta \tilde{S} = \Delta \tilde{H} + T \frac{\partial \Delta \tilde{G}}{\partial T}.
\end{aligned} \tag{3.32}$$

$\Delta \tilde{F}(T, V)$ and $\Delta \tilde{G}(T, V)$ are the maximum work and are, for example, accessible by electrochemical measurements, as Nernst did. Moreover, these thermodynamic functions are the key quantities for the calculation of thermodynamic equilibria.

Equation (3.32) can be rewritten as

$$\begin{aligned}
T^2 \frac{\partial}{\partial T}\left(\frac{\Delta \tilde{F}}{T}\right) + \Delta \tilde{U} &= 0 \\
T^2 \frac{\partial}{\partial T}\left(\frac{\Delta \tilde{G}}{T}\right) + \Delta \tilde{H} &= 0.
\end{aligned} \tag{3.33}$$

Equation (3.33) is an inhomogeneous equation. The homogeneous equation with, e.g., $\tilde{U} = 0$, has the solution $\Delta \tilde{F}(T, V) = C(\tilde{V})T$. A particular solution of the inhomogeneous equation is

$$\Delta \tilde{F}(T, V) = -T \int_0^T \frac{\Delta \tilde{U}(T, \tilde{V})}{T^2} dT.$$

Thus the general solution is

$$\Delta \tilde{F}(T, V) = -T \int_0^T \frac{\Delta \tilde{U}(T, \tilde{V})}{T^2} dT + C(\tilde{V})T. \tag{3.34}$$

---

[14] Walther Hermann Nernst, born Jun. 25, 1864, in Briesen, Prussia, died Nov. 18, 1941, in Muskau, Germany.

Obviously $C(\tilde{V})$ is not accessible by caloric measurements. This quantity was addressed by *Haber*[15] as the *thermodynamically not accessible constant* [21, 22]. Moreover, the integral in Eq. (3.34) should be convergent. Here is the basic problem in the determination of the free energies and the free enthalpies of reaction.

Nernst observed experimentally that by plotting $\Delta \tilde{F}$ against the temperature T, $\Delta \tilde{F}$ approaches with a zero slope zero temperature, i.e.,

$$\lim_{T \to 0} \frac{\partial \Delta \tilde{F}}{\partial T}.$$

For this reason, he concluded that the thermodynamically not accessible constant $C(\tilde{V})$ must be zero.

We can rewrite Eq. (3.32) as

$$
\begin{aligned}
\Delta \tilde{F}(T, \tilde{V}) &= \Delta \tilde{U}_0 + \int_0^T \Delta \tilde{C}_v \, \mathrm{d}T - T \left(\Delta \tilde{S}_0 + \int_0^T \frac{\Delta \tilde{C}_v}{T} \, \mathrm{d}T \right) \\
\Delta \tilde{G}(T, p) &= \Delta \tilde{H}_0 + \int_0^T \Delta \tilde{C}_p \, \mathrm{d}T - T \left(\Delta \tilde{S}_0 + \int_0^T \frac{\Delta \tilde{C}_p}{T} \, \mathrm{d}T \right).
\end{aligned}
\tag{3.35}
$$

Here the index 0 refers to the quantities at zero temperature. From the foregoing,

$$\Delta S_0 = 0.$$

In addition, Eq. (3.35) clearly reflects the importance of the heat capacities in the discussion. We will discuss this topic in Sect. 3.4.1.

Planck actually generalized this statement even further stating that the entropy itself (not only the change in entropy) approaches zero as the temperature approaches zero. This statement demands some special assumptions from the view of quantum mechanics, on the degeneracy of states.

Nernst's formulation was later (1937) refined by *Simon*[16]: The contribution to the entropy of a system from each subsystem which is in internal thermal equilibrium disappears at the absolute zero.

It is also often stated that according to the third law of thermodynamics, absolute zero is unattainable by a suitable process through a finite number of steps. This statement is often presented as an alternative formulation of the third law of thermodynamics.

However, other authors think that this theorem is due to the second and the third law simultaneously. This means that the theorem of the unattainability of absolute zero temperature is not a consequence of the third law exclusively. If this is valid, with the statement of the unattainability of absolute zero we cannot trace back the third law of thermodynamics. Nowadays, there are various formulations of the third law of thermodynamics.

---

[15] Fritz Haber, born Dec. 9, 1858, in Breslau, now Wrocław, Poland, died Jan. 29, 1934, in Basel, Switzerland.

[16] Franz Eugen Simon, born Jul. 2, 1893, in Berlin, died Oct. 31, 1956, in Oxford, England.

- As the absolute temperature approaches to zero, the entropy S tends to a constant S_0.
- The entropy of a perfect crystal of an element at the absolute zero of temperature is zero. For a pure crystalline substance, the zero entropy $S_0 = 0$.
- It is impossible to cool a body to absolute zero by any finite reversible process. Thus, we can approach absolute zero infinitely closely, but not reach this limit in fact.
- $\lim\limits_{T \to 0} S(T, \ldots) = 0$.

Question 3.2. Why we cannot reach absolute zero temperature? In principle it is easy to attain absolute zero temperature, isn't it? Consider an ideal gas. Here, at constant entropy the relation $T V^{\kappa-1} = C(S, n)$. To absolute zero temperature, expand the gas reversibly to infinite volume. The problem is only to have infinite space, to keep it as an ideal gas near absolute zero, and in addition at nonvanishing entropy S. Even, when there would be infinite space available, we could watch this experiment only from a position outside this infinite space.

3.4.1 Heat Capacity

3.4.1.1 Classical View

Before the advent of low-temperature physics it was generally accepted that the heat capacity would be independent of the temperature according to the law of *Dulong*[17] and *Petit*.[18] Only a few remarkable exceptions were known, such as the heat capacity of diamond.

Classical statistical mechanics yields for the energy U of N atoms:

$$U = 3 N k_b T,$$

where k_b is the Boltzmann constant. Obviously, the heat capacity is independent of temperature.

3.4.1.2 Lattice Vibrations

For the thermal properties of solids, *Einstein*[19] developed an equation that could predict the heat capacity of solids in 1907. This model was then refined by *Debye*[20] in 1912. Both models predict a temperature dependence of the heat capacity. At

---

[17] Pierre Louis Dulong born Feb. 12, 1785, in Rouen, France, died Jul. 18, 1838, in Paris.

[18] Alexis Thérèse Petit, born Oct. 2, 1791, in Vesoul, died Jun. 21, 1820, in Paris.

[19] Albert Einstein, born Mar. 14, 1879, in Ulm, Württemberg, Germany, died Apr. 18, 1955, in Princeton, NJ, USA.

[20] Petrus Josephus Wilhelmus Debije, born Mar. 24, 1884, in Maastricht, Netherlands, died Nov. 2, 1966, in Ithaca, NY, USA.

high temperatures, the models yield in the limiting case the law of Dulong and Petit. However, in the low-temperature region, even when both models predict a zero heat capacity, the Debye model agrees more with the experimental behavior.

We will not discuss these models in detail [23], but we will rather investigate certain simple functions of the temperature dependence of the temperature. We will present only the results of Einstein's model

$$C_v = 3Nk_b \left(\frac{\theta}{T}\right)^2 \exp(-\theta/T) \tag{3.36}$$

and *Debye*'s model

$$C_v = \frac{12\pi^4}{5} Nk_b \left(\frac{T}{\Theta}\right)^3$$

at low temperatures. Here θ is the Einstein temperature, whereas Θ is the Debye temperature. In modern texts the derivation of Debye's law is finished after a few lines. Actually, the Debye model is based on the theory of elasticity and in the book of *Schaefer*[21] some 20 pages deal with this topic [24, pp. 571–591].

3.4.1.3 Electronic Contributions

In fact, there is still another contribution to the heat capacity, the electronic contribution to the heat capacity [25]. This type of contribution is also referred to as a Fermi system [26, Chap. 15]. The electronic contribution at low temperatures computes as

$$C_v^{el} = \frac{\pi^2}{2} \frac{N_f k_b^2}{E_f} T,$$

where N_f is total number of free electrons in the crystal, E_f is the Fermi energy, and k_b is the Boltzmann constant. In contrast, the lattice energy of the phonons is rewritten as

$$C_v^{ph} = \frac{12\pi^4}{5} Nk_b \left(\frac{T}{\Theta}\right)^3$$

with $R = nk_b$.

With regard to these issues, the total heat capacity is the sum of the lattice contribution and the electronic contribution:

$$C_v = \alpha T^3 + \gamma T. \tag{3.37}$$

---

[21] Clemens Schaefer, born Mar. 24, 1878, in Remscheid, Germany, died Jul. 9, 1968, in Cologne, Germany.

Question 3.3. Check the formula of Eq. (3.37) in Eq. (3.34). Does the integral converge?

Equation (3.37) can be rearranged into

$$\frac{C_v}{T} = \alpha T^2 + \gamma.$$

Thus a plot of C_v/T vs. T^2 should yield a straight line in the low-temperature approximation.

3.4.2 The Principle of Infinite Steps

The third law is sometimes formulated as the absolute zero can be reached by any process only in infinite steps or asymptotically. There is also a controversy in literature [27]. It has been pointed out that the impossibility of reaching the absolute zero temperature is already a consequence of the second law of thermodynamics. We describe now the process of how to cool down. We inspect the scheme as depicted in Figure 3.2. To achieve cooling, we must change another extensive thermodynamic variable that we address as X. In the first step, we change $X \rightarrow X + \Delta X$ at constant temperature. So the system to be cooled down must be in contact with a thermal reservoir. In the next step, we change $X + \Delta X$ back into X, i.e., $(X + \Delta X \rightarrow X)$, but now we keep the entropy constant. The second step runs adiabatically. After the second step we have restored the variable X, but we have exhausted entropy from the system and thus cooled down. The upper inclined line in Fig. 3.2 represents the envelope of the entropy associated to X, whereas the lower inclined line represents the envelope of the entropy associated to $X + \Delta X$. If the envelope looks like in the left scheme of Fig. 3.2, the absolute zero can be reached in a finite number of steps. In contrast, if the envelope looks like in the right scheme of Fig. 3.2, the absolute zero cannot be reached in a finite number of steps. We have now a closer look on the situation. In the first two steps we achieve a cooling down of ΔT_1. In the next two steps we achieve a cooling down of ΔT_2. So the total cooling down from a start

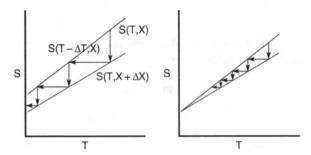

Fig. 3.2 The cooling process (schematically). *Left*: The absolute zero can be reached in a finite number of steps. *Right*: The absolute zero cannot be reached in a finite number of steps

temperature T_s to an end temperature T_e is

$$T_s - T_e = \Delta T_1 + \Delta T_2 + \dots \tag{3.38}$$

Observe that $T_s - T_e < 0$ and consequently $\Delta T_i > 0$ in our notation. If the end temperature is zero ($T_e = 0$) and it cannot be reached in a finite number of steps, then the right-hand side in Eq. (3.38) must be an infinite series that converges to a finite value. This is another formulation of the principle of infinite steps. There are various standard tests for convergence. One of the simplest tests is the ratio test. The series in Eq. (3.38) will converge, if

$$\lim_{k \to \infty} \frac{\Delta T_{k+1}}{\Delta T_k} < 1.$$

More straightforward in the present case is the integral test. If the terms of the series are replaced by a continuous function $f(x)$ that would emerge, when k is replaced by x in the formula for ΔT_k with

$$\lim_{x \to \infty} f(x) \to 0,$$

then the integral

$$\int_{x=t}^{x=\infty} f(x)\mathrm{d}x$$

diverges or converges in the same manner as the series diverges or converges. For the application of this test we note that with increasing x the temperature tends to zero. We can use the formula as reciprocals, i.e.,

$$\int_{T=T_s}^{T=T_e=0} f(T)\mathrm{d}T. \tag{3.39}$$

We try to find out an expression for the terms of ΔT_i. The temperature step ΔT is implicitly given by the entropy equation in Fig. 3.2, namely during the isentropic cooling process we have $S(T - \Delta T, X) = S(T, X + \Delta X)$ or $S(T, X + \Delta X) - S(T - \Delta T, X) = 0$. Taylor expansion gives

$$S(T, X + \Delta X) - S(T - \Delta T, X) = \frac{\partial S(T, X)}{\partial T}\Delta T + \frac{\partial S(T, X)}{\partial X}\Delta X + \dots$$

As a first-order approximation we get

$$\Delta T = -\frac{\frac{\partial S(T,X)}{\partial X}}{\frac{\partial S(T,X)}{\partial T}}\Delta X. \tag{3.40}$$

Clearly, we can force a convergence of the ΔT_i, if we adjust the ΔX_i appropriately. However, this is not what is intended. Instead, we must check if the quotient in Eq. (3.40) as a function of temperature and the further variable is such that the series in Eq. (3.38) converges. Alternatively, we start with a relation that is used to derive the adiabatic gas expansion. We set up the energy at some constant mol number ($dn = 0$) either as a function of temperature and volume $U(T, V)$ or as a function of entropy and volume $U(S, V)$. Both functions must be equal, if the corresponding arguments are inserted into the function declaration:

$$dU(T, V) = C_v dT + \frac{\partial U(T, V)}{\partial V} dV = T dS - p dV = dU(S, V). \qquad (3.41)$$

We annotate that the consideration can be formulated more generally, using a general extensive variable X instead of the volume V. In this case, Eq. (3.41) would look like

$$dU(T, X) = C_X dT + \frac{\partial U(T, X)}{\partial X} dX = T dS + \frac{\partial U(S, X)}{\partial X} dX = dU(S, X).$$

Since we are familiar with the volume energy we do not use the general equation, but exemplify our consideration with the volume energy as given in Eq. (3.41), even when in practice magnetic effects play a major role to achieve deep temperatures. By equating the inner equation of the set of equations in Eq. (3.41), we get

$$\left[\frac{\partial U(T, V)}{\partial V} dV + p \right] dV = -C_v dT + T dS. \qquad (3.42)$$

By setting $dT = 0$ in Eq. (3.42), we move vertically in Fig. 3.2 and by setting $dS = 0$, we move horizontally in Fig. 3.2. Recall that

$$\frac{\partial U(T, V)}{\partial V} dV + p = T \frac{\partial p(T, V)}{\partial T},$$

thus

$$dT = -\frac{T \frac{\partial p(T,V)}{\partial T}}{C_v} dV.$$

On the other hand, replacing the general variable X in Eq. (3.40) by V and using Maxwell's relations for the enumerator we obtain

$$\Delta T = -\frac{\frac{\partial S(T,V)}{\partial V}}{\frac{\partial S(T,V)}{\partial T}} \Delta V = -\frac{\frac{\partial p(T,V)}{\partial T}}{C_v/T} \Delta V, \qquad (3.43)$$

which shows the equivalency of the approach via Eqs. (3.40) and (3.41). Application of the integral convergence test of Eq. (3.39) shows that the order of the enumerator and the nominator in Eq. (3.43) must be at least of the same order to achieve convergence of the series in Eq. (3.38). For instance, if the heat capacity runs with the third power of the temperature ($C_v \propto T^3$), then the pressure should run with temperature with the same power law or with higher power.

3.4.3 Approaching Zero

The Nernst postulate in our nowadays common form states that the entropy of a single phase compound should become zero as the temperature approaches to absolute zero. Let us look how zero is approached for various forms of energy. Table 3.2 lists certain forms of energy, the first differentials, and the second differentials.

Table 3.2 Conditions to run into zero

Energy form	$U(X)$	$\frac{\partial U(X,\dots)}{\partial X} = \xi$	$\frac{\partial^2 U(X,\dots)}{\partial X^2} = \frac{\partial \xi(X,\dots)}{\partial X}$	Symbols
Kinetic energy	$U = \frac{p^2}{2m}$	$\frac{p}{m} = v$	$\frac{\partial v}{\partial p} = \frac{1}{m}$	p: momentum v: velocity m: mass
Plate condenser	$U = \frac{Q^2}{2C_{el}}$	$\frac{Q}{C_{el}} = U_{el}$	$\frac{\partial U_{el}}{\partial Q} = \frac{1}{C_{el}}$	Q: charge U_{el}: potential C_{el}: capacity
Thermal energy		$\frac{\partial U}{\partial S} = T$	$\frac{\partial T}{\partial S} = \frac{T}{C_v}$	S: entropy T: temperature C_v: capacity

Table 3.2 reveals that kinetic energy and the electrical energy of a plate condenser behave in a very related manner. Obviously, in kinetic energy, the mass plays the role of a capacity.

Actually, as the velocity approaches zero, the first differential also approaches zero and we have the situation $dU(0) = 0 \times dp$ in the limiting case of zero velocity. This could be interpreted that in an approximation of first order the energy at zero velocity cannot be increased by transferring momentum to the system. However, at the same time as the momentum increases, the velocity increases with the same order of one.

In both examples mentioned, i.e., kinetic energy and plate condenser, the extensive variable approaches zero as the intensive variable approaches zero. The situation appears to be in a different manner for thermal energy. If we expect some analogy to the forgoing energy forms, we would expect that the heat capacity would not be a constant.

3.4.3.1 Energy Near Absolute Zero

Consider the function

$$U(T) = \frac{a}{4}T^4 = U(S) = \frac{3}{4}\left(\frac{3}{a}\right)^{1/3} S^{4/3}.$$

Observe the relationship to the Stefan – Boltzmann equation. It can be easily verified that

$$\frac{\mathrm{d}U(T)}{\mathrm{d}T} = aT^3 \quad \text{and} \quad \frac{\mathrm{d}U(S)}{\mathrm{d}S} = T.$$

Taylor expansion of $U(S)$ results in

$$U(S + \Delta S) = U(S) + U'(S)\Delta S + \frac{1}{2}U''(S)\Delta S^2 + \dots$$

with

$$U'(S) = \left(\frac{3}{a}\right)^{1/3} S^{1/3}; \quad U''(S) = \frac{1}{3}\left(\frac{3}{a}\right)^{1/3} S^{-2/3}.$$

As S approaches to zero, the first derivative approaches zero, with a lower power than the second derivative approaches infinity. In fact, the series expansion at $S = 0$ is not possible.

References

1. Bruhn, G.W.: Zu T. Beardens Schwierigkeiten mit dem Energiesatz bei destruktiver Interferenz. [electronic] http://www.mathematik.tu-darmstadt.de/~bruhn/interference.html (2008)
2. Cutnell, J.D., Johnson, K.W.: The Principle of Linear Superposition and Interference Phenomena, 7th edn. Chap. 17, pp. 511–536. Wiley, Hobocken, NJ (2007)
3. Enz, C.P.: No Time to be Brief: A Scientific Biography of Wolfgang Pauli, 2nd edn. Oxford University Press, Oxford (2004)
4. Winter, K.: History. In: Neutrino Physics, *Cambridge Monographs on Particle Physics, Nuclear Physics and Cosmology*, vol. 14, 2nd edn. Chap. 1, pp. 4–5. Cambridge University Press, Cambridge (2000). Paperback Edition 2008
5. Harremoeës, P., Topsøe, F.: Maximum entropy fundamentals. Entropy 3(3), 191–226 (2001) [electronic] www.mdpi.org/entropy/
6. Carathéodory, C.: Untersuchungen über die Grundlagen der Thermodynamik. Math. Ann. **67**, 355–386 (1909)
7. Titulaer, U.M., van Kampen, N.G.: On the deduction of Caratheodory's axiom from Kelvin's principle. Physica **31**, 1029–1032 (1965)
8. Macdonald, A.: A new statement of the second law of thermodynamics. Am. J. Phys. **63**, 1122–1132 (1995)
9. Galgani, L., Scotti, A.: Remarks on convexity of thermodynamic functions. Physica **40**(1), 150–152 (1968)
10. Galgani, L., Scotti, A.: Further remarks on convexity of thermodynamic functions. Physica **42**(2), 242–244 (1969)
11. Prestipino, S., Giaquinta, P.V.: The concavity of entropy and extremum principles in thermodynamics. J. Stat. Phys. **111**(1–2), 479–493 (2003). URL http://www.springerlink.com/content/n411802313285077/fulltext.pdf
12. Evans, D.J., Cohen, E.G.D., Morriss, G.P.: Probability of second law violations in shearing steady states. Phys. Rev. Lett. **71**(15), 2401–2404 (1993)

13. Mittag, E., Searles, D.J., Evans, D.J.: Isobaric-isothermal fluctuation theorem. J. Chem. Phys. **116**(16), 6875–6879 (2002)
14. Baranyai, A.: The kinetics of mixing and the fluctuation theorem in ideal mixtures of two component model fluids. J. Chem. Phys. **119**(4), 2144–2146 (2003)
15. Wang, G.M., Sevick, E.M., Mittag, E., Searles, D.J., Evans, D.J.: Experimental demonstration of violations of the second law of thermodynamics for small systems and short time scales. Phys. Rev. Lett. **89**(8905), 601–601 (2002)
16. Prestipino, S., Giaquinta, P.V.: The concavity of entropy and extremum principles in thermodynamics. J. Stat. Phys. **111**(1–2), 479–493 (2003). URL http://www.springerlink.com/content/n411802313285077/fulltext.pdf
17. Hiebert, E.N.: Walther Nernst and the application of physics to chemistry. In: R. Aris, H.T. Davis, R.H. Stuewer (eds.) Springs of Scientific Creativity: Essays on Founders of Modern Science, Chap. 8, pp. 203–231. University of Minnesota Press, Minneapolis, MN (1983)
18. Kox, A.: Confusion and clarification: Albert Einstein and Walther Nernst's heat theorem, 1911–1916. Stud. Hist. Philos. Sci. B Stud. Hist. Philos. Mod. Phys. **37**(1), 101–114 (2006)
19. Nernst, W.: Über die Berechnung chemischer Gleichgewichte aus thermischen Messungen. Nachr. Kgl. Ges. Wiss. Gött. (1), 1–40 (1906). [electronic] http://dz-srv1.sub.uni-goettingen.de/cache/toc/D58106.html
20. Nernst, W.: Über die Beziehungen zwischen Wärmeentwicklung und maximaler Arbeit bei kondensierten Systemen. Ber. Kgl. Pr. Akad.Wiss. **52**, 933–940 (1906). Sitzung 20. Dezember 1906
21. Haber, F.: Thermodynamik technischer Gasreaktionen. Sieben Vorlesungen. Edition Classic. VDM Verlag Dr. Müller Aktiengesellschaft & Co. KG, Saarbrücken (2006). Originally published in 1905
22. Szöllösi-Janze, M.: Fritz Haber, 1868–1934: Eine Biographie. C. H. Beck, Munich (1998)
23. Kompaneeëtìs, A.S.: Theoretical Physics. Dover Phoenix Editions, Mineola, NY (2003). Translated by George Yankovsky
24. Schaefer, C.: Theorie der Wärme, Molekular-kinetische Theorie der Materie, *Einführung in die theoretische Physik*, vol. 2, 3rd edn. De Gruyter, Berlin (1958)
25. Razeghi, M.: Fundamentals of Solid State Engineering, 2nd edn. Springer, New York (2006)
26. Landsberg, P.T.: Thermodynamics and Statistical Mechanics. Dover Publications, New York (1990)
27. Abriata, J.P., Laughlin, D.E.: The third law of thermodynamics and low temperature phase stability. Prog. Mater. Sci. **49**(3–4), 367–387 (2004)

Chapter 4
Equations of State

There is interplay between the fundamental forms and the equations of state. The equations of state can be obtained from any fundamental form. On the other hand, from a certain set of equations of state, the fundamental form can be obtained.

In the strict sense of thermodynamics, an equation of state is a differential of the fundamental form with respect to its natural variables as a function of these natural variables.

For example, using the energy $U(S, V, n)$ as fundamental form, then

$$-\frac{\partial U(S, V, n)}{\partial V} = p(S, V, n) = f(S, V, n)$$

that turns out as the dependence on the pressure on a function of the set S, V, n is an equation of state.

Often, the energy in terms of temperature, volume, and mol number $U(T, V, n)$ is addressed as the caloric state equation, whereas the volume in terms of temperature, pressure, and mol number $V(T, p, n)$ is addressed as the thermal state equation. On the other hand, the energy exclusively expressed in terms of the corresponding natural variables $U(S, V, n)$ belongs to the type of a fundamental equation or fundamental form.

Obviously, the thermal state equation $V(T, p, n) = f(T, p, n)$ is a thermodynamic state equation arising from the Gibbs free enthalpy, because

$$\frac{\partial G(T, p, n)}{\partial p} = V(T, p, n) = f(T, p, n).$$

4.1 The Ideal Gas

If we would know any of the fundamental forms valid for the ideal gas, we could readily derive both the ideal gas law and the *Poisson*[1] equation, namely the adiabatic

---

[1] Siméon Denis Poisson, born Jun. 21, 1781, in Pithiviers, France, died Apr. 25, 1840, in Sceaux, France.

J.K. Fink, *Physical Chemistry in Depth*,
DOI 10.1007/978-3-642-01014-9_4, © Springer-Verlag Berlin Heidelberg 2009

equation of change. Unfortunately, initially we do not have such a knowledge or a feeling how the fundamental form could look like.

In this section, we will show how to derive a fundamental form from the state equations. Further, we will show how to get from certain assumptions on the fundamental from the state equations. The latter approach allows getting information on the thermodynamic properties of the ideal gas.

4.1.1 Energy of the Ideal Gas from State Functions

We recall the thermal state equation for 1 mol of the ideal gas,

$$p\tilde{V} = RT. \tag{4.1}$$

In Eq. (4.1), we express the pressure and the temperature as a function of the energy $\tilde{U}(\tilde{S}, \tilde{V})$ and arrive at the partial differential equation (4.2):

$$-\frac{\partial \tilde{U}(\tilde{S}, \tilde{V})}{\partial \tilde{V}} \tilde{V} = R \frac{\partial \tilde{U}(\tilde{S}, \tilde{V})}{\partial \tilde{S}}. \tag{4.2}$$

This equation has the solution

$$\tilde{U}(\tilde{S}, \tilde{V}) = f\left(\tilde{V}e^{-\tilde{S}/R}\right). \tag{4.3}$$

This means the energy can be any function of $\tilde{V}e^{-\tilde{S}/R}$. Therefore, the ideal gas law as such is not sufficient to get the energy as a function of entropy and volume. The function may contain only $\tilde{V}e^{-\tilde{S}/R}$ as argument. For example, $f(\tilde{V}e^{-\tilde{S}/R}) = \left(\tilde{V}e^{-\tilde{S}/R}\right)^2 + 1$ is a valid function, but $f(\tilde{V}e^{-\tilde{S}/R}) = \tilde{V}\left(\tilde{V}e^{-\tilde{S}/R}\right)^2 + \tilde{S}$ is not a valid function. We will shortly see that the function will be $f(x) \propto x^{1-\kappa}$. As a further state function, use the adiabatic law:

$$p(\tilde{S}, \tilde{V})\tilde{V}^\kappa = f(\tilde{S}) = g(\tilde{S})^{1-\kappa},$$

where $g(\tilde{S})$ is some function of the entropy only. We have put in advance the exponent $1 - \kappa$. Again, we insert for the pressure and arrive at

$$-\frac{\partial \tilde{U}(\tilde{S}, \tilde{V})}{\partial \tilde{V}} \tilde{V}^\kappa = g(\tilde{S})^{1-\kappa}.$$

This differential equation has the solution

$$\tilde{U}(\tilde{S}, \tilde{V}) = -\frac{(g(\tilde{S})\tilde{V})^{1-\kappa}}{1-\kappa} + h(\tilde{S}). \tag{4.4}$$

We can now compare Eqs. (4.3) and (4.4) and seek a solution. A possible way is that we set $h(\tilde{S}) = 0$. Then we set

$$f(x) = -\frac{x^{1-\kappa}}{1-\kappa}$$

and we have a possible solution. Therefore,

$$\tilde{U}(\tilde{S}, \tilde{V}) = -\frac{\left(\tilde{V}e^{-\tilde{S}/R}\right)^{1-\kappa}}{1-\kappa} + C_1.$$

Here we have added C_1 as an additive constant. Since $R = \tilde{C}_p - \tilde{C}_v$ and $\kappa = \tilde{C}_p/\tilde{C}_v$ we come to

$$\tilde{U}(\tilde{S}, \tilde{V}) = C_2 \frac{e^{\tilde{S}/\tilde{C}_v}}{\tilde{V}^{R/\tilde{C}_v}} + C_1. \tag{4.5}$$

In the constant C_2 we have built in $1 - \kappa$, but also other terms to ensure that the expression has the physical dimension of an energy.

4.1.1.1 Energy Dependent on the Number of Moles

We generalize now the considerations and introduce the energy as a function of the entropy, volume, and mol number $U(S, V, n)$. Now volume and entropy are no longer the quantities of 1 mol of gas but the volume and entropy as such, and n is the number of moles of gas we are dealing with. Equation (4.2) changes now into

$$-\frac{\partial U(S, V, n)}{\partial V} V = nR \frac{\partial U(S, V, n)}{\partial S}$$

and we obtain as solution

$$U(S, V, n) = f\left(V e^{-S/(nR)}\right) \times q(n). \tag{4.6}$$

The function $q(n)$ is now new here. From the adiabatic expansion

$$p(S, V, n)V^{\kappa} = h(S, n)$$

we get

$$U(S, V, n) = -\frac{V^{1-\kappa}h(S, n)}{1-\kappa} + C(S, n). \tag{4.7}$$

Combining Eqs. (4.6) and (4.7), using the same arguments as above, we arrive at

$$U(S, V, n) = nC' \frac{e^{(S/n)/(\tilde{C}_v)}}{(V/n)^{R/\tilde{C}_v}} + nC''. \tag{4.8}$$

When we normalize the energy to zero at infinite volume, $C'' = 0$. At constant molar entropy $\tilde{S}$ and constant molar volume $\tilde{V}$, the energy reaches zero at zero material. However, it is not possible to normalize the energy to zero at zero entropy. Wrong equations of state would emerge in this case. Thus, the fundamental form serves as approximation for the ideal gas that admittedly does not exist at zero entropy.

Equation (4.8) could be more directly derived from Eq. (4.5), by presuming that entropy, volume, and mol numbers are extensive variables as such. We use the homogeneity property of the energy, which demands that

$$U(\lambda S, \lambda V, \lambda n) = \lambda U(S, V, n).$$

Note that we could have achieved this result by replacing all the variables S and V by their respective molar variables $S/n = \tilde{S}$, $V/n = \tilde{V}$ and the energy $U(S, V, n)$ by $U(S/n, V/n)/n = \tilde{U}(\tilde{S}, \tilde{V})$. We may now show that when the homogeneity is fulfilled, the energy can be expressed in terms of the products of intensive and extensive variables, i.e.,

$$U(S, V, n) = T(S, V, n)S - p(S, V, n)V + \mu(S, V, n)n. \tag{4.9}$$

We can verify this by forming the partial derivatives of the energy, e.g., $T(S, V, n) = \partial U(S, V, n)/\partial S$, from Eq. (4.8) and inserting them into Eq. (4.9). Note that this procedure does not work with Eq. (4.5), even when we omit the chemical potential. This means that homogeneity is fulfilled only when we scale the entropy, the volume, and the mol number simultaneously. In other words, when we use only entropy and volume, then this set of variables is not sufficient to fulfill the homogeneity conditions.

Summarizing, we needed to obtain for Eq. (4.8) totally three equations, namely the thermal equation of state, the adiabatic equation, and the scaling law. The latter turned out to be equivalent with the *Gibbs – Duhem* equation. This means that for the three variables S, V, n we needed three equations to get the fundamental form.

Example 4.1. We extend Eq. (4.8) for a two-component system of components (1) and (2) by setting

$$\tilde{C}_v = \frac{n_1 \tilde{C}_{v,1} + n_2 \tilde{C}_{v,2}}{n_1 + n_2}.$$
$$n = n_1 + n_2$$

Therefore, the energy is now a function of S, V, n_1, n_2, thus $U(S, V, n_1, n_2)$. Show that the ideal gas equation turns out now to be

$$-R \frac{\dfrac{\partial U}{\partial S}}{\dfrac{\partial U}{\partial V}} = \frac{RT}{p} = \frac{V}{n_1 + n_2}.$$

Further, show that in analogy to Eq. (4.9)

$$U(S, V, n_1, n_2) = T(S, V, n_1, n_2)S - p(S, V, n_1, n_2)V + \\ + \mu_1(S, V, n_1, n_2)n_1 + \mu_2(S, V, n_1, n_2)n_2.$$

This means here also the *Euler* homogeneity relation holds. □

4.1.1.2 Free Energy

When the ideal gas law is written in terms of energy, a partial differential equation is obtained. The situation becomes more simple, if we want to get the free energy as fundamental form. Once we have an expression for the free energy, we can use the Legendre transformation to get the ordinary energy. We illustrate now this type of approach.

The free energy is otherwise known as *Helmholtz*[2] energy. The free energy is the Legendre transformation of the energy with respect to the entropy, in molar quantities:

$$\tilde{F} = \tilde{U} - T\tilde{S}.$$

We omit now the transformation procedure from the free energy with $\tilde{S}$ and $\tilde{V}$ as independent arguments. The molar free energy uses T and $\tilde{V}$ as independent variables and the total differential is

$$d\tilde{F}(T, \tilde{V}) = -\tilde{S}(T, \tilde{V})dT - p(T, \tilde{V})d\tilde{V}.$$

Inserting into the ideal gas equation we get a more simple equation than Eq. (4.2), namely

$$-\frac{\partial \tilde{F}(\tilde{T}, \tilde{V})}{\partial \tilde{V}}\tilde{V} = RT.$$

Integration with respect to $\tilde{V}$ results in

$$\tilde{F}(\tilde{T}, \tilde{V}) = -RT \ln \tilde{V} + C_1(T).$$

The first derivative with respect to T is

$$\frac{\partial \tilde{F}(\tilde{T}, \tilde{V})}{\partial T} = -\tilde{S} = +\frac{dC_1(T)}{dT}.$$

---

[2] Hermann Ludwig Ferdinand von Helmholtz, born Aug. 31, 1821, in Potsdam, Prussia, died Sep. 8, 1894, in Charlottenburg, Berlin.

and the second derivative with respect to T is

$$\frac{\partial^2 \tilde{F}(\tilde{T}, \tilde{V})}{\partial T^2} = -\frac{d\tilde{S}}{dT} = -\frac{\tilde{C}_v}{T} = +\frac{d^2 C_1(T)}{dT^2}. \tag{4.10}$$

Assuming that $\tilde{C}_v$ is a constant, from Eq. (4.10) the most right equation can be integrated to result in

$$\tilde{F}(T, \tilde{V}) = \tilde{C}_v T \ln\left(\frac{C}{\tilde{C}_v T}\right) - RT \ln(\tilde{V}) + \tilde{C}_v T \tag{4.11}$$

besides an integration constant. The derivative of the free energy $\tilde{F}(T, \tilde{V})$ with respect to the temperature T is

$$\frac{\partial \tilde{F}(T, \tilde{V})}{\partial T} = -\tilde{S}(T, \tilde{V}) = \tilde{C}_v \ln\left(\frac{C}{\tilde{C}_v T}\right) - R \ln(\tilde{V}). \tag{4.12}$$

We rewrite Eq. (4.12) to find a more familiar form

$$\exp\left(\frac{\tilde{S}}{R}\right)\left(\frac{C}{\tilde{C}_v}\right)^{\left(\frac{\tilde{C}_v}{R}\right)} = T^{\left(\frac{\tilde{C}_v}{R}\right)}\tilde{V}. \tag{4.13}$$

Equation (4.13) is denoted often as the adiabatic equation for an ideal gas, hereby the left-hand side is simply put as a constant:

$$Const = T^{\left(\frac{\tilde{C}_v}{R}\right)}\tilde{V} = T^{(\kappa - 1)}\tilde{V}.$$

In fact, this constant is not really a constant. If it would be a constant, then there would be only a single adiabatic curve for a fixed amount of gas. However, the more informative statement in comparison to Eq. (4.14) is Eq. (4.13). The derivative of the free energy $\tilde{F}(T, \tilde{V})$ from Eq. (4.11) with respect to the volume $\tilde{V}$ is

$$\frac{\partial \tilde{F}(T, \tilde{V})}{\partial \tilde{V}} = -p(T, \tilde{V}) = -\frac{RT}{\tilde{V}}. \tag{4.14}$$

Of course, again the ideal gas equation emerges. We visualize that for the equation of state the use of the Legendre transformation is more useful than the direct use of the energy equation.

4.1.1.3 Legendre Transformations

From Eq. (4.8), with $C'' = 0$, the derivatives

$$T(S, V, n) = \frac{\partial U(S, V, n)}{\partial S} = \frac{1}{n\tilde{C}_v}U(S, V, n)$$

$$-p(S, V, n) = \frac{\partial U(S, V, n)}{\partial V} = -\frac{R}{\tilde{C}_v V}U(S, V, n)$$

can be formed. From these derivatives, immediately the ideal gas equation follows by division. The equations can be resolved as

$$S(T, V, n) = n\tilde{C}_v \left(\ln \frac{\check{C}_v T}{C'} + \frac{R}{\tilde{C}_v} \ln \frac{V}{n} \right)$$

$$S(T, p, n) = n\tilde{C}_v \left(\ln \frac{\check{C}_v T}{C'} + \frac{R}{\tilde{C}_v} \ln \frac{RT}{p} \right)$$

$$V(S, p, n) = n \exp \frac{S}{n(\tilde{C}_v + r)} \left(\frac{\tilde{C}_v p}{RC'} \right)^{-\check{C}_v/(\check{C}_v + R)}$$

$$V(T, p, n) = \frac{nRT}{p}$$

and inserted in the energy equation. With these equations, analytical forms of the Legendre transformations, e.g., $F(T, V, n) = U(T, V, n) - T S(T, V, n)$, can be obtained. We do not perform these calculations, which are straightforward, in detail.

4.1.1.4 Change of Variables

By changing in the energy the independent variable from $S \to T$ it turns out that

$$\tilde{U}(T, \tilde{V}) = \check{C}_v T. \tag{4.15}$$

Inspection of Eq. (4.15) reveals now an interesting fact. The volume disappears as variable. This means that the energy is not dependent on the volume, if we cross over from the entropy as the independent variable to the temperature as independent variable.

4.1.1.5 Free Enthalpy

Here we still show another method on how to get the fundamental equations, without the need of integration of the equations of state.

In Sect. 1.13.1, the formalism of Taylor series in operator notation is explained. Here, we follow this formalism. We apply Eq. (1.69) for the calculation of the molar Gibbs free energy, using

$$\tilde{G}(T_0 + \Delta T, p_0 + \Delta p) = \exp \left(\Delta T \frac{\partial}{\partial T_0} \right) \exp \left(\Delta p \frac{\partial}{\partial p_0} \right) [\tilde{G}(T_0, p_0)]. \tag{4.16}$$

We equate at first the second part of the operator of Eq. (4.16), i.e.,

$$\exp \left(\Delta p \frac{\partial}{\partial p_0} \right) [\tilde{G}(T_0, p_0)] =$$

$$\tilde{G}(T_0, p_0) + \tilde{V} \Delta p + \frac{1}{2!} \tilde{V}_p \Delta p^2 + \frac{1}{3!} \tilde{V}_{pp} \Delta p^3 + \dots. \tag{4.17}$$

$\tilde{V}_p$, $\tilde{V}_{pp}$, ... are the first, second, ..., etc. derivatives of the molar volume with respect to pressure. We exemplify the special case of an ideal gas and insert $\tilde{V} = RT/p$ at $p = p_0$ and $T = T_0$. Inserting in Eq. (4.17) we get, using $z = \Delta p/p_0$

$$\exp(\Delta p \frac{\partial}{\partial p_0})[\tilde{G}(T_0, p_0)] = \tilde{G}(T_0, p_0) + RT_0 \left(z - \frac{1}{2!}z^2 + \frac{2!}{3!}z^3 - \ldots \right). \quad (4.18)$$

Note that the Taylor expansion of $\ln x$ is

$$\ln(x + \Delta x) = \ln(x) + \frac{1}{x}\Delta x - \frac{1}{2!x^2}\Delta x^2 + \frac{2!}{3!x^3}\Delta x^3 - \ldots$$

Therefore, the term in brackets of Eq. (4.18) is

$$\left(z - \frac{1}{2!}z^2 + \frac{2!}{3!}z^3 - \ldots \right) = \ln \frac{p_0 + \Delta p}{p_0}.$$

In resolving the sum, we have silently done an integration. Thus,

$$\exp \left(\Delta p \frac{\partial}{\partial p_0} \right) [\tilde{G}(T_0, p_0)] = \tilde{G}(T_0, p_0) + RT_0 \ln \left(\frac{p_0 + \Delta p}{p_0} \right). \quad (4.19)$$

Now we proceed to evaluate the first operator of the result of the second operator, i.e.,

$$\exp \left(\Delta T \frac{\partial}{\partial T_0} \right) \left[\tilde{G}(T_0, p_0) + RT \ln \left(\frac{p_0 + \Delta p}{p_0} \right) \right].$$

Expansion of the first term $\tilde{G}(T_0, p_0)$ gives

$$\exp(\Delta T \frac{\partial}{\partial T_0})[\tilde{G}(T_0, p_0)] = \tilde{G}(T_0, p_0) - \tilde{S}\Delta T - \frac{1}{2!}\frac{\partial \tilde{S}}{\partial T_0}\Delta T^2$$

$$- \frac{1}{3!}\frac{\partial^2 \tilde{S}}{\partial T_0^2}\Delta T^3 - \frac{1}{4!}\frac{\partial^3 \tilde{S}}{\partial T_0^3}\Delta T^4 - \ldots. \quad (4.20)$$

For an ideal gas with constant molar heat capacity, $\tilde{S} = \tilde{C}_p \ln T + C'$, we have

$$\frac{\partial \tilde{S}}{\partial T} = \frac{\tilde{C}_p}{T}; \quad \frac{\partial^2 \tilde{S}}{\partial T^2} = -\frac{\tilde{C}_p}{T^2}; \quad \frac{\partial^3 \tilde{S}}{\partial T^3} = 2!\frac{\tilde{C}_p}{T^2}, \text{etc.,}$$

and similar for $T = T_0$. Then the right-hand side of Eq. (4.20) turns into

$$\tilde{G}(T_0, p_0) - \tilde{C}_p \Delta T \left(\ln T + C'' + \frac{1}{2!} \frac{\Delta T}{T_0} - \frac{1}{3!} \frac{\Delta T^2}{T_0^2} + \frac{2!}{4!} \frac{\Delta T^3}{T_0^3} - \cdots \right).$$

We annotate that

$$(x + \Delta x) \ln(x + \Delta x) - (x + \Delta x) = x \ln(x) - x + \ln(x) \Delta x$$
$$+ \frac{1}{2!} \frac{1}{x} \Delta x^2 - \frac{1}{3!} \frac{1}{x^2} \Delta x^3 + \frac{2!}{4!} \frac{1}{x^3} \Delta x^4 - \frac{3!}{5!} \frac{1}{x^4} \Delta x^5 + \cdots$$
$$= x \ln(x) - x + \Delta x \left(\ln(x) + \frac{1}{2!} \frac{\Delta x}{x} - \frac{1}{3!} \frac{\Delta x^2}{x^2} + \frac{2!}{4!} \frac{\Delta x^3}{x^3} - \frac{3!}{5!} \frac{\Delta x^4}{x^4} + \cdots \right).$$

For this reason, we can simplify Eq. (4.20) into

$$\exp(\Delta T \frac{\partial}{\partial T_0})[\tilde{G}(T_0, p_0)] = \tilde{G}(T_0, p_0) - \tilde{C}_p \left((T_0 + \Delta T) \ln(T_0 + \Delta T) \right.$$
$$\left. - T_0 \ln T_0 - \Delta T \right) - \tilde{C}_p C'' \Delta T. \tag{4.21}$$

Finally, the second term in Eq. (4.19) expands into

$$RT_0 \ln \left(\frac{p_0 + \Delta p}{p_0} \right) + R \ln \left(\frac{p_0 + \Delta p}{p_0} \right) \Delta T.$$

We can now set $\Delta p = p - p_0$ and $\Delta T = T - T_0$ and get from Eq. (4.16)

$$\tilde{G}(T, p) = \tilde{G}(T_0, p_0) + C''' - \tilde{C}_p T (\ln T - C^{iv} T) + RT \ln p - RT \ln p_0.$$

Here we have rearranged the constants into a more compact notation. The application of the method becomes even more simple, when the Taylor expansion ends after some terms, as is the case for the incompressible body, where $\tilde{V} = \tilde{V}_0(1 + \alpha(T - T_0))$ and $\tilde{V}_p = 0$.

We annotate that the method we have shown is to find an approximate solution of a differential equation near a fixed point via Taylor expansion. Here, as we could solve all terms of the expansion, we have found an analytical solution of the differential equation.

4.1.2 State Functions from Fundamental Forms

There is no need to have the complete fundamental form as such in an analytical equation, to get some information on the state functions. We may postulate some properties of the fundamental forms.

For example, we assume that the molar energy and the molar enthalpy are functions of the temperature alone. When we expand the energy $\tilde{U}$ as a general function of temperature and volume $\tilde{U} = \tilde{U}(T, \tilde{V})$

$$d\tilde{U}(T, \tilde{V}) = \tilde{C}_v dT + \frac{\partial \tilde{U}(T, \tilde{V})}{\partial \tilde{V}} d\tilde{V}$$

and the enthalpy $\tilde{H}$ as a function of temperature and pressure $\tilde{H} = \tilde{H}(T, p)$

$$d\tilde{H}(T, p) = \tilde{C}_p dT + \frac{\partial \tilde{H}(T, p)}{\partial p} dp$$

and postulate $U = U(T)$ and $H = H(T)$ as stated, then

$$\frac{\partial \tilde{U}(T, \tilde{V})}{\partial \tilde{V}} = 0; \quad \frac{\partial \tilde{H}(T, p)}{\partial p} = 0.$$

Further, it turns out that

$$\tilde{C}_v = f(T); \quad \tilde{C}_p = g(T),$$

where $f(T)$ and $g(T)$ are some functions of temperature. The Legendre transformation of the energy into the enthalpy is by its definition

$$\tilde{H} - \tilde{U} = p\tilde{V}. \tag{4.22}$$

In order to get the equation of the ideal gas, we must further postulate that $g(T) - f(T) = R$, since $d(p\tilde{V}) = (\tilde{C}_p - \tilde{C}_v)dT$. However, this assumption is not sufficient to obtain the adiabatic equation properly, we must more strictly assume that both $\tilde{C}_p$ and $\tilde{C}_v$ are constants. Further, we have to adjust the constant of integration properly.

In this way, we can obtain equations of states using certain assumptions on the fundamental form.

Before we leave this topic, we mention that we could not use also the free enthalpy and the free energy

$$\tilde{G} - \tilde{F} = p\tilde{V}$$

instead of Eq. (4.22), because an identity is obtained. On the other hand, in Eq. (4.22) we have made use of both the Legendre transformation and the change of the variable, when entropy is replaced by temperature which is not a Legendre transformation.

4.1.2.1 Other Approaches

We can choose another approach avoiding assumptions concerning the functional dependence of the heat capacities. From thermodynamics, we get the general relations

$$\frac{\partial \tilde{H}(T, p)}{\partial p} = \tilde{V}(T, p) - T \frac{\partial \tilde{V}(T, p)}{\partial T}$$
$$\frac{\partial \tilde{U}(T, \tilde{V})}{\partial \tilde{V}} = -p(T, \tilde{V}) + T \frac{\partial p(T, \tilde{V})}{\partial T},$$

(4.23)

which are by postulate equal to zero. The solution of the partial differential equations Eq. (4.23) yields

$$\tilde{V} = T C_1(p)$$
$$p = T C_2(\tilde{V}).$$

(4.24)

Here $C_1(p)$ and $C_2(\tilde{V})$ are integration constants, which are functions of p and $\tilde{V}$, respectively. We form the quotient of both equations in Eq. (4.24),

$$\frac{\tilde{V}}{p} = \frac{C_1(p)}{C_2(\tilde{V})}$$

rearrange and find

$$V C_2(\tilde{V}) = p C_1(p) = R.$$

(4.25)

Since the left-hand side in Eq. (4.25) is dependent only on $\tilde{V}$ and the right-hand side is dependent only on p, both equations must be equal to a constant, which we address as R. This constant is even not dependent on the temperature, because the dependence on the temperature was established already in Eq. (4.24). This can also be verified by substituting back $p(T, \tilde{V}) = R(T)/\tilde{V}$ into Eq. (4.24), resulting in $d R(T)/d T = 0$. Therefore, $C_1(p) = R/p$ and $C_2(\tilde{V}) = R/\tilde{V}$. Thus,

$$p \tilde{V} = RT.$$

In order to find out what is R, we form for an ideal gas

$$d \tilde{H} - d \tilde{U} = d(p \tilde{V}) = (\tilde{C}_p - \tilde{C}_v) d T.$$

This is the derivation of the ideal gas law, using purely thermodynamic arguments. We summarize once more what we have done. We have assumed that the energy of the ideal gas at constant temperature does on depend on the volume, as well as the enthalpy of the ideal gas does not depend on the pressure.

In still another approach, we derive now the enthalpy as function of temperature and pressure, i.e.,

$$\frac{\partial \tilde{H}(T, p)}{\partial p} = \tilde{V}(T, p) + \frac{\partial \tilde{V}(T, p)}{\partial p} \left[p + \frac{\partial \tilde{U}(T, \tilde{V}(T, p))}{\partial \tilde{V}} \right].$$

Observe carefully the difference to the first equation in Eq. (4.23). For an ideal gas, we get the differential equation

$$0 = \tilde{V}(T, p) + p \frac{\partial \tilde{V}(T, p)}{\partial p},$$

which solves to

$$p \tilde{V}(T, p) = f(T),$$

where $f(T)$ is some undetermined function of the temperature.

To get this unknown function, we have to make additional assumptions. An analogous procedure, by forming the derivative with respect to the temperature results in

$$\tilde{C}_p(T) = \tilde{C}_v(T) + p \frac{\partial \tilde{V}(T, p)}{\partial T},$$

which solves with constant difference of $\tilde{C}_p(T) - \tilde{C}_v(T) = R$ to

$$p \tilde{V}(T, p) = RT + g(p).$$

Setting $f(T) = RT$, and $g(p)$ we could obtain an expression for the relations of temperature, pressure, and volume of the ideal gas, i.e., the ideal gas equation.

4.1.3 Poisson Equation

To obtain the Poisson equation correctly, we need the assumption that the ratio $\tilde{C}_p / \tilde{C}_v = \kappa$ must be a constant.

In the textbooks usually the derivation of the adiabatic expansion starts with the relation:

$$d\tilde{U}(T, \tilde{V}) = \tilde{C}_v dT = d\tilde{U}(\tilde{S}, \tilde{V}) = T d\tilde{S} - p d\tilde{V}$$

purely on thermodynamic arguments, whereas the textbooks avoid to derive the ideal gas equation based on thermodynamic arguments, but take the ideal gas equation as given.

At constant entropy and inserting the ideal gas law, already derived, we have

$$\tilde{C}_v dT = -\frac{RT}{\tilde{V}} d\tilde{V},$$

which is the differential form of the Poisson equation. By doing the integration, it is usually neglected that the entropy is basically still a variable. Therefore, the constant of integration is not an absolute constant, but a function of the entropy. For this reason, in the $p-V$ diagram an array of curves may appear, depending on the fixed value of the entropy.

4.1.3.1 Related General Relations

We recall the well-known equation of state of adiabatic change of the ideal gas:

$$p(S, V)V^\kappa = C(S).$$

The term $C(S)$ is usually addressed as a constant, but really it is a function of the entropy. Moreover, there is a constant attached with a physical dimension such that the left-hand side and the right-hand side of the equation coincide, i.e., the right hand-side must have the physical dimension of $\text{Pa}\left(\text{m}^3\right)^\kappa$. We ask now if there are analogous equations of this kind. Consider the equation

$$T(S, V)S^\lambda = C(V). \tag{4.26}$$

In fact, this equation holds under some conditions. We start with the argumentation that

$$\frac{\partial S(T, V)}{\partial T} = \frac{C_v(T, V)}{T}.$$

and therefore

$$S(T, V) = \int \frac{C_v(T, V)}{T} dT + C_1.$$

At low temperatures the T^3 law is valid, i.e.,

$$C_v(T, V) = AT^3.$$

Therefore, at low temperatures

$$S(T, V) = \frac{A}{3}T^3 + C_1. \tag{4.27}$$

The integration constant C_1 becomes zero, if we demand that the temperature is zero at zero entropy. From Eq. (4.27) we can establish Eq. (4.26) with $\lambda = -1/3$. In this way, the adiabatic law of compression, valid for ideal gases, corresponds to the T^3 law at low temperatures for ideal solids.

4.1.3.2 Isodynamic Path

We are setting up the equations in a slightly different form. The energy of an ideal gas is a function of temperature alone:

$$\tilde{U} = \tilde{C}_v T + \tilde{U}_0. \tag{4.28}$$

On the other hand, the entropy is

$$\tilde{S} = \tilde{C}_v \ln T + R \ln \tilde{V} + \tilde{S}_0. \tag{4.29}$$

For an ideal gas, $R = \tilde{C}_p - \tilde{C}_v = 8.316 \ \text{J} \, \text{mol}^{-1}$ and $\tilde{C}_p = 5R/2$. Resolving T in Eq. (4.29) and substituting in Eq. (4.28) gives

$$\tilde{U} = \tilde{C}_v \exp\left(\frac{\tilde{S} - \tilde{S}_0}{\tilde{C}_v}\right) \tilde{V}^{\left(\frac{\tilde{C}_v - \tilde{C}_p}{\tilde{C}_v}\right)} + \tilde{U}_0. \tag{4.30}$$

From Eq. (4.30) isodynamics can be drawn in terms of $\tilde{S}$ and $\tilde{V}$, cf., Fig. 4.1. Further, we can discuss the curve along the isodynamic path. Along the isodynamic path $\mathrm{d}\tilde{U} = 0 = T \mathrm{d}\tilde{S} - p \mathrm{d}\tilde{V}$

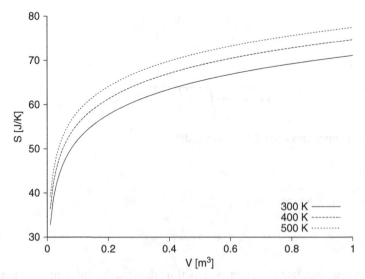

Fig. 4.1 Entropy change and volume change of 1 mol of ideal gas at the isodynamic expansion

$$\frac{\partial \tilde{S}(\tilde{U}, \tilde{V})}{\partial \tilde{V}} = \frac{p}{T}$$

holds. At any tangent along the isodynamic path $d\tilde{S} = (p/T)d\tilde{V}$. The total differential of second order is

$$d^2\tilde{U} = \frac{\partial^2 \tilde{U}}{\partial \tilde{S}^2} d\tilde{S}^2 + 2\frac{\partial^2 \tilde{U}}{\partial \tilde{S}\partial \tilde{V}} dSdV + \frac{\partial^2 \tilde{U}}{\partial \tilde{V}^2} d\tilde{V}^2. \tag{4.31}$$

Inserting $d\tilde{S} = (p/T)d\tilde{V}$ in Eq. (4.31) and using Eq. (4.30) to carry out the differentiation we obtain

$$d^2\tilde{U} = T\frac{\tilde{C}_p - \tilde{C}_v}{\tilde{V}^2} d\tilde{V}^2 = \frac{p}{\tilde{V}} d\tilde{V}^2. \tag{4.32}$$

The second total derivative along the path is positive. This means that if we are moving on a tangent away from the path, the energy will increase. Thus, the isodynamic path is a minimum with respect to a deviation in direction of any tangent touching the path. The total differential of third order is

$$d^3\tilde{U} = \frac{\partial^3 \tilde{U}}{\partial \tilde{S}^3} d\tilde{S}^3 + 3\frac{\partial^3 \tilde{U}}{\partial \tilde{S}^2\partial \tilde{V}} d\tilde{S}^2 d\tilde{V} + 3\frac{\partial^3 \tilde{U}}{\partial \tilde{S}\partial \tilde{V}^2} dSdV^2 + \frac{\partial^3 \tilde{U}}{\partial \tilde{V}^3} d\tilde{V}^3.$$

A similar procedure results in

$$d^3\tilde{U} = -2T\frac{\tilde{C}_p - \tilde{C}_v}{\tilde{V}^3} d\tilde{V}^3 = -\frac{2p}{\tilde{V}^2} d\tilde{V}^3.$$

Of course, we could express $d\tilde{V}$ in Eqs. (4.31) and (4.32) as function of $d\tilde{S}$ at the minimum to obtain

$$d^2\tilde{U} = \frac{T}{R} d\tilde{S}^2$$

and

$$d^3\tilde{U} = -\frac{2T}{R^2} d\tilde{S}^3.$$

For the reasons explained in Sect. 1.11.3, the energy function along the isodynamic path flattens out, as both the entropy and the volume increase.

4.2 Internal Pressure

The internal pressure is reflected by the dependence of the internal energy now as function of temperature and volume with the volume as $\Pi = \partial U(T, V)/\partial V$. The ordinary pressure is $p = -\partial U(S, V)/\partial V$. We can access the internal pressure from Eq. (4.33):

$$T\,dS - p\,dV = C_v\,dT + \frac{\partial U(T, V)}{\partial V}\,dV. \tag{4.33}$$

There are two ways of approach. The common approach is thinking about a process running at constant temperature $dT = 0$. Then we have

$$\frac{\partial U(T, V)}{\partial V} = T\frac{\partial S(T, V)}{\partial V} - p = T\frac{\partial p(T, V)}{\partial T} - p.$$

The last equality follows from Maxwell's relations. The other approach is to think about an isentropic process with $dS = 0$. Then we have

$$\frac{\partial U(T, V)}{\partial V} = -C_v\frac{\partial T(S, V)}{\partial V} - p.$$

The term $\partial T(S, V)/\partial V$ is the temperature change at adiabatic (or more exactly isentropic) expansion. Moreover, Eq. (4.33) is the starting point for the derivation of the isentropic equation of state.

Example 4.2. Using Jacobian determinants, it can be shown that both the isothermal and the isentropic approach to the internal pressure are the same:

$$\frac{\partial U(T, V)}{\partial V} = \frac{\partial(U, T)}{\partial(V, T)} = \frac{\partial(U, T)}{\partial(V, S)}\frac{\partial(V, S)}{\partial(V, T)} =$$
$$\left[\frac{\partial(U, S)}{\partial(V, S)}\frac{\partial(T, V)}{\partial(S, V)} - \frac{\partial(U, V)}{\partial(S, V)}\frac{\partial(T, S)}{\partial(V, S)}\right]\frac{\partial(V, S)}{\partial(V, T)} =$$
$$\frac{\partial(U, S)}{\partial(V, S)} + \frac{\partial(U, V)}{\partial(S, V)}\frac{\partial(S, T)}{\partial(V, T)} = \frac{\partial(U, S)}{\partial(V, S)} - \frac{\partial(U, V)}{\partial(T, V)}\frac{\partial(T, S)}{\partial(V, S)}. \tag{4.34}$$

Resolving the second term in square brackets of Eq. (4.34) just in a different way and taking care that $\partial(x, y) = -\partial(y, x)$, we obtain the two equations in the last line. The left equation resolves in the partial differentials from the isothermal approach and the right equation resolves in the partial differentials from the isentropic approach. □

In Table 4.1 we summarize the most important cases for the internal pressure.

Table 4.1 Appearance of internal pressure

$\Pi =$	Type of body
$\Pi = 0$	Ideal gas
$\Pi = \frac{a}{V^2}$	van der Waals gas
$\Pi = \frac{A}{TV^2}$	Berthelot gas
$\Pi = \frac{C}{V^{2/3}}$	Gravitational cloud
$\Pi \to \infty$	Incompressible body
$\Pi = -p$	Ideally diluted gas

Note: V^2 means l^6

4.3 The van der Waals Gas

We modify now the energy function $\tilde{U}(\tilde{S}, \tilde{V})$ of Eq. (4.5) by replacing $\tilde{V}$ by $\tilde{V} - b$ and adding the term $-a/\tilde{V}$ to the energy equation, as shown in Eq. (4.35):

$$\tilde{U}(\tilde{S}, \tilde{V}) = C \exp(\tilde{S}/\tilde{C}_v)(\tilde{V} - b)^{-R/\tilde{C}_v} - a/\tilde{V}. \qquad (4.35)$$

We check out now if the *van der Waals*[3] equations will appear. In fact,

$$\frac{\partial \tilde{U}(\tilde{S}, \tilde{V})}{\partial \tilde{S}} = \frac{1}{\tilde{C}_v} C \exp(\tilde{S}/\tilde{C}_v)(\tilde{V} - b)^{-R/\tilde{C}_v} = T \qquad (4.36)$$

and

$$\frac{\partial \tilde{U}(\tilde{S}, \tilde{V})}{\partial \tilde{V}} = -\frac{R}{C_v V} C \exp(\tilde{S}/\tilde{C}_v)(\tilde{V} - b)^{-R/\tilde{C}_v} + a/\tilde{V}^2 = -p. \qquad (4.37)$$

Resolving the entropy $\tilde{S}$ in Eqs. (4.36) and (4.37) results in the van der Waals equation:

$$\left(p + \frac{a}{\tilde{V}^2}\right)(\tilde{V} - b) = RT.$$

In comparison to the ideal gas, the entropy term in the van der Waals gas is not modified at all. However, the modifications in the terms concerning the volume influence the temperature. From Eq. (4.36) we learn that the temperature of the van der Waals gas T_{vdW} in relation to the temperature of the ideal gas T_{id} at the same entropy, is

$$\frac{T_{vdW}}{T_{id}} = \left(1 - \frac{b}{\tilde{V}}\right)^{-R/C_v}.$$

The adiabatic equation for the van der Waals gas can be obtained immediately by rearranging Eq. (4.37). An equation resembling closely to that of the ideal gas is obtained, i.e.,

$$\left(p + \frac{a}{\tilde{V}^2}\right)(\tilde{V} - b)^{\kappa} = \frac{R}{C_v} C \exp(\tilde{S}/\tilde{C}_v). \qquad (4.38)$$

---

[3] Johannes Diederik van der Waals, born Nov. 23, 1837, in Leiden, Netherlands, died Mar. 9, 1923, in Amsterdam.

4.3.1 Critical Point

The critical point can be obtained from the van der Waals equation by resolving into
p and setting

$$\frac{\partial p(T, \tilde{V})}{\partial \tilde{V}} = 0; \qquad \frac{\partial^2 p(T, \tilde{V})}{\partial \tilde{V}^2} = 0.$$

Note that we form the differentials at constant temperature. It emerges for the critical
point,

$$\frac{p_c \tilde{V}_c}{RT_c} = \frac{3}{8}; \qquad a = \frac{9RT_c \tilde{V}_c}{8}; \qquad b = \frac{\tilde{V}_c}{3}.$$

If we use the adiabatic van der Waals equation, Eq. (4.38), we can likely form the
first and second derivative of the pressure with respect to the volume at constant
entropy and set them to zero. However, then the constants a and b will emerge in a
different way. For example, $b = \tilde{V}_c(2 - \kappa)/3$.

The reduced adiabatic van der Waals equation looks like

$$\left(\breve{p} + \frac{3\kappa}{\breve{V}^2(2 - \kappa)} \right) \left(3\breve{V} + \kappa - 2 \right)^\kappa = f(\breve{S}).$$

The internal pressure and the co-volume change their signs, when turning from $\kappa =
5/3$ to $\kappa = 7/3$. Thus, for $\kappa > 2$ the $\breve{p} - \breve{V}$ plot with $\breve{S}$ as parameter shows an
unusual behavior. The situation becomes different, when we try to use the values for
a and b obtained from the isothermal van der Waals equation. A drawback is that
the reduced form looks somewhat complicated.

4.3.2 Compressibility Factor

The compressibility factor is defined as

$$\frac{pV}{nRT} = \kappa. \tag{4.39}$$

The compressibility factor is a dimensionless quantity. We can interpret the com-
pressibility factor in terms of entropy, volume, and energy. Since $dU(S, V, n) =
TdS - pdV + \mu dn$, by switching to S as the dependent variable, we get immediately
$\partial S(U, V, n)/\partial V = p/T$. Thus, Eq. (4.39) reads as

$$\frac{pV}{nRT} = \kappa = \frac{V}{nR} \frac{\partial S(U, V, n)}{\partial V}.$$

The form of $S(U, V, n)$ corresponds to the isodynamic expansion of a gas as done in the Gay-Lussac process. It is typical for a gas that it expands deliberately. Therefore, we demand $\kappa > 0$ for any gas.

4.3.3 Relation to Internal Pressure

For a van der Waals gas, the pressure reads as

$$p = \frac{nRT}{V - b} - \frac{a}{V^2}.$$

Thus,

$$\frac{\partial U(T, V)}{\partial V} = \frac{a}{V^2}.$$

On the other hand, for a van der Waals gas,

$$T\frac{\partial p(T, V)}{\partial T} = \frac{nRT}{V - b} = -C_v\frac{\partial T(S, V)}{\partial V}.$$

Thus, the isentropic temperature change of a van der Waals gas on the volume V is dependent solely on the van der Waals parameter b. For example, this relation is relevant for diesel engines or for the impact sensitivity of explosives.

4.4 Incompressible Body

We have to distinguish between an isothermal incompressible body and an isentropic incompressible body. We characterize

- The isentropic incompressible body as $\frac{\partial V(S,p)}{\partial p} = 0$
- The isothermal incompressible body as $\frac{\partial V(T,p)}{\partial p} = 0$.

The fundamental equations for an incompressible body have been derived [1]. In the following, we suppress the mol number as a variable.

4.4.1 Isentropic Incompressible Body

Since for the isentropic incompressible body we have the independent variables S and p as starting point, the enthalpy is a convenient thermodynamic function to handle this problem. We put

$$dH(S, p) = T\,dS + V\,dp \tag{4.40}$$

and further

$$\frac{\partial V(S, p)}{\partial p} = 0 = \frac{\partial^2 H(S, p)}{\partial p^2}. \tag{4.41}$$

Integration yields

$$\frac{\partial H(S, p)}{\partial p} = 0 \times p + C_1(S) = V. \tag{4.42}$$

Therefore, the volume must be a function of the entropy alone, $V = V(S)$. This means also that the volume is only a function of the entropy alone and not a function of pressure:

$$V = C_1(S). \tag{4.43}$$

A second integration yields

$$H(S, p) = pC_1(S) + C_2(S) = pV + C_2(S). \tag{4.44}$$

We arrive immediately with $H - pV = U$ at

$$U = C_2(S).$$

Therefore, the energy is a function of the entropy alone.

4.4.2 Isothermal Incompressible Body

We can repeat the same consideration as shown in Sect. 4.4.1 with the free enthalpy $G(T, p)$. The partial derivative of second order with respect to pressure is

$$\frac{\partial V(T, p)}{\partial p} = 0 = \frac{\partial^2 G(T, p)}{\partial p^2}.$$

Integration yields

$$\frac{\partial G(T, p)}{\partial p} = 0 \times p + C_3(T) = V. \tag{4.45}$$

Therefore, the volume must be a function of the temperature, $V = V(T)$,

$$V = C_3(T). \tag{4.46}$$

A second integration yields

$$G(T, p) = pC_3(T) + C_4(T) = pV(T) + C_4(T). \tag{4.47}$$

We arrive immediately with $G - pV = F$ with

$$F = C_4(T).$$

Therefore, the free energy is a function of the temperature alone. We emphasize that for the isothermal incompressible body the heat capacity at constant volume is not accessible. When we change the temperature, which is needed to measure the C_v then we would also change the volume, due to Eq. (4.46). An exception is the special case $dV(T)/dT = 0$ that occurs in liquid water at the density maximum. Therefore, C_v is not defined in general; however, C_p is readily accessible.

We now make use of the entropy as a function of temperature and pressure, i.e.,

$$dS(T, p) = \frac{\partial S(T, p)}{\partial T} dT + \frac{\partial S(T, p)}{\partial p} dp. \tag{4.48}$$

Now

$$\frac{\partial S(T, p)}{\partial T} = \frac{C_p(T, p)}{T} \quad \text{and} \quad \frac{\partial S(T, p)}{\partial p} = -\frac{\partial V(T, p)}{\partial T},$$

which follows from the definition of the heat capacity and from the second differential (Maxwell's relations) from $dG = -SdT + Vdp$. These are general relations.

We restrict now $C_p = C_p(T)$, the heat capacity should only be a function of the temperature. From

$$\frac{\partial C_p(T)/T}{\partial p} = 0 = -\frac{d^2 V(T)}{dT^2}.$$

Therefore, together with Eq. (4.46), the volume must be of the form $V(T) = AT + B$. Usually we write

$$V = V_0[1 + \alpha(T - T_0)]. \tag{4.49}$$

Here, α is the thermal expansion coefficient. We do not need the special equation, Eq. (4.49), but we use the more general form from Eq. (4.46):

$$\frac{dV(T)}{dT} = \frac{dC_3(T)}{dT} = C_3'(T).$$

We restrict C_p still further to be a constant. Integration of Eq. (4.48) results in

$$S(T, p) = C_p \ln T - pC_3'(T) + C_5 \tag{4.50}$$

Summarizing, we have assumed that the heat capacity at constant pressure C_p is constant and the volume is a function of the temperature alone. Forming the differential of Eq. (4.47) with respect to temperature,

$$\partial G(T, p)/\partial T = -S(T, p),$$

we can compare this equation with Eq. (4.50). We have

$$-\frac{\partial G(T, p)}{\partial T} = -p\frac{dC_3(T)}{dT} - \frac{dC_4(T)}{dT},$$

$$S(T, p) = -p\frac{dC_3(T)}{dT} + C_p \ln T + C_5. \tag{4.51}$$

The right-hand sides of Eq. (4.51) are the differential equation for $C_4(T)$. We emphasize here once more that we did not make any assumption about the special form of $V(T)$, only that it should be not dependent on the pressure, which follows from the condition of isothermal incompressibility, Eq. (4.46).

4.4.3 Isentropic Equation of State

We can calculate from Eqs. (4.48) and (4.49) the isentropic equation of state as

$$\frac{dT}{T} - \frac{V_0\alpha}{C_p}dp = 0$$

or

$$dT = \frac{T\tilde{V}_0\alpha}{\tilde{C}_p}dp. \tag{4.52}$$

We have divided both numerator and denominator by the mol number to introduce now the molar quantities. Here we observe a mysterious and miraculous effect of the energy law. It is interesting to note that the isentropic application of pressure itself is not directly associated with a change of volume, however, with a change of temperature. The change of temperature effects a change in volume. So the change of volume under pressure and consequently the compression work is done in an indirect way.

In the case of a positive expansion coefficient with increasing pressure a rise in temperature is predicted, in the case of a negative expansion coefficient, the temperature decreases, and in the case of a zero thermal expansion coefficient the temperature does not change at all. All these three cases are realized in the case of water. The thermal expansion coefficient is 0 at 4 °C. At 2 °C, the thermal expansion coefficient is $-1.6\,10^{-5}\mathrm{K}^{-1}$ and at 6 °C the thermal expansion coefficient is $1.6\,10^{-5}\mathrm{K}^{-1}$. The effect associated is very small, but has been verified [2, p. 199].

In detail the coefficient in Eq. (4.52) is calculated to be

$$dT = \frac{T\tilde{V}_0\alpha}{\tilde{C}_p}dp = \frac{277\,\text{K} \times 18 * 10^{-6}\,\text{m}^3\text{mol}^{-1} \times 1.6 * 10^{-5}\,\text{K}^{-1}}{75\,\text{J}\,\text{mol}^{-1}\text{K}^{-1}}dp.$$

Thus, the numerical value is $1.06 \times 10^{-9}\ \text{K}\,\text{m}^3\text{J}^{-1}$, which means that for an increase of the pressure of 100 atm, a temperature change of 0.01 K will be observed. The unit m^3J^{-1} is the same as Pa^{-1}.

We discuss finally the case of a thermal expansion coefficient of zero. In this case, a pressure change does not effect a temperature change of the material and also no change in volume. So no input of compression energy is achieved. In contrary to the adiabatic application of pressure in the isothermal case in fact no change in volume will occur. Therefore, the incompressible body should be more accurately addressed as the *isothermal incompressible body*.

4.4.4 Comparison to Ideal Gas

We repeat here once more how to obtain the energy function of the ideal gas. The procedure we utilize now is quite similar to that in Sect. 4.4. We start here with Eq. (4.1) and insert in the second term the respective expression from the ideal gas:

$$\frac{\partial \tilde{V}(T, p)}{\partial T} = \frac{R}{p}$$

and

$$d\tilde{S}(T, p) = \frac{\tilde{C}_p}{T}dT - \frac{R}{p}dp. \tag{4.53}$$

If $\tilde{C}_p$ and R are constants, we can directly integrate Eq. (4.53):

$$\int d\tilde{S}(T, p) = \int \frac{\tilde{C}_p}{T}dT - \int \frac{R}{p}dp = \tilde{C}_p \ln T - R \ln p + C_1.$$
$$\tilde{S}(T, p) = \tilde{C}_p \ln T - R \ln p + C_1. \tag{4.54}$$

In Eq. (4.54) we replace p by $RT/\tilde{V}$ and resolve to T as independent variable:

$$T = C \exp\left(\frac{\tilde{S}}{\tilde{C}_p - R}\right)\tilde{V}^{\frac{R}{R - \tilde{C}_p}},$$

where we have collected several constants into one. Inserting in the expression for the energy and repeating the integration procedure yields finally the expression of the energy for the ideal gas. From the energy equation, all other thermodynamic functions and all the equations of state can be derived.

4.5 Corresponding States

The theorem of corresponding states goes back to van der Waals, who formulated the principle of corresponding states based on his equation of state. In addition, *van der Waals* deduced straightforwardly that in reduced coordinates , the vapor pressure curve and the coexistence curve must be the same for all fluids [3]. An extensive treatment of the corresponding state principle has been published by Xiang [4].

The theorem of corresponding states usually starts with the van der Waals equation. Temperature, pressure, and volume are replaced with the corresponding reduced quantities, e.g., $T = T_c \check{T}$. We arrive finally at an equation with a complete similar form than like that we started before. But all the constants, which are material specific went away, even the gas constant.

The properties of various equations of state have been reviewed in much more detail as presented here [5, pp. 37–53]. In Table 4.2, various gas equations are summarized.

Table 4.2 Various gas equations

Type of equation	Parametric form
Ideal gas equation	$p\tilde{V} = RT$
van der Waals equation	$\left(p + \frac{a}{\tilde{V}^2}\right)\left(\tilde{V} - b\right) = RT$
Berthelot equation	$\left(p + \frac{a}{T\tilde{V}^2}\right)\left(\tilde{V} - b\right) = RT$
Combined van der Waals equation and Berthelot equation	$\left(p + \frac{a}{T^A\tilde{V}^2}\right)\left(\tilde{V} - b\right) = RT$
Redlich – Kwong equation	$\left(p + \frac{a}{T^{0.5}\tilde{V}(\tilde{V}+b)}\right)\left(\tilde{V} - b\right) = RT$
Peng – Robinson equation	$\left(p + \frac{a}{\frac{1}{2}\tilde{V}(\tilde{V}+b)}\right)\left(\tilde{V} - b\right) = RT$
Dieterici equation	$p\left(\tilde{V} - b\right) = RT \exp\left(-\frac{a}{RTV}\right)$

Now we are going the reverse way. We are starting with a reduced equation and we are asking whether there are other forms possible than the common form

$$\left(\check{p} + \frac{3}{\check{V}^2}\right)\left(\check{V} - \frac{1}{3}\right) - \frac{8\check{T}}{3} = 0.$$

We are going the same way as usual, by replacing the constants $3, 1/3, 8/3$ by α, β, ρ:

$$\left(\check{p} + \frac{\alpha}{\check{V}^2}\right)\left(\check{V} - \beta\right) - \rho\check{T} = 0.$$

Resolving to $\check{p}$ and forming the first and second derivative with respect to $\check{V}$, and equating to zero $\partial \check{p}/\partial \check{V}$ and $\partial^2 \check{p}/\partial \check{V}^2$, respectively, we obtain for these three equations at $\check{p} = 1$, $\check{V} = 1$, and $\check{T} = 1$, just and only $\alpha = 3$, $\beta = 1/3$, $\rho = 8/3$. So there is no other possible solution for this set.

The procedure can be extended, if we introduce other functions, e.g.,

$$\left(\check{p} + \frac{A(\check{T})}{\check{V}^2}\right)(\check{V} - B(\check{T})) - \rho\check{T} = 0. \tag{4.55}$$

A similar procedure as indicated above shows that any functions $A(\check{T})$ and $B(\check{T})$ are allowed, provided $A(1) = 3$, $B(1) = 1/3$. So also the equation of *Berthelot*,[4] with $A(\check{T}) = 3/\check{T}$, $B(\check{T}) = 1/3$ is included immediately into the general consideration. Various gas equations in reduced form are shown in Table 4.3. The field is still open to research. A reduced equation of state based on the critical coefficient and still other approaches have been derived [6].

Table 4.3 Various gas equations in reduced form

Type of equation	Reduced form
Ideal gas equation	$\check{p}\check{V} = \frac{8}{3}\check{T}$
van der Waals equation	$\left(\check{p} + \frac{3}{\check{V}^2}\right)(\check{V} - \frac{1}{3}) = \frac{8}{3}\check{T}$
Berthelot equation	$\left(\check{p} + \frac{3}{\check{T}\check{V}^2}\right)(\check{V} - \frac{1}{3}) = \frac{8}{3}\check{T}$
Combined van der Waals equation and Berthelot equation	$\left(\check{p} + \frac{1}{\check{V}q(\check{V}+q)}\right)(\check{V} - q) = 3\check{T}$
	$q = 2^{1/3} - 1$
Peng – Robinson equation	$\left(\check{p} + \frac{1}{\check{T}\check{V}q(\check{V}+q)}\right)(\check{V} - q) = 3\check{T}$
	$q = 2^{1/3} - 1$
Dieterici equation	$\check{p}\left(\check{V} - \frac{1}{2}\right) = \frac{1}{2}\check{T}\exp\left(2 - \frac{2}{\check{T}\check{V}}\right)$
Modified van der Waals equation	$\left(\check{p} + \frac{3F(\check{T})}{\check{V}^2}\right)(\check{V} - \frac{1}{3}) = \frac{8}{3}\check{T}$
	$F(\check{T}) = c + (1 - c)\exp\left(1 - \frac{1}{\check{T}}\right)$
	$c = \frac{e}{e-1}$

Question 4.1. For the van der Waals equation, how the plot of the reduced temperature, $\check{T}$ against the reduced volume $\check{V}$ at constant pressure looks? In addition to the common $p - V$ plot at constant temperature, try a $T - S$ plot at constant pressure. Resolve the reduced energy equation for the van der Waals equation with respect to reduced entropy $\check{S} = S/(n\check{C}_v)$ and reduced temperature and make a parametric plot at constant pressure with the reduced volume $\check{V}$ as parameter. Can an arbitrary constant zero entropy be added to the expression of the entropy?

---

[4] Pierre Eugène Marcellin Berthelot, born Oct. 27, 1827, in Paris, died Mar. 18, 1907, in Paris.

The principle of corresponding states is a special case of the application of scaling laws. In addition to PVT diagrams, transport properties, surface tension, etc., can be modeled [7,8]. Further applications of scaling laws can be found in Sect. 11.4.

References

1. Bechtel, S.E., Rooney, F.J., Wang, Q.: A thermodynamic definition of pressure for incompressible viscous fluids. Int. J. Eng. Sci. **42**(19–20), 1987–1994 (2004)
2. Schaefer, C.: Theorie der Wärme, Molekular-kinetische Theorie der Materie, *Einführung in die theoretische Physik*, vol. 2, 3rd edn. De Gruyter, Berlin (1958)
3. van der Waals, J.D.: On The Continuity of the Gaseous and Liquid States/ edited with an introductory essay by J.S. Rowlinson, *Studies in Statistical Mechanics*, vol. 14. North-Holland, Amsterdam (1988). Edited by J.S. Rowlinson, PhD thesis of van der Waals, Leiden, (1873)
4. Xiang, H.W.: Thermodynamic properties. In: The Corresponding-States Principle and its Practice: Thermodynamics, Transport and Surface Properties of Fluids, 1st edn. Chap. 5, pp. 49–148. Elsevier, Amsterdam (2005)
5. Wei, Y.S.: Prediction of the fluid phase equilibria of binary and ternary mixtures. Ph.D. thesis, Swinburne University of Technology, Melbourne, Australia (1999). URL http://www.swinburne.edu.au/ict/research/cms/documents/disertations/yswChap3.pdf; http://www.swinburne.edu.au/ict/research/cms/dissertations.htm
6. Tian, J., Gui, Y.: An extension of the van der Waals equation of state. Mod. Phys. Lett. B **18**, 213–220 (2004)
7. Dee, G.T., Sauer, B.B.: The principle of corresponding states for polymer liquid surface tension. Polymer **36**(8), 1673–1681 (1995)
8. Xiang, H.W.: The Corresponding-States Principle and its Practice, 1st edn. Elsevier, Amsterdam (2005)

Chapter 5
Thermodynamic Processes

We discuss now a few elementary thermodynamic processes. Traditionally, the thermodynamic processes are classified according to the variable of function that remains constant in the course of the process. Thus, an isothermal process occurs at constant temperature, an isochoric process occurs at constant volume, an isodynamic process[1] occurs at constant energy, etc.

5.1 Mechanical Heat Equivalent

Joule[2] determined in his famous experiment the mechanical heat equivalent. The experiments of Joule in obtaining the mechanical energy equivalent are a landmark in thermodynamics, as it has been demonstrated that heat is a form of energy.

In the experiment, a falling weight stirred water in a tank, and by stirring heat was generated. We set up the energy equation: $dU(T, h) = C_v dT + mgdh = 0$. Thus

$$\frac{dT}{dh} = \frac{mg}{C_v}.$$

The basic problem is that the thermal units and the mechanical units were different at this time. We do not use the old units, but restrict to the MKS system for the mechanical units and to calories and Kelvin for the thermal units. In the experiment, all the numerical values are known. Say a weight of 1 kg was used that fell down 1 m and 1 kg of water was used. An increase of temperature of 2.34 mK was observed. Note that in fact Joule used a much more heavy machinery, probably a very expensive device at this time. We insert now the numbers in the equation:

---

[1] This term has been originally used by Gibbs and contemporary scientists.

[2] James Prescott Joule, born Dec. 24, 1818, in Salford, Lancashire [now in Greater Manchester], England, died Oct. 11, 1889, in Sale, Cheshire.

J.K. Fink, *Physical Chemistry in Depth*,
DOI 10.1007/978-3-642-01014-9_5, © Springer-Verlag Berlin Heidelberg 2009

$$\frac{2.34\,10^{-3}\,\text{K}}{1\,\text{m}} = \frac{9.81\,\text{J}\,\text{m}^{-1}}{10^3\,\text{cal}\,\text{K}^{-1}}.$$

Note that the effect is very small, when an apparatus is used as in the numeric example. Actually, the units in the equation do not cancel, but we can introduce $cal = \lambda * J$. We should obtain $\lambda = 4.18$. Joule recovered this ratio in all his experiments. Already *Carnot*[3] gave in his posthumous published work a remarkable accurate numerical value for the mechanical heat equivalent and even before Joule, Robert Julius Mayer succeeded in this problem.

We quote Joule's report on the experiment literally. It is delivered that Joule did his experiments on the mechanical equivalent of heat by measuring the temperature increase of water, when the energy of a falling weight was dissipated into the water. However, from his original paper it becomes clear that he performed also several other experiments of this type.

5.1.1 On the Mechanical Equivalent of Heat

By James P. Joule [Brit. Assoc. Rep. 1845, trans. Chemical Sect. p. 31. Read before the British Association at Cambridge, Jun. 1845.]

The author gave the results of some experiments, in order to confirm the views he had already derived from experiments on the heat evolved by magneto-electricity, and from experiments on the changes of temperature produced by the condensation and rarefaction of elastic fluids. He exhibited to the Section an apparatus consisting of a can of peculiar construction filled with water. A sort of paddle wheel was placed in the can, to which motion could be communicated by means of weights thrown over two pulleys working on the contrary directions. He stated that the force spent in revolving the paddle wheel produced a certain increment in the temperature of the water; and hence he drew the conclusion that when the temperature of a pound of water is increased by one degree of *Fahrenheit*'s[4] scale, an amount of vis viva is communicated to it equal to that acquired by a weight of 890 pounds after falling from the altitude of one foot.

On the Existence of an Equivalent Relation Between Heat and the Ordinary Forms of Mechanical Power

By James P. Joule, Esq. [In the letter to the Editors of the 'Philosophical Magazine.'] series 3, vol. xxvii, p. 205

Gentlemen,

The principal part of this letter was brought under the notice of the British association at its last meeting at Cambridge. I have hitherto hesitated to give it further publication, not because I was in any degree doubtful of the conclusions at which I had arrived, but because I intended to make a slight alteration in the apparatus calculated to give still greater precision to the experiments. Being unable, however, just at present to spare time necessary to fulfill

---

[3] Nicolas Léonard Sadi Carnot, born Jun. 1, 1796, in Paris, died Aug. 24, 1832, in Paris

[4] Daniel Gabriel Fahrenheit, born May 24, 1686, in Danzig, now Poland, died Sep. 16, 1736, in Den Haag, The Netherlands

this design, and being at the same time most anxious to convince the scientific world of the truth of the positions I have maintained, I hope you will do me the favor of publishing this letter in your excellent Magazine.

The apparatus exhibited before the Association consisted of a brass paddle wheel working horizontally in a can of water. Motion could be communicated to this paddle by means of weights, pulleys, &c., exactly in the matter described in a previous paper (Phil. Mag. Ser. 3, Vol. xxiii, p. 436.).

The paddle moved with great resistance in the can of water, so that the weights (each of four pounds) descended at the slow rate of about one foot per second. The height of the pulleys from the ground was twelve yards, and consequently, when the weights had descended through that distance, they had to be wound up again in order to renew the motion of the paddle. After this operation had been repeated sixteen times, the increase of the temperature of the water was ascertained by means of a very sensible and accurate thermometer.

A series of nine experiments was performed in the above manner, and nine experiments were made in order to eliminate the cooling or heating effects of the atmosphere. After reducing the result to the capacity for heat of a pound of water, it appeared that for each degree of heat evolved by the friction of water a mechanical power equal to that which can raise a weight of 890 lb. to the height of one foot had been expended.

The equivalents I have already obtained are; – 1st, 823 lb., derived from magneto-electrical experiments (Phil. Mag. Ser. 3 Vol. xxiii. pp. 263, 347); 2nd, 795 lb., deduced from the cold produced by the rarefaction of air (Ibid. May 1845, p. 369); and 3rd, 774 lb. from experiments (hitherto unpublished) on the motion of water through narrow tubes. This last class of experiments being similar to that with the paddle wheel, we may take the mean of 774 and 890, or 832 lb., as the equivalent derived from the friction of water. In such delicate experiments, where one hardly ever collects more than one another than that above exhibited could hardly have been expected. I may therefore conclude that the existence of an equivalent relation between heat and the ordinary forms of mechanical power is proved; and assume 817 lb., the mean of the results of three distinct classes of experiments, as the equivalent, until more accurate experiments shall have been made.

Any of your readers who are so fortunate as to reside amid the romantic scenery of Wales or Scotland could, I doubt not, confirm my experiments by trying the temperature of the water at the top and at the bottom of a cascade. If my views be correct, a fall of 817 feet will course generate one degree of heat, and the temperature of the river Niagara will be raised about one fifth of a degree by its fall of 160 feet.

Admitting the correctness of the equivalent I have named, it is obvious that the vis viva of the particles of a pound water at (say) 51° is equal to the vis viva possessed by a pound of water at 50° plus the vis viva which would be acquired by a weight of 817 lb. after falling through the perpendicular height of one foot.

Assuming that the expansion of elastic fluids on the removal of pressure is owing to the centrifugal force of revolving atmospheres of electricity, we can easily estimate the absolute quantity of heat in matter. For in an elastic fluid the pressure will be proportional to the square of the velocity of the revolving atmosphere, and the vis viva of the atmospheres will also be proportional to the square of their velocity; consequently the pressure of elastic fluids at the temperatures 32° and 33° is 480:481; consequently the zero of temperature must be 480° below the freezing point of water.

We see then what an enormous quantity of vis viva exists in matter. A single pound of water at 60° must possess $480° + 28° = 508°$ of heat. In other words, it must possess a vis viva equal to that acquired by a weight of 415036 lb. after falling through the perpendicular height of one foot. The velocity with which the atmosphere of electricity must revolve in order to present this enormous amount of vis viva must of course be prodigious, and equal probably to the velocity of light in the planetary space, or to that of an electric discharge as determined by the experiments of *Wheatstone*.

The paddle wheel used by Rennie in his experiments on the friction of water (Phil. Trans. 1831, plate xi, fig, 1) was somewhat similar to mine. I have employed, however, a greater number of floats, and also a corresponding number of stationary floats, in order to prevent the rotatory motion of the can.

I remain, Gentlemen,

Yours Respectfully,

James P Joule.

Finally we should mention that the determination of the mechanical heat equivalent is nowadays an experiment for undergraduate students. Controlled release of the heat generated by friction is used, by winding a thin rope around a moving drum, filled with water, as shown in Fig. 5.1. The drum is rotated as fast that the weight is maintained in a fixed position. Thus, the friction force between the ribbon and drum compensates the gravitational force. The total friction work is the force times the total path applied. If the drum is turned n times, then the path is n circumferences of the drum. The friction work is converted into heat. The heat capacity of the drum may be neglected or extrapolated with different fillings with water. Heat losses may be compensated by starting the experiment at a certain temperature $-\Delta T$ below the room temperature and stopping the experiment at $+\Delta T$. You can easily construct such a device at home.

Fig. 5.1 Determination of the mechanical heat equivalent in the laboratory

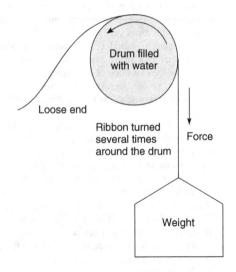

Drum filled with water

Loose end

Ribbon turned several times around the drum

Force

Weight

If the rope is fixed on the drum one turn of the drum would lift the weight by just the circumference $d\pi$ of the drum, if d is the outer diameter of the drum. The gravitational work is then

$$W_g = d\pi n,$$

where n is the number of turns. If the drum is rotated so that the weight remains at constant height, then all the gravitational W_g work appears as frictional work, W_f, i.e., $W_g = W_f$. However, the frictional work is immediately converted into heat.

Thus,

$$f\, mgd\pi n = \tilde{C}_{H_2O} n_{H_2O} \Delta T. \tag{5.1}$$

The specific heat is given in $\text{cal}\,\text{mol}^{-1}\text{K}^{-1}$, so the right-hand side of Eq. (5.1) is in the units of cal. The mechanical work on the left-hand side of Eq. (5.1) is given in J. The conversion factor f has then the physical dimension $\text{cal}\,\text{J}^{-1}$. Note that formally the conversion factor must be chosen (or inserted) in Eq. (5.1) to validate the equal sign.

5.2 Buoyancy

Buoyancy plays a role during the derivation of the sedimentation equilibrium and also in the broader sense with the ultracentrifuge. Someone said once with respect to the buoyancy that the most difficult problems in physical chemistry are the hydrodynamic problems. Now, first buoyancy has to do with hydrostatics rather than with hydrodynamics.

Question 5.1. Place a toy ship into a bathtub. Does the water level rise now the toy ship or does sink? Is the force of gravity shielded by the surrounding liquid, for instance similar as the magnetic field in nuclear magnetic resonance (shielding effect)? In nuclear magnetic resonance, the field is not really shielded, but rather overlaid.

Because buoyancy problems are often misunderstood, we want it to be concerned with the problem. In fact, buoyancy is associated with the coupling of two systems.

5.2.1 Work in the Field of Gravity

If a body in a gravity field is in a height of h then it has the energy

$$U_{grav} = mgh.$$

Acceleration due to gravity g acts downward, thus toward falling height. The above equation applies strictly in vacuum. That can serve thus also as a characteristic of the vacuum.

5.2.2 Force in the Field of Gravity

We obtain the force, which acts on the mass, by the derivative with respect to the height, here

$$K_{grav} = -\frac{\mathrm{d}\,mgh}{\mathrm{d}\,h} = -mg.$$ (5.2)

It is evident that the mass and acceleration due to gravity are accepted here as constant. Further, it is remarkable that the equation is formally alike with the equation of the inert mass. However, here we are concerned with the so-called *heavy mass*.

5.2.3 Force Acting on a Slow-Moving Mass

To the force on a slow-moving mass, the law applies

$$F = \dot{p} = m\dot{v}.$$ (5.3)

The force is any, which acts on the mass. It acts, by changing its motion. The first formulation is in the equation of Newton (the change of the momentum with time), and in the second formulation, the mass is taken as constant and extracted from the differential.

In the field of gravity, the mass takes up energy when it changes its height in positive direction, so

$$\Delta U = mg\,\Delta h.$$ (5.4)

In fact, gh is a scalar product of vectors.

In vacuum actually only this happens, although this is actually incomprehensible. A volume of the ether is moved to the place of the mass, which shifted in the medium. Obviously, the vacuum does not have mass. On the other hand, the vacuum has nevertheless electrical characteristics, in addition, characteristics, which permit a transmission to the gravitation as field effect. The concept must be wrong thus eo ipso.

5.2.4 Motion in a Medium

In a medium now with the change of the position of a body, another change of position of the medium takes place. When the body moves, in the previous position the surrounding medium fills the previous position of the body. The situation is illustrated in Fig. 5.2. The entire work, which was carried out here, results in

$$E = mgh - m'gh.$$ (5.5)

The mass which was actually moved can easily be calculated via the density. In general

Fig. 5.2 Exchange of bodies
in the gravity field: the heavy
body (*black*) goes upward.
Where it rests, a same volume
of liquid must come
downward

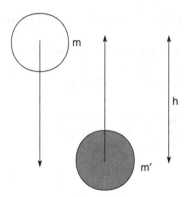

$$\rho = \frac{m}{V} \tag{5.6}$$

holds. The same relation is valid also for m'. We rearrange Eq. (5.5) into

$$E = mgh\left(1 - \frac{m'}{m}\right).$$

Since the volume is clearly the same for the body and the medium, we arrive at

$$E = mgh\left(1 - \frac{\rho'}{\rho}\right).$$

Depending on the ratio of the densities $\frac{\rho'}{\rho}$, there is a gain or a loss of energy if the body moves upward. Therefore, to arrive at a state of minimal energy, if the body is more dense than the surrounding medium, it moves deliberately in the direction of gravity, i.e., down. If the body is less dense than the surrounding medium, it moves deliberately against the direction of gravity, i.e., up. It is important to note that the shape and the size of the moving particle must be the same.

5.3 Processes in the Electric Field

The conservation of energy in the presence of fields such as gravity, electric, and magnetic fields is sometimes not easy to perceive. Here we discuss processes associated with the electric plate condenser.

5.3.1 Electric Plate Condenser

If a particle moves through the plates of an electric plate condenser, it is deflected. It is suggestive that the particle gains kinetic energy at the cost of the electric energy of the plate condenser.

Fig. 5.3 Motion of a particle across the plates of a plate condenser

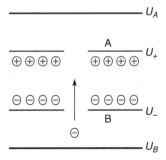

Consider the motion of a particle across the plates, as shown in Fig. 5.3. We presume that outside of the condenser in the regions (B) and (A) there is no electric field. (B) means before and (A) means after. So the particle approaches the condenser with constant kinetic energy. Therefore, the electrode U_B must be at the same electric potential than the electrode U_-. Inside the condenser, the particle is accelerated and leaves the condenser with increased kinetic energy. Therefore, it is expected that the energy of the condenser is decreased by this process and the voltage across the plates drops. For a moment, we pause at this point.

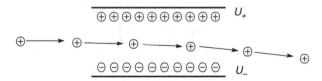

Fig. 5.4 Motion of a particle along the plates of a plate condenser

Consider now the motion of a particle along the plates and initially parallel to the plates, as shown in Fig. 5.4. Again, we presume that outside the condenser there is no field, and the electric potential is zero. It now becomes obvious that when the particle enters the condenser, it struggles against the electric field and loses the kinetic energy. Due to the electric force, the particle is deflected. When the particle leaves the condenser, it moves again into a region of zero electric potential. By this process, the condenser gets back its electric energy and the particle gets back its original initial kinetic energy. When the particle moves in another direction, it means that the component of velocity in the original direction, say x, has decreased and a component normal to the faces of the condenser has appeared. So the condenser acts as an elastic device.

We return now to Fig. 5.3, where we interrupted the discussion. The further fate of the particle when leaving the condenser depends on the electric potential in the region (A) behind the condenser. When the electric potential remains the same as at the plate, then the velocity of the particle remains increased. This can be achieved, when in the region (A) there is another plate kept at the same electric potential as plate U_+. However, the charge of the condenser is unchanged. On the other hand, the electric potential in the region (A) is weakened because a particle of opposite charge has entered the region. These thoughts are reflecting the fact that the charged particle itself is the source of another electric potential. The total potential is superimposed by the potential of the particle and by the potential of the condenser.

If the potential $U_A = U_- = U_B$, then when the particle reaches the plate with the potential U_A, the initial kinetic energy is restored.

We consider now the discharge of a condenser. In Fig. 5.5, a charge moves from one plate to another and finally neutralizes there a charge of opposite sign. In this case, the electric charge at the plates is really decreasing. Thus, we have a drop of the electric voltage across the condenser. In the energy balance, we have to consider the energy of neutralization and maybe the kinetic energy of entropy of the newly created particle.

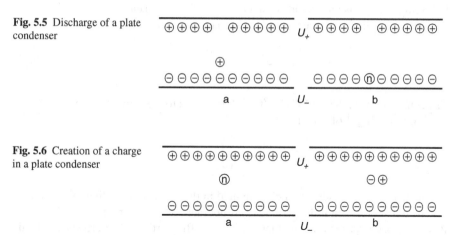

Fig. 5.5 Discharge of a plate condenser

Fig. 5.6 Creation of a charge in a plate condenser

Finally we consider the creation of a charged particle, e.g., by photoionization inside the plate condenser. This process is shown in Fig. 5.6. When a neutral particle is in the condenser, we do not care about the electric potential therein. So we ideally neglect electric polarization of the particle. When the particle dissociates into ions, without any kinetic energy, then these ions are finding themselves in a region of some electric potential. It is suggestive that the work of ionization is dependent on the electric potential in the region where the ions are created.

Instead, consider the process realized in some different way. Let the dissociation take place outside the condenser at zero potential. Next, move both the positively charged ion and the negatively charged ion at the position in Fig. 5.6b. If the negative particle needs work to be done for the placement, the same work is set free by the placement of the positive particle and vice versa. Since the electric work done cancels, the work needed to be done by the dissociation is not dependent on the potential. The situation is different, if a single charge is created.

It should be emphasized that it is not meaningful to split the energy of a particle in a field and the energy of the field itself into separate subsystems. We cannot state that when a particle moves to a higher potential then the field is weakened and the particle gains energy. This appears because the particle interacts with the field by its own field. Otherwise, not any force could emerge. In other words, the charged particle changes the energy simply by its presence at some position.

5.3.2 Electric Potential

After the qualitative discussion exemplified with the electric plate condenser, we can explain the electric potential.

Coulomb[5] invented a torsion balance, which could measure electrostatic forces in relationship to their distance. We start with Coulomb's law that describes the relationship between force, charge, and distance. He is most famous for his discovery in electrostatics. His other fields of interest were friction phenomena.

The force exerted between two electric charges q_1 and q_2 is

$$\mathcal{F}_{1,2} = \kappa \frac{q_1 q_2}{r_{1,2}^2}. \tag{5.7}$$

The constant κ is $8.9876 \times 10^9 \, \mathrm{N \, m^2 C^{-2}}$. It is related to the permittivity $\varepsilon_0 = 8.85 \times 10^{-12} \, \mathrm{C^2 s^2 m^{-3} kg^{-1}}$ of vacuum as

$$\kappa = \frac{1}{4\pi \varepsilon_0}.$$

The particle (1) builds up an electric field in the space surrounding this particle. The field is there, even if the particle (2) is not there. However, to measure the field, it is necessary to probe it by another particle (2). But particle (2) exerts itself a field that disturbs the field of particle (1). Therefore, to probe the field of field strength of particle (1), the charge must be made as small as possible. We can define the field of particle (1) as the specific force $\mathcal{E}_2$ acting on particle (2), at the limit of zero charge of particle (2):

$$\mathcal{E}_2 = \lim_{q_2 \to 0} \frac{\mathcal{F}_{1,2}}{q_2}. \tag{5.8}$$

---

[5] Charles Augustin de Coulomb, born Jun. 14, 1735, in Angoulême, France, died Aug. 23, 1806, in Paris, France.

Now the specific force of Eq. (5.8) is no longer dependent on the charge of particle (2).

The energy of the electric system is obtained by placing the particle from a site of zero force, infinitely far away from the other in a distinct position. So

$$U_{el,1,2} = - \int_{\infty}^{r_{1,2}} \mathcal{F}_{1,2} dr_{1,2} = \frac{\kappa q_1 q_2}{r_{1,2}}. \tag{5.9}$$

If both particles have charges of the same sign, then $q_1 q_2 > 0$. Diminishing the distance increases the energy.

We can deliberately interchange particles (1) and (2). There is some special symmetry in the mathematical formulation. In the same way as done in Eq. (5.8), we can define the potential resulting from particle (1) at the place of particle (2) as

$$\phi_2 = \lim_{q_2 \to 0} \frac{\mathcal{F}_{1,2} r_{1,2}}{q_2}, \tag{5.10}$$

and the potential of particle (2) at the place of particle (1) as

$$\phi_1 = \lim_{q_1 \to 0} \frac{\mathcal{F}_{2,1} r_{2,1}}{q_1}.$$

The electric potentials are governed by the charge of the particles, which may not be equal. Therefore, at the position of particle (1) there may be an electric potential originated by particle (2) that is different from the potential acting on particle (1) at the position of particle (1) originated by particle (2).

The total electric energy of the system may be expressed in a symmetric form as

$$U_{el,1,2} = \frac{1}{2}(q_1 \phi_1 + q_2 \phi_2). \tag{5.11}$$

If we add a third particle to the system, we can again get the work to be done by placing particle (3) into its final position:

$$U_{el,1,2,3} = \kappa \frac{q_1 q_2}{r_{1,2}} + \kappa \frac{q_1 q_3}{r_{1,3}} + \kappa \frac{q_2 q_3}{r_{2,3}}. \tag{5.12}$$

The first term in Eq. (5.12) reflects the energy to build up the system of two particles. When particle (3) is added, the energetic interaction with particle (1) and with particle (2) must be considered. The terms in Eq. (5.12) can be split and rearranged as

$$\frac{\kappa}{2} \left(q_1 \left[\frac{q_2}{r_{1,2}} + \frac{q_3}{r_{1,3}} \right] + q_2 \left[\frac{q_3}{r_{2,3}} + \frac{q_1}{r_{1,2}} \right] + q_3 \left[\frac{q_1}{r_{1,3}} + \frac{q_2}{r_{2,3}} \right] \right). \tag{5.13}$$

Like in Eq. (5.11), we can express the electric energy in terms of electric potentials as

$$U_{el,1,2,3} = \frac{1}{2}(q_1\phi_1 + q_2\phi_2 + q_3\phi_3). \tag{5.14}$$

Clearly, an arbitrary complicated system of multiple particles can be built according to the rules developed, with its electric energy of

$$U_{el,1,2,\dots,n} = \frac{1}{2}\sum_{i=1}^{n} q_i\phi_i. \tag{5.15}$$

Reverting now to Fig. 5.3, the action of a particle that is moving across a condenser becomes more clear. The particle is negatively charged. If the potential as suggested by the charge distributions becomes increasingly positive, when the particle moves upward, the particle, which we may address as the nth particle, decreases the energy of the condenser, simply by the presence of a charge at a certain position.

When two particles disappear at a certain position, due to neutralization, as shown in Fig. 5.5b, then no change in electric energy takes place. Rather the change in energy takes place during the movement of the particles in question to a position close together, as shown in Fig. 5.5a. From the view of electrostatics, if the particles are overlapping, they are literally neutralized. This opens immediately the possiblity to think that even neutral particles may contain charges. As a result, neutral matter may be prone to electric effects; for example, they may exhibit an electric dipole moment, they may decompose into ions.

We emphasize that the same considerations are valid for other kinds of potentials, e.g., gravity.

Example 5.1. We calculate the energy necessary to separate totally 1 mol of a pair of two elementary charges from a distance of 0.150 nm to infinity. Using $N_A = 6.023 \times 10^{23}\,\text{mol}^{-1}$, $\kappa = 8.9876 \times 10^9\,\text{N m}^2\text{C}^{-2}$, the elementary charge $q_e = 1.602 \times 10^{-19}\,\text{C}$, from the formula

$$U_{el} = N_A \kappa \frac{q_e^2}{r},$$

we get $U_{el} = 9.91 \times 10^5\,\text{kg m}^2\text{s}^{-2}\text{mol}^{-1} = 991\,\text{kJ mol}^{-1}$. The latter expression is the Coulomb energy for an isolated system. $\square$

In a crystal lattice, the ions are surrounded by several ions of opposite charge. Therefore, the lattice has a different energy than it would be if the ions were separated into isolated molecules. In a crystal the potential can be calculated, evaluating the distances and charges of the nearest neighboring ions, than of the neighbors in the second shell, etc. In the case of a rock salt type crystal with two types of charges q_+ and q_-, we get

$$U_{el} = N_A \kappa q_+ q_- \left[\frac{6}{r} - \frac{12}{\sqrt{2}r} + \frac{8}{\sqrt{3}r} - \frac{6}{\sqrt{4}r} + \ldots \right]. \qquad (5.16)$$

The lattice structure of rock salt is shown in Fig. 5.8. Each atom of type (A) has six neighbors of atoms of type (B) and vice versa. The series in brackets in Eq. (5.16) converges and is addressed as the *Madelung*[6] constant.

Fig. 5.7 Lattice structure of rock salt

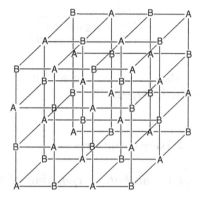

In fact, the series for the cubic lattice of the sodium chloride type arises from the general sum [1]:

$$b_n(2s) = \sum_{k_1,k_2,k_3=-\infty}^{\infty} \frac{(-1)^{k_1+k_2+k_3}}{(k_1^2 + k_2^2 + k_3^2)^s}, \qquad (5.17)$$

which reduces for the interesting case to Benson's formula,

$$b_3(1) = \sum_{i,j,k=-\infty}^{\infty} \frac{(-1)^{i+j+k}}{\sqrt{i^2 + j^2 + k^2}} =$$

$$-12\pi \sum_{m,n=1,3\ldots}^{\infty} \mathrm{sech}^2\left(\frac{1}{2}\pi\sqrt{m^2 + n^2}\right). \qquad (5.18)$$

For some cases of even n, Eq. (5.17) can be expressed in closed forms, involving *Riemann* ζ-functions. The Madelung constant is tabulated for various crystal types in Table 5.1. To get the actual energy of a lattice, the Coulomb terms in addition repulsive terms to the higher order such as in the *Lennard-Jones*[7] potential must be considered.

---

[6] Erwin Madelung, born May 18, 1881, in Bonn, Germany, died Aug. 1, 1972, in Frankfurt/Main, Germany.

[7] Sir John Edward Lennard-Jones, born Oct. 27, 1894, in Leigh, Lancashire, died Nov. 1, 1954, in Stoke-on-Trent, Staffordshire.

Table 5.1 Madelung constant for various crystal types

Crystal type	M[a]	C[b]	Type
Sodium chloride	1.748	6:6	AB
Caesium chloride	1.763	8:8	AB
Wurtzite	1.641	4:4	AB
Zinc blende	1.638	4:4	AB
Rutile	2,408	6	AB_2
Fluorite	2.519	8	AB_2

[a] Madelung constant
[b] Coordination number

The special case of cubic lattice systems has been reviewed in depth suitable for didactic purposes [2].

5.4 Isentropic Process

5.4.1 Adiabatic Expansion of the Ideal Gas

The adiabatic expansion of the ideal gas, c.f., Sect. 4.1.3, is in fact an isentropic expansion in manal equilibrium with the environment. The ideal gas, i.e., the system, is in equilibrium with regard to compression energy. We start again with

$$dU(S, V) = T(S, V)dS - p(S, V)dV$$

and express the internal energy on the other hand as

$$dU(T, V) = C_v(T, V)dT + \Pi(T, V)dV.$$

Here we do not change the amount of matter, therefore $dn = 0$ and we do not introduce the amount of matter in the variable set. Further, we put $dS = 0$. So no entropy may be exchanged with the environment. Or if the gas generates entropy, this entropy must be just removed from the gas into the environment to keep the entropy of the gas constant. It does not matter, how the entropy is kept constant. But if the gas and environment do not generate entropy, this indicates an equilibrium process and that all of the manal energy lost during expansion $-p(S, V)dV$ is taken up by the environment. The gas must be in contact via a piston with an environment having the same pressure as the gas at every stage of expansion. Further, $\Pi = 0$ for an ideal gas. So

$$C_v dT = -pdV,$$

and with the ideal gas law, the law for the adiabatic or isentropic expansion turns out as usual

$$pV^{\kappa} = Const. \tag{5.19}$$

There is usually no additional comment on the constant in Eq. (5.19). In fact, the constant is a function of the entropy: $C_1 n \exp(S/(n\tilde{C}_v))$.

We emphasize that the *Const.* in Eq. (5.19) emerges by a sloppy way of doing the calculation. Subsequently, we illustrate the calculation more accurately. In the first step, we set the total differentials of the energy in the two variable sets equal, i.e., $dU(T, V) = dU(S, V)$. If we have $\Pi = 0$, then according to the law of *Schwarz*, we obtain $\partial C_v/\partial V = 0$, or that $C_v = f(T)$ exclusively. In our calculation, we put more restrictive C_v as a constant. Thus, we obtain

$$C_v dT + p(S, V)dV = T(S, V)dS.$$

We can deliberately change the variable set in p and T as long the values of p and T remain the same. So we get

$$C_v dT + \frac{nRT}{V}dV = TdS,$$

or

$$\frac{dT}{T} + \frac{R}{\tilde{C}_v}\frac{dV}{V} = \frac{dS}{C_v}. \tag{5.20}$$

The integration of Eq. (5.20) with $dS = 0$ is often done by variable separation, thereby forgetting that the constant of integration is a function of the entropy and moreover of the mol number in general.

Next, we examine another method to calculate the isentropic expansion of the ideal gas. Recall that the entropy of an ideal gas is

$$S = nR \ln \frac{V}{n} + n\tilde{C}_v \ln T + C. \tag{5.21}$$

Equation (5.21) gives directly the adiabatic law in full generality. In Eq. (5.21) we can insert the ideal gas law to substitute the volume with the pressure and form the derivative of the pressure with respect to the temperature. The result is

$$\frac{\partial p(S, T)}{\partial T} = \frac{n(C_v + R)}{V} = \frac{nC_p}{V}.$$

5.5 Changes of Energy with Matter

We will investigate what will happen, when we are allowing a gas to escape from a system. The energy of a system containing a single component changes with matter by $\partial U(S, V, n)/\partial n$. This is the chemical potential $\mu(S, V, n)$. However, this is probably not what we have in mind when we are opening the valve of a vessel.

By taking the partial derivative of the energy with the mol number, we have to keep the total entropy and the total volume a constant. In this case, in fact, the entropy and the volume per mol and thus per molecule increases according to

$$S = n\tilde{S}, \quad V = n\tilde{V},$$

when $S = C_1$ and $V = C_2$. To keep the entropy of the system constant, we have to either release the gas from the system without any entropy content or feed the entropy that is lost by the escaping gas back into the system. In the case of the Gay – Lussac experiment, the missing entropy is created in the system itself because the process is irreversible. In fact

$$\frac{\partial U(S, V, n)}{\partial n} = \mu = \frac{G}{n}.$$

The situation becomes different if we want to remove some gas from the system, keeping the volume V constant, and not the total entropy S, but the molar entropy $\tilde{S} = C_3$. Then we have to apply the chain rule for functions with multiple arguments

$$\frac{\partial U(n\tilde{S}, V, n)}{\partial n} = T\tilde{S} + \mu = \frac{H}{n}. \tag{5.22}$$

The last identity in Eq. (5.22) arises from the *Euler* relation of homogeneous functions of first order $U = TS - pV + \mu n$ and the definition of the enthalpy as $H = U + pV$.

When we want to remove the gas in keeping the total entropy constant and the molar volume constant, we get

$$\frac{\partial U(S, n\tilde{V}, n)}{\partial n} = -p\tilde{V} + \mu = \frac{F}{n}.$$

Finally, when we want to keep both the molar entropy and the molar volume constant, we get

$$\frac{\partial U(n\tilde{S}, n\tilde{V}, n)}{\partial n} = T\tilde{S} - p\tilde{V} + \mu = \frac{U}{n}. \tag{5.23}$$

Equation (5.23) means simply that we are cutting away a part of the system. We summarize the results once more in Table 5.2.

It is interesting to observe that the variable that is kept constant will be converted in the respective Legendre transformation, but when the molar quantity is kept constant, it will not turn into the Legendre transformations.

Example 5.2. We expand the procedure of forming certain derivatives keeping certain molar quantities constant. We consider the energy of a system with charged

Table 5.2 Change of the energy at constant molar volume and constant molar entropy

Equation		Constants
$\dfrac{\partial U(n\tilde{S}, V, n)}{\partial n} = T\tilde{S} + \mu$	$= \dfrac{H}{n}$	$\tilde{S}, V$
$\dfrac{\partial U(S, n\tilde{V}, n)}{\partial n} = -p\tilde{V} + \mu$	$= \dfrac{F}{n}$	$S, \tilde{V}$
$\dfrac{\partial U(n\tilde{S}, n\tilde{V}, n)}{\partial n} = T\tilde{S} - p\tilde{V} + \mu =$	$\dfrac{U}{n}$	$\tilde{S}, \tilde{V}$
$\dfrac{\partial U(S, V, n)}{\partial n} = \mu$	$= \dfrac{G}{n}$	S, V

particles of one kind only as $U(S, V, Q, n)$. Such a system bears a total charge. More reasonably, we have a system consisting of an ion with the charge z^+ together with the respective counterion with the charge z^-. Here the positive charges must balance the negative charges. So we would have a two-component system with an energy of $U(S, V, Q^+, n^+, Q^-, n^-)$. We consider now a process in that the some ions are added to the system, the counterions being constant. In this case, we have

$$
\begin{aligned}
n^+ &\to n^+ + \delta n^+ \\
Q^+ &\to Q^+ + \delta Q^+.
\end{aligned}
\tag{5.24}
$$

The two equations in Eq. (5.24) are coupled because the increase of particles in the system is indispensably coupled with an increase of charge. However, we have the molar charge $\tilde{Q}^+$ constant. Therefore, we get for constant S, V, Q^-, n^-

$$
\frac{\partial U(S, V, \tilde{Q}^+ n^+, n^+, Q^-, n^-)}{\partial n^+} = \phi \tilde{Q}^+ + \mu^+ = \mu^{*+}
$$

Here ϕ is the electric potential, i.e., the voltage relative to some reference voltage and μ^+ is the chemical potential for the positive charge. μ^{*+} is addressed as the electrochemical potential for the positive charge. It arises naturally, because the charge is coupled with the mol number. There is no need to define the electrochemical potential, as can be found in some textbooks.

Clearly, the total charge changes for the process in Eq. (5.24). To remain neutral, we would have to introduce a similar process for the counterions. □

From Example 5.2, it would be natural to address the potentials listed in Table 5.2 as thermochemical potential, manochemical potential, etc.

5.6 Isothermal Process

The isothermal process is a process in that $dT = 0$. In the case of an ideal gas, the expansion may achieved in contact with a thermal reservoir to run an isothermal process. This process occurs as a basic step in the Carnot process, and it is a reversible process.

On the other hand, when during the expansion no work is done, the gas remains isothermal. This process is addressed as the Gay – Lussac process. In general, we cannot state for an isothermal process, whether the process is reversible or not.

5.7 Isodynamic Process

In an isodynamic process, the energy remains constant, i.e., $dU = 0$. The motion of a friction-free pendulum is an isodynamic process. A friction-free pendulum is a pure mechanical system, and the system does not have any entropy. So the entropy cannot change at all and the process is a reversible one.

In addition, the Gay – Lussac process is an isodynamic process. We will deal with the Gay – Lussac process in detail in Sect. 5.7.1.

5.7.1 Gay – Lussac Process

We can immediately see from

$$d\tilde{U} = 0 = \tilde{C}_v dT = T d\tilde{S} - p d\tilde{V}$$

that the isodynamic expansion of the ideal gas is an isothermal process because $d\tilde{U} = 0 = \tilde{C}_v dT$. So

$$d\tilde{S}(\tilde{U}, \tilde{V}) = \frac{1}{T} d\tilde{U} + \frac{p}{T} d\tilde{V} = \frac{R}{\tilde{V}} d\tilde{V},$$

and further

$$\tilde{S} = R \ln \tilde{V} + C. \tag{5.25}$$

Equation (5.25) is the condition of the Gay – Lussac experiment. The isodynamic lines must be at the same time isotherms. Conversely, an isothermal process of an ideal gas is an isodynamic process. Such a process is taking place in the Carnot cycle. This means that the amount of energy is taken up from a thermal reservoir as thermal energy is released as expansion energy and vice versa. The situation is different in the Gay – Lussac experiment. Here the gas itself converts internally. When we write the energy as

$$U(S, V, n) = TS - pV + \mu n,$$

we can see that during the isodynamic expansion of and ideal gas, the temperature remains constant. This follows immediately from $dU(T, V, n) = C_v dT$, since $\partial U(T, V, n)/\partial V = 0$ and $dn = 0$.

For this reason, also $pV = nRT$ is constant. When we plot TS viz. pV, then we arrive at a vertical straight line. When the entropy increases, the chemical potential must continuously decrease. A plot in the remaining third dimension, μn viz. TS thus will give a straight line. Note that TS, $-pV$, and μn, all have the dimension of an energy.

Since both the energy U and the term pV are constant, we have $U + pV = H$ and thus the enthalpy in the Gay – Lussac process is constant. Conversely, we can state that for an ideal gas, when we set $pV = C$ as a constant, then the energy remains constant.

5.8 Joule Thomson Process

Before starting, we explicitly state that the *Joule Thomson* process is not an equilibrium process, in contrast to most of the processes dealt with ordinary thermodynamics. The important issue in the Joule Thomson process is that the enthalpy is constant in the course of the process. However, the arguments to justify that the enthalpy is constant are usually somewhat dubious. The flow through the porous plug does not occur in the usual derivation, but rather the initial and the final states are considered. This means that only the states when the gas is entirely on the left side and on the right side are evaluated and we could not stop the experiment before at all.

Fig. 5.8 Scheme of the Joule Thomson Process

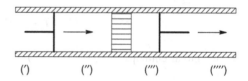

$(')$ $('')$ $(''')$ $('''')$

We will revisit now the conditions of the process. The scheme of the Joule Thomson process is shown in Fig. 5.8. The system consists of totally four subsystems. Subsystems $(')$ and $('''')$ are reservoirs with respect to pressure, p' and p'''', i.e., manostats. The subsystems $('')$ and $(''')$ are gases, adiabatically isolated and separated by a porous plug. The gases may exchange work of compression or expansion with the pressure reservoirs and may loose matter through the porous plug. Of course, the gases contain a certain amount of entropy. We set up now the individual energies for the four subsystems:

$$
\begin{aligned}
dU' &= -p'dV' \\
dU'' &= T''dS'' - p''dV'' + \mu''dn'' \\
dU''' &= T'''dS''' - p'''dV''' + \mu'''dn''' \\
dU'''' &= -p''''dV''''
\end{aligned}
\tag{5.26}
$$

The constraint equations are

$$dV' + dV'' = 0$$
$$dV''' + dV'''' = 0$$
$$dn'' + dn''' = 0 \quad .$$
$$dS'' - \tilde{S}'' dn'' = 0$$
$$dS''' - \tilde{S}''' dn''' = 0$$

(5.27)

Since the total system is isolated, we have to make the total energy stationary, i.e.,

$$dU = 0 = dU' + dU'' + dU''' + dU''''.$$

Using the method of undetermined multipliers, from Eqs. (5.26) and (5.27) the condition for stationary energy turns out as

$$p'' = p'$$
$$p''' = p''''$$
$$\mu'' + T'' \tilde{S}'' = \mu''' + T''' \tilde{S}'''$$

(5.28)

We emphasize that Eq. (5.28) refers to stationary energy and not to thermodynamic equilibrium. Therefore in general, $T'' \neq T'''$ and $p'' \neq p'''$. At this stage, we recall that the molar energy $\tilde{U}$ is

$$\tilde{U} = T\tilde{S} - p\tilde{V} + \mu$$

and the molar enthalpy $\tilde{H}$ is

$$\tilde{H} = \tilde{U} + p\tilde{V} = T\tilde{S} + \mu.$$

5.8.1 Enthalpy

Therefore, the last equation in the set of Eq. (5.28) in fact turns into

$$H'' = H'''$$

which justifies the statement of constant enthalpy.

We remark that the energy of the gas, e.g., in the system (") changes with the loss of matter under the condition of constant molar entropy $\tilde{S}''$ and constant molar volume $\tilde{V}''$. Whenever a certain amount of volume escapes through the porous plug, it takes with it the entropy content in this particular volume and the volume diminishes accordingly, because in addition the plunger moves. Thus, $S = n'' \tilde{S}''$ and $V'' = n'' \tilde{V}''$. Using the chain rule

$$\frac{\partial U''(n''\tilde{S}'', n''\tilde{V}'', n'')}{\partial n''} = T''\tilde{S}'' - p''\tilde{V}'' + \mu''.$$

Therefore, the molar energy $\tilde{U}''$ is constant and the change of the energy U'' with the mol number n'' is $dU'' = \tilde{U}'' dn''$. In fact, all parameters ($\tilde{S}''$, $\tilde{V}''$, T'', p'') are constant during the thermodynamic process.

We rewrite the change of energy of the coupled manostat as

$$dU' = -p'dV' = +p''dV'' = +p''\tilde{V}'' dn'',$$

to get

$$dU' + dU'' = (T''\tilde{S}'' + \mu'')dn'' = \tilde{H}'' dn''. \qquad (5.29)$$

Equation (5.29) states that the molar enthalpy $\tilde{H}''$ is also a constant. Essentially the same holds for the right-hand systems (''') and (''''). Therefore, we have also that $\tilde{H}'''$ is a constant. However, the statement that the molar enthalpies are constants is not sufficient. In order to have them equal, we need the condition of constant energy

$$\sum_n dU^{(n)'} = 0 = \tilde{H}'' dn'' + \tilde{H}''' dn'''.$$

From this condition, with $dn'' + dn''' = 0$, equal enthalpy emerges.

5.8.2 Temperature Drop

The further procedure to get the temperature dependence on the pressure is a standard procedure. The result is

$$\frac{dT}{dp} = \frac{\tilde{V}}{\tilde{C}_p}(T\alpha - 1), \qquad (5.30)$$

where α is the coefficient of thermal expansion. As the pressure decreases, the temperature will decrease only above an initial temperature $T > 1/\alpha$ which is called the inversion temperature. Equation (5.30) can be directly obtained, without using the molar enthalpy by expanding

$$d\left[\mu(T + \Delta T, p + \Delta p) + (T + \Delta T)S(T + \Delta T, p + \Delta p)\right] = 0.$$

5.8.3 Entropy Gain

We form the expression for the total entropy and get

$$dS = dS'' + dS''' = \tilde{S}'' dn'' + \tilde{S}''' dn'''$$

or

$$\frac{dS}{dn''} = \tilde{S}'' - \tilde{S}'''. \tag{5.31}$$

Expanding the entropy as a function of temperature and pressure results in

$$\tilde{S}(T + \Delta T, p + \Delta p) = \tilde{S}(T, p) + \frac{\tilde{C}_p}{T}\Delta T - \alpha \tilde{V} \Delta p.$$

Using for $\Delta T/\Delta p$ the differential quotient of Eq. (5.30) gives

$$\tilde{S}(T + dT, p + dp) = \tilde{S}(T, p) - \frac{\tilde{V}}{T}dp = \tilde{S}(T, p) + \frac{\tilde{C}_p}{T(1 - \alpha T)}dT.$$

This result can be used in Eq. (5.31) to yield

$$\frac{dS}{dn''} = \frac{\tilde{V}'''}{T'''}dp. \tag{5.32}$$

As the gas normally flows from the subsystem ($''$) to the subsystem ($'''$), $dn'' < 0$. The total entropy increases, when $dp < 0$, since the molar volume and the temperature are positive quantities. In Eq. (5.32), $dp = p''' - p'' = p'''' - p'$. We have expanded here the molar entropy of the subsystem ($'''$). We could also obtain the result by expanding the molar entropy of the subsystem ($''$) which results in

$$\frac{dS}{dn''} = -\frac{\tilde{V}'''}{T'''}dp.$$

Now $dp = p'' - p'''$, and therefore the sign turns. On the other hand, we may conclude that in an approximation of first order

$$\frac{\tilde{V}''}{T''} = \frac{\tilde{V}'''}{T'''}.$$

In some other way of consideration, for constant enthalpy $\tilde{H}$ we have

$$d\tilde{H}(\tilde{S}, p) = T d\tilde{S} + \tilde{V} dp = 0 \tag{5.33}$$

and thus

$$\frac{\partial \tilde{S}(\tilde{H}, p)}{\partial p} = -\frac{\tilde{V}}{T}. \tag{5.34}$$

Note that the entropy is produced essentially in the porous plug. If we had an active device, such as a cooler that absorbs reversibly entropy, then the process would be reversible, and systems of different temperature and pressure would be in equilibrium.

Finally we annotate that Eq. (5.34) resembles distantly the *Clausius – Clapeyron* equation. The relationship becomes more obvious, by formally interchanging T and S. Of course, for the Clausius – Clapeyron equation the constraints are completely different and the Clausius – Clapeyron equation deals with equilibrium states.

References

1. Weisstein, E.W.: Madelung constants. From MathWorld – A Wolfram Web Resource: [electronic] http://mathworld.wolfram.com/MadelungConstants.html (2004)
2. Grosso Jr., R.P., Fermann, J.T., Vining, W.J.: An in-depth look at the Madelung constant for cubic crystal systems. J. Chem. Educ. **78**(9), 1198–1202 (2001)

Chapter 6
Equilibrium

In equilibrium, a certain extensive variable may change freely. Further, equilibrium is characterized by $dU = 0$, for a closed system. We will now restrict to the exchange of like energy forms. Standard exchange of an extensive variable X means that $dX_1 = -dX_2$. For example, if two cylinders (1) and (2) with pistons both with equal area are connected as depicted in Fig. 6.2, then the volumes change as $dV_1 = -dV_2$. In this case, in equilibrium the pressures are equal, since

$$dU = -p_1 dV_1 - p_2 dV_2 = (-p_1 + p_2) dV_1 = 0. \tag{6.1}$$

In general, in the case of standard exchange the intensive variables will be equal. $dU = \xi_1 dX_1 + \xi_2 dX_2 = 0$ and so $\xi_1 = \xi_2$. Free exchange of the intensive variables is an essential condition. If the intensive variables are equal, then we may not conclude that the system is in equilibrium. For example, two systems placed together thermally insulated may have exactly the same temperature and the systems are of course not in thermal equilibrium.

6.1 Energy and Entropy Formulation

Both the energy formulation and the entropy formulation serve to obtain the position of equilibria. In general, it is stated that both methods are equivalent. This is basically true. However, there is a subtle difference from the point of view.

In the energy formulation, the entropy is faced as a thermodynamic variable, formally equivalent to the other thermodynamic variables such as volume, mol numbers. The energy is treated as a function of these variables. In contrast to this view, in the entropy formulation, the entropy as such is the function, and the energy is positioned side on side with the other thermodynamic variables, such as volume, mol numbers.

The entropy formulation was introduced by *Massieu* [1]:

In recording, with the aid of the two quantities, energy and entropy, the relations, which translate analytically the two principles, we obtain two relations between the coefficients which occur in a given phenomenon; but it may be easier and also more suggestive

J.K. Fink, *Physical Chemistry in Depth*,
DOI 10.1007/978-3-642-01014-9_6, © Springer-Verlag Berlin Heidelberg 2009

to employ various functions of these quantities. In a memoir, of which some extracts appeared as early as 1869, a modest scholar, M. Massieu, indicated in particular a remarkable function that he termed as characteristic function, and by the employment of which calculations are simplified in certain cases [2].

For a list of some thermodynamic functions, see Table 2.3. We will exemplify the different approaches to get the conditions of equilibrium in Sect. 6.3.6.

6.2 Mechanical and Chemical Equilibrium

We explain at first the concept of mechanical equilibrium. A common illustration is the mechanical equilibrium of two masses on an inclined plane under the action of gravity. The masses are connected with an inextensible string as shown in Fig. 6.1.

Fig. 6.1 Mechanical equilibrium of two masses on an inclined plane under the action of gravity

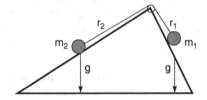

According to *Newton*, equilibrium is established when the resultant forces on the masses, i.e., in the direction of the respective planes, are equal. There is another formulation of equilibrium, however, due to the principle of virtual work. Equilibrium is established in this way, if a small, allowed displacement of the masses will not cause a change in energy.

The change in energy of the system with the displacement is

$$dU = m_1 \mathbf{g} \delta \mathbf{r}_1 + m_2 \mathbf{g} \delta \mathbf{r}_2,$$

with the restriction of the allowed displacement

$$0 = \delta |\mathbf{r}_1| + \delta |\mathbf{r}_2|. \tag{6.2}$$

Equation (6.2) is a constraint equation. In equilibrium, $dU = 0$. We observe the similarity to chemical equilibrium, if we substitute formally

$$dU = \mu' \delta n' + \mu'' \delta n''.$$

However, there is a difference because in chemical equilibrium in most cases the Gibbs energy is used instead of the internal energy. The reason for this difference is that the system under consideration is imbedded into a reservoir, which keeps temperature and pressure constant. However, there is no mysterious difference between the mechanical and thermodynamic equilibrium. We will explain this issue in detail in Sect. 6.5.2.

6.3 Isodynamic and Equilibrium Processes

There is a striking difference between an equilibrium process and a process that runs at constant energy. Both for an isodynamic process and for an equilibrium process in an isolated system $dU=0$. If the energy is a function of a set of variable $X_1, X_2, \ldots, U = U(X_1, X_2, \ldots)$, then these variables may change in some way that the energy remains constant $U = C$, then the condition $dU = 0$ is certainly fulfilled. This is true for an isodynamic process and for an equilibrium process. We emphasize that an isodynamic process is not necessarily an equilibrium process.

6.3.1 Constraints in Equilibrium Processes

Here we discuss qualitatively some aspects of equilibrium. We think of simple coupled systems with k components that can be described with the variables S, V, $n_1, \ldots, n_k$. In equilibrium, the intensive variables will be adjusted dependent on the constraints of the system. For example, if there is no constraint on the volume of a single system, i.e., $dV \neq 0$, the pressure will approach zero, in order to approach an extremum of the energy, $dU \to 0$. The other extreme is a fixed volume, $dV=0$, where the pressure can take any value for $dU \to 0$. The same holds for the other variables in the set needed to describe the system in the sense of thermodynamics.

If two subsystems (') and ('') are coupled so that the total volume is constant, then the constraint equation in differential form is $dV' + dV'' = 0$, which will demand equal pressure in both subsystems.

In the same manner, if two subsystems can exchange matter of kind (1), then $dn_1' + dn_1''=0$, and the chemical potentials become equal, $\mu_1' = \mu_1''$. We have such a situation in an osmotic cell. In such a cell, usually two components are present, and only the exchange of one component is allowed. The semipermeable wall is in a fixed position. For this reason, the pressures in both compartments are different. If we would allow the semipermeable wall to move, then this wall would move in the direction of decreasing pressure. Actually, one of the subsystems would disappear finally.

If there is no wall between the two systems, then the pressures will be equal in equilibrium. We can try to treat the equilibrium in a distillation apparatus, analogous to isopiestics. Actually, there is no thermal equilibrium as the temperatures are different. However, the pressures must be equal because the system would try to expand through the cooler into the other system otherwise. In order that there is no flow of the distilling matter, the chemical potential in the high-temperature region must be equal to the chemical potential in the low-temperature region, even when the temperatures are artificially kept different.

6.3.2 Coupled Systems

Consider two coupled one-component systems ($'$) and ($''$), with the extensive variables S', V', n' and S'', V'', n''. The total energy is the sum of the individual energies:

$$U_{tot} = U'(S', V', n') + U''(S'', V'', n'').$$

If the systems are isolated, we can adjust these six variables independently. Once we have adjusted, we can no longer change by the demand of isolation, so the energies of the subsystems U', U'' remain constant forever. In other words, there is no process that could change the set of variables.

If we allow some process to take place, then the energies of the subsystems U', U'' may change as the extensive variables are changing. However, the total energy remains constant, if the compound system, made up from both subsystems under discussion, is isolated. We discuss now qualitatively some processes, namely the subsystems being connected via a

1. Diathermic wall
2. Adiabatic movable plunger
3. Diathermic movable plunger
4. Porous adiabatic fixed wall.

6.3.2.1 Diathermic Wall

If the subsystems are connected via a diathermic wall, heat may be exchanged. However, the individual mol numbers and the volumes are fixed. So we are left over to vary S' and S'' with the constraint $U_{tot} = C$. A further constraint is needed, which turns out to be the maximum of entropy principle.

By the way, we can easily calculate the isodynamic path by changing the variable S into T. At constant volume, mol number, and total energy, we get

$$C_v' T' + C_v'' T'' = C.$$

6.3.2.2 Adiabatic Movable Plunger

If the subsystems are connected by an adiabatic plunger, the individual volumes may change, say with the standard constraint $V' + V'' = C$. We have further the constraint $U_{tot} = C$. The mol numbers remain fixed. This means that the variation of the entropies S' and S'' and one volume, either V' or V'', may be still varied. We need an additional constraint, otherwise the problem remains ill-posed [3, pp. 321–323].

6.3.2.3 Diathermic Movable Plunger

In contrast to the adiabatic plunger, the diathermic plunger allows the calculation of the equilibrium via the additional condition $T' = T''$.

6.3.2.4 Porous Adiabatic Fixed Wall

In the case of the connection via a porous fixed wall, the individual volumes remain fixed. The total energy $U_{tot} = C$ is fixed. The mol numbers are related by the constraint $n' + n'' = C$. No heat may pass through the porous wall. However, this does not mean that no entropy may move through the wall. Because by the permeation of matter, the entropy attached to the particles is crossing the wall.

The process of transfer of matter may be explained in a rather complicated way. Assume that matter is transferred from the system ($'$) to the system ($''$). We discuss the process in the system ($''$) in some detail. In the first step, the matter n'' being originally in the system ($''$) is somewhat compressed to provide a void space. We subdivide the system ($''$) temporarily into the void space system ($''_v$) and into the compressed region ($''_c$).

Into this void space of the system ($''_v$), matter from the system ($'$) is placed, in such amount to reach the pressure of the compressed region of ($''_c$). Then mixing of the void region with the compressed region takes place. This does not affect the entropy of the system ($''$), if the properties of the matter in both void region ($''_v$) and compressed region ($''_c$) are equal.

However, if the temperature of the transferred matter in ($''_v$) and of the partially compressed matter in ($''_c$) is not equal, a heat exchange between the systems ($''_v$) and ($''_c$) should take place before mixing. Alternatively, the process may be considered as mixing the systems ($''_v$) and ($''_c$) without previous heat exchange and achieving the heat exchange in the mixture. The situation resembles a process that occurs in the Gibbs' paradox. See Sect. 10.1 for the discussion of Gibbs' paradox.

Obviously, by the transfer of matter and thus of the entropy attached to the matter, an exchange of entropy may occur, even when the wall does not conduct heat. In a system composed of two components, separated by a semipermeable wall, as occurs in osmosis, we find just this situation. If the semipermeable wall is not fixed, then the compartment with the impermeable component would try to enlarge its volume due to the pressure difference.

6.3.3 Conditions of Equilibrium

For a more detailed illustration, we consider two chambers filled with a gas. The gas may exchange volume energy via a piston. This is shown in Fig. 6.2. The gas with volume V_h has a pressure of p_h, and the gas with volume V_l has a pressure of p_l. We presume that the lower pressure p_l is smaller than the higher pressure p_h,

Fig. 6.2 Two gases
connected via a piston

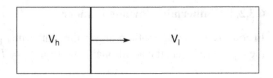

i.e., $p_h > p_l$. Further, in the course of exchange of volume, the total volume remains constant:

$$\mathrm{d}V_{tot} = \mathrm{d}V_h + \mathrm{d}V_l = 0. \qquad (6.3)$$

Therefore, if the system moves toward equilibrium, its total volume energy will decrease, i.e.,

$$-p_l\mathrm{d}V_l - p_h\mathrm{d}V_h < 0. \qquad (6.4)$$

However, the statement $\mathrm{d}U < 0$ cannot be true because we have presumed that the total energy should remain constant, as is demanded in an isolated system. In fact, we are dealing here with the experiment of Gay–Lussac. We have forgotten another thermodynamic variable, namely the entropy. The true and complete expression for the energy reads as

$$\mathrm{d}U(S_l, V_l, S_h, V_h) = +T_l\mathrm{d}S_l + T_h\mathrm{d}S_h - p_l\mathrm{d}V_l - p_h\mathrm{d}V_h = 0. \qquad (6.5)$$

We are dealing here with an isodynamic process, but not with an equilibrium process. Further, since we do not allow the exchange of matter, the mol numbers and thus the corresponding chemical potentials are not relevant in the present consideration.

In an equilibrium process, $\mathrm{d}U = 0$ together with all the constraints in the system. If only the volume may exchange freely, then $\mathrm{d}S_l = 0$ and $\mathrm{d}S_h = 0$, but Eq. (6.3) still holds. Thus, we get by adding $\lambda_V(\mathrm{d}V_{tot} - \mathrm{d}V_h + \mathrm{d}V_l) = 0$ to the energy differential equation (6.5) the condition

$$p_l = p_h. \qquad (6.6)$$

If, in addition, free exchange of entropy is possible between the two gases, we have further the constraint $\lambda_S(\mathrm{d}S_{tot} - \mathrm{d}S_h + \mathrm{d}S_l) = 0$ and

$$T_l = T_h.$$

In particular, in equilibrium also, the total entropy is a conservative quantity. We can take this as the characteristics for equilibrium.

On the other hand, if we measure the pressures and find them equal, as Eq. (6.6) is demanding, then we cannot conclude that there is really equilibrium in the system. Imagine that the plunger is fixed and accidentally equal pressure is established. However, the subsystems are thermally insulated. Then when removing somehow the thermal insulation, equal temperatures will be established, but no longer equal pressures. Therefore, we must take care that the inhibition devices must be unlocked to guarantee free adjustment in direction of equilibrium.

> We mention here what types of pistons are described in the literature. The *Rayleigh*[1] model for *Brown*ian[2] motion is also called the Rayleigh piston. It considers a one-dimensional heavy particle of mass M colliding with light noninteracting particles of mass m, which are always thermalized. According to this model, the moments of time intervals between collision events are finite [4]. The nonequilibrium version of a the Rayleigh piston is known as the adiabatic piston. The absence of a corresponding heat exchange indicates the origin of the name adiabatic piston [5]. Historically, the Rayleigh piston gave rise to sophisticated studies in connection with the second law of thermodynamics.

6.3.4 Equilibrium of Volume Energy

We still consider two gases as shown in Fig. 6.2; however, we will use now the subscripts and superscripts that are common in thermodynamics. We will refer to the gas in the left chamber with ($'$) and the gas in the right chamber with ($''$), and consequently, we will index all the parameters that will refer to these gases in this way.

Again, if only the exchange of volume energy, i.e., the energy of compression and the energy of expansion is in between the two gases, is allowed, we will rewrite Eq. (6.6) in the form

$$p' = p''. \tag{6.7}$$

Equation (6.7) states that the pressure in the left chamber and the pressure in the right chamber must be equal. We have used the condition that the energy must be stationary and the sum of the volumes must be constant to obtain the condition of equilibrium. We can use both requirements to check if there is in fact equilibrium. This means

1. We can change the volume of the gas in the left chamber by dV' and the volume of the gas in the right chamber by $-dV''$. In this way, the condition $V' + V'' = V_{tot} = C$ is fulfilled. If we are doing so, then we should stay at constant energy with respect to the first total differential, i.e., $dU = 0$.
2. We can check if the pressure in the left chamber is the same as the pressure in the right chamber.

The two methods are not completely equivalent. Using the first method, we check, in fact, the condition of equilibrium, according to the original prescription.

---

[1] John William Strutt, 3rd Baron Rayleigh, born Nov. 12, 1842, in Langford Grove, Maldon, Essex, England, died Jun. 30, 1919, in Terling Place, Witham, Essex.

[2] Robert Brown, born Dec. 21, 1773, in Montrose, Scotland, died Jun. 10, 1858, in London, England.

If we are using the second method, we are checking a condition that is derived from the original prescription. In fact, if we are establishing that the pressures are equal, we are ensuring that equilibrium could be established only according to the original prescription. This does not mean, however, that the volume can really exchange without energy loss or gain of the system. We must check that the piston can move freely. Therefore, checking the equality of pressure, this means that equilibrium could be established. However, it does not mean that we can move in fact to a nearby equilibrium state, if we are changing one of the parameters of the gases.

6.3.5 Free Adjustable Parameters

We are dealing now with ideal gases, and here the ideal gas law holds

$$pV = nRT \tag{6.8}$$

for both gases. Inserting the ideal gas law in Eq. (6.7), we obtain for the condition of equal pressures

$$p' = \frac{n'RT'}{V'} = p'' = \frac{n''RT''}{V''}. \tag{6.9}$$

We emphasize now that we have totally six parameters in Eq. (6.9), which are in the function declaration of the pressures, according to Eq. (6.8), i.e.,

$$V', T', n', V'', T'', n''.$$

We can use Eq. (6.9) to calculate one of the parameters from the other remaining five parameters. So we can freely and independently choose totally five parameters and Eq. (6.9) remains valid. We are not restricted longer to the condition that has been used to indicate the equilibrium, i.e., $V' + V'' = V_{tot} = C$. In other words, it is our botheration how we can effect that the pressures remain equal. For example, we can add some gas in the left chamber. In the right chamber, we can allow increasing the volume, in order to move from an equilibrium state to another equilibrium state. Or else, we can keep all parameters in the right chamber constant and we increase both volume and temperature in the left chamber to maintain the pressure. All these things and much more changes are allowed, as long as we do not violate Eq. (6.7).

Therefore, in the course of change of the parameters, we have a wide variety of freedom to change them. However, after reaching a new equilibrium state, we have to check whether equilibrium is established with the original prescription or with the condition of equal pressure. If we use the original prescription, we must again use a thermodynamic process with constant volume. But we have now the possibility to choose another total volume than used previously. However, the previously used total volume is in the same way arbitrary than the total volume after changing some of the parameters. In general, all the possible variations can be expressed as

$$dp'(V', n', T') = \frac{\partial p'}{\partial V'} dV' + \frac{\partial p'}{\partial n'} dn' + \frac{\partial p'}{\partial T'} dT'$$
$$dp''(V'', n'', T'') = \frac{\partial p''}{\partial V''} dV'' + \frac{\partial p''}{\partial n''} dn'' + \frac{\partial p''}{\partial T''} dT''$$

$$(6.10)$$

With Eq. (6.8), Eq. (6.10) turns into

$$dp' = -\frac{p'}{V'} dV' + \frac{p'}{n'} dn' + \frac{p'}{T'} dT'$$
$$dp'' = -\frac{p''}{V''} dV'' + \frac{p''}{n''} dn'' + \frac{p''}{T''} dT''$$

$$(6.11)$$

or

$$d \ln p' = -\frac{dV'}{V'} + \frac{dn'}{n'} + \frac{dT'}{T'}$$
$$d \ln p'' = -\frac{dV''}{V''} + \frac{dn''}{n''} + \frac{dT''}{T''}$$

$$(6.12)$$

In Eqs. (6.11) and (6.12), we have suppressed the list of arguments in the pressures. In the equilibrium condition, Eq. (6.7), we can form the logarithm and the differential to arrive at

$$d \ln p' = d \ln p'',$$

which conforms to the left-hand sides in Eq. (6.12).

Of course, we can add some constraints to Eq. (6.9). If thermal equilibrium in addition to the manal equilibrium is established, then we have the additional condition

$$T' = T''.$$

This condition is derived in the same way as the condition of manal equilibrium, i.e., the total entropy is constant, if we are considering the two gases exclusively. Or else, we could demand that one of the volumes must remain constant, i.e., for example, $V' = C$ or $dV' = 0$.

However, the situation changes, if we want to place a porous diaphragm or a permeable membrane instead of the piston in Fig. 6.2. In order to prevent the transfer of gases, no driving force should be there, which implies that the chemical potentials of the cases in both chambers must be equal. Therefore, in this case, in addition

$$\mu' = \mu'' \tag{6.13}$$

holds. If Eq. (6.13) holds, for a one-component system, we cannot demand a condition like $p' = p''$ together with $T' \neq T''$ or a condition $p' \neq p''$ together with $T' = T''$. This arises because of the Gibbs–Duhem equation

$$SdT - Vdp + nd\mu = 0.$$

We start from a point of equilibrium with $T' = T''$ and $p' = p''$. If we want to vary the temperature T of one of the gases, say $T' \rightarrow T' + dT'$ at constant chemical potential μ', i.e., $d\mu' = 0$, we have to adjust also the pressure p' accordingly. However, in this case, the equilibrium in all aspects is no longer established. For example, if we keep the chemical potential μ'' constant, also μ' must be constant, and we can vary pressure and temperature in the system $(')$ only with the constraint

$$\frac{dp'(T', \mu')}{dT'} = \frac{S'}{V'}. \tag{6.14}$$

In general, the differential $\partial p/\partial T$ at constant chemical potential is positive, since both entropy and volume are positive quantities.

Question 6.1. How we can verify Eq. (6.14)? Look at Fig. 6.3. There are two systems separated by a membrane, permeable only to matter but not to entropy. Consider the left system $(')$ to be connected with a heating element and a piston to change the temperature and the pressure of this subsystem. Vary both temperature and pressure in such a way that no flow of matter can be observed from the right-hand system to the left-hand system and vice versa.

Fig. 6.3 Change of chemical potential with temperature and pressure

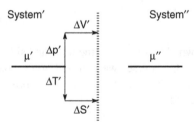

We emphasize that the foregoing discussion is strictly valid only for systems consisting of a single component. In a multicomponent system, the chemical potential may vary in addition to the composition of the system.

We show the situation in Fig. 6.3. An increase in temperature lowers the chemical potential. The entropy flows in the system where the parameters remain unvaried. An increase in pressure increases the chemical potential. In this state, the system wants to expand. Increasing both temperature and pressure can compensate the changes in chemical potential. The system remains in equilibrium with respect to the flow of matter, but it runs out of equilibrium with respect to entropy flow and volume change.

Remarkably, if the temperature and the pressure are increased to adjust the chemical potential back to its original value partly, then the system wants to expand its volume, but at the same time matter wants to flow in.

Summarizing, we have illustrated the way to establish the conditions of equilibrium and to vary the parameters to maintain equilibrium with a simple example.

1. The procedure consists in establishing the conditions at which the process under consideration occurs in equilibrium. The process is the movement of the piston. From this constraint, the condition of equal pressures is derived.
2. Then we develop the pressure as a function of all the possible arguments and perform variations that do not violate the condition of equal pressure. There is no need to vary all the allowed arguments. Instead, we can keep some of the arguments deliberately fixed.

6.3.6 Approaching Equilibrium

We demand now that the two gases in Fig. 6.2 form an isolated system and we want a free exchange of entropy between the two chambers. The respective mol numbers we treat as constant, $dn' = 0$ and $dn'' = 0$. The isolated system means that the total energy is constant. We formulate three equations of constraint:

$$\begin{aligned} U_{tot} &= U'(S', V') + U'(S'', V'') \\ S_{tot} &= S' + S''. \\ V_{tot} &= V' + V'' \end{aligned} \qquad (6.15)$$

In the set of equations of (6.15), we have fixed the total energy U_{tot}, the total entropy S_{tot}, and the total volume V_{tot}. We have three equations and four variables S', V', S'', V''. For this reason, one variable may be fixed deliberately. On the other hand, the variables may move in some way. This means that the system may approach equilibrium. However, we show that the movement to equilibrium may occur only in special cases. We presume that the condition of equilibrium would be equal temperature and equal pressure:

$$T' = T'' \quad p' = p''.$$

These thermodynamic functions are not fixed, but must be simply equal.

Example 6.1. We consider a special example. Allow the general energy functions of the gases to be $U(S, V) = \exp(S)V^{-2/3}$, thus

$$\begin{aligned} U'(S', V') &= \exp(S')V'^{-2/3} \\ U''(S'', V'') &= \exp(S'')V''^{-2/3} \end{aligned}.$$

Here we have suppressed the mol number and the constant that establishes the physical dimension of an energy on the right-hand side. Further, we measure the entropy in units of the heat capacity. The temperature and the pressure would be in general

$$T = \partial U(S, V)/\partial S = U(S, V)$$
$$p = -\partial U(S, V)/\partial V = \frac{2}{3} U(S, V)V$$

and the difference of temperature and pressure would be

$$T' - T'' = U'(S', V') - U''(S'', V'')$$
$$p' - p'' = \frac{2}{3} \left[U'(S', V')V' - U''(S'', V'')V'' \right].$$

Even when the variables of the system may be varied with respect to one variable, the equilibrium conditions can be reached only for equal energies of the subsystems and consequently for equal volume.   □

The demand of constant energy in Eq. (6.15) implies of course $dU_{tot} = 0$. However, $dU_{tot} = 0$ does not mean that the energy is constant at some fixed value, but is stationary with respect to a differential of first order.

To clarify more that we treat now the energy as a function, we should subsequently add the argument list to the function. If we use the condition

$$dU_{tot}(S', V', S'', V'') = 0,$$

then we have a different situation. We are seeking for a distinct set of variables S', V', S'', V'', for which the energy would be stationary, either a minimum or a maximum. The value of the energy is a priori not fixed. This means that we do not deal with an isolated system. However, if a value for the energy that meets incidentally the value of the energy that makes the energy stationary is chosen, then we are dealing with an isolated system by chance.

6.3.6.1 Stationary Energy

If we reformulate Eq. (6.15)

$$U_{tot}(S', V', S'', V'') = U'(S', V') + U''(S'', V'') \rightarrow \text{Stationary}$$
$$S_{tot} = S' + S'' \tag{6.16}$$
$$V_{tot} = V' + V''.$$

The first equation in (6.16) consists, in fact, of two equations, i.e.,

$$\frac{\partial U}{\partial S'} = 0, \quad \frac{\partial U}{\partial V'} = 0,$$

with the other equations in Eq. (6.16) being constraints. We have now four variables and four equations. This means that the system is fixed. If we would have chosen initial values of S', V', S'', V'' so that the system is not in equilibrium, then the system could move only into the equilibrium by changing its energy. This is in contradiction to the concept of an isolated system, however. We perform now the calculation in detail:

$$U_{tot}(S', V', S'', V'') = U'(S', V') + U''(S'', V'') \rightarrow \text{Stationary}$$

means that

$$0 = \frac{\partial U}{\partial S'} = \frac{\partial U'}{\partial S'} + \frac{\partial U''}{\partial S'} = \frac{\partial U'}{\partial S'} - \frac{\partial U''}{\partial S''} = \quad T' - T''$$

and

$$0 = \frac{\partial U}{\partial V'} = \frac{\partial U'}{\partial V'} + \frac{\partial U''}{\partial V'} = \frac{\partial U'}{\partial V'} - \frac{\partial U''}{\partial V''} = -p' + p''.$$

Forming the second derivative yields

$$0 = \frac{\partial(T' - T'')}{\partial S'} = \frac{\partial T'}{\partial S'} + \frac{\partial T''}{\partial S''}; \quad 0 = \frac{\partial(-p' + p'')}{\partial V'} = -\frac{\partial p'}{\partial V'} - \frac{\partial p''}{\partial V''}.$$

According to the conditions of stability, we are facing a minimum. On the other hand, we could fix the total energy and the volume as correct in an isolated system and allow the total entropy to vary. We do not state from where the entropy can be delivered or to where the entropy could disappear.

6.3.6.2 Stationary Entropy

So we set up the equations as

$$S_{tot}(U', V', U'', V'') = S'(U', V') + S'(U'', V'') \rightarrow \text{Stationary}$$
$$U_{tot} = U' + U''$$
$$V_{tot} = V' + V''.$$

An analogous procedure as above results in

$$\frac{1}{T'} - \frac{1}{T''} = 0$$
$$\frac{p'}{T'} - \frac{p''}{T''} = 0$$.

Thus, the same conditions of equilibrium appear as above. Forming the second derivative results in

$$\frac{\partial(1/T'-1/T'')}{\partial U'}=-\frac{1}{T'^2}\frac{\partial T'}{\partial U'}-\frac{1}{T''^2}\frac{\partial T''}{\partial U''}, \text{ and}$$

$$\frac{\partial(p'/T'-p''/T'')}{\partial V'}=+\frac{1}{T'}\frac{\partial p'}{\partial V'}-\frac{p'}{T'^2}\frac{\partial T'}{\partial V'}+\frac{1}{T''}\frac{\partial p''}{\partial V''}-\frac{p''}{T''^2}\frac{\partial T''}{\partial V''}.$$

These equations indicate that a maximum of entropy is achieved at equilibrium. On the other hand, if we demand an isolated system that is even not transparent for entropy, then we must conclude that under such circumstances entropy will be generated, if such as system is moving toward equilibrium. We have used here the principle of maximum entropy under the constraints of constant energy and constant volume.

6.3.6.3 Stationary Volume

For completeness, we investigate the volume as a function of energy and entropy. Equation (6.16) changes now into

$$V_{tot}(S',U',S'',U'')=V'(S',U')+V''(S'',U'')\rightarrow \text{Stationary}$$
$$S_{tot}=S'+S''$$
$$U_{tot}=U'+U''.$$

At the conditions for equilibrium, we get

$$\frac{1}{p'}-\frac{1}{p''}=0, \quad \frac{T'}{p'}+\frac{T''}{p''}=0.$$

The second derivatives become

$$\frac{\partial(\frac{1}{p'}-\frac{1}{p''})}{\partial U'}=-\frac{1}{p'^2}\frac{\partial p'}{\partial U'}-\frac{1}{p''^2}\frac{\partial p''}{\partial U''}, \text{ and}$$

$$\frac{\partial(-\frac{T'}{p'}+\frac{T''}{p''})}{\partial S'}=-\frac{1}{p'}\frac{\partial T'}{\partial S'}+\frac{T'}{p'^2}\frac{\partial p'}{\partial S'}-\frac{1}{p''}\frac{\partial T'''}{\partial S''}+\frac{T''}{p''^2}\frac{\partial p''}{\partial S''}.$$

In a gas, the pressure always points out from the volume. The energy will increase with increasing pressure. Therefore, from this aspect, the volume tends to increase, in order to move in the direction of equilibrium. On the other hand, a solid body may behave similar to a spring. Therefore, in certain regions, the volume of the solid body may tend to decrease, in order to move in the direction of equilibrium.

6.3.6.4 The Stationarity

We can observe that in all the three cases treated here, if either energy U or entropy S of volume V as a function of the remaining variables will become stationary, we arrive at the same conditions for equilibrium. This behavior is caused by the form of the equations. Therefore, from the conditions of equilibrium, we cannot trace back the original equations.

6.4 Setting Up the Equations for Equilibrium

When the equations for the equilibrium are set up, usually this procedure runs or should run through several steps.

1. First, the real system is inspected.
2. A thermodynamic model suitable to establish the thermodynamic relations is set up. For instance, in the case of an osmotic cell, the osmotic pressure in reality is established due to the action of gravity of a liquid column. However, the thermodynamic model reduced the liquid column in that simply a pressure is exerted to the membrane.
3. From the thermodynamic model, the relevant thermodynamic variables and the constraint equations are fixed.
4. The equations of energy, enthalpy, free energy, free enthalpy, are set up, and the suitable thermodynamic function is minimized under the equations of constraint.
5. From the minimum of the thermodynamic function, the equilibrium conditions in terms of the intensive variables are obtained, e.g.,

$$T' = T'', \quad p' = p'', \quad \mu' = \mu''.$$

These equations are the conditions for persistent equilibrium.
6. In order to get the conditions for persistent equilibrium when the variables are varied in the system under investigation under the constraint equations for equilibrium. In this way the mutual dependence of these variables under persistent equilibrium can be obtained. For example, if in the first system, a certain variable was originally n_2, then in the system opposed to the original system, this variable could be

$$n_2 + \delta n_2.$$

Such changes must be compatible with the conditions set up in the original, i.e., the nonvaried system. In the varied system, the conditions for the constraints and for persistent equilibrium still hold as elaborated before.
7. However, the equations of persistent equilibrium transform now for the varied form into

$$T' + \delta T' = T'' + \delta T''$$
$$p' + \delta p' = p'' + \delta p''$$
$$\mu' + \delta \mu' = \mu'' + \delta \mu''.$$

The varied state should be in the same way as before a valid state of equilibrium. There is no need that the varied state can be reached directly from the original system. For instance, if $n_2 \rightarrow n_2 + \delta n_2$, then the original system must be opened and some amount of δn_2 must be added. The system could have to be repressurized, etc.

8. From the varied system, a set of new differential equations is obtained and from integration, the dependence of the equilibrium on the variables under investigation is obtained.

The forgoing prescription may sound somehow abstract, but it will be illustrated by means of several examples to clarify the procedure.

6.5 Thermodynamic Functions

Question 6.2. Why are we using the thermodynamic functions H, F, G, or even other functions derived from the energy U? The answer to this question is that it is sometimes easier to find the conditions for equilibrium by using the thermodynamic functions than using the energy directly.

In general, when we calculate the conditions of equilibrium, we state under which conditions equilibrium is established, instead of giving information whether and how these conditions could be approached from a nonequilibrium state near the equilibrium state.

In order to illustrate the use of thermodynamic functions, we start with the isolated system and then cross to the closed and open system.

6.5.1 Isolated System

We consider an isolated system consisting of two phases (') and ("). The phases are dealt with as subsystems of the isolated system that can exchange freely entropy S, volume V, and matter n among themselves. To indicate to which phase the respective extensive variable belongs, we will add a prime (') or a double prime (") to the variable. Thus, the energy of the subsystem (') is

$$U'(S', V', n') = T'dS' - p'dV' + \mu'dn',$$

and the energy of the subsystem (") is

$$U''(S'', V'', n'') = T''dS'' - p''dV'' + \mu''dn''.$$

We presume that we know the functional dependence of the energy on its variables. U' and U'' may be different functions. The task would not be to find out the nature of the energy as function of entropy, volume, and mol number. The total energy of the combined subsystems is

$$U(S, V, n) = U(S', V', n', S'', V'', n'') = U'(S', V', n') + U'(S'', V'', n'').$$

$U(S, V, n)$ is basically a different function from $U(S', V', n', S'', V'', n'')$, because it is declared with a different set of variables. In the same way as the energy of the system is the sum of the energy of the subsystems, the extensive variables add up

$$S' + S'' = S$$
$$V' + V'' = V \qquad (6.17)$$
$$n' + n'' = n.$$

Likewise, we can identify Eq. (6.17) as a constraint equation, since the system is an isolated one, in which entropy, volume, and mol number must be constant. From the view hitherto presented, the variables of the subsystems are not fixed. However, subsequent considerations will elucidate that these variables can be fixed by the introduction of further equations that are relating these variables.

Since the set of variables S, V, n are identified as constants, also the energy $U(S, V, n)$ must be a constant. The total differential of the energy of this system is

$$dU(S', V', n', S'', V'', n'') = dU'(S', V', n') + dU''(S'', V'', n'') = 0. \qquad (6.18)$$

In the variable list, we have totally six variables, $S', V', n', S'', V'', n''$. We have the three constraint equations Eq. (6.17), relating the variables. The constraint equations in differential form are

$$dS' + dS'' = 0$$
$$dV' + dV'' = 0 .$$
$$dn' + dn'' = 0$$

The energy in Eq. (6.18) will become stationary when

$$\frac{\partial U}{\partial S'} = T' = T'' = \frac{\partial U}{\partial S''}$$
$$\frac{\partial U}{\partial V'} = -p' = -p'' = \frac{\partial U}{\partial V''} \qquad . \qquad (6.19)$$
$$\frac{\partial U}{\partial n'} = \mu' = \mu'' = \frac{\partial U}{\partial n''}$$

Here we have once more three equations relating the six variables. Equation (6.19) states that in both phases the temperatures, the pressures, and the chemical potentials, respectively, are equal, but not a priori constant. However, Eq. (6.17) together with Eq. (6.19) fixes the variables $S', V', n', S'', V'', n''$. Only in special cases, i.e., if the molar entropy and the molar volume of both systems would be equal, ambiguous solutions could be obtained.

On the other hand, a change of, e.g., the total entropy $S \to S + \delta S$, and/or $V \to V + \delta V$, and/or $n \to n + \delta n$ would change the set of the solution of the thermodynamic variables S', V', n', S'', V'', n''. Of course, also the derived functions, i.e., the intensive variables T', p', μ', T'', p'', μ'' would vary. In this way, we have three freedoms, as we could change that total entropy, the total volume, and the total mol number and still we could find solutions with respect to the intensive variables where two phases are present.

If we state that all variables are fixed by the functional form of the energies of the subsystem and the total entropy, volume, and mol number of the combined system, then this should not be confused that there would not be any freedom from the view of the phase rule. In fact, the derivation of the phase rule starts with different prerequisites.

The phase rule is not restricted to isolated systems and we do not ask for the extensive variables, but only for the intensive variables. This means we are not interested with how many matter in the individual phases is, how their volume is and, how their entropy is. Therefore, we are throwing away any information on the size of the phases.

6.5.2 Open System

We are now dealing with two systems. The first system is the system of primary interest, and we will address it with the index "*sys.*" The second system we will call environment, and we will associate it with the index "*env.*" These two systems should be the only in our world. We may gather the two systems into one that we may call the total system "*tot.*" For the combined system, the energies of the two individual systems add up and the total energy is constant:

$$dU_{tot} = dU_{sys} + dU_{env} = 0. \tag{6.20}$$

Further, we point out the constraints for a stand-alone closed system:

$$dV_{tot} = dV_{sys} + dV_{env} = 0$$
$$dS_{tot} = dS_{sys} + dS_{env} = 0.$$

These equations mean that the total energy is constant, but the energies of the individual systems may change. The same is true for the volume and for the entropy.

We imagine an example for the thermodynamic system under consideration: Assume, the system is a gas in a cylinder with freely movable piston. The cylinder is isolated thermally. The environment is the atmosphere. Then the gas may be in equilibrium with respect to compression energy, because it can change freely the volume V_{sys}. Therefore, the pressure of the system p_{sys} is equal to the atmospheric pressure:

$$p_{sys} = p_{env}. \tag{6.21}$$

We may think that the environment is a reservoir with respect to compression energy, in good approximation. This means that the pressure is constant, whatever happens with its volume $dp_{env}/dV_{env} \approx 0$. Since the piston is thermally isolated, an exchange of entropy is not possible $dS_{sys} = 0$. Therefore, the temperature T may not be equal to what is in the environment. $dT_{sys} \neq dT_{sys}$. If the temperatures are equal, then this is purely incidental. If the piston expands, then the system loses energy and

$$dU_{sys} = -p_{sys}\,dV_{sys}.$$

The same formal relation holds for the environment

$$dU_{env} = -p_{env}\,dV_{env}.$$

But the environment gains energy because the system inflates at the costs of the environment. We recall Eq. (6.20) and find for the total energy

$$dU_{tot} = dU_{sys} - p_{env}\,dV_{env} = 0.$$

Here we have inserted the compression energy for the system. Replacing now the environment due to $p_{sys} = p_{env}$ and $dV_{sys} = -dV_{env}$, we have

$$dU_{tot} = dU_{sys} + p_{sys}\,dV_{sys} = 0. \tag{6.22}$$

This is the equilibrium condition for a system being in contact with a reservoir for compression energy, or less precisely with a manostat. Only compression energy may be exchanged with the environment. The equation is in particular appealing, because only variables concerning the system itself are occurring. We do not have to care about the environment. Since we are dealing with a manostat, in equilibrium, $dp_{sys} = dp_{env} = 0$. Therefore, we may safely add to Eq. (6.22) also the term $V_{sys}\,dp_{sys}$, which is zero. Thus, Eq. (6.22) turns into

$$dU_{tot} = dU_{sys} + p_{sys}\,dV_{sys} + V_{sys}\,dp_{sys} = d(U_{sys} + V_{sys}\,p_{sys}) = 0. \tag{6.23}$$

On the other hand, the enthalpy is $H = U + pV$; from Eq. (6.23) we arrive at

$$dH_{sys} = dU_{tot} = 0. \tag{6.24}$$

In Eq. (6.24), on the left-hand side, we are dealing with the enthalpy of the system, whereas on the right-hand side, we are dealing with the energy of the system plus environment, i.e., the combined or total system.

Summarizing, if there is a free exchange of volume energy of the system under consideration with an environment that acts as a manostat, using the enthalpy of the system only describes the equilibrium properly by $dH_{sys} = 0$. We do not need to know anything of the environment.

In other words, if we dealing with the enthalpy (at constant pressure), we are virtually enlarging the pure system with an environment of constant pressure. The condition of equilibrium $dH_{sys} = 0$ is equivalent to the statement that the change total energy dU_{tot} of such a virtually enlarged system is zero.

The procedure seems to be strange at the first glance because we are taking a function where we *add* in fact a certain energy form to the energy of the system. However, the interpretation that we are adding up a virtual environment makes this procedure plausible.

In fact, the same is true when a system is open with respect to other energy forms than compression energy. If a free exchange of entropy is possible, then with the same arguments as pointed out above we may introduce $F_{sys} = U_{sys} - T_{sys} S_{sys}$ and find that $dF_{sys} = 0$ for a process under constant temperature.

However, the most important case is that a system is in contact with a reservoir of constant pressure and constant temperature, and compression energy and thermal energy can be freely exchanged. Here the function $G_{sys} = U_{sys} - T_{sys} S_{sys} + p_{sys} V_{sys}$ is appropriate and $dG_{sys} = 0$ is the condition for equilibrium.

Table 6.1 Energy functions suitable to describe equilibrium

Free exchange of	With	Energy function
—	—	U
S	Thermostat	$F = U - TS$
V	Manostat	$H = U + pV$
S, V	Thermostat and manostat	$G = U - TS + pV$
S, n	Thermostat and chemostat	$B = U - TS - \mu n$
S, V, n	Thermostat, manostat, and chemostat	$I = U - TS + pV - \mu n$

Summarizing, we can trace back the important thermodynamic functions regarding an open system to the simple internal energy of a corresponding gathered closed system. We summarize the situation in Table 6.1. The last entries B and I in Table 6.1 are somewhat uncommon, however straightforward. Similar to a thermostat that keeps the temperature constant and a manostat that keeps the pressure constant, a chemostat is a device that keeps the chemical potential constant, even when matter enters it. The chemostat is a device with the property $d\mu/dn \approx 0$. In fact, commonly the term chemostat is used for a continuous bio-reactor [6, 7]. By the continuous feed of nutrients, the chemical conditions and thus the chemical potential of the contents are kept constant.

Example 6.2. We close with a thermodynamic example of a chemostat. Imagine a stiff plastic bottle filled with brine, with water (1) as system (') and some salt (2) in the sea as system ("). The bottle is permeable to water, but not to the salts dissolved. The bottle is idealized stiff, so no volume change will occur. The bottle is transparent to entropy, so thermal equilibrium can be established. We have the constraint equations

$$dS' = -dS''; \quad dn_1' = -dn_1''; \quad dn_2' = 0; \quad dV' = 0; \quad dV'' = 0.$$

The sea is a reservoir for temperature and the component water (1). If pure water is moving from the bottle into the sea, the salt concentration in the sea will not change. We can express this property of the sea as $dT''/dS'' \approx 0$ and $d\mu_1''/dn_1'' \approx 0$.

Without doing the calculations with the constraint equations and making the energy stationary in detail, we will find

$$T' = T''; \quad \mu_1' = \mu_1''. \tag{6.25}$$

Since the system ($''$) is a chemostat and a thermostat, this means that both T' and μ_1' must be constant. We stress here that these functions are really constant because the system ($''$) is a reservoir and not merely equal as Eq. (6.25) expresses. The energy for the system ($'$) is

$$dU' = T'dS' - p'dV' + \mu_1'dn_1' + \mu_2'dn_2'. \tag{6.26}$$

We seek for the best thermodynamic function which we address as B to describe this case:

$$B'(T', V', \mu_1', n_2') = U' - T'S' - \mu_1'n_1'. \tag{6.27}$$

Equation (6.27) is the Legendre transformation of the energy with respect to entropy, volume, and water. Taking the constraints $dV' = 0$ and $dn_2' = 0$ into consideration, in equilibrium $dB' = 0$. From this condition, immediately $dT' = 0$ and $d\mu_1' = 0$ follow. This is the same, as obtained in Eq. (6.25), however in a more simple way.
□

6.6 Diffusion Equilibria of Gases

6.6.1 One Kind of Gas

A very simple case is the equilibrium of diffusion. Two gases of the same kind are in a container of fixed volume and are separated by a diaphragm. The container itself is placed in a thermostat. The experimental setup is shown in Fig. 6.4. The gas may move freely through the diaphragm. Thus, the mol numbers of the gas in the individual chambers may vary freely, but since the gas may not move outward, the sum of the mol numbers of both chambers is constant. However, the volumes of the individual chambers V', V'' are fixed. Due to the thermostat, the whole system may exchange thermal energy with the environment.

From the experimental setup, as shown in Fig. 6.4, we can develop a thermodynamic model. This model is shown in Fig. 6.5. The thermodynamic model is not unique. We have presumed that the thermal energy may enter from the thermostat into the two chambers. However, across the diaphragm, no thermal energy can be exchanged. If we would allow an exchange of thermal energy across the diaphragm,

Fig. 6.4 Equilibrium of
diffusion

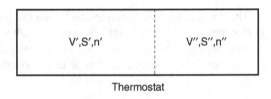

Thermostat

Fig. 6.5 Thermodynamic
model of the equilibrium of
diffusion

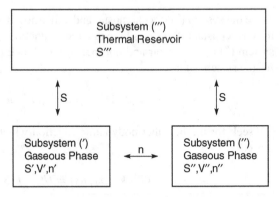

then it would not be necessary to place both chambers into the thermostat. So we
could place one of the vertical double arrows concerning entropy flow arrows in
Fig. 6.5 beside the flow of matter and the model would still do the same job. The
energy of the thermostat U''' is a function of the entropy exclusively, $U''' = U'''(S''')$.
This does not mean that this is true in general. However, we are using the thermostat in the experiment only for exchange of energy, and we demand that during the
experiment nothing else should occur in the thermostat. In our sense, the thermostat
is a big system, say filled with a large amount of water that would not exchange other
energy forms with the system under consideration and moreover with another environment. Because of the large amount of water, we find the property $dT'''/dS''' \approx 0$.
The energy individual chambers U' and U'' are functions that must show the same
character. In particular, $U' = U'(S', n')$, and the same for $U'' = U''(S'', n'')$. Note
that the energy is a function of only the variables that are belonging to the system
itself. For example, $U''' \neq U'''(S')$. We do not address the energy as functions of
the volume, in the course of the experiment, because the volume does not change
at all in the course of the experiment. However, if we would change the volume of
one or both chambers, then we would immediately see that the energy is in fact a
function of the volume. In fact, we will make later use of functional dependence of
the variables on the volume, to elucidate their properties, but this has nothing to do
with the experiment itself.

In equilibrium, the total energy U_{tot} of the system is constant. This means that
exchange processes can occur that do not violate the conservation of total energy.
The we can set up the relations as follows:

$$dU_{tot} = dU' + dU'' + dU''' \quad = 0$$
$$dS_{tot} = dS' + dS'' + dS''' \quad = 0$$
$$dV' = 0$$
$$dV'' = 0$$
$$dV_{tot} = dV' + dV'' \quad\quad\quad = 0$$
$$dn_{tot} = dn' + dn'' \quad\quad\quad = 0.$$

We have included also the constancy of the individual volumes dV' and dV''. From this, it follows also that the sum of the volumes is constant. However, the statement that the individual volumes are constant contains still more information. Moreover, inspection of the functional dependence of the energies reveals that we do not need the volume itself.

In order to find the energy to be stationary, we multiply the constraint equations by an undetermined multiplier and add to the energy equation:

$$dU' + dU'' + dU''' =$$
$$T'dS' + \mu'dn' + T''dS'' + \mu''dn'' + T'''dS''' +$$
$$+ \lambda_S(dS' + dS'' + dS''') + \lambda_n(dn' + dn''). \quad (6.28)$$

Rearrangement of the terms in Eq. (6.28) gives the set of equations

$$T' + \lambda_S = 0$$
$$T'' + \lambda_S = 0$$
$$T''' + \lambda_S = 0,$$
$$\mu' + \lambda_n = 0$$
$$\mu'' + \lambda_n = 0$$

from that we obtain

$$T' = T'' = T''' \quad \text{and} \quad \mu' = \mu''. \quad (6.29)$$

Equation (6.29) states that all temperatures and all chemicals potentials are equal. But we do not obtain any information on the pressure. Intuitively, we feel that the pressures in both chambers should be the same.

6.6.1.1 Pressure

To get some information about the pressure, we set up for both chambers the Gibbs – Duhem equations:

$$0 = S'dT' - V'dp' + n'd\mu'$$
$$0 = S''dT'' - V''dp'' + n''d\mu''. \quad (6.30)$$

From Eq. (6.30), we conclude that the equations have basically the form

$$S dT - V dp + n d\mu = 0.$$

For this reason, we can resolve p as function of μ and T, i.e.,

$$p = p(T, \mu).$$

If the material is the same, then the function calculus for p' and p'' is the same. So for the same arguments $\mu = \mu' = \mu''$ and $T = T' = T''$ also $p = p' = p''$.

Fig. 6.6 Two gases connected by a semipermeable membrane

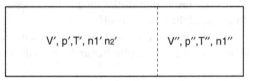

| V', p',T', n1' n2' | V", p",T", n1" |

6.6.2 One Component of Two Permeating

We consider now again two chambers, as shown in Fig. 6.6. In one chamber, there is a mixture of two gases (1) and (2) and in the other chamber there is only one component (1). We assume that the component (1) may move freely from one chamber into the other. The component (2) is restricted to stay in the left chamber. Further, thermal energy may exchange freely between the two chambers. However, manal energy may not exchange. Therefore, we have the conditions

$$S' + S'' = S_{tot}; n_1' + n_1'' = n_{1,tot}. \tag{6.31}$$

From these equations, we derive the conditions of equilibrium expressed in terms of the intensive variables:

$$T' = T'' \tag{6.32}$$
$$\mu_1' = \mu_1''. \tag{6.33}$$

6.6.2.1 Equal Partial Pressures

We start first with a simple argument. If one component can diffuse freely through the barrier, then in both chambers the partial pressures of the component must be equal in equilibrium. Therefore,

$$p_1' = p_1''.$$

We could show this with similar arguments as with one component, treated in Sect. 6.6.1. We note that the total pressure is the sum of the partial pressures, i.e.,

$$p' = p'_1 + p'_2.$$

Further,

$$x'_1 p' = p'_1 = p''_1. \tag{6.34}$$

Here x'_1 and x'_2 are the mole fractions of the component (1) and the component (2) in the left chamber. This is the desired relation between the change of the total pressure and the amount of component (2).

In the chamber ($''$), there is only the single gas component (1). Therefore, $p''_1 = p''$. The difference in pressure is

$$p' - p'' = p'x'_2 = x'_2 \frac{(n'_1 + n'_2)RT'}{V'}. \tag{6.35}$$

We will denote the average molar volume $V'/(n'_1 + n'_2) = \tilde{V}$ and the difference in pressure $p' - p'' = \pi$. Rearrangement of Eq. (6.35) results finally in

$$\pi \tilde{V} = x'_2 RT'.$$

This is completely analogous to the equation in osmosis.

6.6.2.2   Equal Temperatures

Another approach is to use Eq. (6.32). We start with

$$T' = T''.$$

We want to change only T' but not T''. Therefore, $dT' = 0$, and the expansion gives

$$dT' = \frac{\partial T'}{\partial p'} dp' + \frac{\partial T'}{\partial n'_2} dn'_2 = 0. \tag{6.36}$$

We recall the gas law for a gas mixture with two components

$$pV = (n_1 + n_2)RT. \tag{6.37}$$

We resolve now Eq. (6.37) with respect to the temperature and insert the partial differentials into Eq. (6.36):

$$dT' = \frac{V'}{R(n'_1 + n'_2)} dp' - \frac{p'V'}{R(n'_1 + n'_2)^2} dn'_2 = 0. \tag{6.38}$$

This results immediately in

$$dT' = \frac{dp'}{p'} - \frac{dn_2}{(n'_1 + n'_2)} = 0 \tag{6.39}$$

and further in

$$\frac{p'}{n'_1 + n'_2} = C. \tag{6.40}$$

We obtain the constant of integration by the condition

$$\frac{p''}{n'_1} = C.$$

This means that at zero component (2), the pressures are equal $p' = p''$. We arrive now at

$$p'x'_1 = p'', \tag{6.41}$$

which is the same as in Eq. (6.34), with a different philosophy in the approach.

6.6.2.3 Equal Chemical Potentials

We use now the condition of equal chemical potentials, Eq. (6.33). The chemical potential μ''_1 should not change in the course of the process, therefore also the chemical potential μ'_1 should be constant. We expand, as usual $d\mu'_1 = 0$.

$$d\mu'_1 = \frac{\partial \mu'_1}{\partial p'} dp' + \frac{\partial \mu'_1}{\partial n'_2} dn'_2 = 0. \tag{6.42}$$

The chemical potential of the component (1) of an ideal gas mixture of two components is

$$\mu_1 = \mu_{1,0} + RT \ln \frac{p_1}{p} = \mu_{1,0} + RT \ln \frac{n_1}{n_1 + n_2}. \tag{6.43}$$

We get for the partial derivatives

$$\frac{\partial \mu'_1}{\partial p'} = \frac{\partial V'}{\partial n'_1} \quad \text{and} \quad \frac{\partial \mu'_1}{\partial n'_2} = -\frac{RT}{(n'_1 + n'_2)}.$$

Using the ideal gas law for two components, Eq. (6.37), we insert for $\partial V'/\partial n'_1$ in Eq. (6.42) to obtain

$$\frac{RT'}{p'} dp' - \frac{RT'}{(n_1' + n_2')} dn_2' = 0. \tag{6.44}$$

Equation (6.44) is substantially the same as Eq. (6.39). We have now shown that the application of the condition of equal temperature yields the same results as the condition of equal chemical potentials. In the derivation of the osmotic pressure, the condition of chemical potentials is used.

6.6.2.4 Mixing Two Gases

We inspect now the expression

$$\frac{\partial p'}{\partial n_2'}$$

as such. Using the ideal gas law, Eq. (6.37), we find

$$\frac{\partial p'(T', V', n_1', n_2')}{\partial n_2'} = \frac{RT'}{V'} = \frac{p'}{n_1' + n_2'}.$$

We have shown the parameters in the pressure function explicitly to make more clear which parameters stay constant upon building the differential. Integration yields

$$\frac{p'}{n_1 + n_2'} = C. \tag{6.45}$$

Again, Eq. (6.45) is completely analogous to Eq. (6.40). However, we did not make use of the condition of equilibrium at all. In fact, we are describing here a different process, as shown in Fig. 6.7. In a chamber, there is the compound (1) and the compound (2) completely isolated with respect to movement form one chamber into the other. However, there is no thermal insulation. The temperature of the component (2) should be the same as the temperature of the component (1).

In the course of the process, we add the compound (2) to the compound (1) by removing the barrier in between the compounds and then pressing the piston from the left side to the right side as far, so that the original initial volume is retained.

This process, however, is the same as if there would be a semipermeable membrane for compound (1) to still another chamber as shown in Fig. 6.6. We can understand this by the fact that the addition of compound (2) would not change the partial pressure p_1' of the compound (1) in the ideal case. In other words, by performing such a process, there would not be any flow of the compound (1) through the membrane. If we think of the analogous process in the field of osmosis, by such a process, we would not need a membrane at all to find out the increase in pressure by adding a solute with the properties of compound (2).

Fig. 6.7 Addition of a
compound (2) without
membrane equilibrium

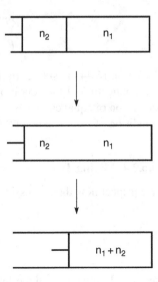

6.7 A Glance on the Constraint Equations

We can imagine some apparatus in which the phase transfer from the liquid state
into the gaseous state occurs. We do not show here the real apparatus, but we start
with a thermodynamic model, as a set of systems that is relevant for the thermodynamic description. In Fig 6.8, a thermodynamic model for gas–liquid equilibrium,
involving a single component is shown. The total system is an isolated system and
consists of four subsystems. Left down we have the liquid phase ($'$) with the set of
extensive variables of S', V', n', i.e., the entropy of the liquid phase, the volume of
the liquid phase, and the mol number of the liquid phase. Right down we have the
gaseous phase ($''$) with the set of extensive variables of S'', V'', n'', i.e., the entropy
of the liquid phase, the volume of the gaseous phase, and the mol number of the
gaseous phase. Liquid phase and gaseous phase can exchange entropy, volume, and
matter, as indicated by the double arrows.

Fig. 6.8 Thermodynamic
model for gas–liquid
equilibrium

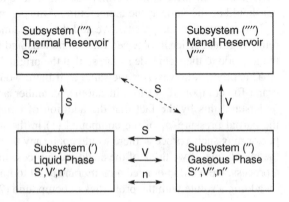

If we would confine ourself that the total system consists of these two subsystems, then no other exchange reactions would be possible. In equilibrium, we would have the total energy, the total entropy, the total volume, and the total mol number to be constant, i.e.,

$$
\begin{aligned}
dU_{tot} &= dU' + dU'' &&= 0 \\
dS_{tot} &= dS' + dS'' &&= 0 \\
dV_{tot} &= dV' + dV'' &&= 0 \\
dn_{tot} &= dn' + dn'' &&= 0
\end{aligned}
$$

If we add the left up system to the latter system, then we have in addition an exchange of entropy from the thermal reservoir to the liquid–gas system, and we have

$$
\begin{aligned}
dU_{tot} &= dU' + dU'' + dU''' &&= 0 \\
dS_{tot} &= dS' + dS'' + dS''' &&= 0 \\
dV_{tot} &= dV' + dV'' &&= 0 \\
dn_{tot} &= dn' + dn'' &&= 0
\end{aligned}
$$

This situation conforms approximately to a pressure cooker, where the sum of the liquid phase and the gaseous phase has constant volume, but from outside heat is transferred. Observe that in Fig. 6.8 there is a dashed double arrow pointing from the thermal reservoir to the gaseous phase. This indicates that entropy could move from the thermal reservoir directly to the gaseous phase. In fact, this type of exchange is not necessary for the description of the thermodynamic system, because entropy from the thermal reservoir can reach the gaseous phase via the liquid phase. In fact, moving along the flow paths of the entropy, including the dashed path would involve a cycle. However, in the physical system, there may be such a contact. In the pressure cooker, the pressure at equilibrium is not fixed because there is no contact to a manal reservoir (manostat).

We proceed now to the last step and connect the system also with a manostat by including the right up subsystem in the total system. Then the relations among the extensive variables will become

$$
\begin{aligned}
dU_{tot} &= dU' + dU'' + dU''' + dU'''' &&= 0 \\
dS_{tot} &= dS' + dS'' + dS''' &&= 0 \\
dV_{tot} &= dV' + dV'' + dV'''' &&= 0 \\
dn_{tot} &= dn' + dn'' &&= 0
\end{aligned}
$$

The same what was stated for the dashed arrow for entropy flow would be valid for an additional value exchange from the manal reservoir to the liquid phase.

In order to get the relations of the intensive variables, we have to seek for a stationary value of dU_{tot} under the respective constraint equations that are written below the energy equation.

6.8 Barometric Formula

The barometric formula relates the change of atmospheric pressure with altitude,

$$p(h) = p(0) \exp\left(-\frac{Mgh}{RT}\right). \qquad (6.46)$$

In Eq. (6.46), $p(h)$ is the pressure at altitude h, $p(0)$ is the pressure at reference level zero altitude, M the molar mass of the gas, g the earth acceleration, R the gas constant, and T the absolute temperature. In spite of its simplicity, it is valid for heights up to 6 km with an error of less than 5%.

The history of the barometric formula is described in the literature [8]. Famous scientists laid the foundations to the barometric formula, among them *Galilei*,[3] *Torricelli*,[4] *Pascal*, and *Boyle*.[5] The explicit formula goes back to *Laplace*. For this reason, the barometric formula is sometimes called Laplace formula.

There are connections to other fields of science. In 1909, *Perrin*[6] showed that small particles suspended in water exhibit a similar relation, where the pressure is replaced with the concentration.

There are several different approaches to derive the barometric formula, i.e., based on hydrostatic considerations, statistical considerations, and energetic considerations [8]. We show here certain methods to derive the barometric formula.

6.8.1 Hydrostatic Derivation

Consider a column of gas on that the gravity is acting. The situation is depicted in Fig. 6.9, left. At a certain height h, a pressure p is acting. Going a small distance lower say at $h - \Delta h$, the pressure increases, because the weight of the gas being in between h and $h - \Delta h$ adds to the pressure. If the column has an area of A, the mass of this gas is $\rho A \Delta h$ and the pressure increase is

$$\Delta p = -\rho g \Delta h. \qquad (6.47)$$

---

[3] Galileo Galilei, born Feb. 15, 1564, in Pisa, Italy, died Jan. 8, 1642, in Arcetri, Italy.

[4] Evangelista Torricelli, born Oct. 15, 1608, in Faenza, Italy, died Oct. 25, 1647, in Florence, Italy.

[5] Robert Boyle, born Jan. 25, 1627, in Lismore, Waterford, Ireland , died Dec. 30, 1691, in London, England.

[6] Jean Baptiste Perrin, born Sep. 30, 1870, in Lille France, died Apr. 17, 1942, in New York City, USA.

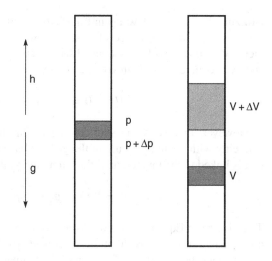

Fig. 6.9 Pressure and volume variation in a gas under the influence of gravity

For an ideal gas, the density ρ is

$$\rho = \frac{m}{V} = \frac{nM}{V} = \frac{Mp}{RT}. \tag{6.48}$$

Inserting Eq. (6.48) into Eq. (6.47), and allowing the differential $\Delta p \to dp$, $\Delta h \to dh$, and integrating results, we get

$$\ln p = -\frac{Mgh}{RT} + C,$$

which can be resolved into Eq. (6.46). The hydrostatic derivation of the barometric formula is based on the equilibrium of the forces acting on the gas.

6.8.2 Energetic Derivation

For the energetic derivation of the barometric formula we consider Fig. 6.9, right sketch. We inspect an exchange of a certain gas from a lower altitude to a higher altitude. This thought experiment is done by removing both portions of gas from the column.

The gas from the lower altitude is placed up and the gas from the higher altitude is placed down. In the course of the change of position, gravitational work has to be done. This work is obtained from an additional work of compression or expansion that the gas is forced to do, in order that the network is balanced. Therefore, we have for a small volume ΔV

$$-p\Delta V + Mng\Delta h = 0.$$

Utilizing the ideal gas law, again Eq. (6.46) is obtained. In the previous discussion, we talked about the balance of work, but we did not fix this concept clearly. In thermodynamics, work is associated with energy. So we mean that the gravitational energy is compensated with volume energy. In other words,

$$dU = 0 = -p dV + Mng dh. \tag{6.49}$$

However, there is some shortcoming in this treatment. Since we are dealing explicitly with the temperature of the gas, we have to consider also the energy form that is linked to the temperature, thus more correctly

$$dU = 0 = T dS - p dV + \mu dn + Mng dh. \tag{6.50}$$

To arrive from Eq. (6.50) at Eq. (6.49), we have to take $dS = 0$. However, this is a contradiction to the assumption of an isothermal process. On the other hand, if we consider the energy as a function of the temperature instead of entropy, then the partial derivative with respect to volume is no longer the pressure.

In the case of an isentropic process, we have to use the adiabatic equation instead of the ideal gas law. Thus to allow an isothermal process, we have to place the column into a thermostat. In this case, not the energy is equated to zero, but rather the free energy

$$dF = 0 = -S dT - p dV + \mu dn + Mng dh.$$

The situation is similar as in the case of the derivation of Kelvin's equation for the surface tension of soap bubbles.

We emphasize that the process is not completely discussed. In a gas column at rest, we cannot simply shift a gas and place it to another position because this position is already occupied with another amount of gas. In fact, we have to shift two portions of gas. One portion is moved upward, thereby being expanded, and the other portion of gas is moved downward, thereby being compressed. However, the work of compression and expansion of the two portions of gas is not done mutually, but at the expense of gravitational work of each gas individually. This is detailed in the discussion on constraints in Sect. 2.10.8.

Question 6.3. In practice, in the atmosphere, from where should a gas take the thermal energy needed for isothermal expansion? The answer is partly by radiation. On the other hand, even in elementary school it is taught that air will cool when it is forced to move upward, when the foehn is discussed. So the barometric formula has a restricted field of application.

The adiabatic process instead of the isothermal process using Eq. (6.50) with $dS = 0$ and $dn = 0$ results in

$$p = p_0^{\frac{C_v}{C_v - C_p}} \left[\frac{C_v p_0 V_0 - Mghn(C_p - C_v)}{C_v V_0} \right]^{\frac{C_p}{C_p - C_v}}. \tag{6.51}$$

Fig. 6.10 Container with gas
sliding on rails in vacuum
under the influence of gravity

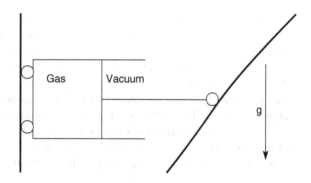

We can imagine the situation by a model sketch as shown in Fig. 6.10. We can
rewrite Eq. (6.51) in reduced variables using the molar volume $\tilde{V}_0$ and a reduced
height z

$$\tilde{V}_0 = \frac{V_0}{n}; \quad z = \frac{Mgh}{p_0\tilde{V}_0} = \frac{Mgh}{RT_0}.$$

$$p_r = \frac{p(h)}{p_0} = \left[1 + (1 - \kappa)\frac{Mgh}{p_0\tilde{V}_0}\right]^{\frac{\kappa}{\kappa-1}} = [1 + z(1 - \kappa)]^{\frac{\kappa}{\kappa-1}}. \qquad (6.52)$$

We can rewrite the isothermal barometric formula (Eq. 6.46) as

$$p_r = \frac{p}{p_0} = \exp\left(-\frac{Mgh}{p_0\tilde{V}_0}\right) = \exp(-z). \qquad (6.53)$$

The difference of Eqs. (6.53) and (6.52) is small in the usual range. This can be seen
by expanding both formulas in Taylor series.

6.8.3 Laps Rate

When the volume is substituted with the temperature in the adiabatic gas law, $pV^\kappa = C = p(nRT/p)^\kappa$, the temperature dependence with height is obtained from Eq. (6.53).
The temperature varies with altitude as

$$T_r = \frac{T(h)}{T_0} = 1 - \frac{Mgh}{T_0\tilde{C}_v}. \qquad (6.54)$$

So Eq. (6.54) predicts a linear temperature drop with altitude. The calculated temperature drop is somewhat higher than observed in meteorological phenomena.

Example 6.3. We can estimate for a gas like air under atmospheric conditions

$$\frac{Mg}{5/2RT_0} = \frac{28.8 \times 10^{-3}\,\text{kg mol}^{-1}\,9.81\,\text{kg m}^{-1}\text{s}^{-2}}{5/2 \times 8.316\,\text{J mol}^{-1}\text{K}^{-1}\,300\,\text{K}} = 4.52 \times 10^{-5}\,\text{m}^{-1}.$$

The number of freedoms for a diatomic molecule is 5/2. So we get $-1.36\,°C$ per 100 m for perfect adiabatic expansion when moving upward. In fact, there are radiation and heat conductivity effects that can feed heat into the gas. Nonpolar diatomic gases such as nitrogen and oxygen cannot absorb radiation, but water vapor and carbon dioxide can absorb radiation.

In meteorology, it is experienced that the dry adiabatic rate is $-1\,°C$ per 100 m change in altitude. In reality, the atmosphere laps rate can reduce from -0.6 to $-0.7\,°C$ per 100 m. The moist adiabatic rate is around $-0.6\,°C$ per 100 m of change in altitude. When the water condenses out as it rises, it adds latent heat due to evaporation. This is why its lapse rate is below that of dry air. □

We can obtain Eq. (6.54) even more simply, by setting the energy as a function of temperature, volume, and height:

$$dU(T, V, h) = C_v dT + \frac{\partial U(T, V, h)}{\partial V} dV + nMg\,dh. \tag{6.55}$$

This is valid for a thin shell, where the height of the system is all the same. Observe that thermodynamic similarity is not established for h, but rather of nh as variable. For an ideal gas, $\partial U(T, V, h)/\partial V = 0$. Integration of Eq. (6.54), again results in Eq. (6.54).

6.8.3.1 Derivation from Chemical Potential

We try now another approach to calculate the temperature gradient of an ideal gas in the gravitational field. The energy of an ideal gas is given by its fundamental form Eq. (4.8). We have to introduce the gravitational energy to arrive at

$$U(S, V, n, h) = n \left[C' \frac{e^{(S/n)/(\tilde{C}_v)}}{(V/n)^{R/\tilde{C}_v}} + C'' \right] + nMgh + C'''. \tag{6.56}$$

To get the condition of zero flow in the direction of gravity, the chemical potential must be constant $\mu = \mu_0$. However, the chemical potential is not obtained by simply forming the derivative of the energy with respect to the mol number $\partial U(S, V, n, h)/\partial n$, but there are some additional constraints. When a gas leaves a shell of a certain height, it takes its entropy with it. Moreover, the shell shrinks in volume. Thus, the correct chemical potential is obtained from the derivative

$$\mu = \frac{U(n\tilde{S}, n\tilde{V}, n, h)}{\partial n}, \tag{6.57}$$

where the molar entropy $\tilde{S}$ and the molar volume $\tilde{V}$ of the gas left in the shell remain constant. When the differential is performed from Eq. (6.56) in this way, and the volume is eliminated from the relation $T = \partial U/\partial S$, again Eq. (6.54) is obtained.

The question, whether a temperature gradient may be possible in thermal equilibrium is of historical interest [9]. *Maxwell* stated that the temperature must be the same in all heights of a column. Otherwise, a perpetuum mobile could be constructed. In contrast, *Loschmidt* believed that the equilibrium temperature decreases with height.

Question 6.4. Is the form $U(S, V, n, h)$ thermodynamically similar, or in other words, is it Eulerian homogeneous of first order?

$$U(\lambda S, \lambda V, \lambda n, \lambda h) = \lambda U(S, V, n, h)$$

Try to reformulate it with the gravity term Mgq with $q = nh$ and set $C''' = 0$. For comparison, recall the momentum as extensive variable $p = mv$. The derivation of Eq. (6.57) works because we do not make there explicit use of the thermodynamic similarity.

The thermodynamics of matter under gravity was derived by *Gibbs* [10, pp. 144–147], even for a multicomponent system. Gibbs started with the relation

$$\delta \int dU + \delta \int ghdm \geq 0, \tag{6.58}$$

where U is the intrinsic energy of an element of mass dm. The total mass is composed of some individual components as $m = \sum_i m_i$. Note that Gibbs did not use molar quantities, but rather mass quantities. Equation (6.58) expresses, using the integral, that the variation of the total intrinsic energy emerges from the total variation of the gravitational energy. The intrinsic energy is setup in the usual way as function of entropy, volume, and mass components. Then the derivation proceeds, similar to the method of undetermined multipliers with the condition $\int \delta dS = 0$ to state that $\int T\delta dS = 0$, and thus that the temperature in each element must be equal. Summarizing, the following conditions for equilibrium emerge:

$$T(h) = C_1; \quad \mu_i(h) + gh = C_2; \quad p(h) + \rho gh = C_3.$$

We emphasize that Gibbs, in his text is talking about a fluid in a tube, when he describes the system. So it is suggestive that he had in mind rather a laboratory system than the earth atmosphere.

The derivation by Gibbs was criticized by *Bohren* and *Albrecht* [11], who suggested for atmospheric systems a temperature gradient, which in turn was objected by *DeCaria* [12], who insists that *Gibbs* is correct.

We may add a comment: the condition of equal temperature arises from the statement

$$\int \delta dS = 0, \tag{6.59}$$

in the derivation by Gibbs. Thus, the total entropy is constant, but may exchange among the individual elements under consideration. On the other hand, we can still keep the total entropy constant, when we demand $dS = 0$. In this case, Eq. (6.59) is still valid, however does not allow for each individual element any exchange of entropy. The condition $dS = 0$ is more restrictive than Eq. (6.59) and demands an adiabatic process, whereas Eq. (6.59) demands an isothermal process. This is ultimately correct, since an ideal thermal insulation does not exist.

In the earth atmosphere, there are usually certain fluxes of heat, therefore both approaches that treat the gas column as a closed thermodynamic system are not correct. The adiabatic case is more suitable, if the respective process of movement of a gas to some height is fast in comparison to the rate of heat conduction, thus entropy exchange of the gas. It was pointed out by *Schaefer* that the isothermal assumption in the barometric formula is highly unrealistic, and the adiabatic equation should be used [13].

6.8.4 Stability of Stars

We may annotate here that the adiabatic expansion in the gravity field plays a certain role for the stability of stars. The Schwarzschild criterion for stability is based on the adiabatic expansion of a gas in the interior of a star under the influence of gravity.

If the star is in its radiative equilibrium, the temperature gradient from the center to its surface will be just the radiative temperature gradient. If an adiabatic motion of a gas region in the star would cause a temperature gradient less than the temperature gradient by radiation, the star would become instable. In other words, if the radiative temperature gradient is less than the adiabatic temperature gradient, a convective motion of a certain region will not occur and the energy transport to the surface of a star will take place by radiation.

6.9 Phase Transition Points

With the term *phase transition point*, we summarize melting point, boiling point, etc. Actually, the phase transition point is not a point, but a certain temperature, at which melting, boiling, etc., occurs. In organic chemistry, often a temperature range for the melting point of certain substances is given for characterization.

It is in order to ask, whether basically melting occurs in a certain temperature range, or whether the range arises due to inaccurate thermometric measurement.

We focus now on a substance, consisting of a single component. The thermodynamic criterion for a phase transition, or in other words that two phases (') and ('') may coexist, is that the chemical potentials of both phases in question are equal. We express the chemical potential as a function of temperature and pressure, $\mu = \mu(T, p)$. The chemical potential as an intensive function may not depend on the size of the system. Since we are dealing with a single component substance, the chemical potential may not be a function of a composition. Therefore, temperature and pressure are sufficient to characterize the chemical potential of a single component substance. In equilibrium with respect to temperature and pressure,

$$p' = p'' \quad \text{and} \quad T' = T'',$$

therefore we can omit the primes in the further discussion. The condition of equilibrium with respect to substance flow from one phase into the other is

$$\mu'(T, p) = \mu''(T, p).$$

More general,

$$\mathrm{d}G(T, p, n) = \mu'\mathrm{d}n' + \mu''\mathrm{d}n'' = (\mu' - \mu'')\mathrm{d}n'. \tag{6.60}$$

The free enthalpy as condition of equilibrium $\mathrm{d}G = 0$ arises naturally, if the system is in contact with a thermal reservoir and a pressure reservoir. The constraint $\mathrm{d}n'' + \mathrm{d}n'' = 0$ arises from the conservation of mass and the assumption that chemical reactions are not allowed. Near equilibrium, a plot of the chemical potentials looks like that in Fig. 6.11.

Fig. 6.11 Plot of the chemical potentials near equilibrium

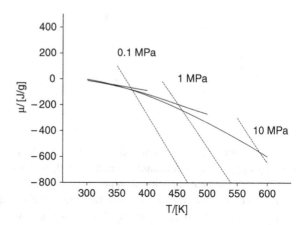

The data are taken from steam tables [14, p. 447]. Traditionally, in the steam tables, the specific quantities are given rather than the molar quantities. Moreover, the entropy is normalized not to absolute zero, but to 0 °C. The full curves are the chemical potentials for the liquid phase at fixed pressure as a function of the temperature. The curves are accessible by measurement only at temperatures below the boiling point. The dashed curves represent the chemical potentials for the gaseous phase at fixed pressure as a function of the temperature. These data are accessible by measurement only at temperatures above the boiling point. Both types of chemical potential are extrapolated in the region that is not accessible to measurement.

The experimental accessible points appear to make a kink at the phase transition. Thus, the chemical potentials of the two phases are running in a way that they can cut only in a point. Therefore, notably for a single substance, the condition of equal chemical potential is fulfilled not in a temperature range, but at a single temperature. Therefore, the term transition point is justified. At a temperature below the intersect, $\mu' < \mu''$ and $dn' > 0$ to get $dG < 0$. Therefore, the phase (') is stable in this region. The situation is reverse at a temperature beyond the intersect. An infinitesimal deviation of the temperature from the temperature at equal chemical potential will cause the less stable phase completely to disappear.

At temperatures above the critical point, the kink in the chemical potentials will disappear. The kink in the chemical potentials causes the entropy, which is $-\tilde{S} = \partial \mu / \partial T$ to change abruptly. Ideally, the entropy jumps in the way like a step function. This behavior, in turn makes the heat capacity to show a Dirac delta function peak at the transition point. The situation is shown schematically in Fig. 6.12.

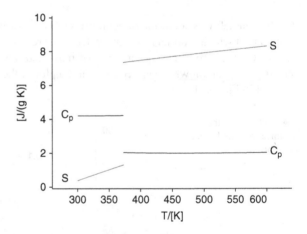

Fig. 6.12 Specific entropy and specific heat capacity of water at a phase transition at 0.1 MPa

In the limit of the critical point, this unusual behavior disappears. Then the entropy approaches a continuous function without a step inside, and the delta function peak in the heat capacity disappears.

If in the course of a phase transition, there is a discontinuity in the entropy, other variables also show a discontinuity. For a homogeneous system, a phase, with one component, the Gibbs – Duhem equation holds:

$$S dT - V dp + n d\mu = 0.$$

Dividing by the mol number, i.e., normalizing the phase to one mole, results in $\tilde{S} dT - \tilde{V} dp + d\mu = 0$. This holds for any phase (', '', ...). If two phases are in equilibrium, the conditions of equilibrium hold:

$$p' = p'' = p; \quad T' = T'' = T; \quad \mu' = \mu'',$$

and we can set up the Gibbs – Duhem equations for each phase and subtract. The result is

$$(\tilde{S}' - \tilde{S}'')dT - (\tilde{V}' - \tilde{V}'')dp. \tag{6.61}$$

Equation (6.61) is the starting point for the Clausius – Clapeyron equation. The equation states that if in the limit the entropies for both phases become equal, also the volumes of the phases must become equal, or the freedom to change the pressure will be lost, $dp = 0$. Otherwise, the equation does not hold longer. In the critical point, the entropies for both phases become equal. Therefore, also the volumes become equal. Of course, this statement holds also vice versa.

From the view of thermodynamics, a temperature range of transition would be only possible, if the curves would coincide in a certain range of temperature, than to intersect in a point as shown in Fig. 6.11. This would imply that the curves should have in this range a common slope.

In practice, the situation is different. Impurities will cause a melting range, apart from an inhomogeneous temperature distribution in a macroscopic system. This causes an apparent temperature range of the phase transition. We emphasize that the slope of the plot is always negative. This arises from the fact that

$$\frac{\partial \mu(T, p, n)}{\partial T} = -\frac{\partial S(T, p, n)}{\partial n} = -\tilde{S}. \tag{6.62}$$

Further, Eq. (6.62) explains which phase is more stable at elevated temperatures. It is the phase with the larger entropy.

The transitions such as melting point, boiling point are addressed as first-order transitions. There are other types of transitions, where the heat capacity in the transition region does not change abruptly, but has the shape of a Greek Λ. These types of phase transitions are called Λ-transitions and belong to the class of higher order transitions. The basic behavior of transitions is shown in textbooks, e.g., [15, p. 145].

From the view of a higher order transition, the first-order transition is obtained, when the lambda region is ideally compressed in such a manner so that the lambda shape approaches a Dirac delta function. Since the thermodynamic data are obtained ultimately by experimental measurement, one can be never sure, whether there is really a first-order transition or a lambda transition in a small region that cannot be differentiated by the experimental precision.

It should be recalled that the experimental data are created by the researcher rather than by nature. If the researcher observes a phase transition, then he adds the latent heat ΔH to his enthalpy table, and adds $\Delta S = \Delta H / T$ to the entropy table. Then when plotting the free energy and finds that the data are coinciding, as shown in Table 6.11, he finds out surprisingly: Hey, there is a first-order transition! However, some statements can be made on the existence of a first-order transition.

- The phases must consist of essentially one component.
- The phases must be strictly homogeneous.

- No superimposed structures even on the microscale should exist or be formed.
- Obviously, a macroscopic phase separation is necessary.

6.10 Colligative Properties

From the equilibrium conditions, we get for a system that is embedded in a thermostat ($dT = 0$) and connected to a manostat ($dp = 0$) the free enthalpy being constant $dG(T, p, \ldots) = 0$ for the process under consideration. Here we are dealing with a process running at constant temperature and at constant pressure.

Question 6.5. How can we come to the ebullioscopic equation, in which we have a variation of the temperature itself? The same open question holds for the other common equations, like osmosis, etc.

6.10.1 Ebullioscopy

6.10.1.1 The Real System

First, we describe the real system. An ebullioscopic apparatus resembles closely to an apparatus suitable for chemical operations done under reflux, but it has a very sensitive thermometer. In fact, the boiling point of a solvent in presence of a nonvolatile solute is measured, more precisely the dependence of the boiling point on the concentration of the nonvolatile solute. The boiling point is the temperature at which equilibrium of liquid and gaseous phase at standard pressure is established.

6.10.1.2 The Thermodynamic Model

The real system disagrees largely with the idealized thermodynamic model, which we set up now. First of all, usually the apparatus is not operating at standard pressure, but at the atmospheric pressure that is available at the time of the experiment. To keep the liquid boiling, a constant heat input from the heating device is compulsory. The condenser acts as a barrier for the vaporized liquid in order to keep the concentration of the nonvolatile solute constant. Further, it is a sink for the continuous heat flow that arises from the heating device.

The instrument is operated under the influence of gravity that will be totally neglected in the thermodynamic model, because the influence of gravity can be justified to be not of importance. However, probably the device would not work at all in an environment lacking gravity. So, in summary the instrument is far away in design from the idealization we are doing now.

Figure. 6.13 represents the thermodynamic model abstracted from the real apparatus. Note that this model resembles closely to that presented in Fig. 6.8.

Fig. 6.13 Thermodynamic model abstracted from a real ebullioscopic apparatus

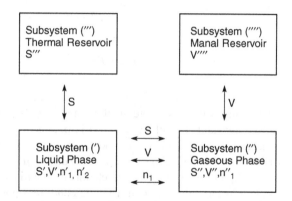

6.10.1.3 Constraint Equations

In fact, we can obtain in a completely similar manner the differential for the total energy

$$dU_{tot} = dU' + dU'' + dU''' + dU'''' = 0$$

and the constraint equations written already in differential form

$$
\begin{aligned}
dS_{tot} &= dS' + dS'' + dS''' &&= 0 \\
dV_{tot} &= dV' + dV'' + dV'''' &&= 0 \\
dn_{tot,1} &= dn'_1 + dn''_1 &&= 0 \\
dn_{tot,2} &= dn'_2 &&= 0
\end{aligned}
\tag{6.63}
$$

Again, here we have assumed that in the environment that consists of the reservoir for thermal and manal energy nothing else will happen as the interplay with the ebullioscopic apparatus. So the total system is an isolated system. The thermal reservoir has the property that the temperature does not change with entropy flow, i.e., $dT'''/dS''' \approx 0$, and the manual reservoir has the property that the pressure does not change with a change in volume, i.e., $dp''''/dV'''' \approx 0$.

As usual we make the total energy U_{tot} stationary by the multiplying the constraint equations by an unknown factors λ_S, λ_V, λ_{n_1}, λ_{n_2} and add to the total energy. This results in the equation for the energy

$$
\begin{aligned}
dU = {}&(T' + \lambda_S)dS' - (p' + \lambda_V)dV' + (\mu'_1 + \lambda_{n_1})dn'_1 + (\mu'_2 + \lambda_{n_2})dn'_2 + \\
&(T'' + \lambda_S)dS'' - (p'' + \lambda_V)dV'' + (\mu''_1 + \lambda_{n_1})dn''_1 + (T''' + \lambda_S)dS''' \\
&\qquad\qquad\qquad\qquad\qquad\qquad\qquad\qquad\qquad - (p'''' + \lambda_V)dV''''.
\end{aligned}
$$

Every term is equated to zero which results in

$$T' = T'' = T''' \quad = \lambda_S$$
$$p' = p'' = p'''' \quad = \lambda_V$$
$$\mu_1' = \mu_1'' \quad\quad = \lambda_{n_1} \tag{6.64}$$
$$\mu_2' \quad\quad\quad = \lambda_{n_2}$$

Actually, we cannot state anything about μ_2' because it is equal to an arbitrary chosen constant. Usually Eq. (6.64) is obtained from the free energy because it is argued that the temperature and the pressure is constant during the process in focus, therefore $dG(T, p, n_1, n_2) = \mu_1' dn_1' + \mu_1'' dn_1''$ with $dn_1' + dn_1'' = 0$ and $dn_2' = 0$, and therefore also the chemical potentials must be equal. Note that the situation is basically different from the conditions pointed out in Eq. (6.25). Recall that $d\mu' - d\mu'' = 0$ is different from $d\mu' = 0$.

We evaluate now once more the number of freedoms in the system. We have the set of extensive variables of S', S'', S''', V', V'', V''', n_1', dn_1'', n_2', so totally nine variables. The constraint equations, Eq. (6.63), contribute to four equations, and the equilibrium conditions, Eq. (6.64), contribute to five relations among the extensive variables. Note that the last equation in Eq. (6.64) is not a relation among the variables. So we have totally nine equations and nine variables, and all the variables can be fixed by the equations.

6.10.1.4 Variation of the Amount of Solute

We repeat the initial question: How we can obtain information on the temperature dependence or the pressure dependence of the equilibrium? We may now add some solute to the liquid phase. Then $n_2' \rightarrow n_2' + \Delta n_2'$. Next, we repeat the experiment. Again, the thermodynamic process is the transfer of water from the liquid phase to the gaseous phase, but now with slightly more solute in the liquid phase.

Again, we expect an equilibrium, but now with different numerical values of temperature, and chemical potential. We fix pressure at the standard pressure. The equilibrium conditions are now

$$T' + \Delta T' = T'' + \Delta T''$$
$$p' = p'' = p'''$$
$$\mu_1' + \Delta \mu_1' = \mu_1'' + \Delta \mu_1''$$

Now we inspect the difference between the first experiment and the second experiment: Forget the conditions of constant temperature and constant pressure. We will simply ask how we are coming from the first experiment to the second experiment. We will focus on the chemical potential and expand $\Delta \mu$ into a Taylor series. Then we obtain

$$\frac{\partial \mu_1'}{\partial n_2'} \Delta n_2' + \frac{\partial \mu_1'}{\partial T'} \Delta T' = \frac{\partial \mu_1''}{\partial T''} \Delta T''. \tag{6.65}$$

Now $\Delta T' = \Delta T'' = \Delta T$ and in general $\partial \mu / \partial T = -\partial S / \partial n = -\tilde{S}$. Therefore, we get from Eq. (6.65)

$$\frac{\partial \mu_1'}{\partial n_2'} \Delta n_2' = -(\tilde{S}'' - \tilde{S}') \Delta T. \qquad (6.66)$$

We recall that in an ideally diluted solution

$$\mu_1'(T', n_1', n_2') = RT' \ln \frac{n_1'}{n_1' + n_2'}.$$

Thus, we finally obtain from Eq. (6.66)

$$RT' \frac{\Delta n_2'}{n_1' + n_2'} = (\tilde{S}'' - \tilde{S}') \Delta T.$$

Since at $\Delta n_2' \to 0$ also $\Delta T \to 0$, we can write $\Delta n_2' = n_2'$ and further with $\tau = \Delta T$

$$RT' x_2' = \Delta \tilde{S} \tau. \qquad (6.67)$$

The entropy and the temperature are energy-conjugated variables, in the same way as the volume and the pressure. Thus, Eq. (6.67) resembles somehow the osmotic equation in terms of energy-conjugated variables.

6.10.1.5 General Variation of the Arguments

There is a still more elegant and more general way to find variations of the equilibrium with the parameters or arguments of the equilibrium conditions. We emphasize that the embedding of the system into a thermostat and into a manostat does not mean that the temperature and the pressure are fixed forever. The embedding means that we have to conduct a single experiment at constant pressure and temperature, but afterward we can readjust the thermostat and the manostat.

We can find all possible variations by using the equations that fix the equilibrium conditions in Eq. (6.64) which we repeat here in differential form:

$$\begin{aligned} dT' &= dT'' = dT \\ dp' &= dp'' = dp \\ d\mu_1' &= d\mu_1'' = d\mu_1 \end{aligned} \qquad . \qquad (6.68)$$

We have not given the connection of these equations with the external values prescribed by the thermostat and the manostat. If a certain variation is required, we must adjust the thermostat and the manostat appropriately to ensure the condition of equilibrium. Otherwise, there would not be equilibrium.

We form the Gibbs–Duhem equations for the system. However, to be independent on the size of the system, we divide them by the total number of moles of the respective phase. So the Gibbs – Duhem equations are

$$0 = \tilde{S}'(T', p')\mathrm{d}T' - \tilde{V}'(T', p')\mathrm{d}p' + x_1'\mathrm{d}\mu_1' + x_2'\mathrm{d}\mu_2'$$
$$0 = \tilde{S}''(T'', p'')\mathrm{d}T'' - \tilde{V}''(T'', p'')\mathrm{d}p'' + x_1''\mathrm{d}\mu_1'' + x_2''\mathrm{d}\mu_2''. \tag{6.69}$$

We did not take care on the equilibrium conditions. Thus, the set in Eq. (6.69) is valid in general for each phase. We have for the phase (')

$$\tilde{S}' = \frac{S'}{n_1' + n_2'}, \quad \tilde{V}' = \frac{V'}{n_1' + n_2'}.$$

Phase (")

$$\tilde{S}'' = \frac{S''}{n_1''}, \quad \tilde{V}'' = \frac{V''}{n_1''}.$$

holds similarly. So in the phase ("), we are facing the molar entropy and the molar volume of the pure components, whereas in the phase (') we are dealing with the molar quantities per mol mixture. Note that in the phase (") $x_1'' = 1$ and therefore $x_2'' = 0$.

Subtracting both equations in Eq. (6.69) and using Eq. (6.68), we obtain

$$0 = \Delta\tilde{S}\mathrm{d}T - \Delta\tilde{V}\mathrm{d}p + \mathrm{d}\mu_1' - x_1'\mathrm{d}\mu_1' - x_2'\mathrm{d}\mu_2'. \tag{6.70}$$

In Eq. (6.70), we have abbreviated $\Delta\tilde{S} = \tilde{S}'' - \tilde{S}'$ and $\Delta\tilde{V} = \tilde{V}'' - \tilde{V}'$. Further, we have suppressed the index of the phase for temperature and pressure.

We form now the derivative with respect to x_1' and comparing with Eq. (2.45), we observe that the last two terms in Eq. (6.70) cancel:

$$0 = \Delta\tilde{S}\frac{\partial T}{\partial x_1'} - \Delta\tilde{V}\frac{\partial p}{\partial x_1'} + \frac{\partial\mu_1'}{\partial x_1'}. \tag{6.71}$$

Note that for an ideal behavior

$$\frac{\partial\mu_1'}{\partial x_1'} = \frac{RT}{x_1'}.$$

So we can rewrite Eq. (6.71) as

$$0 = \Delta\tilde{S}\mathrm{d}T - \Delta\tilde{V}\mathrm{d}p + \frac{RT}{x_1'}\mathrm{d}x_1'. \tag{6.72}$$

Equation (6.72) is now very general. If we keep x_1' as constant, we get a generalized Clausius – Clapeyron equations. If we keep the temperature constant, we get the Raoult's law, and if we keep the pressure constant, we get the ordinary law of ebullioscopy, from where we started.

Example 6.4. We start with the derivation of Raoult's Law. At constant temperature and if $\tilde{V}'' \gg \tilde{V}'$ and if the gaseous phase ($''$) behaves like an ideal gas, we can transform Eq. (6.72) into

$$RT \frac{dp''}{p''} = RT \frac{dx_1'}{x_1'},$$

which integrates to

$$p''(T, x_1') = C(T)x_1'.$$

We find the integration constant that is actually a function of the temperature by setting $x_1' = 1$ to be $C(T) = p''(T, 1)$. Thus, Raoult's law emerges as

$$p''(T, x_1') = x_1' p''(T, 1). \tag{6.73}$$

$p''(T, 1)$ is the vapor pressure of two coexisting phases consisting of pure solvent (1). We can deduce the temperature dependence of the vapor pressure from Eq. (6.72), equating at constant composition, which yields

$$\Delta \tilde{S} dT = \Delta \tilde{V} dp.$$

Again, assuming that the liquid phase is negligible to the gaseous phase and that the gaseous phase is an ideal gas, we come to

$$\Delta \tilde{S} dT = \frac{RT}{p''} dp''.$$

We complete the usual derivation by introducing the molar heat of vaporization $\Delta \tilde{H}_{vap} = T \Delta \tilde{S}$:

$$\frac{\Delta \tilde{H}_{vap}}{RT^2} dT = \frac{dp''}{p''}.$$

Integration results finally in the *Clausius – Clapeyron*[7] equation:

$$p''(T, 1) = C \exp\left(-\frac{\Delta \tilde{H}_{vap}}{RT}\right). \tag{6.74}$$

---

[7] Benoit Paul Emile Clapeyron, born Feb. 26, 1799, in Paris, France, died Jan. 28, 1864, in Paris, France.

Combination of Eqs. (6.73) and (6.74) gives the full general relation between vapor pressure, mole fraction, and temperature of coexistence of a two-component liquid – gas system, the solute being not volatile at all:

$$p''(T, x_1') = x_1' C \exp\left(-\frac{\Delta \tilde{H}_{vap}}{RT}\right). \tag{6.75}$$

If we do not restrict ourselves on a gaseous phase and a liquid phase, we could also get the laws of cryoscopy, etc.   □

6.10.2 Steam Distillation

When in a system the components are essentially immiscible, then the vapor pressures of the individual components add up, and both components distill in the ratio of their vapor pressures. Most common is the use of water as main component; therefore, this particular process is addressed as steam distillation. Steam distillation is a common method for the purification of low volatile substances. Of course, the components should not react in the distillation process. The vapor pressures of some components are shown in Fig. 6.14.

Fig. 6.14 Vapor pressures of some components

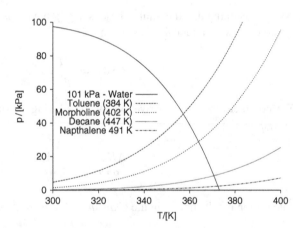

Since water is the mainly used component for volatilization, in Fig. 6.14 the vapor pressure of water is plotted as the complement to atmospheric pressure, i.e. $101 \text{ k Pa} - p_{vap}$. In this way, when the complement line for the water pressure cuts the line of the vapor pressure of the respective substance, the vapor pressures add up to atmospheric pressure [16, 17, p. 709].

6.10.3 Vapor Pressure Osmometry

In vapor pressure osmometry, the system (′) is a droplet of a volatile liquid (1) in which a nonvolatile solute (2) is dissolved. The system (′) can exchange energy of compression with the gas phase (″) that consists entirely of the solvent. Further, solvent (1) may move freely from the liquid phase into the gas phase. The gas phase is in contact with a thermostat and a manostat. We address this as a combined system (‴). Thus, the conditions of equilibrium are

$$\mu_1'(T', p', n_1', n_2') = \mu_1''(T'', p'')$$
$$p' = p'' = p'''$$
$$T'' = T'''$$

The temperature of the thermostat T''' and the pressure of the manostat p''' are constant. We have expressed the chemical potential as a function of intensive variables. Since the chemical potential of the system (″) consists of a single component, it can no longer be dependent on its composition. Further, we want to keep the temperature and the pressure of the system (″) constant, therefore

$$\mu_1''(T'', p'') = C.$$

We cannot make any statement on the temperature of the liquid phase T'. We expand μ_1' as a function of temperature, pressure, and mole fraction and get

$$d\mu_1'(T', p', n_1', n_2') = \frac{\partial \mu_1'}{\partial T'} dT' + \frac{\partial \mu_1'}{\partial p'} dp' + \frac{\partial \mu_1'}{\partial n_1'} dn_1' = 0,$$

and further

$$d\mu_1'(T', p', n_1', n_2') = -\frac{\partial S'}{\partial n_1'} dT' + \frac{\partial V'}{\partial n_1'} dp' - RT' dx_2' = 0.$$

Recall that according to Eq. (1.17), $\partial \ln x_1(n_1, n_2)/\partial n_1 = n_2/(n_1(n_1 + n_2))$ and $\partial x_1/\partial n_1 = n_2/(n_1 + n_2)^2$, further $dx_1/x_1 = d \ln x_1 = d \ln(1 - x_1) \approx -dx_2$.
   Integration at small temperature changes results in

$$-\Delta T' \frac{\partial S'}{\partial n_1'} = RT x_2'. \tag{6.76}$$

We have to make an assumption what $\frac{\partial S'}{\partial n_1'}$ is. This term is the increase of entropy, when 1 mol of solvent enters into the liquid phase. This process is associated with the phase transfer from gaseous to liquid.
   If we think that all the entropy is set free on condensation in the liquid phase, then this expression is the total entropy of condensation, which is just the reverse

to the entropy of vaporization. In fact, thermodynamics cannot answer the question, whether all of the entropy of condensation would be donated to the liquid phase. But we may think that the process of condensation is occurring near the surface and the entropy is distributed according to the ratio of the thermal conductivities of gaseous phase and the liquid phase. Since gases have a low thermal conductivity in comparison to liquids, we may justify the assumption. Again, Eq. (6.76) is very similar to Eq. (6.67) derived form ebullioscopy. With

$$\frac{\partial S'}{\partial n_1'} = -\frac{\Delta \tilde{H}_{vap}}{T},$$

we obtain

$$\Delta T' = \frac{RT^2}{\tilde{H}_{vap}} x_2'.$$

Note that $\Delta T'$ is the increase in temperature, when solute is added up to a mole fraction of x_2', but this temperature increase is also the difference in temperature to the gaseous phase.

In the derivation of colligative properties, it is common to state that the chemical potentials of the phases those components are in equilibrium are equal, i.e., $\mu' = \mu''$. The next step of the derivation, a change of some external parameters is performed. During the change, still equilibrium must be maintained. This is mathematically expressed in terms of the total differential, namely as $d\mu' = d\mu''$. Then, the total differential is expanded into an appropriate series of partial differentials and final adjustments are performed.

Here we show an alternative derivation of the equations of vapor pressure osmometry, using the approach of Taylor expansion of the chemical potential.

All the assumption in the main section apply. However, we will make a small change in the notation. The temperature of the gas phase, previously T'', be will denoted simply as T and the temperature of the liquid phase, previously T', be will denoted as $T + \Delta T$. Thus ΔT is the difference in temperature of liquid phase and gas phase. Note that the liquid phase is warmer than the gas phase.

We consider once more the phase transition of a molecule of the volatile component in detail. The explanation is rather pictorial, as in terms of thermodynamics, the molecules are not known. Moreover, we ignore that single molecules have individual properties and the bulk properties are averages of the individual properties of a molecule.

When the volatile component molecule is in the bulk gas phase, it has the temperature of the bulk gas phase. When the molecule approaches to the phase border, actually there is a temperature gradient. The molecule heats up to the temperature of the liquid phase and condenses just at this temperature. Therefore, just at the process of condensation, the chemical potential of the gaseous phase at the temperature of the liquid phase is equal to the chemical potential of the liquid phase at the bulk temperature of the liquid phase.

Using this model of phase transfer, we have in the general notation $\mu(T, p, n_1, n_2)$ for the chemical potential in equilibrium:

$$\mu_1''(T + \Delta T, p, n_1'', 0) = \mu_1'(T + \Delta T, p, n_1', n_2').$$

We expand now in a Taylor series to get

$$\mu_1''(T + \Delta T, p, n_1'', 0) = \mu_1''(T, p, n_1'', 0) - S'' \Delta T$$

and

$$\mu_1'(T + \Delta T, p, n_1', n_2') = \mu_1'(T, p, n_1', 0) - S' \Delta T + RT \ln x_1'.$$

Thus we have normalized the process to the bulk temperature of the gas phase and a single component (1). Importantly, the chemical potentials at this normalized state refer still to an equilibrium state and they will cancel therefore. Note that the chemical potential for a single component system is not dependent on the mol number, or in other words on the size of the phase.

Combining and using the usual approximation for small x_1, we get

$$\Delta T = \frac{RT}{\Delta S_{vap}} x_2' = \frac{RT^2}{\Delta H_{vap}} x_2'.$$

The alternative derivation shows that it is important to set up the actual conditions in that equilibrium is established. Using in the actual model of phase transfer, using the bulk temperature of the gas phase would give unusual equations. It should be emphasized that the conditions of vapor phase osmometry introduce some questionable conditions, such as the forbidden heat transfer between liquid and gas phases.

Actually, some constraints are introduced that will not hold in general. In fact, we would expect that due to heat conduction, the liquid phase would be cooled down. As a consequence, the chemical potential is changed in this way that additional gas phase will condense in order to heat the droplet. However, the droplet will become more diluted and bigger in size. Eventually, the droplet will fall down from the sensor. The discussion is not as serious. For instance, in membrane osmometry, we have constraints of similar kinds. Actually we cannot fabricate membranes that are really semipermeable walls and no walls that can withstand any pressure.

We emphasize that vapor pressure osmometry is highly analogous to membrane osmometry from the view of thermodynamics. The difference to membrane osmometry is that in the latter case instead of a barrier of thermal energy, a barrier of volume energy is built up. Further, the transfer of the solvent is not accompanied with a phase transfer.

Fig. 6.15 Design for
membrane osmometry. The
bent arrow suggests a transfer
of solvent via the vapor phase
instead through the membrane

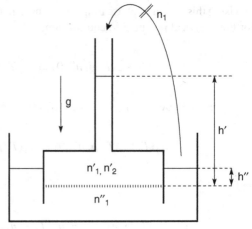

6.10.4 Membrane Osmometry

Usually treatments of membrane osmometry do not refer to the height of a liquid
column that brings the pressure to the membrane under gravity. We take care of
this aspect. However, we will not deal with capillary effects. We show a possible
experimental design for membrane osmometry in Fig. 6.15. The apparatus consists
of two subsystems: System (′) is a solvent (1) with a solute (2) that cannot pass the
membrane, and system (″) consists of a pure solvent (1). The systems are separated
by a semipermeable membrane. The whole design is embedded in a thermostat and
operated in vacuum. We will assume that the solvent has a vapor pressure that does
not contribute to the total pressure that is exerted to the membrane by gravity that is
pointing downward.

The pressure exerted due to gravity by the system (′) is $p' = \rho'gh'$, where ρ' is
the density of the solution. The back pressure due to the system (″) is $p'' = \rho''gh''$,
where ρ'' is the density of the pure solvent. We have the equilibrium conditions

$$T' = T'' = T'''$$
$$\mu_1' = \mu_1'' = C$$

Equal temperatures guarantee that there is no entropy flow, and equal chemical
potentials guarantee that there is no mass flow across the membrane. We did not
make any statement on the pressure. If we would fix the volumes of the two subsystems, then even in the absence of gravity a pressure difference would develop.
We must fix the position of the membrane. If we would not fix the position of the
membrane, a pressure difference would tend to move the membrane and the volume
of the solution would try to increase in the cost of the volume of the pure solute.
This process would take place, even when the chemical potentials on both sides of
the membrane are equal. Thus, a pressure difference causes a bulk expansion and

an associated flow through the membrane, if the membrane is not fixed. This phenomenon is different from a molecular flow caused by a difference of the chemical potential.

In a quite similar procedure as shown in the section vapor osmometry, we can show that the osmotic pressure Δp is

$$\Delta p \tilde{V} = x_2' RT. \tag{6.77}$$

Equation (6.77) is known as the *van't Hoff*[8] equation, who pointed out already the analogy between solutions and gases, because the form resembles much to the ideal gas law [18, 19]. By the way, in this highly interesting paper, *van't Hoff* points out relations between osmotic pressure, vapor pressure depression, freezing point depression, and boiling point elevation, i.e., what we address nowadays in short as colligative properties.

In the limiting case of $\rho' = \rho'' = \rho$, i.e., a highly diluted solution, the pressure expressed as the difference of the heights $\Delta h = h' - h''$ will be modifying Eq. (6.77) as

$$\Delta h \rho g \tilde{V} = x_2' RT. \tag{6.78}$$

We can rewrite Eq. (6.78) also as

$$\Delta h M g = x_2' RT, \tag{6.79}$$

where M is the molecular weight of the solvent.

We will now investigate a volatile solvent, and we ask whether it is necessary to block the contact of the solvent of system ($''$) with the solution in system ($'$) via the vapor phase as shown by the bent arrow in Fig. 6.15.

The vapor pressure of the pure solvent is p_v'', and the vapor pressure of the solution is $p_v' = (1 - x_2')p_v''$. We emphasize that we are talking about the vapor pressure at the phase border. Thus more precisely, p_v'' appears at height h'' and p_v' appears at height h'. Or else, we rewrite the vapor pressure as a function of the height, i.e.,

$$p_v'(h') = (1 - x_2')p_v''(h''). \tag{6.80}$$

Further, we have introduced the subscript (v), in order not to confuse the vapor pressure with the osmotic pressure. Due to the reduction in the vapor pressure of the solution, pure solvent should condense into the column from the bath, system ($''$). This would be sound if the system would operate in the absence of gravity. However, due to gravity, the pressure is decreasing with increasing altitude. The barometric formula says that the pressure reduction as a function of height is

---

[8] Jacobus Henricus van't Hoff, born Aug. 30, 1952, in Rotterdam, Netherlands, died Mar. 1, 1911, in Berlin, Germany.

$$p_v''(h') = p_v''(h'')\exp\left(-\frac{Mg\Delta h}{RT}\right) \approx 1 - p_v''(h'')\frac{Mg\Delta h}{RT}. \tag{6.81}$$

Recall that $p_v'(h')$ in Eq. (6.80) is the vapor pressure of the solution at height h', and $p_v''(h')$ in Eq. (6.81) is the vapor pressure of the pure solvent or the other osmotic half cell corrected to the same height h' by the barometric formula. In equilibrium, just these vapor pressures should be equal. Actually, by the relation

$$p_v'(h') = p_v''(h'),$$

we arrive again at the van't Hoff equation in the form of Eq. (6.79). Mg/RT is typically $0.1 \times 9.81/8.13 \times 300 = 4 \times 10^{-4}\,\mathrm{m}^{-1}$, for a gas with $M = 100$ g mol^{-1}. For air, we would have $\Delta p/\Delta h \approx 0.09$ tor/m. We want to call Mg/RT the altibaric constant. This is a reciprocal length and certainly in general not a constant, because it depends on the molecular weight of the solvent and because the earth acceleration g is dependent on the altitude. So we can call it the altibaric yardstick.

Eventually, we summarize the main results of this section. If the levels of the liquids are at the same height, then the barometric pressure is the same. But the vapor pressure of the solution is less. Therefore, the solution will be diluted by condensation and the level increases. In this way, the vapor pressure increases and the barometric pressure decreases. This goes on until the pressure is in equilibrium. This is at

$$x_2' = \frac{Mg\Delta h}{RT}. \tag{6.82}$$

Equation Eq. (6.82) conforms to Eq. (6.79). Thus, we can conclude that via the vapor phase the same equilibrium would be established than via the membrane. In other words, the membrane is not necessary, if the solvent is sufficiently volatile. We call this procedure the altibaric method. In practical measurements, the establishment of the equilibrium via the membrane is much faster than the establishment via the gas phase. Therefore, classical membrane osmometry is still preferred.

6.10.5 Donnan Equilibrium

The *Donnan*[9] equilibrium is related to osmosis. However, it deals with charged particles. Therefore, not the chemical potential, but rather the electrochemical potential is the starting point to calculate the equilibrium.

---

[9] Frederick George Donnan, born 1870 in Colombo, Ceylon, died 1956 in Sittingbourne, Kent, England.

6.11 Isopiestics

An apparatus for isopiestic experiments is easy to construct. Solutions of a volatile solvent with different concentrations of a nonvolatile solute are placed in an apparatus looking like a receiver of a distillation equipment as shown in Fig. 6.16.

Fig. 6.16 Basic setup of an isopiestic experiment

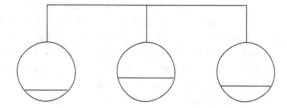

The apparatus is put in a thermostat. The solvent will be transferred via the gaseous phase until all the chemical potentials are equal. A more sophisticated experimental technique of the isopiestic method is the following [20, 21]. Sample cups with solutions are placed in a closed chamber where all solutions share the same vapor phase. The vapor space of the chamber is evacuated to contain only water vapor. Solvent is transferred through the vapor phase. The chamber containing the solutions is kept at isothermal conditions at specific temperature until no more change in the concentration of the solution is observed; thus, thermodynamic equilibrium is reached. When the solutions are in thermodynamic equilibrium, then

$$\mu_1'' = \mu_1' = \mu_1''' = \ldots .$$

Here μ_1'' is the chemical potential of the volatile solvent in the vapor phase, $\mu_1', \mu_1''', \ldots$ are the chemical potentials of the solvent in the various cups, filled with the samples, and the standard solutions, respectively. After establishment of the isopiestic equilibration, the solutions are reweighed. From the equilibrium concentrations, the activity coefficients can be calculated.

6.12 Equilibrium of Distillation

Consider a system consisting of a distillation flask, a condenser, and a receiving flask. In the distillation flask, there is a volatile component (1) that is subject to the distillation process and a nonvolatile component (2) dissolved in the volatile component at a mole fraction x_2. We have $1 = x_1 + x_2$. The distillation flask is kept at some temperature $T + \Delta T$.

In the condenser, a temperature gradient is maintained from $T + \Delta T$ to T. When the volatile component moves from the distillation flask to the receiving flask, it is ideally cooled down to temperature T of the pure volatile liquid component in the receiving flask and vice versa.

The distillation apparatus is not connected to the atmosphere. For this reason, the vapor pressure in the gas phase is that of the vapor pressure of the gas phase above the liquids in both compartments, i.e., the distillation flask and the receiving flask.

Clearly, the distillation process will start when the temperature in the distillation flask will be raised in such a manner that the vapor pressure of the volatile component in which the nonvolatile component is dissolved will exceed the vapor pressure of the same but pure volatile component in the receiver flask. Otherwise, a flow back from the receiver flask to the distillation flask will be expected. The condition of equal vapor pressure arises from the fact that otherwise the subsystems will expand or shrink through the cooler, which is the natural border of the subsystems. We emphasize that there is no free exchange of entropy or heat in the system. The condenser forces a temperature gradient.

In the distillation flask, the vapor pressure is p_D at a mole fraction of x_1 of the volatile component and a temperature of $T + \Delta T$. In the receiver flask, the vapor pressure is p_R at a mole fraction of the volatile component of $x_1 = 1$ and a temperature of T. In equilibrium, both vapor pressures will be equal, i.e.,

$$p_D(x_1, T + \Delta T) = p_R(1, T). \tag{6.83}$$

The vapor pressure is a function of mole fraction and temperature. Using Raoult's law, Eq. (6.83) can be converted into

$$p_D(1, T + \Delta T)x_1 = p_R(1, T). \tag{6.84}$$

The equation of Clausius and Clapeyron in integrated form for the dependence of the vapor pressure p on temperature T reads as

$$p = C \exp\left(-\frac{\Delta \tilde{H}_v}{RT}\right).$$

Here, $\Delta \tilde{H}_v$ is the enthalpy of vaporization that is treated as a constant for a small range of temperature, and C is some constant of integration. Thus Eq. (6.84) turns into

$$x_1 = \exp\left(-\frac{\Delta \tilde{H}_v}{RT}\right)\exp\left(+\frac{\Delta \tilde{H}_v}{R(T + \Delta T)}\right).$$

Taylor expansion of the exponential terms around $\Delta T = 0$ yields

$$x_1 = 1 - x_2 = 1 - \frac{\Delta \tilde{H}_v}{RT^2}\Delta T.$$

Further rearrangement results in

$$\frac{\Delta \tilde{H}_v}{T} \Delta T = \Delta \tilde{S}_v \Delta T = x_2 RT. \tag{6.85}$$

Here, $\Delta \tilde{S}_v$ is the molar entropy of vaporization. Equation (6.85) seems to be closely related to the ebullioscopy law and the law emerging in vapor pressure osmometry. Note that the way of derivation here runs via *Raoult*'s law and the Clausius – Clapeyron equation, whereas ebullioscopy is derived usually via the chemical potential. Moreover, recalling *van't Hoff*'s law of osmometry, $\Pi \tilde{V} = x_2 RT$, we can relate Eq. (6.85) easily to osmometry, arriving at

$$\Delta \tilde{S}_v \Delta T = \Pi \tilde{V}. \tag{6.86}$$

Numerical evaluation of Eq. (6.86) reveals why measurements of the osmotic pressure Π are more sensitive than measurements of the elevation of the boiling point.

There is still another analogy of van't Hoff's law of osmometry and Raoult's law. According to Raoult's law, we can express the reduction of the vapor pressure of a volatile solvent (1) Δp by the dissolution of a nonvolatile component (2) proportional to the vapor pressure of the pure solvent p_0 and the mole fraction of the nonvolatile component x_2, as

$$\Delta p = p_0 x_2.$$

If the gas phase follows the ideal gas law, we can replace p_0 by $RT/\tilde{V}$ and get

$$\Delta p \tilde{V} = x_2 RT. \tag{6.87}$$

In contrast to van't Hoff's law of osmometry, in Eq. (6.87), $\tilde{V}$ is the molar volume of the gas.

6.12.1 Ebullioscopic Constant

The ebullioscopic constants may be derived from Eq. (6.85). For small x_2

$$x_2 = \frac{n_2}{n_1 + n_2} \approx \frac{n_2}{n_1} = \frac{n_2}{w_1} M_1 = m_2 M_1.$$

Here, M_1 is the molecular weight of the solvent (1), in the standard units kg mol^{-1}, and m_2 is the molality of the solute (2) in mol solute per kg solvent.

Thus, Eq. (6.85) turns into

$$\Delta T = \frac{M_2 RT}{\Delta \tilde{S}_v} m_2 = \frac{RT}{\Delta \hat{S}_v} m_2 = K_b m_2. \tag{6.88}$$

$\Delta \hat{S}_v$ is the specific entropy of vaporization, i.e., the entropy of vaporization per kg solvent and K_b is the ebullioscopic constant.

If the molar entropy of vaporization can be treated as a constant, as *Trouton*'s rule suggests, then the relative elevation of the boiling point, $\Delta T / T$, is universally related to the mole fraction x_2 according to Eq. (6.85).

On the other hand, Eq. (6.88) allows to calculate the ebullioscopic constant by probing the elevation of the boiling point with a known solute with the aid of Trouton's rule. The dimensionless quantity $M_1 T_b / K_b$ in Table 6.2 equals to $\Delta \tilde{S}_v / R$.

Table 6.2 Ebullioscopic constant and Trouton's rule

Solvent	$T_b/[K]$	$M_1/[\,kg\,mol^{-1}]$	$K_b/[\,K\,kg\,mol^{-1}]$	$M_1 T_b / K_b$
Acetic acid	390	0.06005	3.08	7.6
Carbon disulfide	319	0.07613	2.34	10.4
Carbon tetrachloride	350	0.15382	5.03	10.7
Cyclohexane	354	0.08416	2.79	10.7
Benzene	353	0.07811	2.53	10.9
Chloroform	334	0.11938	3.63	11.0
Diethyl ether	308	0.07412	2.02	11.3
Camphor	477	0.15223	5.95	12.2
Water	373	0.01802	0.51	13.1
Ethanol	351	0.04607	1.07	15.1

Actually, the ebullioscopic behavior of various solvents including acetic anhydride and acetic acid has been determined at various pressures using benzil and diphenylamine as solute [22].

The results of these determinations have been compared with other data, particularly the degree of association of these solvents. Acetic anhydride shows a slight association, which is in accordance with the slight divergence from the normal value as shown by Trouton's rule.

Eventually, we annotate that the experimental technique of ebullioscopy was greatly improved by the invention of the Beckmann thermometer [23].

6.13 Unlike Energy Forms

In general, equilibrium between unlike energy forms involves an exchange $dU = \xi\,dX + \upsilon\,dY$. For example, in the case of a soap bubble, $dU = \sigma\,dA - p\,dV$. It is not possible to put $\sigma = p$. The two quantities have different physical dimensions.

It is possible to find directly equilibrium relations, if we introduce transformation factors with a certain physical dimension. We show here a slightly complicated, but more general, method as necessary for two variables. We transform the extensive variable so that it will become dimensionless. Then the intensive variable will be transformed to an energy. Taking the example introduced above, $dU = \sigma\,dA - p\,dV$ changes in

$$dU = \sigma \lambda_A d(A/\lambda_A) - p \lambda_V d(V/\lambda_V),$$

with $[\lambda_A] = [\text{m}^2]$ and $[\lambda_V] = [\text{m}^3]$. We introduce new names for the dimensionless extensive variables

$$dU = \sigma \lambda_A d\check{A} - p\lambda_V d\check{V}.$$

We choose the transformers so that $d\check{V} = d\check{A}$. So in equilibrium, $\sigma = p\frac{\lambda_V}{\lambda_A}$. The ratio $\frac{\lambda_V}{\lambda_A}$ has the physical dimension of a length and is dependent on the process, here in particular on the geometry of the system. In fact, it is more easy to derive the relation for a soap bubble, which is a sphere by using

$$dV = 4r^2\pi\, dr \quad \text{and} \quad dA = 16r\pi\, dr$$

(inner and outer surface) and substituting directly in the energy equation. Similarly, gravitational energy can be exchanged with surface energy in the case of a weightless lamella that changes its surface by a moving weight under gravity. Here

$$dU = \sigma\, dA + gd(mh) = 0.$$

Here we find the transformation factors $[\lambda_A] = [\text{m}^2]$ and $[\lambda_{mh}] = \text{kg\,m}$. The ratio of these scale factors has the physical dimension of

$$\frac{[\lambda_{mh}]}{[\lambda_A]} = [\text{kg\,m}^{-1}].$$

This is obviously the mass of the weight and the width of the lamella.

6.13.1 Stiff Systems

Equilibrium is established only when the intensive variables corrected by the transformation factors are equal. We may not conclude what will happen if the system is not in equilibrium, whether it will move to an equilibrium state or rather away. We may think that a system is more far from equilibrium as the difference of intensive variables is more large. In the example above, we may take σ, the width of the lamella w, the mass m, and g as constants. In this case, equilibrium is at $\sigma w = gm$. Under laboratory conditions, we may easily change the width of the lamella w and the hanging mass m, but the other parameters are difficult to change. In this case, equilibrium is established only in one singular point. If only one parameter does not fit, equilibrium is not established. On the other hand, in the case of the soap bubble, inside there is a compressible gas. We have the well-known Kelvin's equation (for a double surface)

$$\Delta p = \frac{\sigma}{4r}.$$

Imagine a soap bubble pending on a drinking straw. Certainly you can reduce the pressure, by allowing some gas to escape. We expect intuitively, if the pressure is decreased, then the bubble will become smaller. We will now inspect Kelvin's equation. Let again σ be constant. The system could move in equilibrium again in this case, if the radius is increased. So the bubble will become bigger. But there is another possibility, if the bubble is very small, smaller than the diameter of the pipe, again the radius increases. There is another pitfall. If we allow escaping certain amounts of the system, then also the mol numbers and the associated chemical potential change.

References

1. Massieu, M.F.: *Sur les fonctions des divers fluides*. Compt. Rendus Acad. Sci. **69**, 858–862 (1869)
2. Poincaré, L.: The New Physics and its Evolution. The International Scientific Series. D. Appleton and Company, New York (1909). Translation of *La physique moderne, son évolution*, Project Gutenberg eBook, [eBook #15207]
3. Callen, H.B.: Thermodynamics. John Wiley, New York (1961)
4. Barkai, E., Silbey, R.J.: Fractional Kramers equation. J. Phys. Chem. B **104**, 3866–3874 (2000)
5. Meurs, P., van den Broeck, C., Garcia, A.: Rectification of thermal fluctuations in ideal gases. Phys. Rev. E **70**(051109), 1–15 (2004)
6. Dejak, C., Pastres, R., Pecenik, G., Polenghi, I.: Classical thermodynamic-statistical approach of open systems deviating from maximum entropy. In: C. Rossi, S. Bastianoni, A. Donati, N. Marchettini (eds.) Tempos in Science and Nature: Structures, Relations, and Complexity, *Annals of the New York Academy of Sciences*, vol. 879, pp. 400–405. Academy of Sciences, New York (1999)
7. Smith, H.L., Waltman, P.: The Theory of the Chemostat: Dynamics of Microbial Competition, *Cambridge Studies in Mathematical Biology*, vol. 13. Cambridge University Press, Cambridge (1995)
8. Berberan-Santos, M.N., Bodunov, E.N., Pogliani, L.: On the barometric formula. Am. J. Phys. **65**(5), 404–412 (1997)
9. Trupp, A.: Energy, entropy: On the occasion of the 100th anniversary of Josef Loschmidt's death in 1895: Is Loschmidt's greatest discovery still waiting for its discovery? Phys. Essays **12**(4), 614–628 (1999)
10. Gibbs, J.W.: On the equilibria of heterogeneous substances. In: The Collected Works, vol. 1. Yale University Press, New Haven, CT (1948 [reprinted])
11. Bohren, C.F., Albrecht, B.A.: Atmospheric Thermodynamics, pp. 164–171. Oxford University Press, Oxford (1998)
12. DeCaria, A.: Atmospheric thermodynamics lesson 10 – the energy minimum principle. ESCI 341 [electronic] http://snowball.millersville.edu/~adecaria/ (2005)
13. Schaefer, C.: Theorie der Wärme, Molekular-kinetische Theorie der Materie, *Einführung in die theoretische Physik*, vol. 2, 3rd edn. De Gruyter, Berlin (1958)
14. Lide, D.R., Kehiaian, H.V.: CRC Handbook of Thermophysical and Thermochemical Data. CRC Press, Boca Raton, FL (1994)
15. Atkins, P.W.: Physical Chemistry, 4th edn. Oxford University Press, Oxford (1990)
16. Badger, W.L., McCabe, W.L.: Elements of Chemical Engineering, *Chemical Engineering Series*, 2nd edn. McGraw-Hill Book Company, Inc., New York, London (1936)
17. Vauck, W.R.A., Müller, H.A.: Grundoperationen chemischer Verfahrenstechnik, 8th edn. VEB Deutscher Verlag für Gundstofindustrie, Leipzig (1989)

18. van't Hoff, J.H.: Die Rolle osmotischen Drucks in der Analogie zwischen Lösungen und Gasen. Z. Phys. Chem. **1**(9), 481–508 (1887)
19. van't Hoff, J.H.: The function of osmotic pressure in the analogy between solutions and gases. Philos. Mag. 5 **56**(159), 81–105 (1888). Translated by W. Ramsay
20. Park, H., Englezos, P.: Osmotic coefficient data for Na2SiO3 and Na2SiO3 —NaOH by an isopiestic method and modeling using Pitzer's model. Fluid Phase Equilib. **153**(1), 87–104 (1998)
21. Zhou, J., Chen, Q.Y., Zhou, Y., Yin, Z.L.: A new isopiestic apparatus for the determination of osmotic coefficients. J. Chem. Thermodyn. **35**(12), 1939–1963 (2003)
22. Beckmann, E., Liesche, O., von Bosse, J.: Ebullioscopic behaviour of solvents at different pressures. 11: Acetic anhydride and acetic acid. Z. Phys. Chem. **88**, 419–427 (1914)
23. Beckmann, E.: Modification of the thermometer for the determination of molecular weights and small temperature differences. Z. Phys. Chem. **51**, 329–343 (1905)

Chapter 7
The Phase Rule

The phase rule by the mathematician and physicist *Gibbs*[1] was commented by *Fuller*[2] in this way:

> The chemist Willard Gibbs developed the phase rule dealing with liquid, gaseous, and crystalline states of substances, apparently not realizing that his phase rule employed the same generalized mathematics as that of Euler's topological vertexes, faces, and edges [1].

7.1 A Schematic Phase Diagram

In Fig. 7.1, the phase diagram of water is presented schematically. We emphasize that we are dealing with a single compound, or as addressed in thermodynamics, we are dealing with a system composed of one component. At low pressures, there is a region where water exists only as gas. At low pressures and low temperatures, water appears as ice. Between ice and gas, there is the liquid region. Every two regions are separated by border lines.

In the inner part of a region, we may change pressure and temperature arbitrarily, i.e., independently, and the phase does not disappear. But there is only one phase, that is, solid ice, liquid water, or water vapor, respectively.

If there should be two phases, we must choose a pair of temperature and pressure so that this tuple lies on a border line. If we want now to change either the temperature or the pressure, the two phases being still there, then we must change the temperature together with the pressure in such a way that the tuple remains on the border line.

Thus, we cannot change temperature and pressure arbitrarily independently. Otherwise, if we are moving away from the border line, one phase will disappear. There is still another unique point, the triple point. At the triple point there are ice, liquid

---

[1] Josiah Willard Gibbs, born Feb. 11, 1839, in New Haven, CT, USA, died Apr. 28, 1903, in New Haven, CT, USA.

[2] Richard Buckminster Fuller, born Jul. 12, 1895, in Milton, M., USA, died Jul. 1, 1983, in Los Angeles CA, USA.

J.K. Fink, *Physical Chemistry in Depth*,
DOI 10.1007/978-3-642-01014-9_7, © Springer-Verlag Berlin Heidelberg 2009

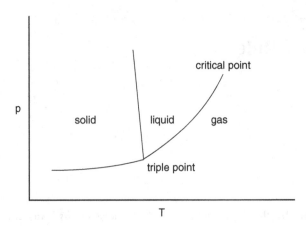

Fig. 7.1 Phase diagram of water

water, and water vapor in equilibrium. At the triple point neither temperature nor pressure may be changed.

7.1.1 Liquid – Gas Curve and Critical Point

If we are moving along the liquid – gas border line in increasing temperature and pressure, then the properties of the liquid and the gas are becoming increasingly similar. The gas becomes increasingly more dense than the liquid. Finally we arrive at the critical point. At the critical point there is no longer a difference the between liquid water and water vapor. The two phases will disappear.

If we are adjusting temperature and pressure accordingly, to move along the border line, then two phases can be watched because of their optical properties. A meniscus can be observed. At the critical point the meniscus disappears.

7.1.2 Solid – Liquid Curve

In the two-phase region of ice and liquid water, the border line is inclined to the ordinate. The curve has a negative slope. This is an exceptional behavior that is exhibited by water, and a few other substances only, e.g., glass ceramics [2] and organic polymers.

For most other compounds the solid – liquid curve has a positive slope. A consequence of this negative slope is the exceptional behavior of water, i.e., ice has a smaller density than liquid water, so ice swims on water. This is highly important for the development of life. If ice would be heavier than liquid water, ice would sink down, making the oceans full of ice.

The phase diagram of water as shown in Fig. 7.1 is not complete. It has been found out that at higher pressures there are further modifications of ice besides of ice I, up to ice VII. However, the ice V, found initially could not be confirmed by other experimenters.

In the contrary to the critical point in the case of the liquid – gas equilibrium, in the case of the solid – liquid equilibrium a critical point has never been observed at all. It is suspected, however, that there should be a critical point also for the solid – liquid equilibrium. At very high pressures, it is believed that the matter will be converted into a metallic state.

7.1.3 Geometrical Interpretation

In a system, consisting of only one component, there are two intensive variables sufficient to describe it. This is usually temperature and pressure. In general, this is not true. For example, if the system consists of large surfaces, then more variables are needed.

However, in most common situations two variables can describe the behavior of a system. For this reason, we can take a two-dimensional plot for a one-component system, as shown in Fig. 7.1. All the information can be shown on a two-dimensional plane.

7.1.3.1 Two-Phase Region

The plane is divided into two parts by a line, with geometrical dimension one. Assume that on one side of the plane is the solid state, and on the other side is the liquid state. If we want to establish a state, where both phases are present, then it is very natural that we can move along the line with the variables temperature and pressure. This line provides a functional dependence of the variables in question, i.e., pressure and temperature

$$p = f(T).$$

We can hardly imagine another situation, for geometrical reasons. If there is still another region besides solid and liquid, then we have to place another line on the plane to separate these regions. Along this new line, the same is true as for the solid – liquid border. To establish the existence of two phases, we can move along the new line.

7.1.3.2 Three-Phase Region

Now two lines, if they are not parallel, are cutting in a point. Here we think of the triple point. To establish the existence of three phases for a one-component system, we have the equations of the two lines, in the two-dimensional space

$$p = f(T), \quad p = g(T).$$

These two equations uniquely determine a point in the two-dimensional space. So for geometrical reasons, in the three-phase region it is not possible to change both temperature and pressure. Otherwise, a phase will disappear.

7.2 The Concept of Freedom

The concept of freedom plays an important role in order to understand the derivation of the phase rule. Therefore, we repeat a few terms concerning this aspect here.

7.2.1 Set of Mathematical Equations

If we have a set of mathematical equations, say

$$x + y + z = 3; \quad x - y = 0; \quad x - y + z = 1,$$

then we can solve for all the variables x, y, z unambiguously. Obviously, for each variable one equation is needed. If there are less equations then variables, say one equation is missing, then still a solution can be obtained, but we can fix one variable arbitrarily. Or we can say, we have the *freedom* to fix one variable deliberately. So the number of freedoms is $V - E = F$. Here V is the number of variables, E is the number of equations, and F is the number of freedoms, i.e., the number of variables that can be chosen by the user as he likes.

The phase rule is deduced from rather formal arguments of linear algebra pointed out above. For the individual phases, there are certain relations, i.e., equations among the intensive variables in equilibrium. These include equal temperatures, equal pressures, and equal chemical potentials. Further, there are the variables that describe the system. We chose the intensive variables and subdivide these into the chemical potentials and other intensive variables, like temperature, pressure. For each component there is a chemical potential. So the number of variables is K for the chemical potential and I for the other intensive variables in a single phase.

7.3 Derivation of the Phase Rule

There are several ways to derive the phase rule. Subsequently, we present some of the methods to derive the phase rule.

7.3.1 Original Derivation

We show here the derivation of the phase rule as done by *Gibbs* in his original paper [3, pp. 55–353]. In a multicomponent system for a homogeneous phase the

Gibbs – Duhem equation is

$$SdT - pdV + \sum_{i=1}^{i=K} n_i d\mu_i = 0. \tag{7.1}$$

We emphasize here that a prerequisite for the validity of the phase rule is that the thermodynamic system can be described fully by Eq. (7.1). This means in detail that entropy, volume, and mol numbers are sufficient as independent variables. The components should be independent as such, which implies no chemical reactions or other transformations. It is interesting to note that the occurrence of phases implies the existence of phase borders, which contribute to surface energy. This energy form is neglected. In the literature, explicitly situations are discussed, where the phase rule fails [4–6].

If there are K components, then there are $K + 2$ variables, i.e., all the μ_i and temperature T, and pressure p. However, Eq. (7.1) restricts the free variation of these variables to $K + 1$. So we have $K + 1$ independent variables. For each phase there is one equation like Eq. (7.1), thus for P phases there are totally $P(K + 1)$-independent variables. The temperatures and the pressures of the individual phases must be equal in equilibrium, so we have in addition $P - 1$ equations relating the temperatures and $P - 1$ equations relating the pressures. Also the chemical potentials must be equal in equilibrium, so for K components we have another $K(P - 1)$ equations. Totally we have $(P - 1)(K + 2)$ equations. The number of freedoms is the difference between variables and equations:

$$P(K + 1) - (P - 1)(K + 2) = K + 2 - P = F.$$

We emphasize once more that the foregoing arguments are based entirely on intensive variables. In other words, we do not worry about the size of the phases, they must be simple there.

7.3.2 Inductive Method

We use now the inductive method. Assume that we have P phases. Each component is present in each phase. Then the number of total independent intensive variables is $V(P)$. If we add one phase, then the independent intensive variables increase by $K + I - 1$. The term -1 arises because we have already merged one equation, i.e., the Gibbs – Duhem equation.

$$V(P + 1) = V(P) + K + I - 1. \tag{7.2}$$

On the other hand, if we add one phase, then we must add for each independent intensive variable one equation to couple the new phase with respect to equilibrium. So the number of equations E increases by

$$E(P+1) = E(P) + K + I. \tag{7.3}$$

We form now the freedoms $V - E$ from the Eqs. (7.2) and (7.3) and obtain

$$F(2) = F(1) - 1$$
$$F(3) = F(2) - 1$$
$$\cdots\cdots\cdots$$
$$F(P+1) = F(P) - 1$$

From this set we immediately observe

$$F(2) = F(1) - 1$$
$$F(3) = F(1) - 2$$
$$\cdots\cdots\cdots \tag{7.4}$$
$$F(P) = F(1) - (P-1)$$

Finally we must know how many freedoms has a system with one phase: $V(1) = K + I - 1$, $E(1) = 0$. We insert this in the last equation of Eq. (7.4) and are finished:

$$F(P) = K + I - P.$$

So the usual "2" in the phase rule is dismantled as the number of intensive variables needed to describe the properties other than the chemical potentials.

Note that in the usual derivation, we state that we have one variable less than the number of components because we are using the mole fractions.

On the other hand, more logically, we could state that we have for one phase $K + I$ variables and one equation, the Gibbs – Duhem equation. Then by adding a new phase, the variables increase by $K + I$ and the equations increase by $K + I + 1$. For one phase, we have then $K + I$ variables and one equation. The result would be the same.

- There may be further equations, in particular, if chemical reactions are allowed.
- The derivation assumes that each component is present in each phase.
- For each phase, there is the same number of variables needed for description.
- For each phase, there is the same number of equations. This means that all the equilibria among the phases are established.

$P + F = K + 2$ holds in particular if thermal, manal, and chemical equilibriums are established. There may be certain cases that need a modification, or positively expressed, the phase rule can be extended to other cases under which the conditions are not valid, e.g., surface energy. We show possible sets of solutions for phases, freedoms, and components in Table 7.1.

Table 7.1 Possible sets of solutions for phases, freedoms, and components

Phases P	Freedoms F	Components K	Independent variables V	Equations E
		$P + F = K + 2$	$F = V - E$	
1	2	1	2	0
2	1	1	4	3
3	0	1	6	6
1	3	2	3	0
2	2	2	6	4
3	1	2	9	8
4	0	2	12	12
1	4	3	4	0
2	3	3	8	5
3	2	3	12	10
4	1	3	16	15
5	0	3	20	20

7.4 Maximum Number of Phases

From the phase rule, we can easily obtain the maximum number of coexisting phases as we set the number of freedoms equal to zero. Thus, the maximum number of phases is

$$P_{max} = K + 2.$$

We must have certain properties to distinguish the phases, e.g., density, composition. These properties must differ for different phases. Otherwise, we could not state that there are certain phases present. We consider the case of one component, and we have two intensive parameters, the temperature T and the pressure p to characterize such properties that we address as Π. There should exist a state function of the form

$$\Pi = \Pi(T, p).$$

Since at maximum three phases should coexist with properties different from each other, these phases should have the properties Π_1, Π_2, Π_3 for a certain pair of T, p. Therefore, the state function has a solution of

$$[\Pi_1 - \Pi(T, p)] [\Pi_2 - \Pi(T, p)] [\Pi_3 - \Pi(T, p)] = 0.$$

So, if we could expand the state function in a polynomial, then the degree of the polynomial is at minimum three.

7.5 Euler's Polyhedral Formula

In topology

$$p + f = k + 2 \tag{7.5}$$

is known as the Euler's polyhedral formula [7]. To avoid confusion with the phase rule, we will now write lower case letters. The polyhedron law states that in a polyhedron there are p vertices, f faces, and k edges. In German there is even a striking similarity with the symbols because faces are Flächen, vertices are Punkte, and edges are Kanten.

For example, in a cube there are 8 vertices, 6 faces, and 12 edges. Imagine an Egyptian pyramid. It has a quadratic bottom face. At the top, four faces are coming together. If we cut the top, then a new face is created. We are loosing the top vertex and we are gaining four additional vertices and four additional edges. At the vertices gained, only three faces are in touch. We will call a vertex, whose n faces are coming together a vertex of order n. If we cutting away a vertex p of order n, we find for the vertices as function of the faces f

$$p(f + 1) = p(f) + n - 1. \tag{7.6}$$

For the edges k as function of the faces f we find

$$k(f + 1) = k(f) + n. \tag{7.7}$$

Therefore, Eq. (7.6) is analogous to Eqs. (7.2) and (7.7) is analogous to Eq. (7.3).

7.5.1 Euler – Poincaré Formula

Euler's formula was generalized by *Schläfli*[3] [8]. The *Euler – Poincaré* formula is an extension of Euler's polyhedral formula. It takes into account more complicated polyhedra with penetrating holes g that are addressed as the genus of the body, internal voids (chambers) with shells s, and loops l that are closed paths surrounding the edges. For this case

$$p + f = k + (l - f) + 2(s - g). \tag{7.8}$$

A conventional polyhedron has as many loops as surfaces, no holes but one inner void which is the volume itself that give rise to only one outer shell. Equation (7.8) reduces then to Eq. (7.5).

---

[3] Ludwig Schläfli, born Jan. 15, 1814, in Grasswil, Switzerland, died Mar. 20, 1895, in Bern, Switzerland.

7.5.2 Simplex

In Fig. 7.2, we show now the situation for dimensions till three. We denote the number of vertices for the simplex in the dimension d as $n(d, 1)$, the number of edges as $n(d, 2)$, the number of faces as $n(d, 2)$, etc. Then we observe the following scheme as given in Table 7.2. The scheme in Table 7.2 is looking like a Pascal triangle, but in the column "nil" the "1" is missing. In fact, from the construction of the simplex from dimension zero upward the rule for the number of elements is

Fig. 7.2 The simplex in zero[th], first, second and third dimension

Table 7.2 Components of the simplex in various dimensions

Dimension	nil	Vertices(p)	Edges (k)	Faces (f)	Volumes (v)
0	–	1	0	0	0
1	–	2	1	0	0
2	–	3	3	1	0
3	–	4	6	4	1

$$n(d + 1, k) = n(d, k) + n(d, k - 1). \tag{7.9}$$

In detail, we are putting always one vertex into the next higher dimension. For example, going from the line, connected with two vertices, we are coming to a triangle, if we are putting one point into the second dimension and we are drawing lines from the two end points of the one-dimensional simplex to the point being in the second dimension. By this construction, the elements are generated by

$$(1 + x)^{d+1}. \tag{7.10}$$

The coefficients of x^1 are the number of vertices (p) in the simplex of dimension d, the coefficients of x^2 are the number of edges (k), etc. On the other hand

$$(1 - x)^{d+1} = 0 \quad \text{at} \quad x = 1 \tag{7.11}$$

holds, and this is the proof [9] that

$$1 - p + k - f + \ldots = 0. \tag{7.12}$$

This holds at least for the simplex. Note that there is only one element of the highest dimension under consideration, e.g., one face in the triangle and one volume in the tetrahedron.

Note that in Eq. (7.11) the point is associated with x^1, the edge with x^2, the plane with x^3 and the volume with x^4. However, in common opinion the point is associated with zeroth dimension, the edge with the first dimension, etc.

Into the simplex, we can introduce some vertices, edges, and surfaces by certain geometrical operations, as exemplified in Fig. 7.3.

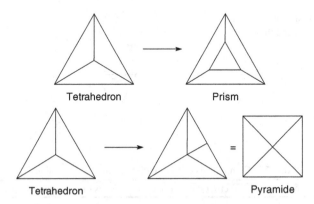

Fig. 7.3 Introduction of vertices, edges, and surfaces into a tetrahedron

We annotate that Eqs. (7.2) and (7.3) correspond geometrically in cutting off a vertex as shown in Fig. 7.3, bottom. The graphs shown in this figure are Schlegel graphs or else addressed as the skeleton of the polyhedron.

We may ask what mathematics wants to teach us here that we do not understand. Actually, remember the famous sentence *In the beginning was the Word* ... and associate the term at x^0 with the ultimate time, which precedes the creation of space.

7.5.3 Inequalities

In a polyhedron, we may denote the number of faces surrounded by three edges, as f_3, the number of faces surrounded by four edges, say f_4, etc. The total number of faces is

$$f = f_3 + f_4 + f_5 + \dots . \tag{7.13}$$

Now every edge shares two faces, therefore the number of edges is

$$k = \frac{3f_3 + 4f_4 + 5f_5 + \dots}{2}.$$

We repeat the same considerations with the vertices. A vertex connect with three edges is p_3, a vertex connected with four edges is p_4, etc. So the total number of vertices is

$$p = p_3 + p_4 + p_5 + \dots$$

and since every vertex connects two edges; the total number of edges is

$$k = \frac{3p_3 + 4p_4 + 5p_5 + \dots}{2}. \qquad (7.14)$$

Note the general inequality $3x_3 + 4x_4 + 5x_5 \dots \geq 3x_3 + 3x_4 + 3x_5 = x \dots$ for positive real x. Therefore, from Eqs. (7.13) and (7.14) we get the inequalities

$$2k \geq 3f \quad \text{and} \quad 2k \geq 3p.$$

From these inequalities together with the polyhedral formula, we get

$$p + f \geq 8.$$

The number of possible polyhedra is shown in Table 7.3. The minor diagonals represent polyhedra of equal number of edges. Note that the minor diagonal is symmetric.

Table 7.3 Number of polyhedra [10, 11]

N	$\downarrow f, p \rightarrow$	4	5	6	7	8	9	10	11	12	13
1	4	1									
2	5		1	1							
7	6		1	2	2	2					
34	7			2	8	11	8	5			
257	8			2	11	42	74	76	38	14	
2606	9				8	74	296	633	768	558	219

7.6 Polyhedral Formula and Phase Rule

Observe the similarity of the polyhedral formula with the phase rule. We emphasize that the analogy to the phase rule to the polyhedral formula is emerging just for the three-dimensional case.

7.6.1 Problem of Physical Interpretation of the Analogy

The problem of the analogy of the phase rule and the polyhedral formula can be addressed as the difficulty in finding reasonable and meaningful relations of a physical and geometrical interpretation what will happen when a new phase will appear and how a corresponding polyhedron will change accordingly to this reorganization.

Comparatively few papers are dealing with the similarity of polyhedral formula and phase rule [12–18]. We mentioned already that Eq. (7.6) is analogous to Eqs. (7.2) and (7.7) is analogous to Eq. (7.3). However, the strict analogy to the phase rule ends here because not all polyhedrons as demanded by Table 7.1 can be constructed. The minimum number of faces in the three-dimensional space is four $f = 4$ for the tetrahedron, whereas the number of freedoms in the phase rule can go down to zero.

Therefore, the analogy of the phase rule to Euler's polyhedron law is not perfect. Moreover, if we think that the edges would correspond to the equations and the faces to freedoms, or something similar, then a problem arises in that the equations would connect quantities of different physical units. Of course, by making these variables dimensionless, a solution can be pointed out.

A phase diagram of a one-component system can be plotted readily in two dimensions, with the pressure and the temperature as coordinates. If we connect the outer lines in Fig. 7.1 mutually, then we could identify a pulped tetrahedron in the figure. However, this model is not sufficient even in a binary system, as besides temperature and pressure the composition emerges as an additional variable. Mostly, common binary phase diagrams are an isobaric section of a general three-dimensional phase diagram with the mole fraction as third variable. However, the mole fractions may be different in the various coexistent phases. In the sense of intensive variables, instead of the mole fraction, the chemical potential μ is the logical pendant to temperature T and pressure p.

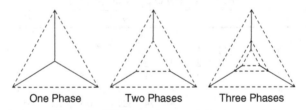

| One Phase | Two Phases | Three Phases |

Fig. 7.4 Geometrical interpretation of phases

There is a way to visualize the phases geometrically. In Fig. 7.4 a possible geometrical representation of a one-component system is given. The dashed lines, which correspond to the edges of the polyhedron, associate the parameters belonging to a certain phase. The full lines represent relations between the phases. Thus for three phases, we have three dashed triangles. The vertices connected by the full lines may be interpreted as the set of T', T'', T''; p', p'', p''; and μ', μ'', μ''. The situation is straightforward for two phases. For a single phase, the full lines are coming together into one point, which we have to define as an unreasonable case. Actually in this case there are no relations among the variables. A geometrical representation of this kind involves two types of lines, here full and dashed lines. From the philosophy of the polyhedral formula, there is simply an edge, which is a drawback. On the other hand, it is possible to construct graphs with arbitrary many phases, which is a shortcoming from the view of the phase rule.

7.6.2 Thermodynamic Surfaces

We can rather restrict ourselves to deal with the phase rule in geometry, which results from a thermodynamic surface. We may think that the thermodynamic variables

$$T', \; p', \; \mu'_1, \; \mu'_2, \; \ldots$$
$$T'', \; p'', \; \mu''_1, \; \mu''_2, \; \ldots$$
$$\ldots\ldots\ldots$$

open a multidimensional space. However, it will become clear immediately that we need only one set of $T, p, \mu_1, \mu_2, \ldots$, where we have suppressed the primes ($'$).

Within this space, a vector is pointing to a thermodynamic state. For example, for a one-component system, Eq. (7.1) reduces to

$$S dT - V dp + n_1 d\mu_1 = 0. \tag{7.15}$$

We recall that the Gibbs – Duhem equation is the complete Legendre transformation of the energy equation, $U(S, V, n)$. A relation between the variables T, p, μ_1 emerges. Therefore, the set of the independent variables consists of any two variables in the set T, p, μ_1. For example, we could rearrange Eq. (7.15) into

$$d\mu_1 = -\tilde{S} dT + \tilde{V} dp. \tag{7.16}$$

Here we have divided by n_1 and introduced the molar quantities for entropy and volume. Obviously in Eq. (7.16) no term that gives information on the size of the system is seen. Moreover, the chemical potential is expressed as a total differential of the independent variables T, p. Thus there should be a function of the form $\mu(T, p)$.

For a certain phase, e.g., gas, in the three-dimensional plot of T, p, μ_1 a two-dimensional surface is obtained. For another phase, say the liquid phase, another two-dimensional surface can be constructed. Coexistence of these two phases appears in these regions where the two surfaces are cutting mutually. This occurs in a line in the three-dimensional space. Similarly, for the solid state we can construct still another surface. Coexistence of all three phases, i.e., solid, liquid, and gaseous phase occurs in the common intersect of the three surfaces, which is in general a point. This is the situation of zero freedom.

The idea can be easily generalized. For a thermodynamic system with K components and two additional variables the dimension of the hypersurface is $K - 1$. Coexistence of P phases gives a hyperline of dimension $K - 1 - (P - 1) = K + 2 - P$, which is interpreted as the freedoms. This geometrical interpretation of the phase rule is completely different from the polyhedral formula.

Question 7.1. How is the shape of the chemical potential near the critical point? Discuss the case, if still more energy forms than thermal energy and volume energy besides chemical energy are needed for the proper description of the thermodynamic system.

References

1. Fuller, R.B., Applewhite, E.J.: Synergetics. Explorations in the Geometry of Thinking. Macmillan Publishing Co. Inc., New York (1975)
2. Tauch, D., Russel, C.: Glass-ceramics with zero thermal expansion in the system bao/al2o3/b2o3. J. Non-Cryst. Solids **351**(27–29), 2294–2298 (2005)
3. Gibbs, J.W.: On the equilibria of heterogenous substances. In: The Collected Works, vol. 1. Yale University Press, New Haven, CT (1948 [reprinted])
4. Johnson, W.C.: On the inapplicability of Gibbs phase rule to coherent solids. Metall. Mater. Trans. A **22**(1), 1093–1097 (1991)
5. Li, D.: Curvature effects on the phase rule. Fluid Phase Equilib. **98**, 13–34 (1994)
6. Shapiro, A.A., Stenby, E.H.: Thermodynamics of the multicomponent vapor-liquid equilibrium under capillary pressure difference. Fluid Phase Equilib. **178**(1–2), 17–32 (2001)
7. Cromwell, P.R.: Polyhedra. Cambridge University Press, Cambridge (1999)
8. Schläfli, L.: Gesammelte mathematische Abhandlungen. Birkhäuser, Basel (1950). Edited by Steiner-Schläfli-Komitee der Schweizerischen Naturforschenden Gesellschaft
9. Tverberg, H.: How to cut a convex polytope into simplices. Geometriae Dedicata **3**(2), 239–240 (1974)
10. Michon, G.P., Anderson, S.E.: Counting polyhedra. [electronic] http://home.att.net/~numericana/data/polyhedra.htm (2001)
11. Weisstein, E.W.: Polyhedral graph. From MathWorld–A Wolfram Web Resource: [electronic] http://mathworld.wolfram.com/PolyhedralGraph.html (2004)
12. Klochko, M.A.: Analogy between phase rule and the Euler theorem for polyhedrons. Izvest. Sectora Fiz.-Khim. Anal., Inst. Obshei. i Neorg. Khim., Akadd. Nauk SSSR **19**, 82–88 (1949)
13. Levin, I.J.: The phase rule and topology. J. Chem. Educ. **23**, 183–185 (1946)
14. Mindel, J.: Gibbs' phase rule and Euler's formula. J. Chem. Educ. **39**, 512–514 (1962)
15. Morikawa, T., Newbold, B.T.: Analogous odd-even parities in mathematics and chemistry. Himi_ (Chemistry) **12**(6), 445–450 (2003)
16. Radhakrishnan, T.P.: Euler's formula and phase rule. J. Math. Chem. **5**, 381 (1990)
17. Rüdel, O.: The phase rule and the boundary law of Euler. Z. Elektrochem. **35**, 54 (1929)
18. Wildeboer, G., Plath, P.J.: Euler'sche Polyederformel und Gibbs'sche Phasenregel. Commun. Math. Comput. Chem. (MATCH) **7**, 163–175 (1979)

Chapter 8
Trouton's Rule

The rule of *Trouton*[1] states that the entropy of vaporization at the boiling point is for all substances approximately

$$\Delta \tilde{S}_v = \Delta \tilde{H}_v / T_b \approx 90 \, \mathrm{J \, mol^{-1} K^{-1}}.$$

Trouton's rule was published in 1883 and after the 100 years anniversary, several papers with historical remarks appeared [1–3].

  Guggenheim has collected data and expressed phase equilibria in terms of corresponding states [4]. The modeling of the coexistence equations with regard to the liquid – gas equilibrium is still an object of research [5].

8.1 Dimensionless Variables

We start with the well-known relations

$$\frac{\mathrm{d}p}{\mathrm{d}T} = \frac{\Delta \tilde{S}_v}{\Delta \tilde{V}_v}.$$

(8.1)

Here $\Delta \tilde{S}_v$ is the entropy of vaporization and $\Delta \tilde{V}_v$ is the volume change on vaporization. Using now

$$\mathrm{d}p = p \, \mathrm{d} \ln p, \quad \mathrm{d}T = -T^2 \, \mathrm{d} \left(\frac{1}{T} \right), \quad \breve{T} = \frac{T}{T_c}, \quad \breve{p} = \frac{p}{p_c},$$

we obtain from Eq. (8.1)

$$\frac{\mathrm{d} \ln \breve{p}}{\mathrm{d} 1 / \breve{T}} = -\frac{T \Delta \tilde{S}_v}{p \Delta \tilde{V}_v} \breve{T}.$$

(8.2)

---

[1] Frederick Thomas Trouton, born Nov. 24, 1863, in Dublin, Ireland, died Sep. 21, 1922, in Downe, Kent.

J.K. Fink, *Physical Chemistry in Depth*,
DOI 10.1007/978-3-642-01014-9_8, © Springer-Verlag Berlin Heidelberg 2009

T_c is the critical temperature and $\check{T}$ is the reduced temperature. The same notation is used for the pressure.

In Eq. (8.2), only reduced, dimensionless variables appears, and this equation should no be longer dependent on the individual properties of the materials. Note that $T\Delta S/p\Delta V$ is some ratio of thermal energy to volume energy.

In fact, a close similarity for the various substances is monitored; however, it can be seen that the monoatomic gases are separated from the polyatomic gases. The difference is not very much pronounced, however, and we neglect this difference here. Further, we deal with the plot as a straight line as a first approximation. Under these conditions, immediately

$$\frac{p\Delta \tilde{V}_v}{T\Delta \tilde{S}_v} = \check{T}\Theta(\check{T}) \tag{8.3}$$

follows. $\Theta(\check{T})$ is a function of $\check{T}$. At the left-hand side of Eq. (8.3), there appears a ratio of energies. This is the energy of expansion when the liquid expands on vaporization, and some sort of thermal energy that must be delivered on vaporization. In fact, it is the enthalpy of vaporization.

In the present form, Eq. (8.3) resembles the law of *Wiedemann*,[2] *Franz*,[3] and *Lorenz*[4]. We can also interpret the left-hand side as a degree of efficiency with regard to mechanical and to thermal work. In the region of low pressures, the vapor approaches an ideal gas. Here Eq. (8.2) becomes

$$\frac{d \ln \check{p}}{d\frac{1}{\check{T}}} = -\frac{\Delta \tilde{H}_v}{RT_c}. \tag{8.4}$$

A plot of the logarithm of the reduced pressure against the reciprocal reduced temperature is shown in Fig. 8.1. The equation of *Clausius – Clapeyron* reads as

$$\frac{d \ln p}{d1/T} = -\frac{\Delta \tilde{H}_v}{R}. \tag{8.5}$$

If we rewrite the equation of *Clausius – Clapeyron* (Eq. 8.5) in reduced quantities, we obtain simply Eq. (8.4).

For substances that obey the theorem of corresponding states, the term $\Delta \tilde{H}_v/RT_c$ should be a universal constant, or more generally a function of the reduced temperature only. It has been pointed out by *Guggenheim* [4] that $\Delta \tilde{H}_v/T$ at corresponding temperatures, i.e., temperatures being the same fraction of the critical temperature, has always the same numerical value.

---

[2] Gustav Heinrich Wiedemann, born Oct. 2, 1826, in Berlin, Germany, died Mar. 23, 1899, in Leipzig, Germany.

[3] Rudolph Franz, born Dec. 16, 1826, in Berlin, Germany, died Dec. 31, 1902, in Berlin, Germany.

[4] Ludvig Lorenz, born Jan. 18, 1829, in Helsingør, Dutch, died Jun. 9, 1891.

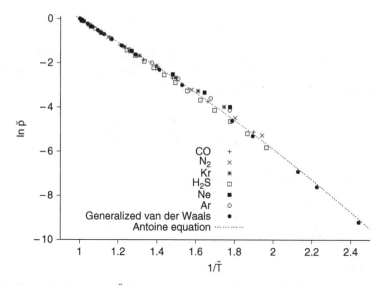

Fig. 8.1 Plot of $\ln \breve{p}$ against $1/\breve{T}$ for various substances [6, p. 74–84]. Theory: Points calculated from Eqs. (8.7) and (8.8)

Further, according to the rule of *Hildebrand*, the boiling point is often at two-thirds of the critical point $T_b/T_c = 2/3$, so that

$$\frac{\Delta \tilde{H}_v}{RT_c} \approx \frac{\Delta \tilde{S}_v T_b}{RT_c} = \frac{90 \times 2}{8.32 \times 3} \approx 7.1. \tag{8.6}$$

A plot of Eq. (8.2) is shown in Fig. 8.1 for various substances.

8.2 Phase Equilibrium

8.2.1 Generalized van der Waals Equation

We try now to obtain an expression for Eq. (8.2) from the generalized van der Waals Equation, Eq. (4.55). In our treatment, we use $A(\breve{T}) = 3/\sqrt{\breve{T}}$ and $B(\breve{T}) = 1/3$, to obtain Eq. (8.7). Observe that this equation is also related to the *Redlich Kwong* equation:

$$\left(\breve{p} + \frac{3}{\sqrt{\breve{T}}\breve{V}^2} \right) \left(\breve{V} - \frac{1}{3} \right) - \frac{8\breve{T}}{3} = 0. \tag{8.7}$$

The thermodynamic equilibrium of two phases of one component is governed by the equations

$$T_l = T_g, \quad p_l = p_g, \quad \mu_l = \mu_g.$$

The same is true for the reduced quantities. The pressure is calculated by resolving Eq. (8.7) with respect to the pressure and inserting either the reduced volume of the liquid phase $\breve{l}$ or the volume of the gas phase $\breve{g}$, to get

$$\frac{8\breve{T}}{3\breve{g} - 1} + \frac{3}{\sqrt{\breve{T}}\breve{g}^2} = \frac{8\breve{T}}{3\breve{l} - 1} + \frac{3}{\sqrt{\breve{T}}\breve{l}^2}. \tag{8.8}$$

The chemical potential is calculated by

$$\int_{\breve{V}=\breve{l}}^{\breve{V}=\breve{g}} \breve{V}\,d\breve{p} = \int_{\breve{V}=\breve{l}}^{\breve{V}=\breve{g}} \breve{V}\frac{\partial \breve{p}}{\partial \breve{V}}\,d\breve{V} = \breve{\mu}_g - \breve{\mu}_l = 0.$$

The result is

$$-\frac{8\breve{T}}{3}\ln\frac{3\breve{g} - 1}{3\breve{l} - 1} - \frac{8\breve{T}}{3(3\breve{g} - 1)} - \frac{6}{\breve{g}\sqrt{\breve{T}}} + \frac{8\breve{T}}{3(3\breve{l} - 1)} + \frac{6}{\breve{l}\sqrt{\breve{T}}} = 0. \tag{8.9}$$

Equation (8.9) can be approximated by assuming $\breve{g} \gg 1 > \breve{l} \approx 1/3$. Further, $\breve{l} = (z + 2)/3$ and the expansion is done for small z, and for $\breve{g}$ the reduced ideal gas equation is inserted. A direct formula for $\ln \breve{p} = \dots$ can be obtained. The approximation is not very satisfactory, however.

Table 8.1 Coexistence curves from the generalized van der Waals equation, Eq. (8.7)

$\breve{p}$	$\breve{T}$	$\breve{g}$	$\breve{l}$
0.0001	0.40974	10922.	0.364268
0.0005	0.449489	2393.87	0.370005
0.001	0.469901	1249.9	0.373214
0.005	0.528107	279.035	0.383495
0.01	0.559416	146.784	0.389832
0.05	0.654298	33.0045	0.413641
0.1	0.708816	17.204	0.431658
0.2	0.775263	8.80953	0.460329
0.3	0.821158	5.85506	0.486751
0.4	0.857526	4.32007	0.513774
0.5	0.888212	3.3667	0.543057
0.6	0.915057	2.70728	0.576344
0.7	0.939105	2.21429	0.616231
0.8	0.961007	1.8193	0.667704
0.9	0.981202	1.4733	0.744082
0.95	0.990759	1.29665	0.806514
0.99	0.998176	1.11535	0.904477
0.995	0.999089	1.07914	0.930727
0.999	0.999818	1.03404	0.96792

The numerical solution of the coexistence curve is calculated from Eqs. (8.8) and (8.9). In these two equations are three unknown variables, i.e., $g, l, \check{T}$. In the first step, a certain value for the reduced volume of the gas phase is prescribed. Next, the reduced volume of the liquid phase and the reduced temperature are calculated from these two equations. Finally, by inserting into Eq. (8.7), the reduced pressure is found. The results are given in Table 8.1 and plotted in Fig. 8.1, marked as theory.

A graphical method for calculating the equilibrium values consists of the direct calculation of the Legendre transformation of the molar-free Gibbs enthalpy, which is essentially the chemical potential.

In Fig. 8.2, leftdown the ordinary plot of the reduced pressure $\check{p}$ against the reduced volume $\check{p}$ is shown. Left up of the integral shows the molar reduced *Helmholtz* energy or some kind of chemical potential. The *Legendre* transformation of this function is the molar reduced *Gibbs* enthalpy, or the reduced chemical potential as function of temperature and pressure. Recalling that

$$\check{\mu}(\check{T}, \check{p}) = \int \check{V}(\check{T}, \check{p}) \mathrm{d}\check{p} = \int \check{V}(\check{T}, \check{p}) \frac{\mathrm{d}\check{p}}{\mathrm{d}V} \mathrm{d}\check{V} = \check{p}\check{V} - \int \check{p}(\check{T}, \check{V}) \mathrm{d}\check{V},$$

we find that $-\check{F}(\check{T}, \check{V}) - \check{p}\check{V}$ is the negative Legendre transformation of $\check{V}(\check{p})\mathrm{d}\check{p}$ and thus the negative reduced chemical potential. More strictly, in the integral, there is an integration constant $C(\check{T})$, which is a function of the reduced temperature. We have skipped this constant in the treatment above.

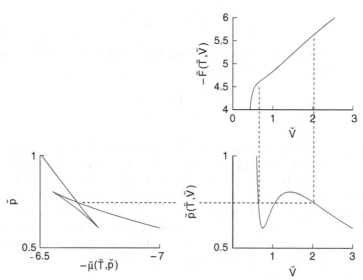

Fig. 8.2 Generalized van der Waals gas, Eq. (8.7): Helmholtz energy, pressure, and chemical potential at $\check{T} = 0.95$

In addition, learn that the edges of the triangle in the left plot of Fig. 8.2 represent the spinodal points. The procedure is explained in more detail in Sect. 1.8.2.

A plot for argon in comparison to that based on the equations of van der Waals, *Carnahan* and *Starling* [7], *Dieterici*, and a generalized equation has been shown in *Sadus* [8, 9]. Another technique proposed is the integration along a path that coincides with the saturation line [10].

8.2.2 Maxwell's Criterion

In elementary text it is often stated that in the plot of pressure against volume of a van der Waals gas (Fig. 8.2), the equilibrium pressure of the coexisting phases is the horizontal line with equal area above and below equilibrium pressure in the two-phase region. This behavior is referred to as Maxwell's criterion.

However, Maxwell's criterion is thrown into the discussion as such and the relations to other criteria of phase equilibria remain in the dark. Here we go into some depth of this topic. For simplicity, we turn back for a moment to ordinary quantities in favor of reduced quantities. Of course, in terms of reduced quantities, the same statements will hold.

First recall that due to the Eulerian homogeneity, the energy can be split into

$$U(S, V, n) = TS - pV + \mu n.$$

Consequently, the molar energy $\tilde{U}(\tilde{S}, \tilde{V})$ becomes

$$\tilde{U}(\tilde{S}, \tilde{V}) = T\tilde{S} - p\tilde{V} + \mu.$$

The Helmholtz energy is the Legendre transformation of the energy with respect to entropy: $\tilde{F}(T, \tilde{V}) = \tilde{U} + p\tilde{V}$. Now

$$-\frac{\partial \tilde{F}(T, \tilde{V})}{\partial \tilde{V}} = p.$$

For this reason, in the $p - \tilde{V}$ diagram, $\tilde{F}(T, \tilde{V}_2) - \tilde{F}(T, \tilde{V}_1)$ is the definite integral between any points V_1 and V_2. In particular, $-\tilde{F}(T, \tilde{V}_g) + \tilde{F}(T, \tilde{V}_l)$ is the area in the $p - \tilde{V}$ diagram from the liquid phase to the gaseous phase. We form now a Legendre transformation, $\mu(T, p) = \tilde{F}(T, \tilde{V}) + p\tilde{V}$. In the $\tilde{F}, V$ diagram, points of the same pressure are on tangents with the same slope. Moreover, since the ordinate in this diagram is the Legendre transformation itself, μ, points of phase equilibrium are those with common tangents, since phase equilibrium demands $\mu_l = \mu_g$. In other words, the chemical potential of the gas phase, μ_g, is equal to the chemical potential of the liquid phase, μ_l. So we have

$$-\tilde{F}(T, \tilde{V}_g) - p\tilde{V}_g + \tilde{F}(T, \tilde{V}_l) + p\tilde{V}_g = -\mu_g + \mu_l = 0.$$

Actually, $p\tilde{V}_g - p\tilde{V}_l$ is just the rectangle of the intersecting points of equilibrium. This completes the illustration of Maxwell's criterion. Eventually, we emphasize that the relationships illustrated in this section are more general as exemplified in the phase equilibrium of coexisting phases of a van der Waals gas.

8.2.3 Plank – Clausius Equation

We emphasize here that *Planck* [11] has already calculated the vapor pressure – temperature coexistence curve by using an equation, Eq. (8.10), which goes back to *Clausius*:

$$\left(p + \frac{a}{(V-b)^2} \right) (V-c) - RT = 0. \tag{8.10}$$

The equation used by Clausius contains four parameters, however, namely a, b, c, R. In order to use the concept of corresponding states, we have three conditions only, i.e., the equation at the critical point itself, the conditions that the first derivative and the second derivative of the pressure with respect to the volume must be zero. Therefore, we can deliberately chose one of the parameters a, b, c, R to fix it.

8.2.4 Redlich Kwong Soave Equation

The *Redlich Kwong* equation was extended by *Soave* [12] to get

$$\left(p + \frac{a\alpha(T, \omega)}{\tilde{V}\left(\tilde{V} + b\right)} \right) \left(\tilde{V} - b\right) = RT. \tag{8.11}$$

The function $\alpha(T, \omega)$ takes into account the molecule's shape. With this equation, phase equilibria were calculated.

8.2.5 Antoine Equation

Whereas for theoretical considerations, the *Clausius – Clapeyron* equation plays a fundamental role, practitioners favor the *Antoine* equation [7]. This equation is very similar to the *Clausius – Clapeyron* equation:

$$\ln p = A + \frac{B}{T+C}. \tag{8.12}$$

A lot of values of parameters A, B, and C have been fitted for various substances and tabulated. The temperature range of validity is carefully attached to the data. Turning to dimensionless variables, one parameter can be omitted by the condition that the reduced pressure approaches one at the reduced temperature of one: $\check{p} \to 1; \check{T} \to 1$. Thus, Eq. (8.12) becomes

$$\ln \check{p} = \check{A}\left(1 - \frac{1 + \check{C}}{\check{T} + \check{C}}\right). \tag{8.13}$$

In Fig. 8.1, the reduced *Antoine* equation with $\check{A} = 4.6$ and $\check{C} = -0.11$ is shown.

8.2.6 Master Curve and Trouton's Rule

It has been pointed out that Trouton's rule is not a strict law, but a remarkable good approximation [13]. The rule works best for nonpolar, quasi-spherical molecules.

If the plot of the reduced vapor pressure against the reciprocal reduced temperature fits a master curve, it does not mean that the rule of *Trouton* is valid. To meet the rule of Trouton, also the rule of *Hildebrand*[5] must be valid. In this case, all the boiling points will be found at $T_c/T = T_c/T_b = 3/2 = 1.5$. On the other hand, if the corresponding states are sound, then all substances should have a common critical pressure, because $\ln p_b/p_c = C$ in this case (since $p_b = 1$ atm).

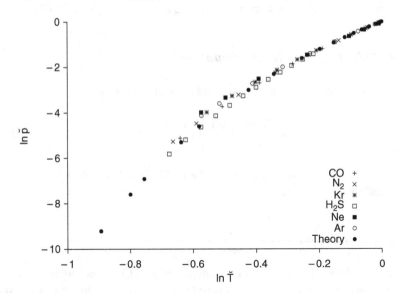

Fig. 8.3 Coexistence curve as double logarithmic plot of the reduced pressure against the reduced temperature

We can see this fact also of the coexistence curve in the double logarithmic plot of the reduced pressure against the reduced temperature. The slope of this plot becomes, in the limiting case, that of an ideal gas phase, and the volume of the liquid

---

[5] Joel H. Hildebrand, born Nov. 16, 1881, in Camden, NJ, USA, died Apr. 30, 1983, in Kensington, CA, USA.

phase being small in comparison to the gas volume also a dimensionless quantity, i.e., $\Delta \tilde{S}_v / R$. This is shown in Fig. 8.3. If the entropy of vaporization is constant, then the curve is a straight line, at least in the region where the law of Trouton is valid. This is obviously not the case, but the curve is not bent very much. Alternatively, the boiling point of all the substances should collapse in one point of the plot. However, in this case, since the pressure at the boiling point is 1 atm, again as argued above the critical points of all substances should be the same. Therefore, the rule of Trouton can be only an approximate rule, or likewise the theorem of corresponding states. To check out, a plot of $\ln p_c$ against $\check{T} / T_b$ is indicated.

References

1. Nash, L.K.: Trouton and T-H-E rule. J. Chem. Educ. **61**(11), 981–984 (1984)
2. Shinoda, K.: Entropy of vaporization at the boiling point. J. Chem. Phys. **78**, 4784 (1983)
3. Wisniak, J.: Frederick Thomas Trouton: The man, the rule, and the ratio. Chem. Educat. **6**(1), 55–61 (2001)
4. Guggenheim, E.A.: The principle of corresponding states. J. Chem. Phys. **13**, 253–261 (1945)
5. Vorob'ev, V.S.: Equation of state for substances that satisfy the corresponding states law. Int. J. Thermophys. **26**, 1409–1421 (2005)
6. Lide, D.R. (ed.): CRC-Handbook of Chemistry and Physics, 71st edn. CRC Press, Boca Raton, FL (1990–1991)
7. Carnahan, N.F., Starling, K.E.: Equation of state for nonattracting rigid spheres. J. Chem. Phys. **51**(2), 635–636 (1969)
8. Sadus, R.J.: Equations of state for fluids: The Dieterici approach revisited. J. Chem. Phys. **115**(3), 1460–1462 (2001)
9. Sadus, R.J.: New Dieterici-type equations of state for fluid phase equilibria. Fluid Phase Equilib. **212**(1–2), 31–39 (2003)
10. Kofke, D.A.: Direct evaluation of phase coexistence by molecular simulation via integration along the saturation line. J. Chem. Phys. **98**(5), 4149–4162 (1993)
11. Planck, M.: Die Theorie des Sättigungsgesetzes. Ann. Phys. **13**, 535–542 (1881)
12. Soave, G.: Equilibrium constants from a modified Redlich-Kwong equation of state. Chem. Eng. Sci. **27**(6), 1197–1203 (1972)
13. Nash, L.K.: Trouton and T-H-E rule. J. Chem. Educ. **61**(11), 981–984 (1984)

Chapter 9
Thermodynamic Cycles

9.1 Thermodynamic Cycles Have Edges

In thermodynamic cycles the states of a system change in such a way that a final
state is reached, which is equal to the initial state. We want to call such a system
the *machine*. The machine takes up a certain form of energy, converts a part of this
energy into another form of energy, and delivers the remainder of the energy taken
up again. We want now to describe these general statements by the example of the
paddle wheel.

9.1.1 The Paddle Wheel

We idealize the paddle wheel, so that we must consider only the gravitational
energy. In the first step we fill a paddle, which is up, with water. This water has
the energy mgh_2. In the second step the paddle moves downward. The energy of the
water changes, as the height changes. Down, thus at the height h_1 in the third step
the entire water is emptied. The water leaves now the paddle wheel with the energy
mgh_1. In the fourth step the empty paddle goes again upward.

If we consider the conservation of energy, then the paddle wheel must have con-
sumed a certain amount of energy. We assume that the paddle wheel delivered this
energy quantity via some energy output, so that it can arrive after the end of the
fourth step in the same condition as started. The paddle wheel nowadays typically
delivers rotational energy, to feed an electric generator. The potential energy of the
water is eventually transformed to some extent into electricity. We idealized the
paddle wheel in such a way that all energy, which loses the water, was converted
from gravitational energy into electric energy. We assumed that no frictional energy
is in play and we have assumed that we do not have to consider the kinetic energy
of the water. In general it is unimportant whether the paddle wheel produces only
electricity, whether it produces surface energy, by grinding grain, or heat energy and
deformation energy, or if it forges steel.

It is essential that the difference of the energy brought in and the energy exhausted
is loss-free transformed. Of course, this is not possible in practice, but the efforts of
engineers go all into this direction, e.g., by bearing the wheel as loss free as possible.

J.K. Fink, *Physical Chemistry in Depth*,
DOI 10.1007/978-3-642-01014-9_9, © Springer-Verlag Berlin Heidelberg 2009

We emphasize a further point: For the execution of a cyclic process a machine is essential and an environment, which consists of three subsystems. From the subsystem (R_2) energy is withdrawn and delivered into the other subsystem (R_1). The third subsystem (NA) serves to absorb the transformed energy. It collects the transformed energy from the machine and uses this energy in some way. It may be necessary that at some stage of the process, the subsystem (NA) will give back some energy to the machine.

We also assume that both the energy transfer from the subsystem (R_2) to the paddle wheel, the machine, and the delivery of energy from the paddle wheel into the other subsystem (NA), occurs loss free. The two mentioned subsystems (R_2), (R_1) serve as reservoirs and are idealized. So that we mean that the intensive variable associated with the energy form under consideration does not change, if we remove from or add energy into the reservoir. In the case of paddle wheel a reservoir is required that does not change the water level, if we take or add water.

Actually, it is not of interest whether the energy is transformed free of loss or not into the subsystem (NA). It is important that the energy goes away from the machine. If we want to allow the process run in the reverse direction, then it is also important that the transfer into the subsystem (NA) takes place loss free from the paddle wheel and in reverse direction. We require thus with the machine works in both direction of the process equally; thus we want a complete reversibility of the process.

We treat now again the paddle wheel more exactly and represent conditions in an $m - gh$ diagram. The partial steps of the paddle wheel in the $m - gh$ diagram are shown in Fig. 9.1.

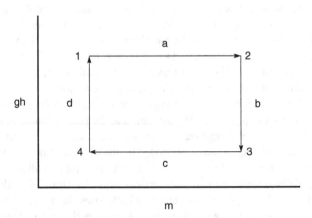

Fig. 9.1 Partial steps of the paddle wheel in the $m - gh$ diagram

(a) The paddle wheel take up in the first step (a) mass at constant height. Here we move parallel to the m-axis (ordinate) to the right.
(b) In the second step (b) the paddle descends, without water drip-out. Therefore, we move vertically downward in the diagram.

(c) By emptying with constant height of (c) we move horizontally to the left in the diagram.
(d) The movement of the empty paddle wheel corresponds to a movement vertically upward in the diagram. We arrive again at the beginning.

We note that the product of abscissa and ordinate is an energy. The surrounded surface in steps (a)–(d) (1-2-3-4) is the net energy, which delivers the paddle wheel within a revolution to the subsystem (NA). The surface in the rectangle of the points (1,2) down to the abscissa is the energy, which delivers the reservoir (R_2). The surface in the rectangle of the points (4, 3) down to the abscissa is the energy, which takes up the reservoir (R_1). So the rectangle (1-2-3-4) is the energy which is transferred by the paddle wheel to the utilizable energy system. It is substantial that the energy in another energy form can be delivered, i.e., transformed by the engine. Otherwise, we could not successfully accomplish the thermodynamic cycle.

Question 9.1. Why do we make at all processes, which deliver only a certain part of the energy? It would be certainly better to convert the entire energy brought into the machine from the reservoir (R_2) into the desired productive energy. This is sound. In fact this is desired. For the paddle wheel along a river, there is a series of such power plants, one building behind the other. That stops, however, as the river reaches the sea. Here is a physical border. I can build at the mouth of a river a power station that discharges the water below the sea level, but I do not bring the water any longer without loss of energy into the sea inside.

It is the gravitation energy, which flows by the action of the machine, so we address this special cyclic process a gravidynamic process. Gaining usable work requires that the water not simply freely gutters down, but the possible work is caught by a machine. As we will see alike, we can understand the process in addition as a chemo+dynamic process in the broader sense, although thereby no chemical reaction occurs.

9.1.1.1 Chemical Potential

Even the paddle wheel has a chemical potential. We deal with the energy as function of height and mol number,

$$U(n, h) = Mngh,$$
$$dU(n, h) = \frac{\partial U(n, h)}{\partial h} dh + \frac{\partial U(n, h)}{\partial n} dn = \Gamma(n, h)\, dh + \mu(n, h)\, dn.$$

We have formed the total differential in a formal manner. However, we observe that Γ is not a function of the height and the chemical potential is not a function of the mol number. So, with more exact analysis,

$$dU(n, h) = Mngdh + Mghdn = \Gamma(n)\, dh + \mu(h)\, dn.$$

We form the second derivatives and find

$$\frac{\partial \Gamma(n)}{\partial n} = \frac{\partial \mu(h)}{\partial h} = mg.$$

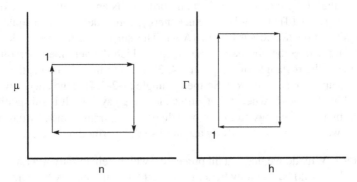

Fig. 9.2 Representation of the cyclic process for the paddle wheel in the $\Gamma - h$ diagram and in the $\mu - n$ diagram. The representation is complementary. The way is in both cases in the clockwise direction, but is once the beginning in the diagram left above and once, with the number 1. The surface of the surrounded rectangle is alike with both representations, although the edge lengths may not be alike

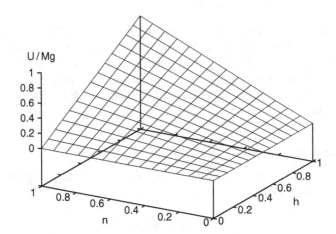

Fig. 9.3 Energy diagram for the paddle wheel process: z-axis U/Mg, other axes h and n. The *lines parallel* to h- and n- axes are *straight lines* opposite to the surface area, which is stretched by the *straight lines h and n*

We made here again use of the law of *Schwarz*. We can represent the process now in refined representation either in $\Gamma - h$ diagram or in $\mu - n$ diagram, cf. Fig. 9.2. As the process of the paddle wheel can be represented with two different energy conjugated variables, also, for example, the Carnot process can be represented in the $p-V$ diagram or in the $T-S$ diagram. In both representations the surrounded surfaces are alike. The individual curves can be, as in the case of the $p-V$ diagram of the Carnot process for an ideal gas, not necessarily straight lines. The energy diagram for the process is shown in Fig. 9.3.

9.1.2 Carnot Process

The Carnot process played an outstanding role for the development of thermodynamics. Usually the Carnot process is exemplified for an ideal gas, in the $p-V$ diagram. We now show that the Carnot is a rectangular process.

The Carnot process is an isothermal–adiabatic cyclic process. With the Carnot process it is the heat energy which flows through the machine. The heat is taken from a thermal reservoir (R_2). The machine itself is an ideal gas that expands isothermally. In the second step the gas expands adiabatically, i.e., isentropically. In the third step, the gas compresses again isothermally, whereby it delivers heat energy to the reservoir (R_2). The fourth step is again an isentropic compression. Here the gas returns into the initial state. We realize that we have two isentropic steps and two isotherm steps. The product of temperature and entropy is an energy, the thermal energy. In Fig. 9.4, we represent now the Carnot process in the $S-T$ diagram. Without repeating the arguments with the paddle wheel in detail, we realize that the Carnot process can be represented completely similarly. Due to the rectangle form we observe that the entropy released to (R_1) has the same value, like the entropy taken up from (R_2). The rectangle must result simply from the demand that the isothermal processes run in a straight line parallel to the abscissa and the isentropic processes run parallel to the ordinate. Otherwise, we could not arrive in a different closed path at the same point (1) back at all.

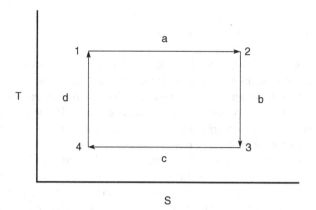

Fig. 9.4 Carnot process in the $S-T$ diagram

In textbooks, often the Carnot cycle is explained based on a thermal reservoir at the higher temperature, at the working gas, and a thermal reservoir at a lower temperature. At the latest, when the isothermal compression is dealt with, the alert student may ask, from where the energy of compression is delivered. It is neither delivered from the thermal reservoirs nor from the working gas. Actually, the energy of compression is delivered from still another auxiliary system what we have termed as (NA) in the discussion of the paddle wheel, Sect. 9.1.1. This system appears already in the first step of the cycle, since the working gas gives the energy obtained from the thermal reservoir to this auxiliary system in a transformed energy form. For this reason, the auxiliary system (NA) is an essential part of the Carnot cycle.

Example 9.1. We try as an exercise the Carnot process in the $T - Q$ diagram, i.e., in Fig. 9.5 we plot the amount of heat over the temperature. The machine transfers now in the first step (a) isothermally heat energy from the reservoir (R_2). Thus, we must vertically in the diagram lift off. In the second step without transfer of heat the temperature is lowered. The heat content of the gas thus remains constant. That means we are forced to go horizontal to the left. Next, we must go down until the initial temperature is reached.

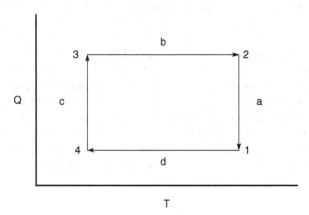

Fig. 9.5 The carnot process in the $T - Q$ diagram

Otherwise, we could not come back to the starting state in an isothermal process. However, this means that in step (d) the same amount of heat is given to the reservoir (R_1) as withdrawn from reservoir (R_2). The individual processes are thus compelling, but the total process does not balance out itself energetically in this representation. If the same energy is taken from the reservoir as the other reservoir is transferred, then no more energy remains left over for work.

We see thus that the representation in the $T - Q$ diagram exhibits an energetic lack and so we reject this representation. In addition the surrounded rectangle is not an energy at all. This consideration shows the meaning of the entropy for the Carnot process. We want to address the Carnot process as a thermodynamic cyclic process in the closer sense, because it is the heat energy, which flows through the machine. The thermodynamic process does not allow has heat energy simply to flow from a warmer to a colder body, but uses the energy due to the inserted machine. We did not restrict the process presented here to an ideal gas at all.   □

Example 9.2. We draw now the Carnot process for an ideal gas in an energy diagram. We draw the steps in the $TS-pV$ diagram. First, observe that $pV = nRT$ for an ideal gas. Thus, we may use a parametric plot with $x = nRT$ and $y = TS$. At constant mol number, in the isothermal steps, the temperature is constant and we have a vertical straight line. In the isentropic steps, the entropy S is constant, and we have a straight line passing the origin.

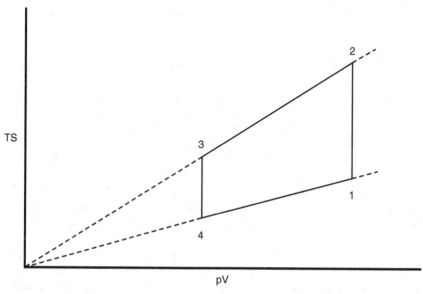

Fig. 9.6 The carnot process in the $TS-pV$ diagram

The process is represented in Fig. 9.6. The area in this diagram does not make sense as an energy in contrast to the $p-V$ diagram and to the $T-S$ diagram. Instead, the axes are themselves energies. Moving from state (1) to state (2), since the temperature is constant the change in energy of the reservoir is $\Delta U = T\Delta S$. This ΔU is not the same as for the gas, even when pV does not change. Recall that the energy is $U = TS - pV + \mu n$, according to the thermodynamical similarity. Thus, we have to consider also the contribution of the term μn, when considering the energy of the gas. We have, with the three basic terms, in square brackets

$$U = [TS] - [nRT] + [n\tilde{C}_v T + nRT - TS].$$

So the interesting term $n\tilde{C}_v T$ rests in the term of the chemical energy. We show for the isentropic step that the energy change is not ΔpV. For the gas itself, in the isentropic step, the change in energy is $\Delta U = -\int p\,dV$, where $pV^\kappa = C$, at constant entropy and mol number. Performing the integration and substituting back, the change in energy is, e.g., $(p_3 V_3 - p_2 V_2)/(\kappa - 1)$, for step (2)→(3). So the term pV does not reflect the change of energy itself in general.

The representation shows clearly that as more steps (3) and (4) are moving toward the origin, less thermal energy moves to the reservoir at the lower temperature. □

There is still another approach to clarify the representation in Fig. 9.6. For the isothermal process, $pV = nRT$, which is a constant. Thus, the process must move parallel to the ordinate. For the isentropic process, we may consider the parametric representation ($x = pV = nRT, y = TS$), where S is constant. Even for an isotropic process, the ideal gas law holds. On resolving the parametric representation, $y/x = \tilde{S}/R$, which indicates that the extrapolated lines cut the origin.

9.1.3 The Process of the Energy Transfer

We discuss now the question how we can achieve that the machine can take a certain amount of energy out of the reservoir. We discuss both the paddle wheel and the Carnot process. Subsequently we want to generalize the treatment.

In order to allow the water run from the reservoir into the machine, thus in the paddle of the water wheel, the height of the paddle must be somewhat lower than the height of the reservoir, or in the limiting case both heights must be equal. During filling the paddle moves down to be filled still more with water. Here not the position of the paddle is the important parameter, but the water level in the paddle. The level of water in the paddle must be slightly lower than the water level in the reservoir, or equal in the limit.

Similar to this process, in the Carnot cycle the temperature of the machine (the gas) must be slightly lower or in the limit equal to the temperature of the reservoir. As in the case of the paddle wheel, the machine must change some thermodynamic variable to give way so that more heat energy can enter into the gas inside. A change of this variable must lower the temperature, because by the take up of heat energy alone the temperature would rise. We keep the temperature constant by the change this mysterious variable while the heat flows from the reservoir into the gas.

We know of course the thermodynamic variable is the volume of the gas. If we would expand the gas adiabatically, then the temperature would become lower. As heat can flow in during expansion, the temperature is kept constant.

We concealed which variable plays the corresponding role with the paddle wheel: In the Carnot process the volume change is used to keep the temperature constant and further to eject the transformed energy as energy of expansion. Just in the same way in the case of the paddle wheel at the iso-hypsic process angular momentum is delivered through the wheel, whereby rotational energy is released. Thus, we see, what is responsible that the energy can flow from the reservoir with constant intensive variable into the machine: It is the change of that energy, which is uncoupled as power energy.

In principle we can imagine the energy transfer in a different manner. We make use of the Maxwell relationship,

$$\frac{\partial S(T, V)}{\partial V} = \frac{\partial p(T, V)}{\partial T}.$$

The change of the entropy with the volume is just as much as the change of the pressure with the temperature at constant volume,

$$\Delta S = \frac{\partial p\,(T,\,V)}{\partial T}\Delta V. \tag{9.1}$$

Relationship (9.1) is applied to the machine in the Carnot process. During the isothermal expansion step we must deliver that quantity entropy, in the course of expansion. If we would not do so, then we could not hold the temperature in the machine. The machine requires thus the subsequent delivery of the entropy. The entropy becomes in any case larger after Maxwell's equation, if we increase the volume.

If we would accomplish the process without work, then entropy would be generated. This is undesirable, of course, and contradicts the demand after reversibility. In equilibrium, the machine takes away from the reservoir the straight necessary entropy. The conclusions developed here are valid also for other machines.

An important prerequisite is that the machine can exchange the energy forms under consideration. Otherwise, such a process is not possible. To illustrate the above statement we want to perform the Carnot process with an incompressible body. The incompressible body has the function of state

$$V = V_0[1 + \alpha(T - T_0)].$$

The volume does not depend on the pressure. Thus, it is not possible to change the volume by application of pressure. A machine using an incompressible body cannot compensate for the entropy that should flow in by an isothermal process by doing expansion work. That is the reason why we use for the Carnot process (compressible) gases and not incompressible solids. Even a compressible body changes its dimensions with temperature by an adiabatic compression only little in comparison to the ideal gas. Therefore, we should apply large pressure changes to achieve appropriate effects. This becomes obvious from the right-hand side in Eq. (9.1). However, this does not mean that a Carnot process is impossible with a compressible solid body.

9.1.4 Chemodynamic Cycles

In certain respects the paddle wheel is already a chemodynamic process. We deal now with a still more obvious chemodynamic process. Basically, in a chemodynamic process the chemical potential and the amount of material change. There is a throughput of matter. An interesting machine consists of a swellable endless ribbon, which reels on two separate large rollers. The rollers are connected rigidly with one another and have unequal diameters. That is the actual trick, which lets the machine run. If the ribbon unreels, then unreeled part from the larger role is larger in length than what is rolled up by the smaller roller. On the other hand, a different tension of the ribbon on both freely hanging sides causes a torque. If we shorten the ribbon on one side and increase on the other side, then the rollers will turn.

The ribbon dips into seawater on one side, and dips into freshwater on the other side. The ribbon can take up only freshwater, but not the salt. With the movement the ribbon stretches and contracts due to the different swelling in the seawater and in the freshwater. The ribbon stretches in freshwater and contracts in seawater. A machine, which exemplifies such a process, is shown in Fig. 9.7.

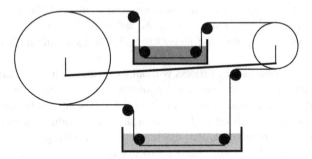

Fig. 9.7 Realization of a chemodynamic process

9.1.4.1 The Steps of the Chemodynamic Process

We idealize the steps: We identify the chemical potential of the water in the volume as μ. As in other cycle processes we index "1" for the lower and "2" for the higher chemical potential. We do that here exceptionally. Otherwise, the index is always used in order to designate the material concerned.

- The ribbon comes first from the seawater. It becomes strained in the step (d) by running along the smaller roller. Here the chemical potential rises. In the first step (a) by immersion into freshwater the ribbon takes up water at constant tension. It lengthens itself thereby at constant chemical potential.
- In the second step (b) the ribbon runs over the larger roller and relaxes. So the chemical potential decreases.
- In the third step (c) the ribbon delivers a part of the stored water in contact with the seawater.
- In the fourth step (d) the ribbon again strained as already described at the beginning. Thus, the chemical potential increases to the starting value. With this change of the tension also a change of the length of the ribbon goes along.

We could point out the essential steps in another way, if we consider a ribbon with a weight hanging on the lower end. It is then immersed in the freshwater. Then we change the weight and dive into seawater. Then we change again the weight and immerse again into water. In this way, we may lift a portion of mass, by an admittedly cumbersome process.

9.1.4.2 Balance of the Process

In the $\mu - n$ diagram again a rectangle results. Altogether the seawater is diluted by freshwater, and some power energy is generated. If we dilute the seawater by simply pouring in freshwater, then only the entropy of the system would increase. With the machine, the dilution becomes possible only under achievement of power work. As a result of the different tension at both rollers different torques arise ωdl and the system delivers mechanical work. In principle, the chemodynamic process dilutes the seawater. The dilution does not take place now simply via the addition of freshwater into the seawater, but that process uses the dilution work. By the way, the reversal of the process effects the production of freshwater.

We examine the energetic situation now more exactly. The energy of the ribbon can be written down as

$$dU(S, V, l, n) = T\,dS - p\,dV - K\,dl + \mu\,dn. \qquad (9.2)$$

The ribbon can exchange thermal energy, expansion work, elastic work (spring energy), and chemical energy. The ribbon behaves similar to a spring on expansion. Therefore, we want to call this form of energy, spring energy. The length is here an extensive variable. Two ribbons of the same kind connected in series result in a double lengthening. The strength is directed opposite to the lengthening, therefore the minus sign.

Question 9.2. Is this explanation sufficient justified, when the expansion work is such as the pressure points in the same direction as the volume changes?

Since the ribbon has a surface, which is once in contact within the water and once in contact with the roller, also surface work could play a role. We neglect the surface work. Further, the process shall take place at constant temperature and at constant pressure. Therefore, it is advantageous to transform the variables. The exchange of water by swelling takes place also at constant tension, or if we do not standardize on the cross section, at constant strength. This suggests transforming the energy forms. However, during the entire cycle process the tension changes, in contrast to the pressure. Therefore, we cannot neglect the expression of the spring energy. The transformed equation reads as

$$dB\,(V, T, l, n) = \mu\,dn + l\,dF. \qquad (9.3)$$

We designate the transformed form of energy as B, which is similar to the free energy of Gibbs, but additionally still the spring energy is transformed. We use the law of *Schwarz*:

$$\frac{\partial \mu\,(n, l)}{\partial F} = \frac{\partial l\,(n, l)}{\partial n}.$$

We omitted the temperature and the pressure in the variable list here, because we do not want to change temperature and pressure during the entire process. However, we cannot exclude a dependence of the thermodynamic functions on temperature and pressure. We assume intuitively that the ribbon becomes longer on swelling. Therefore, the chemical potential must increase with increasing strength of the ribbon. We can set thus the chemical potential, if the strength will change. So we can adjust the strength in such a way that the water in the ribbon has the chemical potential of the water in the tub with freshwater, if it is immersed into the freshwater. On the other hand, water in the ribbon should have the chemical potential of the seawater, if it is immersed into the tub with seawater. In this way, the ribbon under equilibrium conditions can swell or shrink.

9.1.5 Gluodynamic Process

Here we use the energy of adhesion for a cycle process. A tape is partially rolled up on a roller. At one end of the tape a weight is mounted. On the roller a cord is rolled up in the opposite direction. At the end of the cord, also a weight is mounted. The weights must have different masses so that the equilibrium establishes. On unreeling the tape, surface energy is set free. In the arrangement shown in Fig. 9.8, the heights of the weights change equally in opposite direction, thus $dl_1 + dl_2 = 0$. In equilibrium

$$dU = (m_1 - m_2)\, g\, dl + \sigma b\, dl.$$

The change in surface is $dA = b\, dl$, thus width b of the tape times the length variation dl. We carry out the cycle process in such a way that we allow the ribbon

Fig. 9.8 Pressure-sensitive adhesive ribbon on a roller in equilibrium with the force of gravity

to roll up a distance under equilibrium. Then we reduce the surface tension, e.g., by moistening the tape in the roller gap, or by heating. We take away the ballast weight that was before lifted, in order to unreel in equilibrium. Then we dry the tape with a hair dryer and establish the original surface tension. Then again we mount a ballast weight that we can lift then as described in the first step.

9.1.6 The Pump

Also a pump is a chemodynamic machine. The experimental setup is shown in Fig. 9.9. We want to pump an incompressible liquid. We sketch here the steps only briefly:

- Withdrawing the piston. The liquid flows in from container (1) under the pressure p_1.
- The piston stops. The three-way stopcock switches to the other container (2). The pressure actually rises to p_2. Since the fluid is incompressible, simply the pressure rises.
- The piston presses the liquid under pressure p_2 into container (2).
- The piston stops. The three-way stopcock switches back to container (1).

Fig. 9.9 Pump for liquids

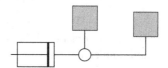

We emphasize that the working system consists only of the cylinder and piston. The machine which moves the pistons back and forward does not belong to the system. This is the same situation as in the paddle wheel, where the shaft, which exhausts the energy as rotational energy, does not belong to the working system. However, this associated auxiliary machine fulfills an important object in the cycle process. It works as buffer storage for energy and in addition as transformer for the form of energy, which is transposed through the machine. The expansion energy, which the piston takes up, must be taken up and stored by the auxiliary machine in the meantime, e.g., as electricity. In the third step the same energy is again used. If the machine works as pump, still energy from the auxiliary machine is added. If the machine works as generator, then in the third step less energy is returned than in the first step was demanded.

We go for a moment back to the paddle wheel. With the paddle wheel this aspect is not visible in all clarity, because the paddle wheel is a well-constructed machine. We therefore invent a paddle wheel that consists of only one paddle. If the paddle with the water moves downward, then rotational energy is delivered into the shaft. The paddle empties down completely. Now the empty paddle must be lifted, and here rotational energy is given back through the shaft.

9.1.7 Combustion Engines

In the age of automotive vehicles it is in order to discuss combustion engines. We deal with a combustion engine that utilizes chlorine and hydrogen as the detonating gases. The reaction is

$$Cl_2 + H_2 \rightarrow 2HCl.$$

The advantage of this treatment is that the total mol numbers are constant. We can then easily deal this reaction as a chemodynamic cycle. If in a combustion process the mol numbers are changing, then it is more advantageous to use the masses instead of the mol numbers. In the view of a chemodynamic cycle a combustion engine is still a four-stroke engine, but the meaning of the individual processes is different in comparison to those what a technician would explain:

1. As the piston moves back the reactants enter with a chemical potential μ_1.
2. The chemical reaction takes place, the chemical potential lowers.
3. The piston moves forward and exhausts the products.
4. The engine changes its state in order to allow entering again the reactants at their chemical potential in equilibrium.

We see, in practice, it is literally impossible to operate such an engine close to equilibrium conditions.

9.1.8 Hysteresis

Hysteresis is another cycle process. In contrast to all other cycles discussed hitherto, in this case no work is generated, but entropy.

9.1.9 Universal Reversible Cycles

Thermodynamic cycles have been generalized. The fundamental equation can be integrated and can be used to derive an expression characterizing the efficiency of a system for any kind of cyclic process by which work is produced [1].

As characteristic for thermodynamic cycles, the working system accesses two reservoirs with a low and a high thermodynamic potential. The thermodynamic potentials, i.e., temperature, chemical potential, hydrostatic pressure, electric potential, etc., show an absolute zero of the lower reservoir, when the efficiency of the cycle $\eta = 1$.

Reference

1. Ginzburg, B.Z., Aharon, I.: The second law of thermodynamics revisited I. The absolute zero of potential and the general expression for the efficiency of universal reversible cycle. Physica A **210**(3–4), 489–495 (1994)

Chapter 10
Thermodynamic Paradoxes

10.1 Gibbs' Paradox

The entropy of mixing ideal gases was treated already by Gibbs. The result is known as the Gibbs' paradox. Several scientists were engaged with this paradox, among them *Rayleigh, Lorentz, Einstein,* and *von Neumann*[1] [1, 2].

The paradox states that the two gases can be as similar as possible, but not completely similar to exhibit an entropy of mixing. Gibbs' paradox was tried to be explained in several ways. For example, it was stated that nature changes the properties in jumps, in other words, in quanta [3, p. 163].

Thus, it has been argued that substances cannot become arbitrarily similar because they consist of isotopes. In contrast, in the case of *o-* and *p-*hydrogen, the difference of these compounds relies not on isotopes, but on their spins. Further, in polymers, an unsaturated end group, with either *cis* or *trans*, can be relevant that two kinds of compounds can be identified. However, the difference can be made essentially arbitrarily small as the molecular weight of the polymer increases.

We point out still another issue. A gas that consists essentially of one kind of matter is still different, because the individual particles exhibit different velocities. By a device that acts as a velocity separator, such as two rotating sectors, we could even separate molecules that move in a certain range of speed. Thus a gas can be separated into two systems that exhibit different velocities. In the macroscopic sense of thermodynamics, these two gases then have different temperatures. If we mix these gases, consisting essentially of the same material, a certain entropy of mixing emerges. The process is the same as if we allow the separated gases to attain thermal equilibrium and then combine them. In the first step, entropy is created as the gases reach equal temperature. In the second step, by combining, the system is simply enlarged and no further entropy is generated.

Gibbs' paradox silently assumes that the components to be mixed have the same temperature.

---

[1] Johann von Neumann, born Dec. 3, 1903, in Budapest, Hungary, died Feb. 8, 1957, in Washington, DC, USA.

Therefore, similarity is not a continuous process as Gibbs believed. A more recent attempt to explain Gibbs' paradox stresses the reversible and irreversible mixing [2]. The chapter also reviews previous views.

The problem was discussed by *Kemp* [4] in a way, as a correction term was introduced. In fact, the molecules present in the system are subdivided into those that are distinguishable and those that are indistinguishable. The process of mixing is split up into several more basic processes in a rather complicated way. The correction term, when subtracted from the entropy of mixing, makes the entropy of mixing zero, when the molecules are indistinguishable. A more general correction term was obtained by MacDonald [5], to which Kemp basically agreed [6].

However, Kemp stresses that the main conclusions in his original paper [4] still remain valid:

- Gibbs' paradox arises from the unsound assumption that the entropy of each gas in an ideal mixture is independent of the presence of the other gases;
- the entropies of ideal mixing cannot be deduced solely from the laws of classical dynamics applied to the ideal gas equation;
- individual entropies cannot be assigned to the constituents of an ideal mixture.

10.1.1 Reasons for the Paradox

On the other hand, it has been pointed out that Gibbs' paradox is not really a paradox [7, 8]. The paradox seems to arise because the conditions of the process of mixing are not fixed properly. Coarsely spoken we can disparage what is happening as follows: two different gases are mixed and the correct entropy of mixing is obtained.

However, after mixing, the scientist states that he was in error, the gases were not different in fact. And he is wondering that the entropy of mixing does not turn out to be zero. However, there was one further process introduced, namely, changing the identity of a component during the process. But in fact, this change of the identity is a further process, which is not included in the consideration.

The conclusion is that we are not allowed to start the process with a certain initial assumption and after doing the process we change the initial assumption.

10.1.1.1 Mixing and Chemical Reaction

To illustrate the problem of Gibbs' paradox more clearly, we introduce now a thermodynamic cycle that involves a chemical reaction. *cis*-2-Butene has a boiling point of 4 °C and *trans*-2-butene has a boiling point of 1 °C.

So both substances are gases at room temperature, even when we must concede that they are not ideal gases. We assume further that we have an apparatus in that we can transform *cis*-2-butene into *trans*-2-butene in a gas mixture of both components, for example, a UV source. Consider the process schematically shown in Fig. 10.1.

At first, we discuss the process qualitatively. We are starting from state (A) with two subsystems, i.e., chambers. Initially, in the left chamber there is pure

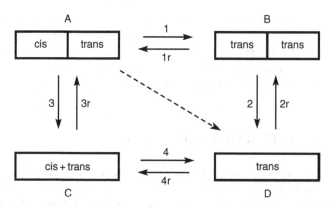

Fig. 10.1 Mixing and chemical reaction

cis-2-butene, and in the right chamber, there is pure *trans*-2-butene. Via process (1) we can transform the *cis*-2-butene in the left chamber into *trans*-2-butene to arrive at state (B). When the wall is removed in process (2) we have the case of mixing of identical gases. This process is in fact simply an enlargement of a homogeneous system. We can come to this state via another series of processes.

Again, starting at state (A), via process (3) we mix *cis*-2-butene and *trans*-2-butene to obtain state (D). Obviously, the process (3) creates entropy of mixing. Then, in the mixture we transform *cis*-2-butene into *trans*-2-butene via process (4). Then, in state (C) we have pure *trans*-2-butene in the combined chamber.

From state (A), the processes (1) and (2) arrive at the same state (C) as the processes (3) and (4). We discuss now the reverse process (4r). The reverse process (4r) can be disassembled as follows:

1. In a certain restricted region of the system containing pure *trans*-2-butene we transform all the material into *cis*-2-butene.
2. Next, we mix the *cis*-2-butene formed in this region with the *trans*-2-butene in the rest of the system. This process creates entropy of mixing.

From this consideration, it turns out that the reverse process (4r) is composed of the reverse process (2r), the reverse process (1r), and the process (3). In other words, in Fig. 10.1 we are running counterclockwise from state (C) to state (D), as it should be in a thermodynamic cycle.

If we think the reverse process (4r) now in the normal (forward) way (4), then we separate first the mixture and then convert one component. By means of the separation process, the entropy of mixing that is created in process (3) is compensated. The separation process creates entropy of de-mixing, which is just the reverse of mixing.

Gibbs' paradox arises from starting the process at state (A) and moving directly to state (C) via the dashed arrow, thereby neglecting the transformation of the distinguishable compounds into undistinguishable compounds, where the entropy of mixing is got back, no matter to which degree the compounds are distinguishable.

So we can resolve the paradox in that the proper process must be chosen depending on whether the components are distinguishable or not. If the components are distinguishable, then the proper process to handle the problem is the process (3), otherwise we must use process (2). We emphasize that the decision whether the components are distinguishable is a yes/no decision.

10.1.2 Processes Concerning the Paradox

Consider an apparatus as sketched in Fig. 10.2. We perform now simple experiments with this apparatus. The left chamber we address with a single prime ($'$) and the right chamber we address with the double prime ($''$). Thus, the left chamber has a volume of V' and a temperature of $T' = T$. The right chamber has a volume of V' and a temperature of $T'' = T$.

Fig. 10.2 Two chambers separated by spool valves M_0, M_1, M_2. M_0 is impermeable, M_1 is permeable to gas (1), and M_2 is permeable to gas (2). The valves can be switched independently. The chambers are placed in a thermostat

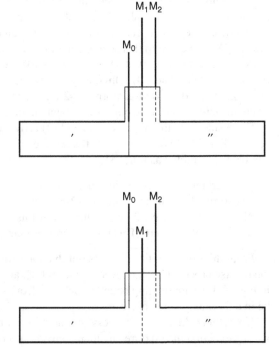

We have derived the energy equation of an ideal gas in Eq. (4.8). From there the entropy of a one-component gas can be calculated as

$$S = \tilde{C}_v n \ln(\tilde{C}_v T) + nR \ln\left(\frac{V}{n}\right) \tag{10.1}$$

besides an arbitrary constant. For processes at constant temperature and constant mol number starting with a volume V_s and ending with a volume V_e, Eq. (10.1) reduces to

$$\Delta S = nR \ln\left(\frac{V_e}{V_s}\right),$$

where ΔS is the change of entropy when the volume changes as $V_s \rightarrow V_e$. We summarize the processes that are discussed in the subsequent sections in Table 10.1.

Table 10.1 Some kinds of mixing processes of ideal gases

Process	Before		After	
	mol	In volume	mol	In volume
Expansion of a one-component gas	n_1'	V'	n'	$V' + V''$
Mixing of identical gases	n_1'	V'	$n_1' + n_1''$	$V' + V''$
	n_1''	V''		
Mixing of different gases	n_1'	V'	n_1'	$V' + V''$
	n_2''	V''	n_2''	$V' + V''$
Mixing of a gas mixture with a pure gas	n_1'	V'		
	n_2'	V'	n_2''	$V' + V''$
	n_1''	V''	$n_1' + n_1''$	$V' + V''$

10.1.2.1 Expansion of a One-Component Gas

Suppose the left chamber is filled with a one-component gas of type (1) and the right chamber is empty. The spool valve is closed. Initially the total entropy is $S_s = S_s'$ since the right chamber is empty ($S_s'' = 0$)m

$$S_s = \tilde{C}_v n_1' \ln(\tilde{C}_v T) + n_1' R \ln\left(\frac{V'}{n_1'}\right). \tag{10.2}$$

If we open the spool valve, the gas is distributed between both chambers. At the end we have an entropy of S_e

$$S_e = \tilde{C}_v n_1' \ln(\tilde{C}_v T) + n_1' R \ln\left(\frac{V' + V''}{n_1'}\right),$$

which is clearly greater than the entropy at the start of the experiment S_s. However, observe that the total mol number remains constant.

10.1.2.2 Mixing of Identical Gases

Now we repeat the experiment if the right chamber is not empty but filled with a certain amount of gas. At the beginning of the experiment the total entropy is

$$S_s = \tilde{C}_v n_1' \ln(\tilde{C}_v T) + n_1' R \ln \left(\frac{V'}{n_1'} \right) + \tilde{C}_v n_1'' \ln(\tilde{C}_v T) + n_1'' R \ln \left(\frac{V''}{n_1''} \right).$$

After opening the spool valve we have just the situation of a single system with increased volume and mol number in comparison to the two separated systems according to Eq. (10.1) with $n \to n_1' + n_1''$ and $V \to V' + V''$. Thus, we get

$$S_e = \tilde{C}_v (n_1' + n_1'') \ln(\tilde{C}_v T) + (n_1' + n_1'') R \ln \left(\frac{V' + V''}{n_1' + n_1''} \right). \qquad (10.3)$$

In case that we adjusted the two chambers at equal pressure and temperature, then the molar volume would not change during the process, and we have

$$\tilde{V} = \frac{V'}{n_1'} = \frac{V''}{n_1''} = \frac{V' + V''}{n_1' + n_1''}. \qquad (10.4)$$

Namely, due to the ideal gas law, we have

$$\frac{RT}{p} = \frac{V'}{n'} = \frac{V''}{n''}.$$

What we here address as molar volume should not be confused with the partial volume of a gas mixture. Under the condition of Eq. (10.4) no increase of entropy is obtained, i.e., $S_e - S_s = 0$ as can be easily verified from Eqs. (10.3) and (10.2)

10.1.2.3 Mixing of Different Gases

Now we proceed to the common mixing of two different gases. This case is treated in the textbooks to explain the entropy of mixing. Suppose that initially in the left chamber is a gas (1) and in the right chamber is a gas (2). Then at the start of the experiment the entropy S_s is

$$S_s = \tilde{C}_v n_1' \ln(\tilde{C}_v T) + n_1' R \ln \left(\frac{V'}{n_1'} \right) + \tilde{C}_v n_2'' \ln(\tilde{C}_v T) + n_2'' R \ln \left(\frac{V''}{n_2''} \right). \qquad (10.5)$$

For simplicity we presume that both gases have the same heat capacity. The basic situation would not change, if we assume

$$\tilde{C}_v = \frac{n_1 \tilde{C}_{v,1} + n_2 \tilde{C}_{v,2}}{n_1 + n_2}.$$

After opening the spool valve the gases will mix and the final entropy is

$$S_e = \tilde{C}_v(n_1' + n_2'') \ln(\tilde{C}_v T) +$$

$$n_1' R \ln \left(\frac{V' + V''}{n_1'} \right) + n_2'' R \ln \left(\frac{V' + V''}{n_2''} \right). \quad (10.6)$$

The difference in entropy from Eqs. (10.6) and (10.5) is

$$\Delta S = S_e - S_s = n_1' R \ln \left(\frac{V' + V''}{V'} \right) + n_2'' R \ln \left(\frac{V' + V''}{V''} \right). \quad (10.7)$$

Now the molar volume of the components of the gases no longer remains constant as in case of Eq. (10.4), but increases. Therefore, the entropy of mixing also increases. Note that we are starting in this section and in Sect. 10.1.2.2 with the same mathematical formula. If we are inserting carefully the starting conditions and the end conditions, then we do not feel the paradox. We emphasize that a necessary condition for the practical application of the formula is that we can perform a mixing experiment, which is possible only if we have the two components already separated. If we cannot purify the components, considerations on the entropy of mixing of these components are useless.

We could visualize the mixing of the gases as a two-step process. In Fig. 10.2, initially the valve M_0 is closed. The mixing could be performed by keeping the valve M_1 closed when the valve M_0 is opened. Since the spool valve M_1 consists of a membrane that is permeable only for gas (1), the gas in the left-hand chamber can start to permeate into the right chamber, thus increasing its volume, whereas gas (2) is still imprisoned in the right-hand chamber. In the second step, the membrane M_2 that is permeable for gas (2) is closed and the membrane M_2 is lifted. So in the second step also gas (2) allows expanding in the whole volume $V' + V''$. In this way, the mixing of the gases consists of a two-step expansion process. It can clearly be seen that for a single-component gas such a process cannot be achieved. To achieve such a process, the membrane must be fabricated in a way that a single gas may not permeate from left to right, but from right to left. This would be a version of Maxwell's demon.

The Gibbs' paradox arises from the idea that in the two chambers are single-component gases and that the different gases could approach asymptotically to an indistinguishable gas. It is opposed from the nature of matter, since it is built from electrons, protons, and neutrons, it is not possible to approach the properties continuously. There can be only one electron added or removed. Thus, the change of the nature of the gases cannot be performed in arbitrary small steps.

However, thermodynamics does not specify what the criteria to make two gases undistinguishable are. For instance, we can distinguish two gas molecules if they bear different energy contents. A gas molecule with high-energy content may start radiating in contrast to a molecule with normal energy content. Or even differences in kinetic energy may lead us to distinguish molecules. Therefore, to define what is a gas mixture or what is a single gas is a somewhat floating problem.

10.1.2.4 Mixing of a Gas Mixture with a Pure Gas

We proceed now to place in the left-hand chamber of Fig. 10.2 a gas mixture of the species (1) and (2) and in the right-hand chamber we place a single-component gas consisting exclusively of the component (1).

In contrast to the problem to make a single gas continuously more and more identical to another gas species, a gas mixture consisting of gases (1) and (2), say with the major component (1), can be nearly continuously made identical to a single gas (1), as the component (2) is decreased.

In other words, instead of making the gas in the left chamber identical to the gas in the right by virtually changing the properties of each individual molecule of the gas in the left chamber in the same manner, we change the overall properties of the gas in the left chamber by changing its composition. At the begin of the experiment the total entropy is

$$
\begin{aligned}
S_s = {} & \tilde{C}_v n_1' \ln(\tilde{C}_v T) + n_1' R \ln \left(\frac{V'}{n_1'} \right) + \\
& + \tilde{C}_v n_2' \ln(\tilde{C}_v T) + n_2' R \ln \left(\frac{V'}{n_2'} \right) + \\
& + \tilde{C}_v n_1'' \ln(\tilde{C}_v T) + n_1'' R \ln \left(\frac{V''}{n_1''} \right).
\end{aligned}
\tag{10.8}
$$

At the end of the experiment, the entropy is

$$
\begin{aligned}
S_e = {} & \tilde{C}_v (n_1' + n_1'') \ln(\tilde{C}_v T) + (n_1' + n_1'') R \ln \left(\frac{V' + V''}{n_1' + n_1''} \right) + \\
& + \tilde{C}_v n_2' \ln(\tilde{C}_v T) + n_2' R \ln \left(\frac{V' + V''}{n_2'} \right).
\end{aligned}
\tag{10.9}
$$

The difference $\Delta S = S_e - S_s$ obtained from Eqs. (10.8) and (10.9) goes in the limit of $n_1' \to 0$ into Eq. (10.7).

We form the differences of the entropies, divide by R and the total sum of moles $n_1'' + n_1' + n_2'$, and substitute

$$
\frac{n_1''}{n_1'' + n_1' + n_2'} = \frac{V''}{V' + V''}
$$

$$
x_{2,e} = \frac{n_2'}{n_2' + n_1' + n_1''}
$$

$$
x_{2,s} = \frac{n_2'}{(n_2' + n_1')}
$$

$$
x_{1,e} = 1 - x_{2,e}
$$

$$
x_{1,s} = 1 - x_{2,s}
$$

where $x_{2,e}$ is the mole fraction of component (2) at the end of the experiment in the combined mixture and $x_{2,s}$ is the mole fraction of component (2) at the beginning of the experiment in the left chamber. We obtain finally

$$\frac{\Delta S}{R(n_1'' + n_1' + n_2')} = x_{1,e} \ln \frac{x_{2,e}}{x_{1,e}} - \ln(x_{2,e}) + \frac{x_{2,e}}{x_{2,s}} \ln(x_{1,s}) + x_{2,e} \ln \frac{x_{2,s}}{x_{1,s}}. \quad (10.10)$$

Equation (10.10) goes in the limit of $x_{2,s} \rightarrow 1$ into the case of mixing of two different gases and in the limit of $x_{1,e} \rightarrow 1$ to zero, which is the case of mixing identical gases.

10.1.3 Some Mixing Paradoxes

The following is rather a pitfall than a paradox:

Question 10.1. When two different ideal gases are mixed, the temperature, the volume, and the pressure remain constant, as well as the total energy. However, the total entropy increases. Thus we have

$$\Delta U = T \Delta S - p \Delta V,$$

with $\Delta U = 0$; $\Delta V = 0$, but $\Delta S > 0$. Resolve the problem. What happens with the chemical potential?

When we start to mix two pure components, the entropy of mixing remains finite. Thus, we believe that we need finite energy in order to separate the mixture back into the pure compounds. However, in practice we never get them clean.

10.2 Maxwell – Boltzmann Paradox

If two systems are in equilibrium, thus in thermal equilibrium, they exhibit equal temperatures. On the other hand, a collection of particles are in equilibrium, if there is a certain velocity distribution. The velocity can serve as a measure of the temperature of a particle. If we break down the collection of particles to thermodynamic systems, the concept of thermal equilibrium is violated. According to the classical view of thermodynamics, we could increase the entropy if we pick out two systems of different temperatures and allow equilibrating. Obviously, the application of thermodynamics is restricted to entities that consist of more than one particle. Actually, the variables in classical thermodynamics rely on averages of large ensembles.

The *Maxwell – Boltzmann* velocity distribution emerged by theoretical arguments in 1866. However, an experimental proof was given only in 1921 by *Stern*[2] [9–11].

---

[2] Otto Stern, born Feb. 17, 1888, in Sohrau, Silesia, died Aug. 17, 1969, in Berkeley, CA, USA.

10.3 Maxwell's Demon

Maxwell[3] invented his demon to raise an objection against the second law of thermodynamics. If a gas is in two chambers separated by a door, a demon placed could open the door when a hot or high energetic molecule wants to come in, and close it when a low energetic molecule wants to come in or a high energetic molecule wants to escape. In this way, cooling of one chamber and the heating of the other chamber could be achieved. The hot chamber could then drive a thermal engine to get mechanical energy.

To overcome this situation, it was suggested that the demon needs more energy to do his basic job than can be obtained by the thermal engine. Nowadays, scientists agree that in small-scale regions the second law can be violated [12–14].

The concept of Maxwell's demon plays a role in the thermodynamics of computation [15]. The views concerning the role of Maxwell's demon are still controversial.

10.4 Loschmidt's Paradox

Loschmidt's paradox states that if there is a motion of a system that leads to a steady increase of entropy with time, then there is certainly another allowed state of motion of the system, found by time reversal, in which the entropy must decrease.

The principle of time reversal is well established in mechanics, as the second law of thermodynamics is well established in thermodynamics. By Loschmidt's considerations, both principles are opposed in a conflicting way.

The paradox can be resolved in assuming that in cosmology there is an arrow of time. On the other hand, by fluctuations, the second law of thermodynamics may be violated [14].

10.5 Thermodynamic Inequalities

The inequality between geometric mean and arithmetic mean turns out by entropy considerations [16, 17].

We consider n identical heat reservoirs each having the same constant heat capacity C. However, heat reservoirs should be initially at different absolute temperatures $T_i, i = 1, \ldots, n$.

When the heat reservoirs are allowed to come into thermal contact they will reach finally an equilibrium temperature T_e. From the energy conservation it follows

---

[3] James Clerk Maxwell, born Jun. 13, 1831, in Edinburgh, Scotland, died Nov. 5, 1879, in Cambridge, England.

$$\sum_{i=1}^{i=n} \Delta U_i = C \sum_{i=1}^{i=n} (T_i - T_e) = 0.$$

In the special case of equal heat capacity, T_e is the geometric mean of the initial temperatures, i.e.,

$$T_e = \frac{1}{n} \sum_{i=1}^{i=n} (T_i).$$

We calculate now the increase of entropy to

$$\Delta S = \sum_{i=1}^{i=n} \int_{T_i}^{T_e} \frac{C}{T} \, dT = nC \ln \left[\frac{T_e}{[T_1 \times T_2 \times \ldots \times T_n]^{1/n}} \right] \geq 0. \qquad (10.11)$$

Now, if we start from a nonequilibrium state the total entropy must increase, or in the limiting case of initial equilibrium be equal to zero. From Eq. (10.11) it follows that

$$\frac{1}{n} \sum_{i=1}^{i=n} (T_i) \geq (T_1 \times T_2 \times \ldots \times T_n)^{1/n}.$$

10.6 Chemical Vapor Deposition

It was argued that the most celebrated violation of thermodynamics is the fact that CVD (chemical vapor deposition) diamond forms under thermodynamically unstable pressure – temperature conditions. A careful analysis of thermodynamic data led to the conclusion that certain equilibria are taking part in the process, so that the process no longer appears as a thermodynamic paradox [18]. The key idea was to assume that CVD diamond formation is a chemical process consisting in accretion of polymantane macromolecules. Thus, violations of thermodynamic principles could be avoided.

10.7 The Thermodynamics of Black Holes

The theory of black holes in general relativity suggests a fundamental relationship between gravitation, thermodynamics, and quantum theory. The cornerstone of this relationship is the thermodynamics of black holes. Certain laws of black hole mechanics can be set up by applying the basic laws of thermodynamics to such a system [19].

In the classical theory a black hole is a region where gravity is so strong that nothing can escape. Therefore, black holes are perfect absorbers. Furthermore, they do not emit. Their temperature is at the absolute zero.

Even in classical general relativity, there is a serious difficulty with the ordinary second law of thermodynamics when a black hole is present [19]. If matter is swallowed in a black hole, the entropy initially present in the matter is lost. No compensating gain of entropy occurs elsewhere. Therefore, the total entropy in the universe decreases. The resort is to continue to count the entropy in a black hole. However, the second law cannot be verified experimentally, because there is no access to the black hole.

In contrast, according to quantum theory, black holes emit a *Hawking* radiation. This radiation exhibits the thermal spectrum. The generalized second law of thermodynamics is the connection of the laws of black hole mechanics to the ordinary laws of thermodynamics.

Bardeen, *Carter*, and *Hawking* [20] published a general proof of the four laws of black hole mechanics. These laws correspond to the four laws of thermodynamics. These laws apply to stationary black holes. For example, in analogy to the second law of thermodynamics, the area of the event horizon A never decreases, thus $dA \geq 0$. Further, in analogy to the third law, it is impossible to reduce the surface gravity to zero by any process in finite steps.

In 1975, *Hawking* stated that the temperature of a black hole is not absolute zero. Due to quantum effects, a black hole radiates all species of particles with a black body radiation [21]. The mean temperature is proportional to the surface gravity of the black hole.

Therefore, an isolated black hole with no feed of matter will evaporate completely within a finite time. If the correlations between the inside and outside of the black hole are not restored during the evaporation process, then when the black hole has evaporated completely, the information with respect to these correlations is lost. Alternatives to this loss of information, e.g., the formation of a high entropy region elsewhere are implausible. This issue is addressed as the black hole information paradox [19].

10.8 The Perpetuum Mobile

There is an excellent compilation concerning the perpetua mobilia by Gramatke [22]. In Table 10.2 a few historical facts are compiled concerning the chronology of the perpetuum mobile [23].

Science distinguishes between a perpetuum mobile of first kind and second kind. A perpetuum mobile of first kind violates the first law of thermodynamics and a perpetuum mobile of second kind violates the second law of thermodynamics. In fact, perpetua mobilia have been invented to deal with literally every known kind of energy. The concept of a perpetuum mobile seems to have fascinated or worried

Table 10.2 Chronology of energy laws and the perpetuum mobile [22]

Date	Event
500 B.C.	*Anaxagoras* states that nothing can emerge from nothing and nothing can be annihilated
1480	Leonardo *da Vinci* is engaged with the perpetuum mobile. He perceives the impossibility of constructing such an engine
1640	*Descartes* formulates a precursor of the energy principle
1660	*Becher* invents a clockwork based on a perpetuum mobile. Finally he realizes that the engine could not work
1719	*Bessler* publishes the book *Perpetuum Mobile Triumphans*
1719	*Droz* invents a perpetual clockwork based on bi-metal. The engine is not a perpetuum mobile, but utilizes temperature fluctuation
1775	The French Academy of Sciences declares to be no longer engaged in the discussion of perpetua mobilia
1842	Julius Robert *Mayer* formulates the mechanical equivalent of heat
1843	James Prescott *Joule* measures the mechanical equivalent of heat
1848	Hermann *Helmholtz* launches the principle of conservation of energy
1880	James Clerk *Maxwell* describes what becomes later famous as Maxwell's demon
1880	Rudolf *Clausius* formulates the second law of thermodynamics
1910	Walther *Nernst* formulates the third law of thermodynamics
1955	*Belousov* discovers the oscillating chemical reaction. This reaction seems to violate the second law of thermodynamics. Belousov's results were not accepted until 1968 *Zhabotinsky* could force the acceptance in the world of chemistry

scientists that they have included the perpetuum mobile in their definitions of the thermodynamic laws. Yet, they rate the perpetuum mobile higher than the laws itself.

Thus, sometimes we find proofs that the statement of impossibility of a perpetuum mobile is tantamount with the first law of thermodynamics, and the same can be found for the second law of thermodynamics. Nevertheless, there are still unperturbed inventors that are launching patents on energy generators, e.g., [24–26]. Patents of this kind can be found as alleged perpetua mobilia, other machines, or engines in the European Patent class F03B17/04 (mechanical), H02K53/00 (electrical).

Interesting for the chemist is that perpetua mobilia are not restricted to mechanics, electricity, but invaded also in the fields of chemistry. For instance, eternal lamps received the attention of alchemists. Another kind of perpetuum mobile consists of two coupled distillation apparatus. The distillate of one apparatus is flowing down in the other apparatus. This instrument is called circulatorium.

10.9 Negative Heat Capacities

Negative heat capacities have been discussed in context of cosmology and with black holes [17, 27]. Since the late nineteenth century astronomers realized that by adding energy to a star, it would expand and cool down. However, this was not pointed out as a paradox in thermodynamics until *Thirring* [28] emphasized that there is a theorem that specific heats are positive.

When the binding energy increases more rapidly than linearly with the number of particles, negative specific heat capacity for some energies are exhibited. In larger systems, a negative heat capacity creates instability.

The physical origin of the apparent paradox of an object cooling while absorbing thermal energy is the impossibility of defining a thermodynamical limit for systems interacting via nonsaturating forces. First-order phase transitions may be viewed as caused by negative heat capacities [29]. Also in the theory of ionization, chemical dissociation, and in the van der Waals gas such issues are found. Therefore, this concept is applicable even outside the realm of stars, star clusters, and black holes.

10.10 Expansion of an Incompressible Body

If we apply pressure to an incompressible body in the course of an adiabatic process, the body may unexpectedly expand. The increase of pressure leaves even an incompressible body literally not cool. The body can accept energy and can warm up when thermally isolated. Since the pressure has no effect on the dimensions, but the temperature has effect, the body may expand on heating.

More precisely we must differentiate between an adiabatically incompressible boy and an isothermally incompressible body. Since the generalized susceptibilities must be positive for the sake of stability of matter, we have

$$\frac{\partial V}{\partial p} \geq 0.$$

Here the dependent variables of the volume must be extensive. On the other hand,

$$\frac{\partial V}{\partial T}$$

can be either positive or negative or zero, as exemplified with water near the freezing point.

10.11 Fixing the Extensive Variables

We deal now with a substance that can be thermodynamically characterized by its entropy, volume, and mol number, (S, V, n). If we place this substance in a container of fixed volume V' which is now a constant, then we can no longer form the partial derivative of the energy $U(S, V', n)$ with respect to the fixed volume V'. In other words, we cannot access the pressure of the substance. In the language of mathematics, the expression

$$p(S, V', n) = -\frac{\partial U(S, V', n)}{\partial V'}$$

is not a valid one. However, we suspect that the substance would have a certain pressure, even when the volume is fixed. If we know the function $U(S, V, n)$ in detail, then we could formally perform the calculus

$$p(S, V, n) = -\frac{\partial U(S, V, n)}{\partial V}. \qquad (10.12)$$

Next, we can insert $V = V'$ in the explicitly calculated function $p(S, V, n)$ and we arrive at $p(S, V', n)$, which is the correct expression for the pressure.

In the train of thoughts above we have mingled the concept of energy as a function of the variables as such and the constraint of fixed volume. The constraint is not a general constraint in that way that it is an inherent property of the substance itself. Thus, we could imagine that the substance can be placed in a container in that the volume can be varied to some extent. In that situation the partial derivative to the volume is associated with an admissible real physical process.

In the case of fixing the volume we face the situation of a virtual process. Even when the volume is fixed, basically we could change the volume. However, we must have some previous knowledge on how the function $p(S, V, n)$ looks like. There is one way to get the fundamental equation $U(S, V, n)$ purely via theoretical considerations. But in order to measure $U(S, V, n)$, variations of the respective extensive variables are necessary.

Thus, the mathematics of thermodynamics suggests that an intensive variable such as the pressure is only experimentally accessible via a physical variation of its associated extensive variable, here the volume V. This means that we cannot construct a pressure sensor that is completely stiff in its volume.

We can express this situation also from the view of the system itself. Any measurement of a property of the system would change the state of the system itself, due to the variation of a certain extensive variable. Of course, in practice the change of the state of the system due to measurement may be negligible.

10.12 Changes in Chemical Potential and Pressure

The chemical potential of a one-component system depends on temperature and pressure. If two vessels (') and (") are connected via a capillary, a flow would occur, if the chemical potentials are not equal, $\mu' \neq \mu''$. We start from a condition where the chemical potentials are equal. For this reason, also the pressures and temperatures are equal:

$$\mu' = \mu''; \quad T' = T''; \quad p' = p''.$$

In such a system, no net flow of entropy and mass occurs. If in one of the subsystems the temperature would change, then the change of the chemical potential can be prevented by changing also the pressure, $\mu'(T', p') = \mu''(T'' + \Delta T, p'' + \Delta p)$. Namely, in general,

$$\mu(T + \Delta T, p + \Delta p) = \mu(T, p) + \frac{\partial \mu(T, p)}{\partial T} \Delta T + \frac{\partial \mu(T, p)}{\partial p} \Delta p.$$

Now, the paradox is that across the capillary there is a pressure drop, but the equal chemical potentials demand that there should be no flow.

We must differentiate between bulk flow and molecular flow. If in the capillary would be a movable plunger or barrier, then the subsystem with the lower pressure would expand at the cost of the other system. However, this process will occur even when there is no barrier. The subsystem moves into the area of the other subsystem. If there is a membrane, not permeable to the flow of entropy, but permeable to matter, then the subsystem will exert to the other system a pressure, but no flow will occur. The situation is similar to that of osmosis. However, when matter moves, it carries entropy. So, under ordinary conditions, it is difficult or impossible to fabricate such a membrane.

It has been found by *Allen*[4] and Jones that at low temperatures in helium, when a temperature difference is produced, a pressure difference arises. This phenomenon is known as the fountain effect, or thermomechanical effect [30]. The superfluid is believed to carry no entropy.

If the two vessels are containing a mixture of two gases, an additional freedom arises, namely the composition. In the two vessels, different compositions may be established when a temperature gradient is forced. Thereby, the pressures in the vessels may be kept equal so that no bulk flow arises. This situation is found in the *Clusius*[5]–*Dickel*[6] effect. A *Clusius–Dickel* column is very sensitive with regard to separation and can be used for isotope separation.

References

1. Rayleigh, J.: The work can be obtained at mixing different gases. In: Scientific Papers, vol. 1, pp. 242–249. Cambridge University Press, Cambridge (1899). Reprint at Dover Publications, New York, 1964
2. Tatarin, V., Borodiouk, O.: Entropy calculation of reversible mixing of ideal gases shows absence of Gibbs paradox. Entropy 1(2), 25–36 (1999). [electronic] www.mdpi.org/entropy/
3. Schaefer, C.: Theorie der Wärme, Molekular-kinetische Theorie der Materie, *Einführung in die theoretische Physik*, vol. 2, 3rd edn. De Gruyter, Berlin (1958)

---

[4] John Frank (Jack) Allen, born May 5, 1908, in Winnipeg, Canada, died Apr. 22, 2001.

[5] Klaus Paul Alfred Clusius, born Mar. 19, 1903, in Breslau, now Wroclaw, Poland, died May 28, 1963, in Zurich, Switzerland.

[6] Gerhard Dickel, born Oct. 28, 1913, in Augsburg, Germany.

4. Kemp, H.R.: Gibbs' paradox for entropy of mixing. J. Chem. Educ. **62**(1), 47–49 (1985)
5. McDonald, J.J.: Gibbs' paradox: Two views on the correction term (lte). J. Chem. Educ. **63**(8), 735 (1986)
6. Kemp, H.R.: Gibbs' paradox: Two views on the correction term (lte). J. Chem. Educ. **63**(8), 736 (1986)
7. Ben-Naim, A.: On the so-called Gibbs paradox, and on the real paradox. Entropy **9**, 132–136 (2007). [electronic] http://www.mdpi.org/entropy/papers/e9030132.pdf
8. Jaynes, E.T.: The Gibbs paradox. In: C.R. Smith, G.J. Erickson, P.O. Neudorfer (eds.) Maximum Entropy and Bayesian Methods, *Fundamental Theories of Physics*, vol. 50, pp. 1–22. Kluwer Academic Publishers, Dordrecht, Holland (1992)
9. Eldridge, J.A.: Experimental test of Maxwell's distribution law. Phys. Rev. **30**(6), 931–935 (1927)
10. Stern, O.: Eine direkte Messung der thermischen Molekulargeschwindigkeit. Z. Phys. A: Hadrons Nucl. **2**(1), 49–56 (1920)
11. Zartman, I.F.: A direct measurement of molecular velocities. Phys. Rev. **37**(4), 383–391 (1931)
12. Baranyai, A.: The kinetics of mixing and the fluctuation theorem in ideal mixtures of two component model fluids. J. Chem. Phys. **119**(4), 2144–2146 (2003)
13. Evans, D.J., Cohen, E.G.D., Morriss, G.P.: Probability of second law violations in shearing steady states. Phys. Rev. Lett. **71**(15), 2401–2404 (1993)
14. Wang, G.M., Sevick, E.M., Mittag, E., Searles, D.J., Evans, D.J.: Experimental demonstration of violations of the second law of thermodynamics for small systems and short time scales. Phys. Rev. Lett. **89**(8905), 601–601 (2002)
15. Bub, J.: Maxwell's demon and the thermodynamics of computation. Stud. Hist. Philos. Sci. B Stud. Hist. Philos. Mod. Phys. **32**(4), 569–579 (2001)
16. Badger, W.L., McCabe, W.L.: Elements of Chemical Engineering, *Chemical Engineering Series*, 2nd edn. McGraw-Hill Book Company, Inc., New York, London (1936)
17. Landsberg, P.T.: Fragmentations, mergings and order: Aspects of entropy. Phys. Stat. Mech. Appl. **305**(1–2), 32–40 (2002)
18. Piekarczyk, W.: Crystal growth of CVD diamond and some of its peculiarities. Cryst. Res. Technol. **34**(5–6), 553–563 (1999)
19. Wald, R.M.: The thermodynamics of black holes. Living Rev. Relat. **4**(6) (2001). [electronic] http://www.livingreviews.org/lrr-2001-6
20. Bardeen, J.M., Carter, B., Hawking, S.W.: The four laws of black hole mechanics. Commun. Math. Phys. **31**(2), 161–170 (1973). URL http://projecteuclid.org/Dienst/UI/1.0/Summarize/euclid.cmp/1103858973?abstract=
21. Hawking, S.W.: Particle creation by black holes. Commun. Math. Phys. **43**(3), 199–220 (1975). URL http://projecteuclid.org/Dienst/UI/1.0/Summarize/euclid.cmp/1103899181?abstract=
22. Gramatke, H.P.: Perpetuum mobile. [electronic] http://www.hp-gramatke.net/perpetuum/index.htm (2003)
23. Michal, S.: Das Perpetuum mobile gestern und heute, 2nd edn. VDI-Verlag, Duüsseldorf (1981)
24. Antonijevic, B. Gravitational alternator. WO Patent 2 005 012 724, assigned to Antonijevic Borisa (Yu), Feb. 10 (2005)
25. Ritscher, H. Einrichtung zur Erzeugung von Elektrischer Energie. DE Patent 8 510 493U, Jul. 11 1987
26. Smeretchanski, M. Mechanical energy generator uses gravity and buoyancy forces acting on variable volume elements controlled by gas cylinders. FR Patent 2 830 575, assigned to Smeretchanski Mikhail (Fr), Apr. 11 2003
27. Lynden-Bell, D.: Negative specific heat in astronomy, physics and chemistry. Phys. Stat. Mech. Appl. **263**(1–4), 293–304 (1999)

28. Thirring, W.: Systems with negative specific heat. Z. Phys. A: Hadrons Nucl. **235**(4), 339–352 (1970)
29. D'Agostino, M., Bougault, R., Gulminelli, F., Bruno, M., Cannata, F., Chomaz, P., Gramegna, F., Iori, I., Le Neindre, N., Margagliotti, G.V.: On the reliability of negative heat capacity measurements. Nucl. Phys. A **699**(3–4), 795–818 (2002)
30. Allen, J.F., Jones, H.: New phenomena connected with heat flow in He II. Nature **141**, 243–244 (1938)

Chapter 11
Dimensional Analysis

Dimensional analysis is rarely taught in pure chemistry, rather in engineering chemistry. Most of the treatises on dimensional analysis content themselves to show how to convert units. Perhaps the most well-known application of dimensional analysis in pure chemistry is one dealing with the conversion of units, that is a part of dimensional analysis in the law of corresponding states [1]. The chapter is intended to show that there are other fields in chemistry that would benefit from dimensional analysis. In particular, it is shown how to establish Stokes' law in a simple way using dimensional analysis. Further, dimensional analysis can be used to find rather unexpected relations. This should be used to motivate the reader to think about the context of various scientific quantities.

There are several papers on dimensional analysis [2–5], but most of them are dealing with the conversion of units that is a part of dimensional analysis. Dimensional analysis in the more advanced sense is presented rather in the older literature [6–8] or in textbooks [9, 10]. One paper is dealing with the theorem of the corresponding states [1], which is a more advanced form of dimensional analysis. In summary, dimensional analysis is used rather in engineering sciences, including technical chemistry, than in pure chemistry.

11.1 Units

In an introductory part, we will discuss the issue and origin of physical units. Some historical facts in particular concerning the origin of units in Europe and the US are published online [11]. The history of the Bureau International des Poids et Mesures is documented elsewhere [12].

11.1.1 The Metric System

The metric system was invented in France, in the course of the French revolution. In 1790 Sir John Riggs Miller and the Bishop of Autum, Prince Talleyrand proposed that Great Britain and France should cooperate to equalize their weights and measures, by the introduction of the metric system.

J.K. Fink, *Physical Chemistry in Depth*,
DOI 10.1007/978-3-642-01014-9_11, © Springer-Verlag Berlin Heidelberg 2009

To focus on the results of these efforts with time, realize that in the UK, only in 1971 the currency was decimalized. The situation complicates that e.g., the US gallon is of a different size in comparison to the UK gallon. So liquid measures of the same name are not of the same size in the US and UK systems. Also that the US ton is 2000 pounds, whereas a UK ton is 2240 pounds. These are also referred to as a short ton and long ton, respectively.

The first step in the creation of the decimal metric system can be derived from the two platinum standards representing the meter and the kilogram. This was introduced on Jun. 22, 1799, in the Archives de la République in Paris. In 1832, Gauß strongly promoted the application of the metric system in Germany. Gauß and *Weber*[1] included also magnetic and electric quantities in the system. In the UK, Maxwell and Thomson pointed out the need for a consistent system of units. In 1874, the British Association for the Advancement of Science (BAAS) introduced the CGS system, based on the units centimeter, gram, and second, prefixes ranging from micro to mega were suggested. However, the use of the CGS system in electricity and magnetism proved to be inconvenient. In the 1880s, the BAAS and the International Electrical Congress suggested further units, among them were ohm, volt, and ampere.

In 1901, *Giorgi*[2] showed that it is possible to combine the CGS system with the electric and magnetic units to create a four-dimensional system by adding the ampere or the ohm, to the CGS units [13].

Before, in the Gaussian system, the electric permittivity and the magnetic permeability were treated as dimensionless quantities. This caused strange situations in case of electric and magnetic units in the Gaussian system. Thus, the equations in electromagnetism were reformulated. In SI units, the vacuum permittivity ε_0 has the unit $A^2 s^2 N^{-1} m^{-2}$ and the magnetic permeability μ_0 has the unit NA^{-2}. Observe that $\sqrt{\varepsilon_0 \mu_0}$ is a reciprocal velocity.

In 1946, the Conférence Générale des Poids et Mesures (CGPM) approved a system, based on meter, kilogram, second, and ampere. Whereas before the cm, a unit with prefix was previously a basic unit, thereafter the kg, also a prefixed unit appeared as basic unit. In 1954, the kelvin and the candela as base units were introduced, and in 1971 the mole appeared as basic unit.

11.1.2 The Units in Detail

11.1.2.1 The Unit of Length

The origins of the meter go back to at least the 18th century. At that time, there were two competing approaches to the definition of a standard unit of length. Some

---

[1] Wilhelm Eduard Weber, born Oct. 24, 1804, in Wittenberg, Germany, died Jun. 23, 1891, in Göttingen, Germany.

[2] Giovanni Giorgi, born Nov. 27, 1871, in Lucca, Italy, died Aug. 19, 1950, in Castiglioncello.

suggested defining the meter as the length of a pendulum having a half-period of one second; others suggested defining the meter as one ten-millionth of the length of the earth's meridian along a quadrant, i.e., one fourth of the circumference of the earth. In 1791, the French Academy of Sciences preferred the meridian definition over the pendulum definition because the force of gravity varies slightly over the surface of the earth, affecting the period of the pendulum. Initially the length was too short by 0.2 mm because the flattening of the earth due to its rotation was neglected.

In 1927, the meter was defined as the distance between the axes of the two central lines marked on the bar of platinum – iridium kept at the Bureau International des Poids et Mesures (BIPM) under certain conditions, such as temperature and pressure. In 1960 a definition came up, based upon the wavelength of ^{86}Kr radiation. In 1983, the CGPM replaced this definition by the following definition [14]:

> The meter is the length of the path travelled by light in vacuum during a time interval of 1/299 792 458 of a second. This definition is based on the constant velocity of light in vacuum at exactly 299792458 m s^{-1}.

11.1.2.2 The Unit of Mass

Around 1800, one kilogram was defined as the mass of a cubic decimeter of water. In 1901, the kilogram was defined as the mass of a certain standard, made of platinum – iridium. This definition is still waiting for a replacement.

11.1.2.3 The Unit of Time

Originally, the second was defined originally as the fraction 1/86,400 of the mean solar day. However, unpredictable irregularities in the rotation of the earth were discovered. In 1960 a definition, based on the tropical year, was presented by the International Astronomical Union. In 1997 the following definition was agreed:

> The second is the duration of 9 192 631 770 periods of the radiation corresponding to the transition between the two hyperfine levels of the ground state of the cesium 133 (^{133}Cs) atom.

11.1.2.4 The Unit of Electric Current

The international electric units were introduced by the International Electrical Congress in Chicago in 1893, and the definitions of the international ampere and the international ohm were confirmed by the International Conference of London in 1908. There were efforts to replace these international units by so-called absolute units. In 1946 a definition that relates to mechanical quantities was proposed:

> The ampere is that constant current which, if maintained in two straight parallel conductors of infinite length, of negligible circular cross section, and placed 1 meter apart in vacuum, would produce between these conductors a force equal to 2×10^{-7} Newton per meter of length.

11.1.2.5 The Unit of Thermodynamic Temperature

The definition of the unit of thermodynamic temperature was given in 1954. The triple point of water was considered as the fundamental fixed point and assigned to the temperature of 273.16 K. In 1967, the notation ° K for Kelvin was replaced by simply K. The unit of thermodynamic temperature was defined as follows:

The Kelvin unit of thermodynamic temperature, is the fraction

$$\frac{1}{273.16}$$

of the thermodynamic temperature of the triple point of water.

The unit of *Celsius*[3] temperature is the degree Celsius, symbol °C. It is by definition equal in magnitude to the kelvin. The numerical value of a Celsius temperature t expressed in degrees Celsius is given by

$$t/^\circ C = T/K - 273.15.$$

The value of -273.15 K is the ice point of water.

11.1.2.6 The Unit of Amount of Substance

After establishment of the fundamental laws of chemistry, units like *gram-atom* or *gram-molecule*, were used to specify amounts of chemical elements or compounds. These units are directly related to atomic weights and molecular weights. These units refer to relative masses. The advent of mass spectrometry showed that the atomic weights arise from mixtures of isotopes. Intermittently two scales, a chemical scale and a physical scale were in use. In 1960, by an agreement between the International Union of Pure and Applied Physics (IUPAP) and the International Union of Pure and Applied Chemistry (IUPAC), this duality was eliminated.

It remained to define the unit of amount of substance by fixing the corresponding mass of carbon 12. By international agreement, this mass has been fixed at 0.012 kg, and the unit of the quantity *amount of substance* was given the name mole (symbol mol). The amount of matter was then defined as follows:

The mole is the amount of substance of a system which contains as many elementary entities as there are atoms in 0.012 kilogram of carbon 12. Its symbol is mol.
When the mole is used, the elementary entities must be specified and may be atoms, molecules, ions, electrons, other particles, or specified groups of such particles.

Avogadro's Number

The history related to *Avogadro*'s number has been compiled by *Murrel* [15]. The measurement of Avogadro's number is close related to the problem of measuring

---

[3] Anders Celsius, born Nov. 27, 1701, in Uppsala, Sweden, died Apr. 25, 1744, in Uppsala, Sweden.

the amount of substance in practice. In Table 11.1 various methods that have been used to measure Avogadro's number are compiled.

Table 11.1 Measurement of the Avogadro number [16–19]

Year	Scientist	Method
1865	Loschmidt	Gas viscosity and internal volume
1870	Kelvin	Expansion of a liquid film to infinite size and comparing with evaporation energy
1870	Kelvin	Inhomogeneity of matter via optical properties
1870	Kelvin	Attraction of electrically charged plates forming an electrochemical element and chemical potential
1871	Rayleigh	Theory of light scattering
1900	Planck	Radiation spectrum
1908	Rutherford, Geiger	Radioactive decay
1910	Einstein, Smoluchowski	Theory of fluctuations
1916	Millikan	Faraday constant and charge of electron
1908	Perrin	Hypsometric law, sedimentation equilibrium
1930	Bearden	X-ray analysis of atomic distances

11.1.2.7 The Unit of Luminous Intensity

Originally, each country had its own, and rather poorly reproducible, unit of luminous intensity. Only in 1909 some efforts to unify the luminous intensity appeared by the international candle based on carbon filament lamps. However, Germany, at the same time, stayed with the *Hefner*[4] candle which is equal to ca. 9/10 of an international candle.

The properties of a black body offered a theoretically perfect solution to define the luminous intensity. In 1933, this principle was adopted to the new candle based on the luminance of the black body radiation at the temperature of freezing platinum. In 1979, because of experimental difficulties in realizing a black body radiator at high temperatures, a new definition of the candela appeared:

> The candela is the luminous intensity, in a given direction, of a source that emits monochromatic radiation of frequency 540 x 1012 hertz and that has a radiant intensity in that direction of 1/683 watt per steradian.

11.1.3 The International System

The International System (Le Systeme international d'Unites) was established in 1960 (11th CGPM, Conférence Générale des Poids et Mesures). There are seven basic units that are given in Table 11.2. The definition of the units was revised from time to time. For instance, originally the unit of length was given as $1/(40 \times 10^6)$ of

---

[4] Friedrich Franz Philipp von Hefner-Alteneck, born Apr. 27, 1845, in Aschaffenburg, Germany, died Jan. 7, 1904, in Biesdorf.

the length of the earth meridian cutting Paris. In Table 11.3 the recent definitions of the basic units are summarized.

Table 11.2 The basic units in the international system [20]

Kind	Unit name	Abbreviation
Length	Meter	m
Mass	Kilogram	kg
Time	Second	s
Electric current	Ampere	A
Temperature	Kelvin	K
Amount of substance	Mole	mol
Luminous intensity	Candela	cd

Table 11.3 Recent definitions of the basic units [21]

Unit	Year[a]	Definition
Kilogram	1901	One kilogram is the mass of the international prototype which is in the custody of the Bureau International des Poids et Mesures at Sévres, France.
Meter	1983	One meter is the length of the path traveled by light in vacuum during a time interval of 1/299 792 458 of a second.
Second	1983	One second is the duration of 9 192 631 770 periods of the radiation corresponding to the transition between the two hyperfine levels of the ground state of the ^{133}Cs atom.
Kelvin	1967	One Kelvin is the fraction 1/273.16 of the temperature of the triple point of water.
Ampere	1948	One ampere is the constant current which, if maintained in two straight parallel conductors of infinite length, of negligible circular cross section, and placed 1 meter apart in a vacuum, would produce a force equal to 2×10^{-7} Newton per meter of length of the wire.
Candela	1970	One candela is the luminous intensity, in a given direction, of a source that emits monochromatic radiation of a frequency 540,1012 Hz and that has a radiant intensity in that direction of (1/683) watt per steradian.
Mole	1970	One mole is the amount of substance of a system which contains as many elementary entities as there are atoms in 0.012 kg of carbon.

[a] Year of revision

11.1.4 Derived Units

From the basic units, other units may be derived, e.g., Newton, Pascal, Joule, Volt To the basic units prefixes may be attached. The prefixes are given in Table 11.4. We ask now whether the seven basic units are literally basic. For example, we can derive the temperature from the ideal gas law

$$pV = nRT.$$

Table 11.4 Prefixes for units

Prefix	Abbreviation	Power	Prefix	Abbreviation	Power
deca	da	10^1	deci	d	10^{-1}
hecto	h	10^2	centi	c	10^{-2}
kilo	k	10^3	milli	m	10^{-3}
mega	M	10^6	micro	μ	10^{-6}
giga	G	10^9	nano	n	10^{-9}
tera	T	10^{12}	pico	p	10^{-12}
peta	P	10^{15}	femto	f	10^{-15}
exa	E	10^{18}	atto	a	10^{-18}
zetta	Z	10^{21}	zepto	z	10^{-21}
yotta	Y	10^{24}	yocto	y	10^{-24}

The ideal gas law provides a relation between temperature and the product of pressure and volume. However, a conversion factor R is needed. Instead of making use of the conversion factor, we can state directly

$$\frac{pV}{n} = T'.$$

In this view, T' has the dimension of energy per mol. We could readily measure the temperature in terms of volume and pressure – if an ideal gas is available. So we have traced back the temperature to the (mean) energy of a particle. The temperature is expressed in terms of mechanical quantities.

In the thermodynamic formulation of the energy as

$$dU = T'dS',$$

the entropy S' has the dimension of an amount of a substance, i.e., mole.

11.2 The Minimal Set of Basic Fundamental Units

In a similar way as shown for the temperature, also other physical quantities even belonging to the so-called basic or fundamental units can be traced back to mechanical physical quantities. It remains to answer how many basic units are really needed at minimum. First, we state that the introduction of a physical unit requires the definition of a physical quantity itself. Without a physical quantity now unit for this quantity is needed. The new physical quantity is introduced by a relation including other previously known physical quantities.

In fact, the question arises whether we need really these seven basic units, or whether the number of basic units can be still further reduced. For example, around 1900, *Ostwald*[5] suggested that the energy should be one of the basic units. Instead

---

[5] Wilhelm Ostwald, born Sep. 2, 1853, in Riga, Latvia, died Apr. 4, 1932, in Großbothen, Germany.

of mass – length – time, there should be energy – length – time. So Ostwald has put the concept of energy over the concept of the substance. At the end of the 19th century, the scientific world was divided into the energists and into the atomists. It is interesting to emphasize that Ostwald, even being a chemist, belonged to the energists. We cite from the literature [22]. The English translation can be found elsewhere [23, 24]:

Auf der Naturforscherversammlung in Lübeck 1895 ist es zwischen den Atomisten und den Energetikern zu Auseinandersetzungen gekommen, die mit großer Leidenschaftlichkeit geführt wurden.

Die Stimmung auf dieser Tagung können wir beim Lesen von *Sommerfelds* Erinnerungen nachempfinden: Von den Energetikern hat der Dresdener *Helm* einen Vortrag gehalten. Er stand unter dem Einfluß *Ostwalds* und beide unter dem Einfluß der Machschen Philosophie, obwohl *Mach* selbst nicht anwesend war. Die Auseinandersetzung zwischen *Boltzmann* und *Ostwald* erinnerte sowohl vom Inhalt als auch von der Form her an den Kampf zwischen einem Stier und einem wendigen Torero. Diesmal blieb jedoch trotz aller Fechtkünste der Torero (*Ostwald*) auf dem Platz. *Boltzmanns* Argumente waren umwerfend. Wir Mathematiker standen alle auf seiten *Boltzmanns*.

Der historischen Wahrheit zuliebe müssen wir hier erwähnen, daß *Ostwald* den vollständigen Triumph der Atomtheorie noch erlebt hat und klug genug war, seine irrigen Stellungnahmen öffentlich zu korrigieren.

Ostwald himself states [25]:

Die Energie ist daher in allen realen oder konkreten Dingen als wesentlicher Bestandteil enthalten, der niemals fehlt, und insofern können wir sagen, daß in der Energie sich das eigentlich Reale verkörpert. Und zwar ist die Energie das Wirkliche in zweierlei Sinn. Sie ist das Wirkliche insofern, als sie das Wirkende ist; wo irgend etwas geschieht, kann man auch den Grund dieses Geschehens durch Kennzeichnung der beteiligten Energien angeben. Und zweitens ist sie das Wirkliche insofern, als sie den Inhalt des Geschehens anzugeben gestattet. Alles, was geschieht, geschieht durch die Energie und an der Energie. Sie bildet den ruhenden Pol in der Erscheinung Flucht und gleichzeitig die Triebkraft, weiche das Weltall der Erscheinungen um diesen Pol kreisen läßt. Wahrlich, wenn heute ein Dichter Ausschau halten wollte nach dem größten Inhalte menschlichen Denkens und Schauens und wenn er klagen wollte, daß keine großen Gedanken mehr die Seelen zu weitreichendem Umfassen aufregen, so könnte ich ihm den Energiegedanken als den größten nennen, den die Menschheit im letzten Jahrhundert an ihrem Horizonte hat auftauchen sehen, und ein Poet, der das Epos der Energie in würdigen Tönen zu singen verstände, wurde ein Werk schaffen, das den Anspruch hätte, als Epos der Menschheit gewürdigt zu werden... Aus diesem Grunde kann auch niemals eine energetische Theorie irgendeines Erscheinungsgebietes durch die spätere Entwicklung der Wissenschaft widerlegt werden, ebensowenig wie der Fortschritt der Wissenschaft jemals die Sätze von der geometrischen Ähnlichkeit der Dreiecke widerlegen wird. Das einzige, was in solcher Richtung geschehen kann, ist eine Erweiterung oder auch eine Verschärfung des Gesetzesinhaltes; in solchen Fällen aber handelt es sich nur um Verschönerungsarbeit am Gebäude, nie aber um ein vollständiges Niederreißen und Neubauen. Letzteres ist dagegen bei den üblichen mechanistischen Hypothesen immer unvermeidlich gewesen. Man denke beispielsweise nur an die Folgenreihe der Lichthypothesen, von der griechischen Vorstellung, daß Bilder in der Form von Häuten sich von den Gegenständen loslösen, um ins Auge zu gelangen, durch Newtons Lichtkügelchen, Huyghens und seiner Nachfolger Ätherschwingungstheorie auf mechanisch-elastischer Grundlage bis zur modernen elektromagnetischen Theorie.

From the nowadays view, some of the ideas of *Ostwald* have been practiced, insofar as now the units are chosen in that way that they are coupled via the energy. For example, the thermal energy is defined in Joule, and the specific heat capacity has the unit $J\,kg^{-1}K^{-1}$. Similarly, the electric units are defined in the way that $1\,V \times 1\,A \times 1\,s = 1\,J$.

Recently the question of the number of basic units necessary to describe any physical system has been revisited and reviewed [26]. In the Gaussian system, three basic dimensions are still necessary and sufficient to express the dimension of any physical quantity. These correspond to space (L), time (T), and matter (M). The number of units do not depend on the number and the nature of fundamental interactions in world. For example the basic units would be consistent with a world without gravity. Even electric and magnetic phenomena can be expressed based on space, time, and matter. In 1870, *Stoney*,[6] associated another meaning to the fundamental units:

$$L : \frac{e\sqrt{G}}{c^2}; \quad T : \frac{e\sqrt{G}}{c^3}; \quad M : \frac{e}{\sqrt{G}}.$$

Here, e is the elementary charge, c the velocity of light, and G the constant in the gravitational law. From this view, the set e, c, G is more fundamental than the set L, T, M. The expression for M can be obtained from Newton's law of gravitation and Coulomb's law of electrostatic forces, which have both the well-known form

$$F = C \frac{q^2}{r^2},$$

where F is the force, C is a constant, q is either the mass or the electric charge, and r is the mutual distance of mass or charge. In the case of gravitational forces, the dimension of the constant C can be determined; however, in the case of electrostatic forces, the constant was set deliberately to 1 and dimensionless. The expressions for L, T are obtained by further dimensional analysis. This system of fundamental units is sometimes addressed as Stoney scale.

Later, the concept of fundamental units was modified by *Planck* to

$$L : \frac{\sqrt{\hbar G}}{c^{3/2}}; \quad T : \frac{\sqrt{\hbar G}}{c^{5/2}}; \quad M : \sqrt{\frac{\hbar c}{G}}.$$

The use of $\hbar, c, G$ as fundamental units is believed to be more suitable for theoretical considerations.

The set of constants $\hbar, c, k$, where k is the Boltzmann constant have been interpreted as conversion factors of the energy E to reciprocal time, mass, and temperature, respectively, according to the well-known equations:

---

[6] George Johnstone Stoney, born Feb. 15, 1826, in Oak Park, Ireland, died Jul. 5, 1911, in Notting Hill, England.

$$E = \hbar\omega; \quad E = mc^2; \quad E = kT.$$

It has been concluded that in terms of quantum string theory, only two fundamental units are needed, i.e., velocity and length. However, these conclusions are subject of a controversy. In addition, the International System of Units has been criticized from the view of basic physics [26]:

> The SI might be useful from the point of view of technology and metrology, but from the point of view of pure physics four out of its seven basic units are evidently derivative ones.

For example, the electric current can be traced back to a number of electrons in motion. The temperature reflects the average energy of a collection of particles. The mol number is simply a number of molecules. The candela could be expressed as the flux of photons.

11.3 The Nature of a Physical Quantity

A physical quantity is composed of a dimensionless number and the unit. If a car is moving with 90 km h, as referred to commonly in newspapers, then 90 is the dimensionless number and *kmh* is the unit. Here we must take care because in the present case km h is an abbreviation for a composed unit. In particular, km h does not mean km $*$ h, but it means km $*$ h^{-1} and what is given in the newspapers is not just wrong, but it is highly misleading. Even in the sciences, we have sometimes such a situation. For example, the unit for the pressure is Pa, i.e., Pascal, in honor to Blaise *Pascal*.[7] On the other hand, Pa may be understood as Peta years, with P for 10^{15} and a as the abbreviation for year, from the Latin word *annum*.

Further, in the older literature, often the pressure is expressed in mm Hg, which is in fact not a length, but the pressure that exerts a column of mercury with the specified length (normalized at 0 °C). Therefore, also in the scientific literature, there are pitfalls concerning units.

11.3.1 Conversion of Units

The conversion of units is easy. We show this for the velocity. Suppose we want to convert km h^{-1} into m s^{-1}–So what is a velocity of 90 km h^{-1} in m s^{-1}? We simply insert the conversion factor in the units itself

$$km = 1000\,m; \; h = 3600\,s.$$

Therefore, 90 km h^{-1} = 90 × 1000 m(3600 s)$^{-1}$. Finally we gather the numbers which arose in the expression and obtain

---

[7] Blaise Pascal, born Jun. 19, 1623, in Clermont, France, died Aug. 19, 1662, in Paris, France.

$$90 \text{ km h}^{-1} = 90 * 1000/3600 \text{ m s}^{-1} = 25 \text{ m s}^{-1}.$$

The conversion of units does not take place in balancing an equation. If we write

$$h = 3600 \, s,$$

then this is an instruction how to substitute the units themselves. We may not insert into the units the numerical value. For example, since 1 hour is 3600 seconds, inserting the numeric values, is wrong, e.g., $1 = 3600 \, 3600$, as can be seen immediately. Also the following procedure is not correct. We want to introduce a factor that would give finally the number to convert the numbers

$$x_h \times f \times [h] = x_s \times [s]. \tag{11.1}$$

Inserting now the converted units into Eq. (11.1)

$$x_h \times f \times [3600 \, s] = x_s \times [s]$$

gives a wrong result for $f = 1/3600$. Instead, the conversion factor is obtained by successive substitution or even attaching the missing units to the expression. If the unit to be converted does not have the physical basic dimensions to appear finally, physical quantities with dimensions must be attached to the original expression.

We show this for the pressure. If we want to convert mm Hg into Pa, then a little bit more effort is needed. The unit mm Hg looks like a length but refers to a pressure that is exhibited by a column of mercury at $0 \,°C$ and standard earth acceleration $g \, [\text{m s}^{-2}]$.

Without thinking about details, how heavy such a column might be, we start with a formal procedure. We regard now mm Hg as a length as such. We convert at first mm Hg in m Hg, because we need the height in meters:

$$h \text{ mm Hg} = h \times 10^{-3} \text{ m Hg}.$$

So we *substitute* for mm Hg now 10^{-3} m Hg, rather than to resolve the equation. We may also suppress the "Hg" because it is rather an index, to indicate that we are dealing with mercury. For example, there is also sometimes a pressure unit given in mmH_2O. We address the height of the column by h.

Now we have already a length in the SI unit meters at the right-hand side. To this we have to multiply a quantity with the physical dimension $\text{kg m}^{-2}\text{s}^{-2}$ to obtain a pressure in the SI units. This might be split up in a density and an acceleration, i.e., $\text{kg m}^{-3} \times \text{m s}^{-2}$. This is in fact the density of mercury at $0 \,° C$ and the standard earth acceleration. Therefore, we arrive finally from mm Hg at

$$h \times 10^{-3} \times \rho \times g = h \times 10^{-3} \times 9.80665 \times 13595.5 \times 10^3 \text{ kg m}^{-1}\text{s}^{-2}.$$

We gather the numbers at the left-hand side as the conversion factor. To convert mmHg into Pascal, multiply by 133. This is because Pascal is the standard unit for the pressure in SI units, i.e., $Pa = kg\,m^{-1}s^{-2}$.

11.3.2 Presentation of Data in Tables and Figures

In tables, lengthy numbers should be avoided. If a certain quantity is $150,000\,kg$, the next entry is $130,000\,kg$, then it is desirable to abbreviate the entry in a suitable way. Several possibilities are shown in Table 11.5. In column 1 of Table 11.5 we have given the original SI unit, i.e., kg. However, the numbers below the head appear to be lengthy. The correct reading of column 2 expects that we have multiplied these $150,000\,kg$ by a factor of 10^{-3} to obtain the entries below. In column 3, the situation is reverse. The entries are given in multiples of 1000 kg. Therefore, to interpret the information correctly, some basic knowledge, how the table was constructed, is necessary. In column 4, we start with the equation

$$Qty = 150,000\,kg.$$

Table 11.5 Presentation of information in tables

1 Qty kg	2 Qty 10^{-3}kg	3 Qty 10^{+3}kg	4 Qty/[10^{+3}kg]	5 Qty/[t]
150, 000	150	150	150	150
130, 000	130	130	130	130

If we divide by 1000 kg, then just $Qty/[1000\,kg] = 150$ is obtained, the right-hand side being a dimensionless number, the left-hand side being that what is in the first row of the table. The presentation of the data as given in column 4 is recommended by IUPAC. Also column 5 follows without problems, because 1 ton is 1000 kg. On the other hand, if we would resolve the denominator in the mathematical sense, again a highly confusing entry would appear, i.e., $Qty\,10^{-3}\,kg^{-1}$.

The same arguments hold for figures. In figures, the axis labels should appear as dimensionless quantities. If the legend of an axis is $10^3\,K/T$, then a label tick of 4 means $10^3\,K/T = 4$, and therefore

$$T = 10^3/4\,K = 250\,K.$$

There are various alternative possibilities to denote the legend, e.g.,

$$\frac{10^3\,K}{T}; \frac{kK}{T}; \frac{10^3}{T/K}.$$

11.3.3 Π-Theorem

There are still other features in dimensional analysis, in particular the Π-theorem invented by *Buckingham*[8] in 1914 [27]. This powerful topic is reviewed by Wai-Kai Chen [28].

The Π-theorem states: If there are n variables in a problem and these variables contain m primary dimensions, then the equation relating all the variables will have $(n - m)$ dimensionless groups. Here a dimensionless group is some suitable product of the physical quantities that appears to have no physical dimension, i.e., a group lacks physical units. The name Π-theorem originates because *Buckingham* referred to these groups as Π groups.

11.3.4 Basic Principles of Dimensional Analysis

Dimensional analysis was introduced by Fourier in 1822. He stressed the importance of dimensional homogeneity and also the concept of a dimensional formula. In any physical (and chemical) equation, the quantity on the left-hand side must have the same physical dimension as the quantity on the right-hand side. This is the only way that retains the validity if the units used are changed, say for example from cm to inch, etc. The basic principle is best explained in an example.

11.3.5 Period of a Mathematical Pendulum

In a pendulum, the period (T) is dependent on the mass (M), the length of the wire (L), and the gravity acceleration (G) that acts on the mass. We presume some relationships $T = M^a L^b G^c$, a, b, c are undetermined exponents. Now the dimension of T is s (seconds), the dimension of L is m (meter), and the dimension of G is $m\,s^{-2}$. The exponents must be chosen in such a way that the right-hand side of the equation will be a time. Inserting the dimensions in the equation, we arrive at $s^1 = kg^a m^b m^c s^{-2c}$.

It is immediately seen that $a = 0$. Further, $1 = -2c$ and $b + c = 0$. So there should not be a dependence of the period on the mass and the relationship should be $T = K(L/G)^{0.5}$. K is a dimensionless constant or function of a dimensionless quantity.

In a more complicated situation, the following method is good: form the derivatives with respect to kg, m, and s. Equate the resulting equations at kg $=$ m $=$ s $= 1$. This method is suitable for a symbolic calculator and subsequently a session with the program DERIVE® [29] is given that does the job for the above example.

---

[8] Edgar Buckingham, born Jul. 8, 1867, in Philadelphia, PA, died Apr. 29, 1940, in Washington, DC, USA.

The alert reader will observe that chemical equations can be balanced with a similar method.

11.3.5.1 A DERIVE® Session

```
InputMode:=Word
CaseMode:=Sensitive
M:=kg
L:=m
G:=m*s^(-2)
T:=s
T=M^a*L^b*G^c
DIF(T=M^a*L^b*G^c,kg)
DIF(T=M^a*L^b*G^c,m)
DIF(T=M^a*L^b*G^c,s)
[DIF(T=M^a*L^b*G^c,kg),
DIF(T=M^a*L^b*G^c,m),
DIF(T=M^a*L^b*G^c,s)]

[0=a*kg^(a-1)*m^(b+c)*ABS(s)^(-2*c),
0=kg^a*m^(b+c-1)*(b+c)*ABS(s)^(-2*c),
1=-2*c*kg^a*m^(b+c)*s^(-2*c-1)*SIGN(s)^(2*c)]
[0=a*1^(a-1)*1^(b+c)*ABS(1)^(-2*c),
0=1^a*1^(b+c-1)*(b+c)*ABS(1)^(-2*c),
1=-2*c*1^a*1^(b+c)*1^(-2*c-1)*SIGN(1)^(2*c)]

[0=a,0=b+c,1=-2*c]
```

In principle, it is allowed to use arbitrary units, even general units such as L for a length.

11.3.6 Pitch of an Organ Pipe

We want to know the pitch of an organ pipe. The pitch is measured as a frequency v with the physical dimension of s^{-1}. We know intuitively that the pitch is somehow dependent on the length l of the flue pipe.

$$v = f(l).$$

To be correct in dimensions, there must be also a velocity in the equation. Therefore, we set up the relationship as

$$v = \frac{kv}{l}.$$

We interpret the velocity v as the velocity of sound. k is dimensionless function, turning out to be 0.5 for an open flue pipe, i.e., half of the wavelength. Even Lord Rayleigh was engaged in the pitch of organ pipes [30].

11.3.7 Motion of Sphere in a Medium

11.3.7.1 Stokes' Law

To come to a more important item for chemists, we derive *Stokes'*[9] law by means of dimensional analysis. A sphere with radius r is moving with constant speed v in a viscous medium with viscosity η subjected to a constant force F. We assume that a steady velocity will be established. This velocity is dependent on the force acting on the sphere, on the radius of the sphere, and on the viscosity that is surrounding the sphere. The physical dimensions of the variables are

$$r = \text{m}; \quad v = \text{m}\,\text{s}^{-1}; \quad F = \text{kg}\,\text{m}\,\text{s}^{-2}; \quad \eta = \text{kg}\,\text{m}^{-1}\text{s}^{-1}.$$

We assume a product law:

$$r v^a F^b \eta^c = K.$$

Using a similar procedure like in the pendulum example, we obtain

$$a = 1; \quad b = -1; \quad c = 1.$$

Note that one variable, here r, can occur without an exponent, since the equation system will be homogeneous. Rearranging, we arrive at

$$F = \frac{1}{K}\eta r v.$$

Further, observe that K may be a function of a dimensionless quantity. Dimensional analysis gives no information about the coefficient K. It is interesting to find that in the limiting cases of $F \to 0$, K seems to be a constant. If there is no force, then we put $F = 1$ in the law. There is no solution available, say a movement with constant velocity.

11.3.7.2 Newton's Law

There may be also other solutions of the problem. We are doing now a rather stepwise heuristic approach. The individual steps are shown in Table 11.6. First,

---

[9] George Gabriel Stokes, born 1819 in Skreen, County Sligo, Ireland, died in Cambridge, Cambridgeshire, England.

we assume that the velocity of the sphere is dependent on the force. We form the expression F/v. In this expression, still the time is does not cancel out. Therefore, we divide again by the velocity to arrive at F/v^2. Now in step 2 again a length is left over. To remove the length, we could easily multiply by the radius of the sphere, to do a reasonable operation. However, then the mass is left over.

We decide intuitively to proceed in another way, i.e., to add more length to the law, by dividing by the square of the radius of the sphere. What should we do alternatively? The weight of the sphere should not enter the presumed law. No matter whether the sphere is inside empty, or filled with a heavy metal, in the stationary state no inert forces may act, and we did not talk about gravity at all. So we are arriving at step 3 in Table 11.6. We discover now that with respect to the dimensions, a reasonable quantity is left over, i.e., a density. So we can finish the law in step 4, by dividing by the density.

Table 11.6 Stepwise approach of the problem of velocity of a sphere

Step no.	Physical law	Dimension
1	F/v	$\mathrm{kg\,m\,s^{-2}m^{-1}s} = \mathrm{kg\,s^{-1}}$
2	F/v^2	$\mathrm{kg\,m^{-1}}$
3	$F/(v^2r^2)$	$\mathrm{kg\,m^{-3}}$
4	$F/(v^2r^2\rho)$	1

It is left over now to interpret the individual terms appearing in the law. This is not a problem with the force that is acting on the sphere, with the radius of the sphere, and with the velocity. The density introduced in the last step could be the density of the sphere or the density of the surrounding medium. We decide to associate it with the latter.

We have now derived Newton's law that is on the other side of the law of Stokes. Whereas the law of Stokes is the limiting law for creeping flow, the law of Newton appears in aerodynamics where the viscosity of the dragging medium is not important. This type of Newton's law should be distinguished from various other laws named after Newton, e.g., the gravity law, cooling law.

11.3.7.3 Dimensionless Numbers

We can define dimensionless numbers as the rations of various kinds of energies or various kinds of forces. A few examples for energy forms are given in Table 11.7.

Table 11.7 Various types of energies

Energy	Expression
Kinetic energy	$mv^2/2$
Gravitational energy	mgh
Viscous energy	$\eta v D^2$
Thermal energy	mC_sT

η Viscosity; m Mass; v Linear velocity; D, h Some characteristic length; g Earth acceleration; C_s Specific heat capacity; T Temperature

The ratio of the kinetic energy to the viscous energy is

$$\frac{mv^2/2}{\eta v D^2}.$$

This can be further expanded to result in

$$\frac{\rho v D}{\eta} = \text{Re}. \tag{11.2}$$

Here Re is the Reynolds number, in honor to *Reynolds*,[10] a pioneer in fluid mechanics.

11.3.7.4 Friction Factors

In a more general approach, we may define a dimensionless friction factor $\check{f}$ [31, pp. 180–192] as

$$F = AK\check{f}. \tag{11.3}$$

The force F acting on a body with a characteristic area A implies a characteristic kinetic energy K per unit volume. Further, a dimensionless friction factor that includes the shape and other properties serves a proportionality factor. The friction factor is thus not constant.

For a sphere, we do not have another choice as to put the characteristic surface as $A = r^2\pi$ and the kinetic energy to $K = \rho v^2/2$. Therefore, for a sphere, Eq. (11.3) transforms into

$$F = \frac{r^2\pi\rho v^2\check{f}}{2}. \tag{11.4}$$

We can now include both the law of Stokes and the law of Newton, using the Reynolds number, Eq. (11.2). The law of Stokes will emerge from Eq. (11.4), if we take the friction factor as $\check{f} = 24/Re$, and the law of Newton will emerge, if we take the friction factor as $\check{f} = 0.44$. The friction factor for spheres as function of the Reynolds number is plotted in Fig. 11.1.

11.3.8 Dimensionless Formulation of Kinetic Equations

We consider here a consecutive reaction scheme consisting of elementary reactions of first order

---

[10] Osborne Reynolds, born Aug. 23, 1842, in Belfast, Ire, died Feb. 21, 1912, in Watchet, Somerset, England.

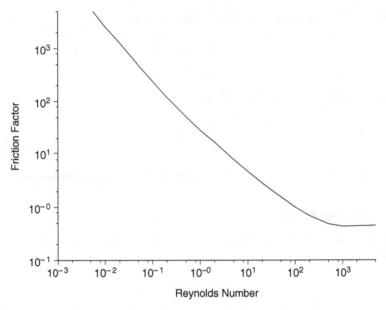

Fig. 11.1 Friction factor for spheres viz. Reynolds number [31, pp. 180–192]

$$A \xrightarrow{k_A} B \xrightarrow{k_B} C.$$

For the change of the concentration of the respective species with time, the following differential equations (11.5) hold:

$$-\frac{d[A]}{dt} = k_A[A]$$
$$-\frac{d[B]}{dt} = k_B[B] - \frac{d[A]}{dt} \qquad (11.5)$$

with the initial condition $t = 0; [A] = [A]_0$. The solution for the time dependence of the concentrations is

$$[A] = [A]_0 \exp(-k_A t)$$

and

$$[B] = \frac{k_A[A]_0}{k_B - k_A} \left[\exp(-k_A t) - \exp(-k_B t)\right].$$

In the course of the calculation, an additional integration constant emerges. Here the integration constant is adjusted in this way that at $t = 0, [B] = 0$. In this solution, there will appear the constants $k_A, k_B, [A]_0$ as parameters.

The parameters can be reduced significantly if reduced quantities are introduced. We introduce

$$\tau = k_A t, \quad \alpha = \frac{[A]}{[A]_0}, \quad \beta = \frac{[B]}{[A]_0}, \quad \text{and} \quad K = \frac{k_B}{k_A}.$$

Inserting these substitutions in Eq. (11.5), a much more simplified system of differential equations arises

$$\begin{aligned} -d\alpha/d\tau &= \alpha \\ -d\beta/d\tau &= K\beta - d\alpha/d\tau \end{aligned} \quad . \tag{11.6}$$

The initial conditions of the reduced quantities are

$$\begin{aligned} \tau = 0; &\quad \alpha = 1 \\ \tau = 0; &\quad \beta = 0 \end{aligned} \quad .$$

The equation can be solved most conveniently by using the *Laplace* transform. The Laplace transform of a function $f(t)$ is simply

$$F(s) = \int_0^\infty f(t)\exp(-st)dt. \tag{11.7}$$

In Table 11.8 we show some properties of the Laplace transform.

Table 11.8 Properties of the Laplace transform

Operation	Function	Laplace transform
Definition	$f(t)$	$F(s) = \int_0^\infty f(t)\exp(-st)dt$
Derivative	$f'(t)$	$sF(s) - f(0)$
Shift	$\exp(-kt)f(t)$	$F(s+k)$
Exponential	$\exp(-kt)$	$1/(s+k)$

Transforming now the equation set (11.6) results in two algebraic equations

$$\begin{aligned} -sA(s) + \alpha(0) &= A(s) \\ -sB(s) + \beta(0) &= KB(s) + A(s) \end{aligned} \quad ,$$

or by substitution with $A(0) = 1$ and $B(0) = 0$:

$$A(s) = \frac{1}{s+1}; \quad B(s) = -\frac{A(s)}{s+K} \quad .$$

The transformation rules given in Table 11.8 are sufficient to get back the relations for $[A]/[A]_0$ and $[B]/[A]_0$ as functions of time into

$$\beta = \frac{1}{1 - K} \left[\exp(-K\tau) - \exp(-\tau) \right].$$

The curves can be plotted in general using only one parameter, i.e., K. Such a plot is given in Fig. 11.2.

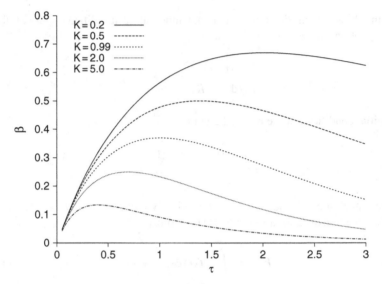

Fig. 11.2 Plot of the reduced concentration $\beta = [B]/[A]_0$ of the intermediate B viz. the reduced time for various ratios of rate constants $K = k_A/k_B$

11.3.9 Solving Differential Equations

There is still another technique, in which dimension analysis is used. The differential equations of transport can be set up sometimes easily, but it is too difficult to solve them under the conditions required. However, if the equations are made dimensionless, then the relevant combinations of dimensionless quantities can be found out.

11.3.9.1 Equation of Motion

We follow here the derivation given by Bird [31, pp. 97–98]. The equation of continuity for a Newtonian fluid with constant density and viscosity is given by

$$\rho \frac{D\mathbf{v}}{Dt} = -\nabla p + \eta \nabla^2 \mathbf{v} + \rho \mathbf{g}.$$

The equation of motion for a noncompressible flow is given as

$$\nabla \cdot \mathbf{v} = 0.$$

This set of equations can be made dimensionless by using the quantities

$$\mathbf{v}^* = \frac{\mathbf{v}}{V}; \quad p^* = \frac{p - p_0}{\rho V^2}; \quad t^* = \frac{tV}{D}; \tag{11.8}$$

$$x^* = \frac{x}{D}; \quad y^* = \frac{y}{D}; \quad z^* = \frac{z}{D};$$

$$\nabla^* = D\nabla; \quad \nabla^{*2} = D^2\nabla^2; \quad \frac{D}{Dt^*} = \left(\frac{D}{V}\right)\frac{D}{Dt}. \tag{11.9}$$

Here we use the star (*), following the notation of Bird [31] to indicate that we are dealing with a reduced and dimensional physical quantity. The result is

$$\nabla^* \cdot \mathbf{v}^* = 0$$

and

$$\frac{D}{Dt^*} = -\nabla^* p^* + \left[\frac{\eta}{DV\rho}\right]\nabla^{*2}\mathbf{v}^2 + \left[\frac{gD}{V^2}\right]\frac{\mathbf{g}}{g}. \tag{11.10}$$

In Eq. (11.10) two dimensionless numbers can be identified, i.e.,

$$Re = \left[\frac{\eta}{DV\rho}\right] \quad \text{and} \quad Fr = \left[\frac{gD}{V^2}\right].$$

Re is the Reynolds number and Fr is the *Froude*[11] number. Numerous flow problems can be solved by applying similarity relations. For the individual problems, the various scaling parameters given in Eqs. (11.8) to (11.9) must be chosen in a proper way. For example, Bird [31, pp. 101–102] exemplifies how the depth of a vortex in an agitated tank can be found out by similarity calculations.

Movement of Ions

This technique has been illustrated by considerations on the movement of ions in hydrated concrete [32]. In this way, the leaching of ionic species and eventually statements concerning the service time of concrete can be obtained.

---

[11] William Froude, born Nov. 28, 1810, in Dartington, Devon, England, died May 4, 1879, in Simonstown, South Africa.

Problems in Fire Research

The equations in problems regarding fire research are so much complicated that a close solution is not possible. However, using dimensional analysis, some valuable insight can be obtained [33].

11.4 Illustrative Examples

11.4.1 Melting Temperature and Coefficient of Thermal Expansion

The melting temperature has naturally the physical dimension of a temperature. To find a relation with other quantities in a dimensionless form, we go through tables and discover that the coefficient of thermal expansion has the dimension of a reciprocal temperature. Now there could be some natural relation between these two quantities. We can fix a zero temperature as reference state or infinite temperature as reference state. It is more reasonable to fix zero temperature as reference state. Therefore, coming from low temperature, it is more likely to choose the coefficients of thermal expansion in the solid state than the coefficients of thermal expansion in the liquid state.

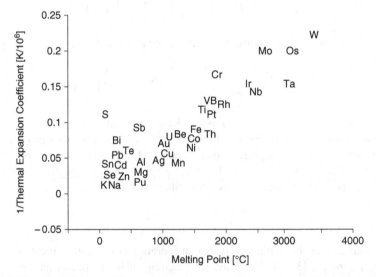

Fig. 11.3 Plot of coefficients of thermal expansion at room temperature of the elements against their melting temperatures. Data from CRC handbook of chemistry and physics [34, p. 12–124]

In Fig. 11.3, the coefficients of thermal expansion at room temperature, as far as available, are plotted against the melting temperatures of the elements. It looks like there is indeed a relation between the coefficients of thermal expansion and the melting temperatures. Elements with a big coefficient of thermal expansion will have a low melting point and vice versa. This finding can be easily explained with the range of action of interaction forces.

Question 11.1. Metals with a low melting point are ductile, whereas metals with a high melting point are brittle. Think about ductile and brittle. Are these contradictory concepts? Choose a physical quantity to characterize this property, e.g., various methods for measuring hardness. Make a plot of melting point against hardness.

11.4.2 Boiling Point of Linear Alkanes

In nearly every textbook on organic chemistry, there is either a plot or a table relating the boiling points of the homologous linear alkanes with the number of carbon atoms. In Fig. 11.4, the boiling point of *n*-alkanes is plotted viz. the number of carbon atoms. We are looking now behind the law governing this relationship. If the substance evaporates then there is an increase in surface. We may even think that the molecules in gas phase have a surface that is the border to the vacuum. Therefore, we may confront the enthalpy of vaporization to the surface energy:

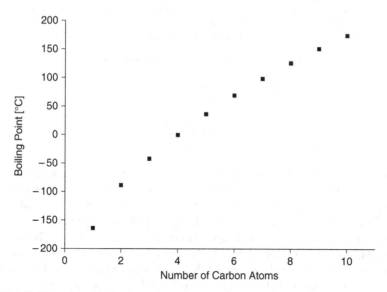

Fig. 11.4 Boiling point of *n*-alkanes against number of carbon atoms

$$\Delta \tilde{H}_v = T_v \Delta \tilde{S}_v = \sigma \tilde{A} = N_A \sigma a. \tag{11.11}$$

In introducing the entropy of vaporization, we indicate that we intend to use Trouton's rule. $\tilde{A}$ is the molar surface, which is related via the Avogadro Number $N\tilde{A}$ to the surface of a single molecule. The gain of area is also dependent on the shape of

the molecule. For geometric similar bodies, there is a relationship of the area to a
characteristic length and also to the volume. For example, for spheres the volume is
related to the area by

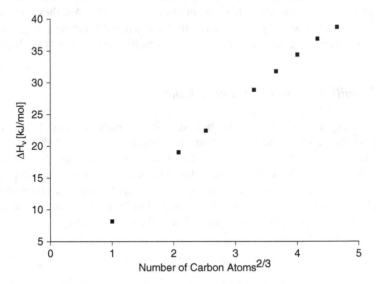

Fig. 11.5 Number of carbon atoms against heat of vaporization of n-alkanes

$$\frac{V^{2/3}}{A} = (36\pi)^{-3} \approx 0.207.$$

Assuming that the molecules in gas phase are collapsed and the density in the
molecules (matter) itself is approximately the same, we obtain $n \propto V$, where n is
the number of carbon atoms. Thus, a relation $\Delta \tilde{H}_v \propto n^{2/3}$ will be expected on the
base of the former considerations. In Fig. 11.5, a plot of the 2/3 power of the number
of carbon atoms against the heat of vaporization of n-alkanes is shown. The linearity
of the plot is better than those on Fig. 11.4, although from a double logarithmic plot
of the boiling point (in Kelvin) against the number of carbons it can be seen that
the slope varies from initially ca. 0.7–0.5 for 10 carbons. It is now a good exercise
for the reader to reason about this shortcoming. In Table 11.9 and in Table 11.10
some data of interest are given. A number of checks can be made. The following
suggested plots are not given here for reasons of space. Plot these items:

- Check whether there is a linear relation between the number of carbons and the
  internal volume.
- Check whether Trouton's rule is sound by plotting the boiling point (in Kelvin)
  against the heat of vaporization.
- Plot the surface tension against the number of carbons.

Table 11.9 Physical data for linear alkanes [34]

Name	n	$\Delta\tilde{H}_v$ [kJmol^{-1}]	T_b [°C]	a [l^{2}at mol^{-2}]	b [l mol^{-1}]
Methane	1	8.17	−164	2.253	0.04278
Ethane	2		−88.6	5.489	0.06380
Propane	3	19.04	−42.1	8.664	0.08445
n-Butane	4	22.44	−0.5	14.47	0.1226
n-Pentane	5		36.1	19.01	0.1460
n-Hexane	6	28.85	69	24.39	0.1735
n-Heptane	7	31.77	98.4	31.51	0.2065
n-Octane	8	34.41	125.7	37.32	0.2368
n-Nonane	9	36.91	150.8		
n-Decane	10	38.75	174.1	48.55	0.2905

Name: Compound name; n: Number of carbon atoms; $\Delta\tilde{H}_v$ [kJ mol^{-1}]: Molar heat of vaporization at the boiling point; T_b [°C]: Boiling point; a[l^{2}atm mol^{-2}]: Internal pressure; b[lmol^{-1}]: Internal volume

Table 11.10 Surface properties [35]

Name	σ [mJ/m^2]	A [mJ/m^2]	B [mJ/(m^{2}K)]
n-Pentane	15.48	18.25	0.11021
n-Hexane	17.91	20.44	0.1022
n-Heptane	19.80	22.1	0.098
n-Octane	21.14	23.52	0.09509
n-Nonane	22.38	24.72	0.09347
n-Decane	23.37	25.67	0.09197

σ[mJ/m^2]: Surface tension at 25 °C; A[mJ/m^2]: Parameter for surface tension equation $\sigma = A - BT$ (Temperature range 0–100 °C): B[mJ m^{-2}K^{-1}]: Parameter for surface tension equation $\sigma = A - BT$ (temperature range 0–100 °C)

- Plot the boiling point against the product of (internal volume)$^{2/3}$ * (surface tension) as suggested from the relation.
- Plot the molar surface area $\tilde{A} = \Delta\tilde{H}_v/\sigma$ against the number of carbon atoms. What is the power law? State about the molecule shape.

Take care on the correct units. Try to explain the findings. There are further tests of interest: Work out similarity theorems for other shapes of the molecules than spheres. Estimate whether the molecule radii obtained from the relation are in the correct range. Take notice that Lord Kelvin established such considerations [17, 18].

Question 11.2. Is there a possibility to obtain a numerical value for the Avogadro number via the surface and the internal volume? Test the relation with other underlying models. Assume that the molecules are glued together by gravitational or electrostatic forces. Calculate the work to be done, if a single molecule is removed far apart from the neighbors, what is in fact an evaporation? What power law would turn out for this case? Try the same assuming dipole interactions. Are the forces in the correct range in comparison to the thermal energy? After doing so much data handling and reasoning, the reader should decide to accept the suggested relationship or to reject it, or to think there may be some truth behind it.

11.4.3 Kelen Tüdös Plot

The copolymerization equation can be written as

$$\frac{m_a}{m_b} = \frac{1 + r_a[M_a]/[M_b]}{1 + r_b[M_b]/[M_a]} \tag{11.12}$$

for small conversions. According to *Fineman* and *Ross* [36], Eq. (11.12) can be linearized into a parametric form

$$y = \left(\frac{m_a}{m_b} - 1\right) \frac{[M_a]}{[M_b]} \left(\frac{m_a}{m_b} - 1\right)^{-1}$$
$$x = \left(\frac{[M_a]}{[M_b]}\right)^2 \left(\frac{m_a}{m_b} - 1\right)^{-1} \quad .$$

According to *Kelen* and *Tüdős* [37], an auxiliary variable is introduced

$$\alpha = \left(\frac{x_{min}}{x_{max}}\right)^{1/2},$$

and the original form is transformed into a dimensionless form, as

$$\tilde{x} = x \, (\alpha + x)^{-1}$$
$$\tilde{y} = y \, (\alpha + x)^{-1} \quad .$$

11.4.4 Life Expectancy and Body Weight of Mammals

Allometry is the science that describes the relationships of properties of living organisms. Usually, allometric laws are expressed as power laws. We will derive some power laws using dimensional analysis.

We know that mammals have a constant temperature. It is essential for the vital activity to maintain this temperature. There is a heat flux through the skin into the surrounding. Big mammals have a lower surface to volume ratio in comparison to small mammals. Therefore, the heat generated by the volume or the mass unit must be less for big mammals. We may think that the biochemical activity is lower for bigger mammals [38, 39]. This means that the biological time will run slower for the bigger varieties. If the cells have a limited capacity to reproduce, this means that bigger mammals will grow older than the smaller ones. Without going into the bio-chemical details of cell division and aging, we try it out first. A double logarithmic plot of life expectancy viz. body weight looks like in Fig. 11.6. The following items should be discussed:

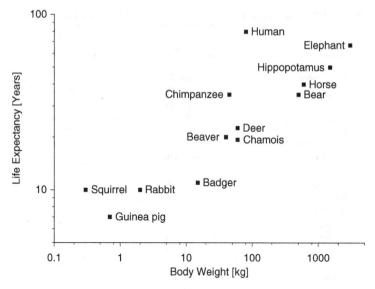

Fig. 11.6 Life expectancy viz. body weight of mammals

- Is the power law like that what we would expect from scaling? This means, balance the heat flow and the heat generation.
- Are there some other relations, like heartbeat frequency on body weight?
- If the consideration is based solely on the energy flow then other animals, like reptiles or insects, should not follow this law. In Table 11.11 some data are given to illustrate this behavior.

We go now a little bit more in detail. Animals are most at rest. In this state, they have to maintain their temperature, as most important task. To achieve this, they must produce thermal energy. The biological cells have now a limited capacity to produce energy. It is the same situation as an electrochemical cell is exhausted, after producing a certain amount of energy, or an engine will be worn out after doing certain strokes. Therefore, the thermal energy that can be delivered per unit weight of tissue is

$$\frac{E}{m} \propto k.$$

If there is a constant energy production $\dot{E}$ during lifetime t_l, we may convert this into

$$\frac{\dot{E} t_l}{m} \propto k. \tag{11.13}$$

Table 11.11 Life expectancy of animals and plants [49, 50] and body weight

Organism	Life expectancy [years]	Body weight [kg]
Deer	15–25	17
Ant (Queen)	10–15	
Badger	10–12	10–20
Bat	12–15	
Bear	30–40	150–780
Beaver	15–25	40
Beech marten	8–10	
Beech tree	600–1000	
Bird of prey	→ 100	
Blackbird	→ 20	
Blindworm	> 30	
Blueberry	→ 28	
Camel	ca. 30	530
Canary	→ 24	
Carp	60–100	
Cat	20–30	3–4
Catfish	60–100	
Cattle	15–25	
Cedar	→ 1000	
Chamois	18–25	60
Chimpanzee	→ 35	40–50
Cockatoo	70–85	
Cypress	→ 2000	
Deer	20–25	40–100
Donkey	40	190
Eagle-owl	→ 70	
Eel	30–50	
Elephant	60–80	1500–4000
Elm	300–600	
Ermine	5–7	
Fir tree	300–500	
Fox	10–12	4
Freshwater crayfish	20	
Frogs	12–20	
Goat	20	30
Gibbon	> 30	
Golden hamster	3	0.1
Goose	40–60	
Guinea pig	7	0.7
Hamster	6–10	
Hare, rabbit	8–12	2
Hedgehog	8–10	0.8
Herring	18	
Hippopotamus	50	1200–3000
Horse	30–50	600
Human	80	80
Ibex	30	
Jaguar	100	22

Table 11.11 (continued)

Organism	Life expectancy [years]	Body weight [kg]
Kangaroo	19	35
Leech	20–27	
Lime tree	600–1020	
Lion	30–35	
Lizard	25–30	
Lobster	50	
Mice	→ 1.5	
Mole	2–3	
Moose	20–25	
Mouse	→ 1.5	
Muskrat	5	
Newt	30–40	
Oak tree	500–1200	
Ostrich	30–40	
Pigeon	30–45	
Pike	60–100	
Pine tree	300–500	
Plaice	60–70	
Poplar	300–500	
Raccoon	13	4.3
Rat	5–7	0.3
Redwood tree	3000–4000	
Rhinoceros	50	
Sheep	15–20	55
Skunk	5–8	
Snail	> 9	
Snakes	25–30	
Spiders	1–20	
Spruce	300–500	
Squirrel	8–12	0.25–0.48
Starling	→ 20	
Stork	→ 70	
Swallow	→ 10	
Tapeworm	25–50	
Turtle	100–300	
Weasel	4–7	
Wolf	16	30–40
Woodchuck	15–18	
Yew tree	900–3000	

Now the energy production must be proportional to the surface of the animal, which is in turn proportional to the power of 2/3 to the body weight m:

$$\dot{E} \propto m^{2/3}.$$

Therefore, we obtain by substituting into Eq. (11.13)

$$\frac{t_l}{m^{1/3}} \propto k.$$

11.4.4.1 Metabolic Rate

The metabolic rate B of an organism with a mass M is related to $B = M^a$, where a was believed to be 3/4. This law goes back to *Kleiber.*[12]

However, simple dimensional analysis suggests $a = 2/3$. In reviewing various data sets [40, 41], it was found that there is little evidence for rejecting $a = 2/3$ in favor of $a = 3/4$ [42]. There is still a controversy concerning the exponent of this particular law [43–45]. In particular, many other physiological variables such as life span, heart rate, radius of aorta, respiratory rate, and so on scale with exponents that are typically simple multiples of 1/4 [46, 47]. For the life expectancy, i.e., the lifespan, a power law of $T = T_0 M^{1/4}$ has been suggested [48]. Brain weight and body weight, life span, gestation time, time of sleeping, and other data have been compiled and related [49].

Summarizing, it has been shown that dimensional analysis can serve as a motivation to become familiar with data, to turn them in various ways. Further, dimensional analysis should give creativity in as far as making curious how some connections on several fields of chemistry could be. The handling of data and plotting should not be a cumbersome task with the advent of spreadsheet calculators. A few examples have been given to illustrate the value of dimensional analysis and also how to deal with the laws in a critical way. It could not be the goal to establish some relationships without critical interpretation of them. Numerous further examples that are not quite obvious can be worked out in this way.

References

1. Hakala, R.W.: Dimensional analysis and the law of corresponding states. J. Chem. Educ. **41**, 380 (1964)
2. Canagaratna, S.G.: Is dimensional analysis the best we have to offer? J. Chem. Educ. **70**, 40 (1993)
3. DeLorenzo, R.: Expanded dimensional analysis: A blending of English and math (TP). J. Chem. Educ. **71**, 789 (1994)
4. Drake, R.F.: Working backwards is a forward step in the solution of problems by dimensional analysis. J. Chem. Educ. **62**, 414 (1985)
5. McClure, J.R.: Dimensional analysis: An analogy to help students relate the concept to problem solving (aa). J. Chem. Educ. **72**, 1093 (1995)
6. Langhaar, H.L.: Dimensional Analysis & the Theory of Models. John Wiley, Chapman & Hall, New York, London (1951)
7. Pankhurst, R.C.: Dimensional Analysis & Scale Factors. Monographs for Students. Chapman, London (1964)
8. Porter, A.W.: The Method of Dimensions, Monographs on Physical Subjects, 2nd edn. Methuen, London (1943)
9. Szirtes, T., Rózsa, P.: Applied Dimensional Analysis and Modeling, 2nd edn. Elsevier, Amsterdam (2007)

---

[12] Max Kleiber, born Jan. 4, 1893, in Zurich, Switzerland, died Jan. 5, 1976, in Davis, California, CA, USA.

10. White, F.M.: Dimensional analysis and similarity. In: Fluid Mechanics, *McGraw-Hill Series in Mechanical Engineering*, 5th edn. Chap. 5, pp. 293–342. McGraw-Hill, Boston, MA (2005)
11. NIST: The NIST reference on constants, units and uncertainty. [electronic] http://physics.nist.gov/cuu/Units/background.html (2005)
12. BIPM: History of the pavillon de breteuil. [electronic] http://www.bipm.org/en/bipm/history/ (1991)
13. Giorgi, G.: Unità razionali di Elettromagnetismo. Atti del'AEI p. 402 (1901)
14. Giacomo, P.: News from the BIPM. Metrologia **20**(1), 25–30 (1984)
15. Murrel, J.: Avogadro and his constant. [electronic] http://www.sussex.ac.uk/chemistry/ ;documents/the_constant_of_avogadro.pdf
16. French, A.P.: Atoms. In: R.G. Lerner, G.L. Trigg (eds.) Encyclopedia of Physics, 2nd edn., pp. 80–84. Wiley VCh, Weinheim (1991)
17. Hawthorne Jr., R.M.: Avogadro's number: Early values by Loschmidt and others. J. Chem. Educ. **47**, 751 (1970)
18. Thomson, W.: The size of atoms. Nature **1**, 551–553 (1870)
19. Virgo, S.E.: Loschmidt's number. Sci. Prog. (St. Albans, United Kingdom) **27**, 634–49 (1933). [electronic] http://gemini.tntech.edu/~tfurtsch/scihist/loschmid.html
20. Taylor, B.N.: Guide for the use of the international system of units (SI). NIST Special Publication 811, Physics Laboratory, National Institute of Standards and Technology, Gaithersburg, MD (1995)
21. Taylor, B.N.: The international system of units (SI). NIST Special Publication 330, Physics Laboratory, National Institute of Standards and Technology, Gaithersburg, MD (2001)
22. Simonyi, K.: Kulturgeschichte der Physik, pp. 395–396. Akadémiai Kiadó, Budapest (1990)
23. Eftekhari, A.: Ludwig Boltzmann (1844–1906). [electronic] http://philsci-archive.pitt.edu/archive/00001717/02/Ludwig_Boltzmann.pdf (no year)
24. Seth, S.: Crisis and the construction of modern theoretical physics. Br. J. Hist. Sci. **40**(1), 25–51 (2007)
25. Ostwald, W.: Die Energie, *Wissen und Können*, vol. 1, pp. 5, 6, 127, 128. Johann Ambrosius Barth Verlag, Leipzig (1908)
26. Duff, M.J., Okun, L.B., Veneziano, G.: Trialogue on the number of fundamental constants. J. High Energy Phys. **3**(23), 31 (2002). [electronic] http://arxiv.org/abs/physics/0110060
27. Buckingham, E.: On physically similar systems: Illustrations of the use of dimensional equations. Phys. Rev. **4**(4), 345Ű 376 (1914). [electronic] http://prola.aps.org/abstract/PR/v4/i4/p345_1
28. Wai-Kai, C.: Dimensional analysis. In: M. Howe-Grant, J.I. Kroschwitz (eds.) Kirk Othmer, Encyclopedia of Chemical Technology, vol. 8, 4th edn., pp. 204–223. John Wiley and Sons, New York (1993)
29. Software House, Inc. Derive, a mathematical assistant, version 3.06. Software House, Inc., Honolulu, Hawaii (1995)
30. Strutt (Lord Rayleigh), J.W.: On the pitch of organ-pipes. Philos. Mag. **13**, 340–347 (1882)
31. Bird, R.B., Stuart, W.E., Lightfoot, E.N.: Transport Phenomena, 2nd edn. John Wiley & Sons Inc., New York (2002)
32. Barbarulo, R., Marchand, J., Snyder, K.A., Prene, S.: Dimensional analysis of ionic transport problems in hydrated cement systems. Part 1. Theoretical considerations. Cem. Concr. Res. **30**(12), 1955–1960 (2000)
33. Thomas, P.H.: Dimensional analysis: A magic art in fire research? Fire Saf. J. **34**(2–3), 111–141 (2000)
34. Lide, D.R. (ed.): CRC-Handbook of Chemistry and Physics, 71st edn. CRC Press, Boca Raton, FL (1990–1991)
35. Riddick, J.A., Bunger, W.B.: Organic Solvents, Physical Properties and Methods of Purification. Wiley, New York (1970)
36. Fineman, M., Ross, S.D.: Linear method for determining monomer reactivity ratios in copolymerization. J. Polym. Sci. **5**(2), 259–262 (1950)

37. Kelen, T., Tüdős, F.: Analysis of the linear methods for determining copolymerization reactivity ratios. I. New improved linear graphic method. J. Macromol. Sci. Pure Appl. Chem. **A9**(1), 1–27 (1975)

38. Thompson, D.W.: On Growth and Form. Cambridge University Press, Cambridge (2000). Edited by John Tyler Bonner, first published in 1917

39. Thompson, D.W.: Über Wachstum und Form. Eichborn, Frankfurt (2006). Edited by Anita Albus and John Tyler Bonner, first published in 1917

40. Bennet, P., Harvey, P.: Active and resting metabolism in birds-allometry, phylometry and ecology. J. Zool. **213**, 327–363 (1987)

41. Heusner, A.A.: Size and power in mammals. J. Exp. Biol. **160**, 25–54 (1991)

42. Dodds, P.S., Rothman, D.H., Weitz, J.S.: Re-examination of the 3/4-law of metabolism. J. Theor. Biol. **209**, 9–27 (2001)

43. Enquist, B.J., Allen, A.P., Brown, J.H., Gillooly, J.F., Kerkhoff, A.J., Niklas, K.J., Price, C.A., West, G.B.: Biological scaling: Does the exception prove the rule? Nature **445**, E9–E10 (2007)

44. Hedin, L.O.: Biological scaling: Does the exception prove the rule? (reply). Nature **445**, E11 (2007)

45. Reich, P.B., Tjoelker, M.G., Machado, J.L., Oleksyn, J.: Biological scaling: Does the exception prove the rule? (reply). Nature **445**, E10–E11 (2007)

46. Schmidt-Nielsen, K.: Scaling: Why Is Animal Size so Important? Cambridge University Press, Cambridge, UK (1984)

47. West, G.B., Woodruff, W.H., Brown, J.H.: Allometric scaling of metabolic rate from molecules and mitochondria to cells and mammals. Proc. Natl. Acad. Sci. USA **99**(Suppl. 1), 2473–2478 (2002)

48. Azbel, M.Y.: Universal biological scaling and mortality. Proc. Natl. Acad. Sci. USA **91**(26), 12453–12457 (1994)

49. Allison, T., Cicchetti, D.: Sleep in mammals: Ecological and constitutional correlates. Science **194**(12), 732–734 (1976)

50. Bogenrieder, A. (ed.): Lexikon der Biologie. Herder, Freiburg. i. Br. (1985)

Chapter 12
Generating Functions

Generating functions are a highly useful tool. Actually, they are rarely used. They can find application within many branches of physical chemistry. Instead of talking long about what is a generating function, we show examples of generating functions. The coefficients of the exponential function are the reciprocal factorials of the natural numbers, because

$$\exp(x) = \frac{1}{0!}x^0 + \frac{1}{1!}x^1 + \frac{1}{2!}x^2 + \cdots .$$ (12.1)

The binomial coefficients are the coefficients of the series expansion of the expression

$$(1+x)^n = \binom{n}{0}x^0 + \binom{n}{1}x^1 + \binom{n}{2}x^2 + \cdots .$$ (12.2)

The Catalan numbers are counting the valid numbers of right and left brackets. For example, there are exactly five legal strings of three pairs of brackets:

$$()()();\quad (())();\quad ()(());\quad (())();\quad ((()))$$

The Catalan numbers are generated by the function $\frac{1-\sqrt{1-4x}}{2x}$

$$\frac{1 - \sqrt{1 - 4x}}{2x} = 1 + x + 2x^2 + 5x^3 + 14x^4 + \cdots .$$

The Hermite polynomials will be received as the coefficients of the series expansion from Eq. (12.3) as

$$\exp\left(-xy - y^2\right) = \sum_{n=1}^{n=\infty} \frac{H_n(x)}{n!} y^n.$$ (12.3)

12.1 Computation of Probabilities

Generating functions work satisfactorily also in the computation of probabilities. If we have either a red or a black ball in a box, then the probability that the box contains a red ball is r and the probability that the box contains a black ball is s. Both probabilities sum up to one, i.e., $r + s = 1$. The probability that in two boxes are k red balls is the coefficient at x^k in the expression

$$(rx + s)(rx + s) = r^2x^2 + (rs + sr)x + s^2 = r^2x^2 + 2rsx + s^2.$$

Actually, we should read $s + rx$ in terms of a generating function rather as $sx^0 + rx^1$. Likewise, s as the coefficient at x^0 is the probability that no red ball is in the box and r as the coefficient at x^1 is the probability that just one red ball is in the box. Since the probability for each box is the same and we have independent events, the generation function for one box is just squared to get the situation for two boxes.

A further example of a generating function is the partition function in statistical thermodynamics. The partition function is

$$Z = \sum_j \exp(-\beta E_j). \tag{12.4}$$

Here β will be identified with $1/kT$, and E_j is the energy level of a possible state j. From the partition function Eq. (12.4), a series of thermodynamic properties can be obtained. For example, the energy and the entropy are

$$U = -\frac{d \ln Z}{d\beta} \quad \text{and} \quad S = k(\ln Z + \beta U).$$

Example 12.1. The energy levels of a simple harmonic oscillator are $E_j = \hbar\omega(j + 1/2)$, $j = 0, 1, \cdots$ Here, ω is some characteristic frequency of the oscillator. We can find a closed expression for the partition function as

$$Z = \sum_j \exp(-\beta\hbar\omega(j + 1/2)) = \frac{\exp(-\hbar\omega/(2kT))}{1 - \exp(-\hbar\omega/(kT))} = \frac{1}{2\sinh(\hbar\omega/(2kT))}.$$

Forming the energy and from the energy, the heat capacity by the derivative with respect to temperature, we get

$$C_v = k\left(\frac{\hbar\omega}{2kT}\right)^2 \frac{1}{\sinh^2\left(\frac{\hbar\omega}{2kT}\right)}.$$

This is the heat capacity of a one-dimensional oscillator according to *Einstein*. The heat capacity deviates at low temperatures. It is not possible to expand into a Taylor series around $T \to 0$. In other words, the function has a pole at zero, which emerges as an essential singular point. A more accurate formula is due to *Debye*. □

12.2 Computation of the Number of Isomers

The problem of counting isomers was accessed starting from graph theory by *Pólya*[1] [1, 2]. Other chemical applications of graph theory are compiled by Balaban [3].

We illustrate the method with a simple example.

With the computation of the number of isomers of different chlorinated benzene molecules, a generating function arises that is connected with the symmetry of the benzene molecule. This function is designated after *Pólya* as the *Zyklenzeiger*. The generating function for chlorine isomers of benzene is

$$F(x) = 1 + x + 3x^2 + \cdots \qquad (12.5)$$

The goal of this section is to show how to calculate the *Zyklenzeiger*.

12.2.1 Isomers of Propionaldehyde

Question 12.1. How many chlorinated isomers has propionaldehyde? We write down the isomers in Fig. 12.1.

We indicate now the formalism, how we can compute the number of isomers: We identify the α-C atom, as a box, and the β-C atom likewise as a box. In the box, which refers to the α-C atom, now two hydrogen atoms, hydrogen atom and one chlorine atom, or two chlorine atoms can be. In the box, which refers to the β-C atom, either three hydrogen atoms can be, or one hydrogen atom and two chlorine atoms, or two hydrogen atoms and one chlorine atom, or three chlorine atoms.

For the α-C atom, there is one possibility to have zero chlorine atoms attached, one possibility to have one chlorine atom attached, and one possibility to have two chlorine atoms attached. So, each coefficient in the generating function is one and we get for the α-C atom $1 + x + x^2$. We repeat the same consideration for the β-C atom and obtain $1 + x + x^2 + x^3$ as the generating function.

The generating function for the entire molecule is simply the product of these two generating functions:

$$F(x) = (1 + x + x^2)(1 + x + x^2 + x^3) = 1 + 2x + 3x^2 + 3x^3 + 2x^4 + x^5.$$

The coefficients at that kth powers of x refer to the amount of that isomers with k chlorine atoms. If we want to know the total number of isomers from $0\,\mathrm{Cl}$ to $5\,\mathrm{Cl}$, then we put $x = 1$, c.f., also Fig. 12.1. The procedure is not restricted to chlorinated isomers, of course.

---

[1] Pólya György, born Dec. 13, 1887, in Budapest, Hungary, died Sep. 7, 1985, in Palo Alto, CA, USA.

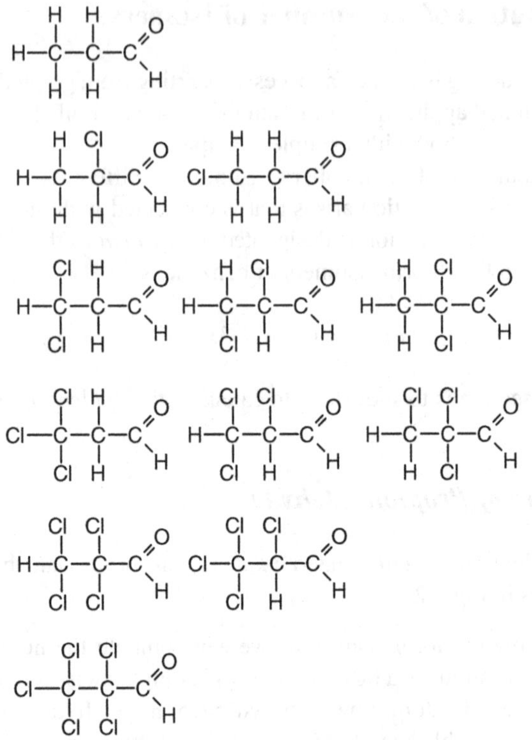

Fig. 12.1 Chlorinated isomers of propionaldehyde

Question 12.2. Why don't we write hydrogen or chlorine for each hydrogen atom either? Because some hydrogen atoms are not distinguishable, indeed three hydrogens at the β-C atom and two at the α-C atom. We could write also for the hydrogen atoms h.

Then the generating function reads as

$$F(h, x) = (h^2 + hx + x^2)(h^3 + h^2x + hx^2 + x^3). \qquad (12.6)$$

Equation (12.6) becomes then a homogeneous function in h and x. However, it is self-understanding that if one hydrogen atom disappears, a chlorine atom replaces the hydrogen atom. So the special form in Eq. (12.6) is superfluous.

The example was carefully selected, because a completely protruding asymmetry exists here. If the group of aldehydes would not be there, then certain isomers would not be distinguishable, if we turn the molecule, and/or mirror it. If a molecule has certain symmetries, then we must consider these symmetries. In this case, isomers created by combinatoric considerations, which are not distinguishable because of certain symmetries, should not be counted twice or multiple times. The issue of counting properly is the tricky problem.

12.2.2 Chlorinated Isomers of Benzene

A more complicated example is to count the isomers of substituted benzene. Here the problem of symmetry becomes sound. We start now to elucidate the symmetries of the planar hexagon.

12.2.2.1 Symmetries of the Planar Hexagon

It is well known that the carbon backbone of benzene forms a regular planar hexagon. The connections to the hydrogen atoms lie in the plane. We label the carbon atoms of the benzene from 1 to 6. We imagine a hexagonal box and place the benzene ring into this box. Only a certain finite number of distinguishable possibilities remain as to how the ring can lie in the box. We dice thus with the benzene ring.

In Fig. 12.2, these possibilities are listed. We bring now some order into the manifold. First, we label the corners of the box. We attach fixed numbers to the corners, just those which have the first benzene ring in Fig. 12.2 (left up). Then we know the possibilities, which are illustrated in Fig. 12.2 and also in Table 12.1.

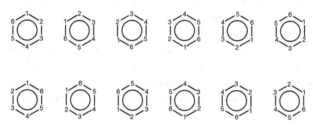

Fig. 12.2 Symmetries of benzene

Table 12.1 Symmetries of benzene

No	box	123456	cycles	Length×numbers
1	ring	123456	(1)(2)(3)(4)(5)(6)	1×6
2		234561	(123456)	6×1
3		345612	(135)(246)	3×2
4		456123	(14)(25)(36)	2×3
5		561234	(153)(264)	3×2
6		612345	(165432)	6×1
7	ring mirrored	165432	(1)(26)(35)(4)	1×2+2×2
8		654321	(16)(25)(34)	2×3
9		543216	(15)(24)(3)(6)	1×2+2×2
10		432165	(14)(23)(56)	2×3
11		321654	(13)(2)(46)(5)	1×2+2×2
12		216543	(12)(36)(45)	2×3

We rotate the ring, by pushing the first number to the rear. In this way, however, not all possibilities are exhausted. We can also mirror the ring. The symmetry operations suggest that the individual relative positions do not change in molecule. The molecule remains intact as such, is not torn apart, and again is thus not assembled. Therefore, a result of dicing with the molecule must be representable as a series of any rotation or mirroring. We introduce a column with cycles in Table 12.1. We find these cycles, by descending from the numbers of the box to the numbers of the ring vertically. From there we notice the number and go from this number in the box again vertically downward. We continue so for a long time, until we come again to the first number, with which we began. Thus, the cycle closes.

We illustrate this procedure with an example. The numbers in the box are 123456. The box is equivalent with the left-up ring in Fig. 12.2. With respect to the box, this ring is the identity transformation. We inspect now the ring with the 3 position up, third from left up in Fig. 12.2, or entry No. 3 in Table 12.1. The numbers of this ring are 345612. We look for the number 1 in the box and go downward; the 1 in the box corresponds to the 3 in the ring. We note thus 13. From the 3 in the box, we go again downward, 3 in the box corresponds to the 5 in the ring. We add the 5 to the sequence of numbers now, thus 135. From the 5 in the box, we go again vertically downward into the ring. To the 5 in the box, the 1 in the ring corresponds. Thus, we are landing again at the number where we started. The cycle is closed now.

We put the cycle brackets to indicate that it is closed: (135). The next higher number, which we did not go through yet, is the 2. We repeat the same procedure and obtain (246). Thus, we get with all positions from 1 to 6 for the special permutation 345612 (135)(246). In fact, we have subdivided the permutation 345612 into two subgroups of the same cycle length, i.e., 3. The whole procedure is a formal analysis, which usefulness we will recognize immediately. In the column length*number, we have classified the permutations. If we finally arrange these permutations, then we obtain

$$3 * [1 * 2 + 2 * 2] + 4 * [2 * 3] + 2 * [3 * 2] + 2 * [6 * 1] + 1 * [1 * 6] =$$
$$3(a)(a)(aa)(aa)+$$
$$4(aa)(aa)(aa)+$$
$$2(aaa)(aaa)+$$
$$2(aaaaaa)+$$
$$1(a)(a)(a)(a)(a)(a) =$$
$$3(a)^2(aa)^2 + 4(aa)^3 + 2(aaa)^2 + 2(aaaaaa) + 1(a)^6. \qquad (12.7)$$

The a in the parentheses refers to the individual positions. If we dice the benzene ring in a certain way, then the same kind of atom must sit at the positions within the parentheses, if we want to have a diced arrangement which cannot be distinguished from the original arrangement. So in the positions in brackets, there may be either exclusively hydrogen atoms or chlorine atoms to have, but no mixture. For example, in a box, (aa) may designate xx inside, or hh, however, not hx. We presume h the case that the position is occupied with hydrogen and x the case that the position is occupied with chlorine. So we obtain from the symbolic notation the generating function

$$3 * [1 * 2 + 2 * 2] + 4 * [2 * 3] + 2 * [3 * 2] + 2 * [6 * 1] + 1 * [1 * 6] =$$
$$3(h + x)^2(h^2 + x^2)^2 + 4(h^2 + x^2)^3 + 2(h^3 + x^3)^2 + 2(h^6 + x^6) + 1(h + x)^6.$$
(12.8)

We must still standardize the generating function. If we set x equal to 0 and set h equal to 1, then a nonsubstituted isomer should be obtained with correct standardization. In this way, we receive the normalization factor.

12.2.3 Other Examples

12.2.3.1 Chlorinated Dibenzodioxins

As further examples, we consider isomers of the chlorinated dibenzodioxins. The molecule is shown in Fig. 12.3. It has the symmetries as given in Table 12.2. We obtain the generating function for the entire number to isomers in Eq. (12.9).

$$F(h, x) = \frac{1}{4}\left[(h + x)^8 + 3\left(h^2 + x^2\right)^4\right]$$
(12.9)

If we set h and x equal to 1, then we obtain 76 isomers. In the literature, the number to chlorinated isomers, often also referred to as congeners, is given as 75. That comes along because here we also take into account completely nonchlorinated isomers. Written out, the generating function reads as

$$F(x) = 1 + 2x + 10x^2 + 14x^3 + 22x^4 + 14x^5 + 10x^6 + 2x^7 + x^8.$$

Fig. 12.3 Dibenzodioxin

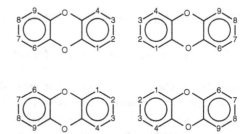

Table 12.2 Symmetries of dibenzodioxin

Box	12346789	Cycles	Length×numbers
	12346789	(1)(2)(3)(4)(5)(6)(7)(8)(9)	1×8
	98764321	(19)(28)(37)(46)	2×4
	67891234	(16)(27)(38)(49)	2×4
	43219876	(14)(23)(69)(78)	2×4

12.2.3.2 Chlorinated Dibenzofurans

We still compute isomers of the chlorinated dibenzofurans. The symmetry here is smaller than with the dibenzodioxins. We obtain therefore a larger number of isomers. The numbering of dibenzodioxin is shown in Fig. 12.4. The molecule has the symmetries given in Table 12.3. We obtain the generating function for the entire number to isomers as

$$F(h, x) = \frac{1}{2} \left[(h + x)^8 + \left(h^2 + x^2 \right)^4 \right].$$ (12.10)

If we put h and x equal to 1, then we find the number of isomers as 136. Written out, the generating function reads

$$F(x) = x^8 + 4x^7 + 16x^6 + 28x^5 + 38x^4 + 28x^3 + 16x^2 + 4x + 1.$$ (12.11)

Fig. 12.4 Dibenzodioxin

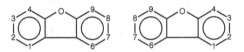

Table 12.3 Symmetries of dibenzofuran

Box	12346789	Cycles	Length×numbers
	12346789	(1)(2)(3)(4)(5)(6)(7)(8)(9)	1×8
	67891234	(16)(27)(38)(49)	2×4

12.3 Calculation of Isotopic Patterns

12.3.1 The Origin of Isotopic Patterns

Most of the chemical elements occur naturally not as a single species, but rather as a mixture of isotopes. The existence of isotopes was predicted already by *de Marignac*[2] and discovered by *Soddy*[3] in 1910 and Todd in 1912, respectively. *Thomson*[4] found that neon forms isotopes with a mass of 20 and 22, when analyzing channel rays. Isotopes can be readily seen by means of a mass spectrometer, which

---

[2] Jean Charles Galinard de Marignac, born Apr. 24, 1817, in Geneva, died Apr. 15, 1894, in Geneva.

[3] Frederick Soddy, born Sep. 2, 1877, in Eastbourne, died Sep. 22, 1956, in Brighton.

[4] Sir Joseph John Thomson, born Dec. 18, 1856, in Cheetham Hall, England, died Aug. 30, 1940, in Cambridge.

was at first designed by *Aston*[5] in 1919. The mass spectra of molecules show often a complicated pattern which is due to fragmentation and due to a convolution of the isotopic patterns of the constituting atoms.

The analysis and interpretation of the isotopic pattern is a valuable tool to identify the nature and structure of molecules, in the mass spectrometer of course converted into gaseous ions, in particular for organic compounds.

According to *Prout*[6], all the atoms are consisting of a cluster of hydrogen atoms. Prout's hypothesis turned out to be wrong in detail because of the discovery of the neutron. For this reason, the atomic weights of the elements are not exact multiples of the atomic weight of hydrogen, but since the weight of the neutron is close to that of the proton, and the mass of the electron is small to the masses of neutron and proton. Further, in an atom or ion, the masses of constituting particles are not simply the sum of isolated particles at rest, but there is a mass defect resulting from the bond strength in the atomic nucleus.

The accurate masses of the ions can be measured by means of precision mass spectrometry. For a variety of applications, it is sufficient to deal with nominal mass numbers. These are the integers close to the accurate masses. The natural abundances of the isotopes of carbon and chlorine are shown in Table 12.4. The abundances of the isotopes are given in % by mole, not in % by weight.

12.3.2 Isotopic Patterns of Molecules

The basic technique, how to calculate relative abundance of isotope clusters, has been given in the literature [4]. Computer programs that calculate the isotopic patterns of molecules are available [5–7].

From the information in Table 12.4, the isotopic patterns of compounds constituted from carbon and chlorine can be easily calculated. We focus on carbon tetrachloride CCl_4. We are constructing now a generating function for the abundance of the respective isotopes, the abundance generating function.

We attach the relative abundance associated with a certain mass number to a dummy variable set to the power of this respective mass number. So the abundance generating function of the isotopic pattern of carbon is

$$C(t) = 0.99t^{12} + 0.01t^{13}$$

and the abundance generating function of the isotopic pattern of chlorine is

$$Cl(t) = 0.76t^{35} + 0.24t^{37}.$$

---

[5] Francis William Aston, born 1877 in Harbourne, Birmingham, Warwickshire, England, died 1945 in Cambridge, Cambridgeshire, England.

[6] William Prout, born Jan. 15, 1785, in Horton, Gloucestershire, England died Apr. 9, 1850, in London, England.

Table 12.4 Abundance of stable isotopes

Isotope	M (nominal)	M (accurate) [g mol^{-1}]	Abundance %
^{1}H	1	1.007825032	99.99
^{2}H	2	2.014101778	0.01
^{12}C	12	12.00000000	98.90
^{13}C	13	13.00335483	1.10
^{14}N	14	14.00307400	99.63
^{15}N	15	15.00010890	0.36
^{16}O	16	15.99491462	99.76
^{17}O	17	16.99913150	0.04
^{18}O	18	17.99916042	0.20
^{35}Cl	35	34.96885272	75.78
^{37}Cl	36	36.96590262	24.22

If a chlorine ion is analyzed in a mass spectrometer, peaks are seen at the mass numbers 35 and 37. The relative intensity of the peaks at mass number n is just the coefficient attached to t^n. The pattern that appears for carbon tetrachloride is a convolution of the abundance generating functions for both carbon and chlorine.

We form now the abundance generating function for the isotopic pattern of carbon tetrachloride:

$$F_{CCl_4} = C * Cl^4 = \left(0.99t^{12} + 0.01t^{13}\right)\left(0.76t^{35} + 0.24t^{37}\right)^4. \tag{12.12}$$

Equation (12.12) may be expanded to result in

$$F_{CCl_4} = 3.310^{-5}t^{161} + 0.0032t^{160} + 4.2 * 10^{-4}t^{159}$$
$$+ 0.041t^{158} + 0.002t^{157} + 0.19t^{156} + 0.0042t^{155}$$
$$+ 0.41t^{154} + 0.0033t^{153} + 0.33t^{152}. \tag{12.13}$$

The coefficients attached to the powers reflect the abundance of the respective masses. For easier look up, these coefficients are collected once more in Table 12.5. Further, the distribution of the isotope pattern can be directly observed in the mass spectrum, cf., Fig. 12.5.

12.3.2.1 General Procedure for Isotopic Patterns

The general procedure to find out the isotopic pattern of a molecule is straightforward.

1. Look up the abundances of the isotopes of the elements needed.
2. Construct the abundance generating functions of the individual atoms by multiplying the abundances of the mass numbers with a dummy variable to the power of the mass number.

Table 12.5 Abundance of the mass numbers for carbon tetrachloride

Mass numbers	Abundance
161	3.3E−5
160	0.0032
159	4.60E−4
158	0.041
157	0.002
156	0.19
155	0.0042
154	0.41
153	0.003
152	0.33

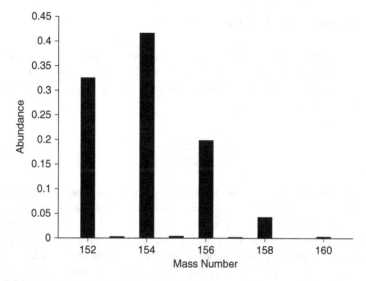

Fig. 12.5 Mass numbers and abundances of CCl_4, the base peak in mass spectrometry

3. Construct the abundance generating function of the molecule by multiplying the abundance generating functions of the constituting atoms. If an atom is occurring n-times, raise the abundance generating function of the atom to the power of n.
4. Expand the abundance generating function of the molecule and read the pattern of the desired mass number, say k as the coefficient of the dummy variable that appears raised to the power of k.

12.3.2.2 De-convolution

It is desirable to trace back the abundance of generating functions of the constituting atoms from a pattern of a molecule, as shown in Fig. 12.5. If the pattern can be resolved into a product of abundance generating functions, then from these functions the atomic composition of the cluster can be found out. The experienced analyst can find this out in one glance.

In fact, a mathematically exact generated pattern, e.g., from Eq. (12.13) can be factorized back without problems. In practise, however, it happens that the pattern found out in a real mass spectrum is incomplete. This means that essential peaks that are weak are in the level of noise. Further, the abundances determined experimentally can slightly differ, depending on the instrumental performance. Therefore, a direct factorization of a pattern obtained from a real experiment is condemned to fail.

There may be methods to construct a pattern from the abundance generating functions of atoms for molecules and to compare these patterns with those obtained by the experiment.

12.3.3 Average Molar Mass

From the abundance generating functions that are used to calculate the isotopic pattern, also the average molar mass of the molecular can be calculated. If the abundance generating function, e.g., for carbon tetrachloride, is as given in Eq. (12.12), then the average molar mass is given as

$$M = \frac{\frac{dF_{CCl_4}(t)}{dt}}{F_{CCl_4}(t)}\Bigg|_{t=1}. \tag{12.14}$$

Forming the derivative effects, the powers will come down. The right-hand side of Eq. (12.14) is then equated at $t = 1$. Division by $F(t)$ is only necessary if the abundances are not normalized to 1. Otherwise, of course, $F(1) = 1$. If the abundance generating function is used as given in Eq. (12.12), then the average based on the nominal mass numbers of the isotopes is obtained. This is often sufficient. For a more accurate average, the powers in the accurate mass numbers must be inserted, e.g., for chlorine:

$$Cl(t) = 0.76t^{34.96885272} + 0.24t^{36.96590262}.$$

12.3.4 Weight Fraction of Atoms

The mass fraction of atoms in a molecule can be also obtained from the abundance generating function. For this purpose, we modify the abundance generating functions somewhat. We show this for carbon:

$$\tilde{C}(t, c) = 0.99(t * c)^{12} + 0.01(t * c)^{13},$$

and oxygen:

$$\tilde{O}(t, o) = 0.99(t * o)^{16} + 0.0004(t * o)^{17} + 0.002(t * o)^{18}.$$

So here we have replaced the dummy parameter t by $t \to t * c$ in the case of the abundance generating function for carbon and t by $t \to t * o$ in the case of the abundance generating function for oxygen.

For carbon monoxide, the abundance generating function expands to

$$
\begin{aligned}
\tilde{F}_{CO} = \tilde{C}(t, c)\tilde{O}(t, o) = \; & 2.0 \times 10^{-5} \times c^{13}o^{18}t^{31} \\
& + 4.0 \times 10^{-6} \times c^{13}o^{17}t^{30} \\
& + 1.0 \times 10^{-2} \times c^{13}o^{16}t^{29} \\
& + 2.0 \times 10^{-2} \times c^{12}o^{18}t^{30} \\
& + 4.0 \times 10^{-4} \times c^{12}o^{17}t^{29} \\
& + 9.9 \times 10^{-1} \times c^{12}o^{16}t^{28}
\end{aligned}
\qquad (12.15)
$$

Inspection of Eq. (12.15) reveals that the exponent of the dummy parameter t informs about the total mass number and the exponents of c and o inform about the particular isotopes constituting the group.

The weight fraction of carbon w_c is obtained by

$$
w_c = \left. \frac{\dfrac{\partial F_{CO}(c, o, t)}{\partial c}}{\dfrac{\partial F_{CO}(c, o, t)}{\partial t}} \right|_{c=o=t=1} = 0.43.
$$

Similarly, the mass fraction of oxygen w_o is obtained by

$$
w_o = \left. \frac{\dfrac{\partial F_{CO}(c, o, t)}{\partial o}}{\dfrac{\partial F_{CO}(c, o, t)}{\partial t}} \right|_{c=o=t=1} = 0.57.
$$

In practice, it is more efficient to calculate the averages from the atoms, to store them and to add the individual averages of the atoms that are constituting a molecule together.

12.3.5 Mixture of Molecules

We introduce the abundance generating function of hydrogen:

$$
\tilde{H}(t, h) = 0.999(t * h)^1 + 0.001(t * h)^2.
$$

From this function and the forgoing function, we calculate the abundance generating function of ethylene:

$$\tilde{F}_{C_2H_4} = \tilde{C}^2(c, t)\tilde{H}^4(h, t).$$

Now it is common that ethylene and carbon monoxide are isotonic. We form the abundance generating functions of a mixture of ethylene and carbon monoxide by multiplying by a factor a and b respectively. By adding we obtain $\tilde{G}$:

$$\tilde{G} = a\tilde{C}^2(c, t)\tilde{H}^4(h, t) + b\tilde{C}(c, t)\tilde{O}(o, t).$$

The expansion gives immediately the expression shown in Table 12.6. Equation (12.16) remains mathematically valid; however, the scheme is arranged in such a way that utmost information can be obtained. In the first column, the mass numbers are given as the exponent of t. In the second column, the relative abundance is given. In the third column, the composition of the atomic cluster is given, and finally a and b indicate to which part of the mixture the peak belongs.

Table 12.6 Expansion of $\tilde{G} = a\tilde{C}^2(c, t)\tilde{H}^4(h, t) + b\tilde{C}(c, t)\tilde{O}(o, t)$

$$
\begin{aligned}
\tilde{G} = \quad & t^{34}(1.0 \times 10^{-16}\ c^{26}h^8 \times a) \\
+ & t^{33}(3.9 \times 10^{-13}\ c^{26}h^7 \times a \\
& +1.9 \times 10^{-14}\ c^{25}h^8 \times a) \\
+ & t^{32}(5.9 \times 10^{-10}\ c^{26}h^6 \times a \\
& +7.9 \times 10^{-11}\ c^{25}h^7 \times a \\
& +9.8 \times 10^{-13}\ c^{24}h^8 \times a) \\
+ & t^{31}(3.9 \times 10^{-7}\ c^{26}h^5 \times a \\
& +1.1 \times 10^{-7}\ c^{25}h^6 \times a \\
& +3.9 \times 10^{-9}\ c^{24}h^7 \times a \\
& +2.0 \times 10^{-5}\ c^{13}o^{18} \times b) \\
+ & t^{30}(9.9 \times 10^{-5}\ c^{26}h^4 \times a \\
& +7.8 \times 10^{-5}\ c^{25}h^5 \times a \\
& +5.8 \times 10^{-6}\ c^{24}h^6 \times a \\
& +4.0 \times 10^{-6}\ c^{13}o^{17} \times b \\
& +1.9 \times 10^{-3}\ c^{12}o^{18} \times b) \\
+ & t^{29}(1.9 \times 10^{-2}\ c^{25}h^4 \times a \\
& +3.9 \times 10^{-3}\ c^{24}h^5 \times a \\
& +9.9 \times 10^{-3}\ c^{13}o^{16} \times b \\
& +3.9 \times 10^{-4}\ c^{12}o^{17} \times b) \\
+ & t^{28}(9.7 \times 10^{-1}\ c^{24}h^4 \times a \\
& +9.8 \times 10^{-1}\ c^{12}o^{16} \times b)
\end{aligned}
$$

(12.16)

The scheme immediately reveals that the mass numbers 32–34 are sensitive exclusively to ethylene, whereas the mass numbers 28–33 are confounded with both compounds. On the otherhand, for example, it can be seen that the mass number 31 is sensitive 100-fold more to component b than to component a. Further, at mass number 28, both compounds have nearly equal response.

The scheme provides fundamental information, which mass spectrometric peaks can be used for quantitative analysis of the two compounds in question.

12.3.6 Alternative Abundance Generating Functions

We mention finally that instead of including the element symbols into the dummy variable t, they can be stand-alone with a different name. We redefine the abundance generating function for carbon $\tilde{C}$ as

$$\tilde{C}(t, {}^{12}C, {}^{13}C) = 0.9 * {}^{12}C * t^{12} + 0.01 * {}^{13}C * t^{13}$$

and the abundance generating function for hydrogen $\tilde{H}$ as

$$\tilde{H}(t, H, D) = 0.999 * t * H + 0.001 * D * t^2.$$

Here we have used rather a chemical notation. Previously we have used lower case letters for the variables. Here ^{12}C and ^{13}C are individual variables, such as H or D. Then the abundance generating function for acetylene $\tilde{A}$ looks like

$$
\begin{aligned}
\tilde{A} = \ & t^{30}(1.0 * 10^{-10}\ {}^{13}C^2 * D^2) \\
& + t^{29}(2.0 * 10^{-8}\ {}^{12}C * {}^{13}C * D^2 \\
& \quad + 2.0 * 10^{-7}\ {}^{13}C^2 * H * D) \\
& + t^{28}(9.8 * 10^{-7}\ {}^{12}C^2 * D^2 \\
& \quad + 4.0 * 10^{-5}\ {}^{12}C * {}^{13}C * H * D \\
& \quad + 2.0 * 10^{-5}\ {}^{13}C^2 * H^2) \\
& + t^{27}(2.0 * 10^{-3}\ {}^{12}C^2 * H * D \\
& \quad + 2.0 * 10^{-2}\ {}^{12}C * {}^{13}C * H^2) \\
& + t^{26}(9.8 * 10^{-1}\ {}^{12}C^2 * H^2)
\end{aligned}
$$

12.4   Forms of Generating Functions

Generating functions do not necessarily occur in the form

$$F(x) = \sum a_k x^k, \tag{12.17}$$

but can appear also in the form

$$F(x) = \int a(k)\, x^k\, dk. \tag{12.18}$$

Further, they can appear also as exponential generating function, Eq. (12.19). In the case of distributions, the moment generating function is

$$F(x) = \sum a_k \exp(kx), \tag{12.19}$$

and the characteristic function is

$$F(x) = \sum a_k \exp(\imath k x).$$ (12.20)

Both functions have corresponding integral forms. Also generating functions with several arguments are permissible and useful.

We already learned a characteristic property of generating functions with the red and black balls. Multiplication of two generating functions again gives a generating function. The coefficient at x collects all probabilities, which generate the probability, looked for. Average values can be obtained with advantage from generating functions. We consider polymers in a pot, in which n-mers with the concentrations c_n are present. The generating function for the concentration is

$$F(s) = c_1 s + c_2 s^2 + c_3 s^3 + \cdots.$$

So, derivatives of the generating function become

$$s \frac{dF(s)}{ds} = c_1 s + 2 c_2 s^2 + 3 c_3 s^3 + \cdots$$

$$s \frac{d\left(s \frac{dF(s)}{ds}\right)}{ds} = c_1 s + 4 c_2 s^2 + 9 c_3 s^3 + \cdots.$$

The number average of the degree of polymerization is

$$P_n = \left. \frac{s \frac{dF(s)}{ds}}{F(s)} \right|_{s=1},$$

and the weight average of the degree of polymerization is

$$P_w = \left. \frac{s \frac{d^2 F(s)}{ds^2}}{\frac{dF(s)}{ds}} + 1 \right|_{s=1}.$$

The polydispersity is

$$U = \frac{P_w}{P_n} - 1 = \frac{s F(s) \frac{d^2 F(s)}{ds^2}}{\left(\frac{dF(s)}{ds}\right)^2} + \frac{F(s)}{s \frac{dF(s)}{ds}} - 1.$$

$$\text{at } s = 1$$

The matter becomes more simple, if the moment generating function is chosen:

$$F(x) = c_1 \exp(x) + c_2 \exp(2x) + c_3 \exp(3x) + \cdots.$$

The first derivative is

$$\frac{dF(x)}{dx} = c_1 \exp(x) + 2c_2 \exp(2x) + 3c_3 \exp(3x) + \cdots .$$

So for the number average of the degree of polymerization in terms of the moment generating function is

$$P_n = \frac{F'(0)}{F(0)} = \frac{c_1 + 2c_2 + 3c_3 + \cdots}{c_1 + c_2 + c_3 + \cdots}.$$

Similarly, for the weight average of the degree of polymerization

$$P_w = \frac{F''(0)}{F'(0)} = \frac{c_1 + 4c_2 + 9c_3 + \cdots}{c_1 + 2c_2 + 3c_3 + \cdots}$$

is obtained.

12.4.1 Fourier Series

We set up a probability generating function

$$W(x) = \sum_{k=0}^{k=\infty} w(k) x^k.$$

The coefficient $w(k)$ is the probability that the system exhibits the property k. For example, we could consider a mixture of polymers. Therein is a certain quantity of particles with the degree of polymerization of k. $w(k)$ is then the probability, or the frequency, with which particles with the polymerization degree of k are present. We replace now x by $\exp(\iota x)$ and obtain the sum

$$F(x) = \sum_{k=0}^{k=\infty} w(k) \exp(\iota k x). \tag{12.21}$$

This is the discrete Fourier transform of the probability density function, also addressed as the characteristic function. We multiply now Eq. (12.21) by $\exp(-\iota l x)$, $l = 1, 2, 3, \ldots$ and integrate over an interval of 2π to get

$$\int_0^{2\pi} F(x) \exp(-\iota l x) dx = \int_0^{2\pi} \left[\sum_{k=0}^{k=\infty} w(k) \exp(\iota k x) \exp(-\iota l x) \right] dx.$$

We take knowledge of

$$\int_0^{2\pi} \exp(ikx)dx = 0$$

$$\int_0^{2\pi} \exp(ikx)\exp(ilx)dx = 0$$

$$\int_0^{2\pi} \exp(ikx)\exp(-ikx)dx = \int_0^{2\pi} dx = 2\pi$$

Therefore,

$$\int_0^{2\pi} F(x)\exp(-\imath kx)dx = 2\pi\,w(k). \tag{12.22}$$

In this way, the probabilities $w(k)$ can be obtained from Eq. (12.22). The definition of the Fourier transform is not unique, but it is historically grown from the different fields of science. Various conventions are in use. In Table 12.7 we show some pairs of Fourier transforms. The last definition in Table 12.7 involves two parameters (a, b) and is the most versatile one. In the integral form, the characteristic function uses $(1, 1)$.

Table 12.7 Pairs of Fourier transforms

$F(k) = \int_{-\infty}^{\infty} f(x)\exp(-2\pi\imath kx)dx$	
	$f(x) = \int_{-\infty}^{\infty} F(k)\exp(+2\pi\imath kx)dk$
$F(\omega) = \int_{-\infty}^{\infty} f(t)\exp(-\imath\omega t)dt$	
	$f(t) = \frac{1}{2\pi}\int_{-\infty}^{\infty} F(\omega)\exp(+\imath\omega t)d\omega$
$F(\omega) = \frac{1}{\sqrt{2\pi}}\int_{-\infty}^{\infty} f(t)\exp(-\imath\omega t)dt$	
	$f(t) = \frac{1}{\sqrt{2\pi}}\int_{-\infty}^{\infty} F(\omega)\exp(+\imath\omega t)d\omega$
$F(\omega) = \sqrt{\frac{|b|}{(2\pi)^{1+a}}}\int_{-\infty}^{\infty} f(t)\exp(-\imath b\omega t)dt$	
	$f(t) = \sqrt{\frac{|b|}{(2\pi)^{1-a}}}\int_{-\infty}^{\infty} F(\omega)\exp(+\imath b\omega t)d\omega$

12.4.2 Z-Transformation

The Z-transformation of a function $f(n)$ is given by

$$\sum_{n=0}^{n=\infty} f(n)z^{-n}.$$

The inverse Z-transformation of $F(z)$ is given by the contour integral

$$\frac{1}{2\pi i}\oint F(z)z^{n-1}dz.$$

Z-transforms are literally discrete analogues of Laplace transforms. They are widely used for solving difference equations, especially in digital signal processing and control theory. They can be thought of as producing generating functions, of the kind commonly used in combinatorics and number theory. For example, the Z-transformation of 2^{-n} is

$$\frac{2z}{2z - 1}.$$

12.4.3 Linear Transformations

We make now, similarly as is common with the different integral transforms, a correspondence table between the stochastic variable and the associated characteristic function. Note, there are several integral transforms. The most well-known integral transformation[7] might be the Fourier transform. Further, we emphasize the Laplace transform, the Mellin transform, and the Hilbert transform. These transformations are useful for the solution of various differential equations, in communications technology, all ranges of the frequency analysis, also for optical problems and much other more. We designate the stochastic variable with X. The associated characteristic function should be

$$F(s) = \int_{x=-\infty}^{x=+\infty} p(x) \exp(\imath s x) \, dx.$$

12.4.3.1 Linear Shift

If to X a constant b is added, i.e., $X \rightarrow X + b$, then the characteristic function changes by $F(s) \rightarrow F(s) \exp(\imath s b)$.

12.4.3.2 Multiplication by a Constant

If X is multiplied by a constant of a, i.e., $X \rightarrow aX$, then the characteristic function changes by

$$F(s) \rightarrow F(as).$$

12.4.3.3 Addition of Two Independent Variables

We add two stochastic variables, which should be independent [8]. $Z \rightarrow X + Y$, the characteristic function for X is $F(x)$ and the characteristic function for Y is $G(x)$. Then for Z the characteristic function is $F(x)G(y)$. We summarize the results in

---

[7] The shortform for an integral transformation is transform.

Table 12.8. We can clearly see the relationship with the correspondence tables of other integral transforms.

Table 12.8 Correspondences of stochastic variables with their appropriate characteristic functions

Transformation	Result
$X \rightarrow X + a$	$F(x) \rightarrow F(x)\exp(\iota a)$
$X \rightarrow Xb$	$F(x) \rightarrow F(xb)$
$X + Y \rightarrow Z$	$F(x), G(y) \rightarrow F(x)G(y)$

12.5 Solution of Systems of Differential Equations

In a decay chain, a system of differential equations may emerge:

$$-\dot{n}_1 = kn_1$$
$$-\dot{n}_2 = kn_2 - \dot{n}_1$$
$$-\dot{n}_3 = kn_3 - \dot{n}_2$$
$$\dots\dots\dots$$

n_j are the number of particles of type j in the system, k is a rate constant, and $\dot{n}_j$ is the rate at which the number is changing. The rate constant is a known quantity. We form the generating function

$$F(x,t) = n_1(t)\exp(x) + n_2(t)\exp(2x) + n_3(t)\exp(3x) + \cdots . \qquad (12.23)$$

Further, we multiply the set of equations by $\exp(lx), l = 1, 2, \dots$. Inserting the generating function into Eq. (12.23) yields

$$-\frac{\partial F(x,t)}{\partial x} = kF(x,t) + kF(x,t)\exp(x).$$

So the whole set of equations can be solved in one stroke. Applications for this technology are the Brunauer – Emmet – Teller adsorption isotherm, various polymerization reactions and depolymerization reactions, and counting problems in statistical thermodynamics.

12.6 Statistical Thermodynamics

Basically, it is believed that the properties of a macroscopic system can be derived from the properties and interactions of the microscopic particles that constitute the macroscopic system by a mechanical theory.

Actually, the number of particles is too large that this can be achieved by strict application of the laws known in mechanics. Therefore, a statistical method has been developed. For example, the pressure of a gas is expressed as the average force per unit area exerted by its particles when they collide with the container walls.

12.6.1 Thermodynamic Probabilities

To each particle in a set of particles, a quantity of energy can be associated. This may be the energy of motion, the potential energy, the chemical energy, etc. This energy remains not for ever the same for a certain particle. The particles may exchange energy among another, as they collide, move in a field, or undergo chemical reaction.

Therefore, even when at a certain initial time the energy of all the particles is the same, after a sufficient long time, the energy will distributed among the particles in a specific manner. We will address this steady state as the equilibrium. Even in equilibrium, the process of energy exchanges continues, but at a macroscopic scale no fluctuations in the distribution of the energy can be observed.

In a system of constant energy, the energy may be distributed in a number of ways. We illustrate this with a small set of four particles.

12.6.1.1 Permutations

First, we will construct the number of permutations of particles. We start with a single particle a. We can place the second particle b before or behind particle a. Thus, we arrive at ba and ab. The next particle c can be placed before the first, in between the two, or at the end. This is the general procedure to construct the permutations of particles. This can be easily done nowadays with an appropriate text editor. The pattern resulting from this procedure is given in Table 12.9.

Table 12.9 Permutations of a set of four particles

1–6	7–12	13–18	19–24
abcd	bacd	cabd	dabc
abdc	badc	cadb	dacb
acbd	bcad	cbad	dbac
acdb	bcda	cbda	dbca
adbc	bdac	cdab	dcab
adcb	bdca	cdba	dcba

The numbers of permutations are calculated straightforward. For one particle, there is one permutation. For the second, there are two. To these two permutations, for each permutation ab and ba three more permutations can be constructed. This is 2×3. In general if the number of permutations for n particles are $P(n)$, then for $n + 1$ particles the number of permutations are $P(n + 1) = P(n)(n + 1)$. Thus, tracing back

$$P(n) = 1 \times 2 \times \ldots \times n = n!.$$

12.6.1.2 Distribution of Particles at Energy Levels

We illustrate the possible distributions of particles among energy levels. We assume that a particle can take up a certain energy and can choose between three different energies: $\varepsilon_1, \varepsilon_2, \varepsilon_3$. We want to associate the following energies to these levels, in arbitrary energy units eu:

$$\varepsilon_1 = 1 \, \text{eu}, \qquad\qquad \varepsilon_2 = 2 \, \text{eu}, \qquad\qquad \varepsilon_3 = 3 \, \text{eu}.$$

The number of particles that can be at all the possible energy units is given in Table 12.10.

Table 12.10 Number of four different particles distributed at three energy levels

$\varepsilon_1 = 1 \, \text{eu}$	$\varepsilon_2 = 2 \, \text{eu}$	$\varepsilon_3 = 3 \, \text{eu}$	Total energy eu
4	0	0	4
3	1	0	5
3	0	1	6
2	2	0	6
2	1	1	7
2	0	2	8
1	3	0	7
1	2	1	8
1	1	2	9
1	0	3	10
0	4	0	8
0	3	1	9
0	2	2	10
0	1	3	11
0	0	4	12

It can be readily seen from Table 12.10 that there are several energy levels that can be established in more than only one distribution of particles. There arises the question, which particular distribution is the preferred, or is there a preferred distribution at all?

The numbers of distributions for certain energy levels in Table 12.10 are obtained by the generating function

$$F(x, y) = \frac{1}{(1 - xy)(1 - xy^2)(1 - xy^3)}. \tag{12.24}$$

The situation is close to that of the money change problem. Taylor expansion of the generating function in Eq. (12.24), either in x or in y and collecting the term at x^4, results in

$$[x^4]\{F(x, y)\} = y^{12} + y^{11} + 2y^{10} + 2y^9 + 3y^8 + 2y^7 + 2y^6 + y^5 + y^4.$$

The coefficients at y^k are just the number of ways, in how many ways an energy of k eu can be achieved. All the particles are indistinguishable, in this case.

Still more detailed information is available, if we modify Eq. (12.24) into

$$F(x, y) = \frac{1}{(1 - xy_1 y)(1 - xy_2 y^2)(1 - xy_3 y^3)}.$$

Again, extracting the term at order x^4, we arrive at

$$y_3^4 y^{12}$$
$$+ y_2 y_3^3 y^{11}$$
$$+ y_3^2 (y_1 y_3 + y_2^2) y^{10}$$
$$+ y_2 y_3 (y_1 y_3 + y_2^2) y^9$$
$$+ (y_1^2 y_3^2 + y_1 y_2^2 y_3 + y_2^4) y^8$$
$$+ y_1 y_2 (y_1 y_3 + y_2^2) y^7$$
$$+ y_1^2 (y_1 y_3 + y_2^2) y^6$$
$$+ y_1^3 y_2 y^5$$
$$+ y_1^4 y^4.$$

The powers at y_1, y_2, y_3 refer to the number of particles in the energy level ε_1, ε_2, ε_3.

The calculation proceeds in still another way. If we are fixing the energy, then we can inspect a given distribution and look up, in how many ways the particles are placed into the boxes with different energy levels. Here we assume that the particles at the same energy levels are not distinguishable, but the particles in different boxes are distinguishable.

Question 12.3. What is the reason for this procedure? If all the particles would be distinguishable, then no distribution would be preferred.

We want now to fix the total energy of these four particles to 8 eu. Inspection of Table 12.10 shows that there are three possible distributions with four particles into three energy levels with 1 eu, 2 eu, 3 eu respectively. These are

1. 2 particles at 1 eu and 2 particles at 3 eu,
2. 1 particle at 1 eu and 2 particles at 2 eu and 1 particle at 1 eu,
3. 4 particles at 2 eu.

The number of ways how the distributions can be established, if the particles at a certain energy level are not distinguishable, can be easily obtained from Table 12.9. For example, for case 1 (2 particles at 1 eu and 2 particles at 3 eu), we have to put a separator after the second particle, like $ab|cd$, to indicate the border of the energy levels. The particles before | have the energy ε_1 and the particles behind | have the energy ε_2. Within a certain energy level, we sort the particles, e.g., $ba|dc$ into $ab|cd$ and sort out equal patterns. So the numbers of realizations of distributions are obtained. In fact, it turns out that if N particles should be distributed among k energy levels, the number of different ways to establish is

$$\bullet \quad M(N, N_1, \ldots) = \frac{N}{N_1 \times N_2 \times \ldots \times N_k}. \tag{12.25}$$

Thus, the number of realization for the distribution is given in Table 12.11.

Table 12.11 Number of ways to establish a distribution for four particles at three energy levels with a total energy of 8 eu as given in Table 12.10

$\varepsilon_1 = 1\,\text{eu}$	$\varepsilon_2 = 2\,\text{eu}$	$\varepsilon_3 = 3\,\text{eu}$	$M(N, N_1, \ldots)$
2	0	2	6
1	2	1	12
0	4	0	1

Shown for a total energy of 8 eu

12.6.1.3 Generating Functions for the Distribution

In the classical derivation, there are somehow mysterious assumptions concerning the distribution, that is, in cases when the particles can be distinguished from each other and when they are indistinguishable.

Here we proceed in a very different way. We presume that the particles may have a certain energy among $\varepsilon_1, \varepsilon_2, \ldots \varepsilon_k$. We can select a particle from stock and we want to know the probability that the particle chosen by random has the energy ε_j.

Here we presume the principle of equal a priori probabilities. This means that we assume that all states corresponding to a given energy are equally probable for a certain molecule. This means that we can find the molecule in a high-energy state or a low-energy state with equal probability. This may be expressed that a molecule may freely move from a low energy to a high energy. There are no strong molecules that can climb easier to a high-energy level than weak molecules. The molecules are perfectly similar with respect what they will do in the next step, when at the same energy level. Therefore, we cannot distinguish them, if they are at the same energy level, but we can distinguish them, if they are at different energy levels. For instance, we could construct an energy dispersive engine that can separate these molecules.

So the probability to catch a molecule with an energy of ε_1 let be ε_1, and so on. We form now the function

$$f = \varepsilon_1 + \varepsilon_2 + \cdots + \varepsilon_k.$$

Here $\varepsilon_1 = 1/k$ in order to get the sum to 1. We address f as the probability generating function for the energy of one particle. Further, we assume the particles are independent. Then for N particles the generating function is

$$F = f^N = (\varepsilon_1 + \varepsilon_2 + \cdots + \varepsilon_k)^N.$$

If we are collecting the term at $\varepsilon_1^{N_1} \varepsilon_2^{N_2} \times \cdots \times \varepsilon_k^{N_k}$, then we arrive at the probability that the set of particles has N_1 particles with energy ε_1, N_2 particles with energy ε_2, etc. This is of course the multinomial coefficient, given in Eq. (12.25). In statistical thermodynamics, the most probable distribution is that where

$$\frac{N}{N_1 \times N_2 \times \cdots} \to Max$$

tends to be a maximum. There are two further conditions:

1. The number of particles remains constant.
2. The total energy remains constant.

These two conditions are fulfilled if the method of undetermined multipliers is used to get the maximum by variational calculus.

12.6.2 Partition Function

For the most probable distribution

$$N_i(\varepsilon_i) \propto \exp\left(-\frac{\varepsilon_i}{kT}\right). \tag{12.26}$$

Here k is the Boltzmann constant and T is the absolute temperature. $N_i(\varepsilon_i)$ is the number of particles with energy ε_i. The partition function is finally obtained as

$$z = \sum_i \exp\left(-\frac{\varepsilon_i}{kT}\right). \tag{12.27}$$

The acronym z is common because the German term is *Zustandssumme*. From Eq. (12.27), the mean energy per particle can be calculated

$$\bar{\varepsilon} = kT^2 \frac{\partial \ln z}{\partial T}.$$

In fact, the partition function is a generating function. We can see this if we put $1/x = -kT$. Then

$$z = \sum_i \exp\left(\varepsilon_i x\right).$$

It can be immediately seen that

$$\frac{1}{z}\frac{\partial z}{\partial x} = kT^2 \frac{\partial \ln z}{\partial T}.$$

Example 12.2. We derive Eq. (12.26) by ordinary thermodynamics. Consider a system of two kinds of particles (1) and (2) that can be mutually transformed one into the other (1) $\rightleftharpoons$ (2). The energy of particles (1) and (2) is $U_1(X_1, n_1)$ and $U_1(X_2, n_2)$, respectively. Here X_1, X_2 is the extensive quantity of a not closer specified energy form. We form now the total differential with the constraints $X_1 = \tilde{X}_1 n_1$, $X_2 = \tilde{X}_2 n_2$, and $dn_1 + dn_2 = 0$. Here the molar quantities $\tilde{X}_1, \tilde{X}_2$ should be constants. Then the total differential of the total energy of the system is

$$\frac{\partial U}{\partial n_1} = \frac{\partial U_1}{\partial X_1}\tilde{X}_1 + \mu_1 - \frac{\partial U_2}{\partial X_2}\tilde{X}_2 - \mu_2 = N_L\varepsilon_1 + \mu_1 - N_L\varepsilon_2 - \mu_2. \quad (12.28)$$

N_L is the Avogadro number. We equate Eq. (12.28) to zero and expand the chemical potentials of a mixed phase into

$$0 = N_L\varepsilon_1 + \mu_{1,0} + RT \ln x_1 - N_L\varepsilon_2 - \mu_{2,0} + RT \ln x_2.$$

This results into

$$\frac{n_2}{n_1} = \frac{\exp(-\mu_{2,0})\exp(-\varepsilon_2/(kT))}{\exp(-\mu_{1,0})\exp(-\varepsilon_1/(kT))} = \frac{g_2\exp(-\varepsilon_2/(kT))}{g_1\exp(-\varepsilon_1/(kT))}. \quad (12.29)$$

So the chemical potential of the pure component acts as the statistical weight. In Eq. (12.29) we replace now the index 2 by i, sum over i and take the reciprocal. This yields finally

$$\frac{n_1}{n_{tot}} = \frac{g_1\exp(-\varepsilon_1/(kT))}{\sum_i g_i\exp(-\varepsilon_i/(kT))}.$$

□

12.7 Adsorption Isotherms

Adsorption phenomena are of interest in several topics and find uses including purification technology, refrigeration technology [9]. The uptake of gases by charcoal was studied by *Scheele*[8] in 1773 and further by *Fontana*[9] in 1777.

> *Dewar*[10] found that charcoal cooled in liquid air was capable of taking up large quantities of such gases as oxygen and nitrogen.

---

[8] Carl Wilhelm Scheele, born Dec. 9, 1742, in Stralsund, Sweden, died May 21, 1786, in Köping.

[9] Félice Fontana, born Apr. 15, 1730, in Pomarolo, Italy, died Mar. 10, 1805, in Florence, Italy.

[10] James Dewar, born Sep. 20, 1842, in Kincardine-on-Forth, in Scotland, died Mar. 27, 1923, in London, England.

Prior to 1910 many different theories of adsorption had been proposed, but none of them had been very successful. In most of these theories the increased concentration of the adsorbed substance near the surface was thought to be analogous to the retention of the earth's atmosphere by the gravitational attraction of the earth. An adsorbed gas was thus regarded as a kind of miniature atmosphere extending out a short distance from a solid substance. In general such theories were called upon to account only for qualitative aspects of the adsorption of gases on solids [10].

Molecules and atoms can attach themselves onto surfaces as physical adsorption or as chemical adsorption. Physical adsorption arises from a van der Waals attraction of the adsorbate to the surface, in the order of $20 \, \mathrm{kJ \, mol^{-1}}$. In chemical adsorption, the adsorbate sticks to the solid by the formation of a chemical bond with the surface. This interaction is much more stronger than physical adsorption.

The situation is shown in Fig. 12.6. Molecules being in the gas phase can be adsorbed on a surface, when striking the surface. Molecules being adsorbed on the surface can be desorbed and move again into the gas phase. In the case of chemical sorption, the free valences of the surface are spent by an adsorption process. Therefore, the concept of chemical sorption is reasonable only for a monomolecular layer (s_1). If physical adsorption takes place, then also multimolecular layers can be formed. This is typical in the region of temperature and pressure when the gas starts to condense. On the other hand, when the surface is not smooth is idealized as shown in Fig. 12.6, the phenomenon of capillary condensation can play a role.

Fig. 12.6 Adsorption and desorption of gas molecules on a surface

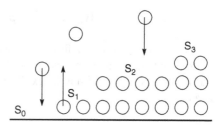

Capillary condensation is the reverse effect from the pressure increase of small droplets. As shown by *Kelvin* from thermodynamic arguments, small droplets with a strongly curved surface have a larger equilibrium vapor pressure than planar surfaces. On the other hand, the Kelvin equation (12.30) predicts that the equilibrium vapor pressure of a surface with a negative radius, this is a hole, is still lower than the equilibrium vapor pressure of a planar surface. This is the reason for capillary condensation. The vapor pressure of a spherical droplet is

$$
\ln\left(\frac{p(r)}{p(r = \infty)}\right) = \frac{\sigma \tilde{V}}{RTr}.
\tag{12.30}
$$

Here $p(r)$ is the vapor pressure of a spherical droplet with radius r, $\tilde{V}$ is the molar volume of the liquid, and σ is the surface tension of the liquid. Eq. (12.30) is derived from the Kelvin Eq. (12.31)

$$\Delta p = \sigma \left(\frac{1}{r_1} - \frac{1}{r_2} \right), \tag{12.31}$$

considering the increase of the radius of the droplet by a condensation from the gas phase, which is assumed to be an ideal gas. Besides adsorption on the surface, molecules from the gas phase may be captured in the inner region of the solid body. This phenomenon is addressed as absorption. If both absorption and adsorption is important, then we address this phenomenon as sorption.

12.7.1 Langmuir Isotherm

The formula of what is now addressed as *Langmuir*[11] Isotherm [10–13] was derived in 1916. We are dealing exclusively with a monomolecular layer. This means, in Fig. 12.6 only the layer labeled s_1 is present.

So we have a bare surface s_0 and a surface that is occupied with a monomolecular layer s_1. This means also that multilayer surfaces are not present, i.e., $s_2 = 0$, $s_3 = 0$, ..., $s_i = 0$. If the total area is A, then this area is $A = s_0 + s_1$.

The derivation of the *Langmuir* isotherm is based on a kinetic model. The rate of adsorption is proportional to the pressure in the gas phase p and to the area s_0 of free sites:

$$ds_{a,1}/dt = k_{a,1} p s_0 = k_{a,1} p (A - s_1).$$

The second index 1 of k indicates that we are dealing with layer 1. Basically, this index is not necessary here, but we will make use in Sect. 12.7.3 in the case of multilayer adsorption of this index. The rate of desorption is proportional to the area occupied:

$$-ds_{d,1}/dt = k_{d,1} s_1.$$

The total rate of change of unit surface $ds_1/dt = ds_{a,1}/dt + ds_{d,1}/dt$ is obtained by combination of both adsorption and desorption:

$$ds_1/dt = k_{a,1} p (A - s_1) - k_{d,1} s_1.$$

Integration with the initial condition $t = 0$, $s_1 = 0$ gives

$$s_1(t) = \frac{A k_{a,1} p}{k_{a,1} p + k_{d,1}} \left[1 - \exp \left(-(k_{a,1} p + k_{d,1}) t \right) \right]. \tag{12.32}$$

---

[11] Irving Langmuir, born Jan. 31, 1881, in Brooklyn, NY, died Aug. 16, 1957, in Falmouth, MA, USA.

The term in square brackets of Eq. (12.32) at $t \to \infty$ tends to 1, so Eq. (12.32) reduces in the stationary state to

$$s_1(\infty) = \frac{A k_{a,1} p}{k_{a,1} p + k_{d,1}}.$$ (12.33)

Equation (12.33) is referred to as the Langmuir isotherm. The Langmuir isotherm explains adsorption for monomolecular layers only. The equation can be generally applied to all cases involving chemisorption. At high pressures and physical adsorption, the Langmuir isotherm may fail to predict experimental results. At high pressures, multilayers may be formed.

12.7.2 Frumkin Isotherm

Frumkin[12] was the pioneer of Russian electrochemistry [14]. *Frumkin* assumed that there is some interaction between the adsorbates. The free enthalpy of adsorption is then $\Delta G = \Delta G_0 + \gamma s$. If the adsorbates attract one another, then $\gamma > 0$. If they repel, then $\gamma < 0$. In the case that $\gamma = 0$, we recover the Langmuir result. The Frumkin isotherm [15] is written as

$$\frac{s}{1-s} = c \exp\left(-\frac{\Delta G_0}{kT}\right) \exp\left(\frac{\gamma s}{kT}\right).$$

γ / kT is the Frumkin function.

12.7.3 BET Isotherm

The paper by Brunauer, Emmet, and Teller [16], entitled *Adsorption of Gases in Multimolecular Layers* was published in 1938. In 2003 the American Chemical Society celebrated its 125th anniversary. At this occasion, a ranking list of the 125 most cited papers of the Journal of the American Chemical Society was published. The paper of *Brunauer*,[13] *Emmet*,[14] and *Teller*[15] is on rank 4 with 4,808 citations. By the way, the most cited paper from this journal among all is that by *Lineweaver* and *Burk* [17] with 10,638 citations.

Starting from Langmuir's equation, we proceed now to derive the equation of Bunauer, Emmet, and Teller. It is interesting to note that in the paper by Brunauer,

---

[12] Александр Наумович Фрумкин, born 1895 in Kishinev, died 1976 in Tula.

[13] Brunauer István, born Feb. 12, 1903, in Budapest, Hungary, died Jul. 6, 1986, in Potsdam, NY, USA.

[14] Paul Hugh Emmett, born Sep. 22, 1900, in Portland, Oregon, died Apr. 22, 1985.

[15] Edward (Ede) Teller, born Jan. 15, 1908, in Budapest, died Sep. 9, 2003, in Stanford, CA, USA.

Emmet, and Teller, the equation of Langmuir is mentioned, but the authors do not cite the original paper of Langmuir. So it seems that the equation of Langmuir was so popular at this time (1938) to the scientists, like logarithm or sinus and US dollar.

The equation of Brunauer, Emmet, and Teller deals with the multilayer adsorption. Here again we have a bare surface s_0, a surface that is occupied exclusively with a monomolecular layer s_1. Further, we have a surface s_2 that consists of two layers of adsorbate, etc. The total surface is the sum of these individual surfaces:

$$A = \sum_{i=0}^{i=\infty} s_i.$$

The total volume adsorbed is simply

$$v = v_0 \sum_{i=0}^{i=\infty} i s_i,$$

where v_0 is the volume of a monomolecular layer on a surface of the unit size, i.e., $1\,m^2$.

The surface of the bare layer s_0 increases by desorption of the first layer and decreases by condensation of a first layer onto the bare layer. Layers of higher order do not have any influence there. Analogous to the equation of Langmuir we proceed to equate the rate of adsorption of a first layer and a second layer above the previous layers and so on. The surface of the first layer may change in four ways:

1. The first layer s_1 increases by condensation onto the bare layer s_0.
2. The first layer s_1 decreases by vaporization of the first layer s_1.
3. The first layer s_1 increases by vaporization of the second layer s_2.
4. The first layer s_1 decreases by condensation onto the first layer s_1.

The situation is completely analogous to the higher layers with the areas $s_2, s_3, \ldots$

$$\frac{ds_0}{dt} = -k_{a,1} p s_0 + k_{d,1} s_1$$

$$\frac{ds_1}{dt} = k_{a,1} p s_0 - k_{d,1} s_1 + k_{d,2} s_2 - k_{a,2} p s_1$$

$$\frac{ds_2}{dt} = k_{a,2} p s_1 - k_{d,2} s_2 + k_{d,3} s_3 - k_{a,3} p s_2 \tag{12.34}$$

$$\frac{ds_3}{dt} = k_{a,3} p s_2 - k_{d,3} s_3 + k_{d,4} s_4 - k_{a,4} p s_3$$

$$\ldots = \ldots$$

In the original paper by Brunauer, Emmet, and Teller, the kinetic constants are given in a slightly different notation

$$k_{a,i} = a_i, \; i = 1, 2, \ldots$$

and

$$k_{d,i} = b_i \exp\left(-\frac{E_i}{RT}\right).$$

For this reason, here we also start counting with 1. In this series of equations (12.34), only the first terms $k_{a,1}$ and $k_{d,1}$ are different from the terms of higher order $k_{a,2}, k_{a,3}$ and $k_{d,2}, k_{d,3}$, respectively. So

$$k_a = k_{a,2} = k_{a,3} = \ldots$$

and

$$k_d = k_{d,2} = k_{d,3} = \ldots.$$

Substituting, we obtain

$$
\begin{aligned}
ds_0/dt &= -k_{a,1}\, ps_0 + k_{d,1} s_1 \\
ds_1/dt &= k_{a,1}\, ps_0 - k_{d,1} s_1 + k_d s_2 - k_a\, ps_1 \\
ds_2/dt &= k_a\, ps_1 - k_d s_2 + k_d s_3 - k_a\, ps_2 \\
ds_3/dt &= k_a\, ps_2 - k_d s_3 + k_d s_4 - k_a\, ps_3 \\
\ldots &= \ldots
\end{aligned}
\tag{12.35}
$$

We introduce the generating function

$$F(x, t) = s_0(t) + s_1(t)x + s_2(t)x^2 + s_3(t)x^3. \tag{12.36}$$

We multiply now the individual equations in Eq. (12.35) successively by x, x^2, x^3, x^4 and insert Eq. (12.36) to obtain.

$$
\begin{aligned}
x\frac{dF(x,t)}{dt} =& \\
& - xk_{a,1}\, ps_0(t) + x^2\left[F(x,t)k_a - k_a\, ps_0(t) + k_{a,1}\, ps_0(t)\right] \\
& - x\left[F(x,t)k_d - k_d s_0(t) - k_{d,1} s_1(t)\right] + x^2\left[k_d s_1(t) - k_{d,1} s_1(t)\right] \\
& + F(x,t)k_d - k_d s_0(t) - x\left[k_d s_1(t)\right] \\
& + xk_a p\left[s_0(t) - F(x,t)\right].
\end{aligned}
\tag{12.37}
$$

In equilibrium, $dF(x,t)/dt = 0$ and $F, s_0, s_1, \ldots$ are no longer functions of the time. We can resolve (12.37) to obtain

$$F(x) = \frac{x\left(k_a\, ps_0 - k_{a,1}\, ps_0 - s_1(k_d - k_{d,1})\right) - k_d s_0}{xk_a p - k_d}. \tag{12.38}$$

From the first equation in (12.35)

$$-k_{a,1}ps_0 + k_{d,1}s_1 = 0,$$

we can express s_1 as s_0

$$s_1 = k_{a,1}s_0/k_{d,1},$$

and by the condition $F(1) = A$, the bare area s_0 is obtained:

$$s_0 = \frac{Ak_{d,1}(k_a p - k_d)}{k_a p k_{d,1} - k_d(k_{a,1}p + k_{d,1})}.$$

Substituting into Eq. (12.38) yields

$$F(x) = \frac{A(k_a p - k_d)\left(x(k_a p k_{d,1} - k_{a,1}p k_d) - k_d k_{d,1}\right)}{\left(k_a p k_{d,1} - k_d(k_{a,1}p + k_{d,1})\right)(xk_a px - k_d)}.$$

This is essentially a geometric distribution. The total amount adsorbed is

$$\frac{v}{A} = v_0 \left.\frac{\frac{dF(x)}{dx}}{F(x)}\right|_{x=1}.$$

References

1. Pólya, G.: Kombinatorische Anzahlbestimmungen für Gruppen, Graphen und chemische Verbindungen. Acta Math. **68**, 145–254 (1937)
2. Pólya, G.: Combinatorial Enumeration of Groups, Graphs, and Chemical Compounds. Springer, New York (1987). Edited by R.C. Read
3. Balaban, A.T. (ed.): Chemical Applications of Graph Theory. Academic Press, London (1976)
4. Gorman, M., DeMattia, D., Doonan, D., Gohlke, R.S.: The calculation of relative abundance of isotope clusters in mass spectrometry. J. Chem. Educ. **47**, 467 (1970)
5. Arnold, L.J.: Mass spectra and the Macintosh: Isotope pattern calculator. A program to calculate isotopic ratios for molecular fragments (CS). J. Chem. Educ. **69**, 811 (1992)
6. Mattson, B., Carberry, E.: A new program for the calculation of mass spectrum isotope peaks. J. Chem. Educ. **50**, 511 (1973)
7. Mattson, B.M., Carberry, E.: Updated student-use programs for the calculation of mass spectral isotope patterns (cs). J. Chem. Educ. **60**, 736 (1983)
8. Fisz, M.: Wahrscheinlichkeitsrechnung und mathematische Statistik. VEB Deutscher Verlag der Wissenschaften, Berlin (1980)
9. Rouquerol, F., Rouquerol, J., Sing, K.S.W.: Adsorption by Powders and Porous Solids. Principles, Methodology and Applications, 1st edn. Academic Press, San Diego (1999)
10. Langmuir, I.: The condensation and evaporation of gas molecules. Proc. Natl. Acad. Sci. USA **3**, 141 (1917)
11. Langmuir, I.: Evaporation, condensation, and adsorption. Phys. Rev. **8**, 149 (1916)
12. Langmuir, I.: The adsorption of gases on plane surfaces of glass, mica and platinum. J. Am. Chem. Soc. **40**, 1361–1403 (1918)

13. Langmuir, I.: Vapor pressures, evaporation, condensation and adsorption. J. Am. Chem. Soc. **54**, 2798–2832 (1932)
14. Parsons, R.: Alexander Naumovich Frumkin. Electrochim. Acta **46**(8), 1095–1100 (2001)
15. Frumkin, A.N.: Surface tension curves of the higher fatty acids and the equation of condition of the surface layer. Z. Phys. Chem. **116**, 466–484 (1925)
16. Brunauer, S., Emmet, P.H., Teller, E.: Adsorption of gases in multimolecular layers. J. Am. Chem. Soc. **60**, 309–319 (1938)
17. Lineweaver, H., Burk, D.: The determination of enzyme dissociation constants. J. Am. Chem. Soc. **56**, 658–666 (1934)

Chapter 13
Stoichiometry

Stoichiometry roots from $\sigma\tau o\iota\chi\varepsilon\alpha$, which means letter or basic matter, and $\mu\varepsilon\tau\rho\nu$, which means measure. This particular topic of chemistry is engaged with the setup of chemical formulas and the description of chemical reactions. Only if the stoichiometry of a chemical reaction is known, the yield can be calculated. Stoichiometry was found by Richter [1]. *Richter*[1] lived in Breslau and was a chemist and mining expert.

Balancing chemical equations is a primary task in basic chemical education. In the course of an examination, it may happen that a student can balance a chemical equation, but he cannot explain the general algorithm how he succeeded. In textbooks this topic is actually unclearly presented, making use of common multiples and the method of successive rectification of the equation with respect to a certain atom, etc.

How to balance chemical equations was always a continuing field because of its importance [2–6]. A critical review has been given by *Herndon* [7]. A general procedure to obtain the stoichiometric coefficient is to solve the system of the homogeneous linear equations that are obtained from the principles of conservation of matter and charge. The earliest chemical paper by *J. Bottomly* in 1878 used this method [8].

Both the oxidation number method and the ion – electron half-reaction method start by establishing the relative proportions of reagents that are taking part in a separate oxidation and reduction procedure.

13.1 Half-Reaction Method

Consider the equation

$$HIO_3 + FeI_2 + HCl \rightarrow FeCl_3 + ICl + H_2O. \tag{13.1}$$

---

[1] Jeremias Benjamin Richter, born Mar. 10, 1762, in Hirschberg, now Jelenia Gora, Poland, died Apr. 14, 1807, in Berlin, Germany.

J.K. Fink, *Physical Chemistry in Depth*,
DOI 10.1007/978-3-642-01014-9_13, © Springer-Verlag Berlin Heidelberg 2009

The application of the half-reaction method starts like the following [7]:

1. It is easily perceived that the ferrous species is oxidized and the iodate is reduced.
2. The equation is split into two separate oxidation and reduction equations. These equations are balanced by inspection resulting in Eq. (13.2).
3. The lowest common multipliers are used to cancel the electron in Eq. (13.2).

$$FeI_2 + 5\,HCl \rightarrow FeCl_3 + 2\,ICl + 5\,H^+ + 5e^-$$
$$4\,H^+ + 4\,e^- + HIO_3 + HCl \rightarrow ICl + 3\,H_2O. \tag{13.2}$$

Finally the balanced Eq. (13.3) emerges from Eq. (13.1):

$$5\,HIO_3 + 4\,FeI_2 + 25\,HCl \rightarrow 4\,FeCl_3 + 13\,ICl + 15\,H_2O. \tag{13.3}$$

13.2 Balancing by Inspection

The balance of chemical equations occurs often by *inspection*, whatever this is. The rules for balancing chemical equations by inspection have been summarized as such [9]:

1. Any elements that appear only once on each side of the equation are balanced first.
2. Terms containing elements that have been balanced with respect to each other become part of a *linked set*. Usually, there will be two different linked sets in an equation.
3. Sometimes compounds that contain elements common to both linked sites are *twinned*. That is, the same compound is written twice in the same equation, so that it can become part of both linked sets. The twin terms are later combined.
4. An element that occurs in both linked sets can be used to tie them together. Often this will be the last element in the equation remaining unbalanced. Determine the increase or decrease in number of atoms of that element as you go from the left to the right of the equation in each linked set. Then use these numbers as factors for balancing the two sets with respect to each other.
5. For ionic equations, first split the equation into linked sets and balance all the elements within each set, using twin terms when needed. Then determine the increase or decrease in charge for each linked set, and use these numbers as factors for balancing the two sets to each other. On occasion it may be simpler to balance the charges before balancing the last several elements.

13.3 A Challenging Balance

In 1995 an article by *Stout* [10] on a chemical equation that is difficult or incredibly challenging to balance appeared. This article caused a series of subsequent contributions on this topic [7, 9, 11–18]. Even from the title of some papers, e.g., *"Letter to the Editor about How Do I Balance Thee?...,"* "Response to Letter to the Editor about ...," "Redox Challenges – Reply," it can be felt that the topic was rather passionately discussed. In 1997 the editor [19] lost his patience in stating

> Because essentially everything that can be said has been said, I am declaring a moratorium on manuscripts about how to balance equations.

The famous equation by Stout to balance is [10]

$$[Cr(N_2H_4CO)_6]_4[Cr(CN)_6]_3 + KMnO_4 + H_2SO_4 \rightarrow$$
$$K_2Cr_2O_7 + MnSO_4 + CO_2 + KNO_3 + K_2SO_4 + H_2O.$$

$$(13.4)$$

13.3.1 Matrix Method

Smith and Missen [20] exemplified the solution to obtain the stoichiometric coefficients of this equation by the matrix method [21]. A reacting system consisting of a set of chemical reaction equations is represented by a formula matrix $A = (a_{ki})$, where the element a_{ki} of this matrix is the subscript of the chemical element k in the compound i occurring in the reaction equation. Consider a matrix $N = (v_{ij})$ of stoichiometric coefficients, where v_{ij} is the coefficient of the species i in the chemical equation j. The matrix N is obtained by solving the matrix equation

$$AN = 0.$$

In Eq. (13.4) we have the compounds

$$\begin{pmatrix} [Cr(N_2H_4CO)_6]_4[Cr(CN)_6]_3 \\ KMnO_4 \\ H_2SO_4 \\ K_2Cr_2O_7 \\ MnSO_4 \\ CO_2 \\ KNO_3 \\ H_2O \end{pmatrix}$$

and the elements (Cr, N, H, C, O, K, Mn, S).

The compound matrix consists of the vectors

$$
\begin{aligned}
[Cr(N_2H_4CO)_6]_4[Cr(CN)_6]_3 &= (7,66,96,42,24,0,0,0); \\
KMnO_4 &= (0,\ 0,\ 0,\ 0,\ 4,1,1,0); \\
H_2SO_4 &= (0,\ 0,\ 2,\ 0,\ 4,0,0,1); \\
K_2Cr_2O_7 &= (2,\ 0,\ 0,\ 0,\ 7,2,0,0); \\
MnSO_4 &= (0,\ 0,\ 0,\ 0,\ 4,0,1,1); \\
CO_2 &= (0,\ 0,\ 0,\ 1,\ 2,0,0,0); \\
KNO_3 &= (0,\ 1,\ 0,\ 0,\ 3,1,0,0); \\
K_2SO_4 &= (0,\ 0,\ 0,\ 0,\ 4,2,0,1); \\
H_2O &= (0,\ 0,\ 2,\ 0,\ 1,0,0,\ 0)
\end{aligned}
\tag{13.5}
$$

In Mathematica®, the matrix A can be composed from the vectors in Eq. (13.5) with the MatrixForm and with the Transpose command. Further, the matrix can be reduced to the unit form with the RowReduce command. The mathematica program spies out then a matrix A^* where the coefficients appear then at the right most column.

13.3.2 Equating the Number of Atoms

Glaister [22] from the Department of Mathematics, University of Reading, showed a simple method to balance this equation by equating the number of atoms on each side.

13.4 Balancing with Generating Functions

The following should be helpful in this regard for clarifying, how equations are to balance properly. It is shown that a deeper penetration of stoichiometry can literally point out also the impossibility of certain reactions for reasons of the balance.

13.4.1 A Fundamental Example

13.4.1.1 Synthesis of Ammonia

The stoichiometric coefficients for the reaction

$$
H_2 + N_2 \rightarrow NH_3
$$

are to be determined. In addition, we provide the compounds with at first unknown stoichiometric coefficients:

$$
x\, H_2 + y\, N_2 \rightarrow z\, NH_3.
\tag{13.6}
$$

The homogeneity of this equation allows that we can select one of these coefficients (x, y, z) arbitrarily, e.g., $z = 1$.

13.4.1.2 Balance Over Atoms

Now the proper balance over Eq. (13.6) consists in adjusting the coefficients in such a way that on the left-hand side and on the right-hand side, the number of the individual atoms are equal. With this rule, we find

$$\text{for H: } 2x = 3$$
$$\text{for N: } 2y = 1$$

The regulation is thus obviously this that we must pull for each atom in each component in the equation the index as coefficient. If a certain atom in a component does not occur, then zero is to be pulled as coefficient.

13.4.1.3 Partial Derivatives with Respect to the Atoms

A more general method is the method of partial derivatives with respect to the atoms. We regard the chemical equation as a mathematical equation and rewrite as

$$x * h^2 + y * n^2 = nh^3.$$

Then we form the derivatives $\partial(\bullet)/\partial h$ and $\partial(\bullet)/\partial n$ and obtain

$$\partial(\bullet)/\partial h : 2xh = 3nh^2$$
$$\partial(\bullet)/\partial n : 2yn = h^3$$

(13.7)

We equate now Eq. (13.7) at $n = h = 1$ with the result

$$x = \frac{3}{2} \quad \text{and} \quad y = \frac{1}{2}.$$

A generalization on more complicated systems should not make problems.

13.4.1.4 Using a Symbolic Calculator

We use the symbolic calculator DERIVE®, in order to find out the stoichiometric coefficients of the reaction (13.4). In Table 13.1 a somewhat abbreviated DERIVE® session in order to calculate stoichiometric coefficients for this reaction is shown.

13.4.2 System of Linear Equations

Next, we exemplify the reaction

$$KOH + CO_2 \rightarrow K_2CO_3 + H_2O.$$

(13.8)

Table 13.1 A DERIVE® session in order to calculate stoichiometric coefficients

```
% This and slightly abbreviated minutes of a session
% Some comments have been inserted, are marked as %
  InputMode:=Word
  CaseMode:=Sensitive
% Typing in the compounds
  (Cr*(N^2*H^4*C*O)^6)^4*(Cr*(C*N)^6)^3 K*Mn*O^4
  H^2*S*O^4
  K^2*Cr^2*O^7
  Mn*S*O^4
  C*O^2
  K*N*O^3
  K^2*S*O^4
  H^2*O
  (Cr*(N^2*H^4*C*O)^6)^4*(Cr*(C*N)^6)^3
  +a*(K*Mn*O^4)+b*(H^2*S*O^4)
  =c*(K^2*Cr^2*O^7)+d*(Mn*S*O^4)+e*(C*O^2)
  +f*(K*N*O^3)
  +g*(K^2*S*O^4)+h*(H^2*O)
% Simplify
% Forming the derivatives
  DIF((Cr*(N^2*H^4*C*O)^6)^4*(Cr*(C*N)^6)^3
  +a*(K*Mn*O^4)+b*(H^2*S*O^4)=c*(K^2*Cr~
  ^2*O^7)+d*(Mn*S*O^4)+e*(C*O^2)+f*(K*N*O^3)
  +g*(K^2*S*O^4)+h*(H^2*O),Cr)
% ..........
% Adding the equations together as vector
  [7*C^42*Cr^6*H^96*N^66*O^24=2*Cr*K^2*O^7*c,
  66*C^42*Cr^7*H^96*N^65*O^24=K*O^3*f,
  96*C^42*Cr^7*H^95*N^66*O^24+2*H*O^4*S*b=2*H*O*h,
  42*C^41*Cr^7*H^96*N^66*O^24=O^2*e,
  24*C^42*Cr^7*H^96*N^66*O^23+4*O^3*(H^2*S*b+K*Mn*a)=
  =2*C*O*e+7*Cr^2*K^2*O^6*c
  +H^2*h+O^2*(4*K^2*O*S*g+3*K*N*f+4*Mn*O*S*d),
  Mn*O^4*a=O^3*(2*Cr^2*K*O^4*c+2*K*O*S*g+N*f),
  K*O^4*a=O^4*S*d,H^2*O^4*b=K^2*O^4*g+Mn*O^4*d]
%Substitute for each chemical element '1'
% and solve the system of equations
  [a=588/5,b=1399/10,c=7/2,d=588/5,e=42,
  f=66,g=223/10,h=1879/10]
% multiply by 10
  [a=588/5,b=1399/10,c=7/2,d=588/5,e=42,
  f=66,g=223/10,h=1879/10]*10
% Simplify
  [10*a=1176,10*b=1399,10*c=35,10*d=1176,10*e=420,
  10*f=660,10*g=223,10*h=1879]
```

Here are involved four chemical compounds with totally four atoms. In Eq. (13.7), we had three compounds with totally two atoms. As we can see, for the solution of the stoichiometric coefficients, a system of linear equations arises. Thus, certain rules from algebra are applicable.

For a solution of a linear system, one equation is necessary for each unknown variable. The equations must not contradict themselves on the one hand; on the other hand, each equation must contain information about the relation of the unknown quantities, which is not derivable from the remaining equations.

If E is the number of equations and U the number of unknown variables, then there are freely selectable $U - E = F$ unknown variables, if F is positive. If more equations than variables are available, then the system is overdetermined and it can be that some equations are redundant, because they can be obtained by a linear combination of the others. Or the equations are such in they lead to a contradiction, e.g., $x + y = 1; x + y = 2$.

The number of equations results from the number of different atoms (A) in the compounds. The number of the unknown coefficients is one less than the number of compounds involved (C). In fact, because of the homogeneity of one coefficient, we can adjust it arbitrarily. It applies thus for

$$C - 1 - A = F.$$

In the case of the ammonia synthesis, the situation is very clear because here $F = 0$. For the reaction of carbon dioxide with caustic potash solution, we obtain $F = -1$. Obviously, here an equation is redundant, or one atom is too much. We do not suspect a contradiction, because the reaction can be found in all textbooks.

Now it is the hydroxide, which is essentially involved in the reaction. All well-known reactions of this kind are formulated also without the counterion (K^+). From the balance, it is suggested purely to formulate the reaction in a simpler way, i.e., one atom can be left off. Of course, pure mathematics does not permit us to draw conclusions from the laws of algebra on the existence of ions. We cannot even say which of the equations is redundant.

Here we may annotate that the phase rule and the polyhedron law are based on similar arguments, c.f., Sect. 7.6, p. 265.

13.4.3 Definition of Pseudoatoms

In the example

$$x \, KOH + y \, CO_2 \rightarrow z \, H_2O + K_2CO_3, \tag{13.9}$$

the atoms are defined in conventional way. It is now possible to define groups in the molecular formula as atom or pseudoatom (atavistically: to gather several atoms into radicals). A well-known example of radicals is the ammonium radical. From the chemical view, a radical is a gather yard of certain atoms in a compound that is left intact in the course of a chemical reaction. Radicals emerge as such on the right-hand side of the chemical equation, i.e, in the products.

Table 13.2 Gathering of atoms into groups at the reactants

Groups	Equation		Atoms + groups	x	y	z
–	$x\,KOH + y\,CO_2$	$\rightarrow z\,H_2O + K_2CO_3$	4	2	1	1
$Q = KO$	$x\,HQ + y\,CO_2$	$\rightarrow z\,H_2O + COQ_2$	4	2	1	1
$Q = KH$	$x\,OQ + y\,CO_2$	$\rightarrow z\,H_2O + K_2CO_3$	5	a	a	a
$Q = OH$	$x\,KQ + y\,CO_2$	$\rightarrow z\,HQ + K_2CO_3$	5	a	a	a
$Q = CO$	$x\,KOH + y\,OQ$	$\rightarrow z\,H_2O + K_2O_2Q$	4	2	1	1
$Q = CO_2$	$x\,KOH + y\,Q$	$\rightarrow z\,H_2O + K_2OQ$	4	2	1	1
$Q = O_2$	$x\,KOH + y\,CQ$	$\rightarrow z\,H_2O + K_2COQ$	5	2	1	1

[a] No solution

So certain atoms in Eq. (13.9) reaction can be combined into groups. For example, as a result of the substitution of Q = CO, Eq. (13.10) is obtained:

$$x\,KOH + y\,OQ \rightarrow z\,H_2O + K_2OQ. \tag{13.10}$$

The stoichiometric coefficients are obtained as described above by partial differentiating (also with respect to Q) and evaluation of the set of Eq. 1. Table 13.2 shows that the stoichiometric coefficients emerge correctly, with two exceptions, however, if we treat OH or KH as pseudoatoms. In these two cases, hydrogen emerges at the product side (right-hand side of the equation) from *nothing*. Obviously, all other substitutions with a *stoichiometrically permitted* reaction mechanism are compatible with the suggested pseudoatoms. This argument is valid in a mathematical sense, and not necessary in a chemical sense. However, the suggested equation by mathematical arguments must be among those equations, suggested by chemical arguments.

For example, the group of Q = KO permits a mechanism, with which Q will be transferred or attached to CO_2, and then from the CO_2 an oxygen atom will be split off, which emerges eventually in the water group. If we introduce the two substitutes Q=OH and Q=KH, then the introduction of these respective radicals is not valid. We can conclude that bonds must be opened within these radicals in the course of the particular reaction.

On the other hand, the gradual enlargement of Q gives a reference to the group of CO up to the reactant CO_2. Obviously, this means that this group ($Q = CO_2$) does not split in the course of the reaction.

By a combination of such statements, we can arrive in the best case conclusively at a restriction of possible reaction mechanisms. Even we may exclude some reaction mechanisms by a critical examination of such a procedure.

13.4.4 Stoichiometrically Unknown Compounds

Stoichiometrically unknown compounds can be examined, by treating the indices as unknown quantities, e.g.,

$$x\,KOH + y\,CO_n \rightarrow z\,H_2O + K_2CO_3,$$

or

$$x \, \text{KOH} + y \, C_m O_2 \rightarrow z \, H_2O + K_2CO_3.$$

Nonlinear equations are obtained in this case. However, we observe that during the solution of the above equations, the values $m = 1$ are forced and/or $n = 2$. If we put further $y \, C_m O_n$, then we obtain still $n = 2m$, as the relative composition of the compound and further $ym = 1$. So the stoichiometric factor y is fixed besides the uncertainty of a hypothetical m-mer of carbon dioxide.

13.4.5 Contradictory Equations

The attempt to produce hydrogen peroxide from caustic potash solution and carbon dioxide fails because of stoichiometry, as we can verify easily:

$$x \, \text{KOH} + y \, CO_2 \rightarrow z \, H_2O_2 + K_2CO_3. \tag{13.11}$$

No solution is obtained for the stoichiometric coefficients in Eq. (13.11). Likewise, carbon monoxide cannot react with caustic potash solution to produce potash according to the following reaction for stoichiometric reasons:

$$x \, \text{KOH} + y \, \text{CO} \rightarrow z \, H_2O + K_2CO_3.$$

13.4.6 Confounded Equations

Consider the equation

$$H_2 + \text{CO} \rightarrow \text{HCHO} + CH_3\text{OH}. \tag{13.12}$$

Equation (13.12) describes both the formation of formaldehyde and methanol from hydrogen and carbon monoxide. There are three atoms and four compounds. However, forming the derivatives of (13.13)

$$x \, h^2 + y \, co \rightarrow z \, hcho + ch^3 oh \tag{13.13}$$

results in

$$
\begin{aligned}
2x &= 2z + 4 \\
y &= z + 1 \\
y &= z + 1
\end{aligned}
\tag{13.14}
$$

This means there are two identical equations. We can set $z = 0$ in (13.14) and remove $H_2 + CO \rightarrow HCHO$ from this equation. Thus, Eq. (13.12) consists of two elementary equations.

13.4.7 Matrix Notation

13.4.7.1 Inhomogeneous Equations

We can solve Eq. (13.8) also by the methods of linear algebra. We place the coefficients

$$x\,KOH + y\,CO_2 \rightarrow z\,K_2CO_3 + H_2O.$$

In matrix notation, a system of linear equations can be written down as

$$Ax = b, \tag{13.15}$$

where A is the coefficient matrix, x is the vector of variables, and b is the right-hand side of the equations. For Eq. (13.8), the matrix notation Eq. (13.15) reads as

$$\begin{pmatrix} 0 & 1 & -1 \\ 1 & 2 & 3 \\ 1 & 0 & 0 \end{pmatrix} \begin{pmatrix} x \\ y \\ z \end{pmatrix} = \begin{pmatrix} 0 \\ 1 \\ 2 \end{pmatrix}.$$

The first row is the derivative with respect to carbon ($\partial \bullet /\partial C$), the second row is the derivative with respect to oxygen ($\partial \bullet /\partial O$), the third row is the derivative with respect to hydrogen ($\partial \bullet /\partial H$). Since there are three unknown coefficients, we need only the information from three equations, which are obtained by forming the derivatives. We could likewise instead of one derivatives take also the derivative with respect to potassium ($\partial \bullet /\partial K$), and we would obtain the same final result.

The coefficients are obtained by forming the inverse matrix

$$x = A^{-1}b,$$

in detail:

$$\begin{pmatrix} 2 \\ 1 \\ 1 \end{pmatrix} = \begin{pmatrix} 0 & 0 & 1 \\ 3 & -1 & 1 \\ 2 & -1 & 1 \end{pmatrix} \begin{pmatrix} 0 \\ 1 \\ 2 \end{pmatrix}.$$

When we apply this method of matrix inversion, we must be sure that the procedure will work. This means that we have chosen the unknowns and the equations appropriate, finally there should be a nontrivial solution, i.e., the unknown variables x, y, z must be different from zero.

13.4.7.2 Homogeneous Equations

We revisit once more the Eq. (13.8) in the form

$$x_1 \, KOH + x_2 \, CO_2 \rightarrow x_3 \, K_2CO_3 + x_4 \, H_2O. \tag{13.16}$$

We have renamed the coefficients, and we have added x_4 as the stoichiometric coefficient for water. We treat now $x_1, \ldots, x_n$ as a vector and the coefficients of the system of the linear equations as a matrix. Thus, we have

$$Ax = 0,$$

with x

$$x = \begin{pmatrix} x_1 \\ x_2 \\ \ldots \\ x_n \end{pmatrix},$$

and A

$$A = \begin{pmatrix} a_{1,1} & a_{1,2} & \ldots & a_{1,n} \\ a_{2,1} & a_{2,2} & \ldots & a_{2,n} \\ \ldots & \ldots & \ldots & \ldots \\ a_{m,1} & a_{m,2} & \ldots & a_{m,n} \end{pmatrix}.$$

Therefore,

$$Ax = \begin{pmatrix} a_{1,1}x_1 + a_{1,2}x_2 + \ldots + a_{1,n}x_n \\ a_{2,1}x_1 + a_{2,2}x_2 + \ldots + a_{2,n}x_n \\ \ldots + \ldots + \ldots + \ldots \\ a_{m,1}x_1 + a_{m,2}x_2 + \ldots + a_{m,n}x_n \end{pmatrix}. \tag{13.17}$$

This is just the system of equations we have to solve. We insert for the coefficients now for Eq. (13.16) and get

$$A = \begin{pmatrix} 1 & 0 & -2 & 0 \\ 1 & 2 & -3 & -1 \\ 1 & 0 & 0 & -2 \\ 0 & 1 & -1 & 0 \end{pmatrix}. \tag{13.18}$$

We calculate now the echelon of the matrix Eq. (13.18) as

$$A = \begin{pmatrix} 1 & 0 & 0 & -2 \\ 0 & 1 & 0 & -1 \\ 0 & 0 & 1 & -1 \\ 0 & 0 & 0 & 0 \end{pmatrix}. \tag{13.19}$$

In fact, the last row of Eq. (13.19) contains exclusively zeros.

13.4.8 Linear Algebra

Linear algebra deals among other topics with systems of linear equations. The results can be readily used in vector and tensor analysis. We explain now a few paradigms of linear algebra.

13.4.8.1 System of Linear Equations

Equation (13.20) shows a system of inhomogeneous linear equations

$$\begin{aligned} a_{1,1}x_1 + a_{1,2}x_2 + \ldots + a_{1,n}x_n &= b_1 \\ a_{2,1}x_1 + a_{2,2}x_2 + \ldots + a_{2,n}x_n &= b_2 \\ &\ldots\ldots\ldots\ldots\ldots \\ a_{m,1}x_1 + a_{m,2}x_2 + \ldots + a_{m,n}x_n &= b_n \end{aligned} \tag{13.20}$$

We can convert this inhomogeneous system always in a homogeneous system, if we put $b_k = -a_{k,n+1}x_{n+1}$ and thinking $x_{n+1} = 1$, i.e.,

$$\begin{aligned} a_{1,1}x_1 + a_{1,2}x_2 + \ldots + a_{1,n}x_n + a_{1,n+1}x_{n+1} &= 0 \\ a_{2,1}x_1 + a_{2,2}x_2 + \ldots + a_{2,n}x_n a_{2,n+1}x_{n+1} &= 0 \\ &\ldots\ldots\ldots\ldots\ldots \\ a_{m,1}x_1 + a_{m,2}x_2 + \ldots + a_{m,n}x_n a_{m,n+1}x_{n+1} &= 0 \end{aligned}.$$

13.4.8.2 Echelon

A matrix is in row echelon[2] form if

1. all rows that consist entirely of zeros are grouped together at the bottom of the matrix; and
2. the first (counting left to right) nonzero entry in each nonzero row appears in a column to the right of the first nonzero entry in the preceding row, if there is a preceding row at all.

---

[2] Echelon refers originally to a combat formation in which the members are arranged diagonally.

13.4.8.3 Rank

The rank of a matrix A is the number of nonzero rows in the reduced row echelon form of A.

13.4.8.4 Linear Independence

We call a system of equations Eq. (13.17) $Ax = 0$ linearly independent, if the only solution of $x_1, x_2, \ldots$ is $x_1 = 0, x_2 = 0, \ldots$. In this case, the echelon of the corresponding matrix is triangular. This means there are leading 1s in the diagonal. We see from the echelon of the matrix Eq. (13.18) that the equations are not linearly independent. Therefore, a nonzero solution of the stoichiometric coefficients exists.

In the matrix Eq. (13.18), we can omit any of the rows, i.e., the first row, the second row, the third row, or the fourth row, and we will arrive at the echelon

$$A = \begin{pmatrix} 1 & 0 & 0 & -2 \\ 0 & 1 & 0 & -1 \\ 0 & 0 & 1 & -1 \end{pmatrix}.$$

The echelon is now one row smaller than the echelon, Eq. (13.19), but the row with the zeros disappeared and the echelon is of the same form. Therefore, we have in the matrix one equation that can be deduced by a combination of the other equations.

13.4.8.5 Independency

We consider once more the impossible reaction

$$x_1 \, KOH + x_2 \, CO_2 \rightarrow x_3 \, K_2CO_3 + x_4 \, H_2O_2.$$

The matrix corresponding to $(K, O, H, C)^T$ reads as

$$A = \begin{pmatrix} 1 & 0 & -2 & 0 \\ 1 & 2 & -3 & -2 \\ 1 & 0 & 0 & -2 \\ 0 & 1 & -1 & 0 \end{pmatrix},$$

and the echelon of this matrix is the diagonal matrix

$$\begin{pmatrix} 1 & 0 & 0 & 0 \\ 0 & 1 & 0 & 0 \\ 0 & 0 & 1 & 0 \\ 0 & 1 & 0 & 1 \end{pmatrix}.$$

Therefore, there is no other solution other than all the coefficients being zero: $x_1 = 0, \ldots, x_4 = 0$.

13.4.8.6 Confounding

We inspect now the equation

$$x_1 \, KOH + x_2 \, CO_2 \rightarrow x_3 \, K_2CO_3 + x_3 \, KHCO_3 + x_5 \, H_2O. \tag{13.21}$$

The coefficient matrix of Eq. (13.21) is

$$\begin{pmatrix} 0 & 1 & -1 & -1 & 0 \\ 1 & 2 & -3 & -3 & -1 \\ 1 & 0 & -1 & 0 & -2 \\ 1 & 0 & -1 & -2 & 0 \end{pmatrix},$$

and the echelon of this matrix is

$$\begin{pmatrix} 1 & 0 & -1 & 0 & -2 \\ 0 & 1 & -1 & 0 & -1 \\ 0 & 0 & 0 & 1 & -1 \\ 0 & 0 & 0 & 0 & 0 \end{pmatrix}.$$

In contrast to the echelon of Eq. (13.19), one row with a leading 1 in column 3 is missing. In other words, the leading zeros jump by two. This means that there are totally two independent variables that can be chosen deliberately. We can conclude further that this chemical equation consists of two confounded elementary equations.

Question 13.1. In the case of a single chemical equation, are there always n chemical elements and $n + 1$ chemical compounds?

13.4.9 Ions in Equations

Of course, balances accomplish also with reactions, in which ions are involved. Beside differentiating with respect to the atoms, it is also differentiated with respect to the charge. In the reaction

$$Cr_2O_7^{2-} + 6J^- + 14H^+ \rightarrow 2Cr^{3+} + 3J_2 + 7H_2O$$

replace the charges by $e_{\pm n}$, e.g., $Cr_2O_7e_{-2}$ and differentiate with respect to the charge. Stoichiometry does not bother whether the charges are in fact positive or negative. Actually, the principle works well if positive and negative charges are exchanged. So in retrospect, we could state that stoichiometry suggests the existence of antimatter.

13.4.10 Homologues

The advantage of partial differentiating emerges more distinctly in systems with infinitely many components. In the following, we discuss another simple example with finitely many components. The expression $(1 - x^{n-1})/(1 - x)$ expands, as well known, into the geometric series

$$1 + x + x^2 + \ldots + x^{n-2}.$$

Using this relationship, we now gain the number of the linear alkane homologues:

$$F(h, c) = h^3 c \frac{1 - (ch^2)^{n-1}}{1 - (ch^2)} ch^3 = h^3 c c h^3 + \ldots + h^3 c (ch^2)^{n-2} ch^3. \quad (13.22)$$

Equation (13.22) is the generating function for the number of the linear alkanes in h and c.

Question 13.2. What is now the ratio of the number of hydrogen atoms and the number of carbon atoms, H/C, in this series of alkanes? We obtain it either on the right side of Eq. (13.22) by counting or on the left side by differentiating. We may build simply the expression

$$\left. \frac{\frac{\partial F(h, c)}{\partial h}}{\frac{\partial F(h, c)}{\partial c}} \right|_{c=h=1} = \frac{2(n+4)}{n+2} = H/C; n = 2, 3, \ldots .$$

Going to the limit is necessary here because of 0/0. Further, we obtain the limit value $H/C = 2$ for large n.

13.4.11 Symbolic Computational Programs

For the formation of the differentials and the solution of the sets of equations, symbolic computational programs can be used in favor, such as DERIVE® or Mathematica®.

13.5 Further Examples

Finally, we present further examples to illustrate the method of balancing chemical equations using generating functions.

13.5.1 Analytical Chemistry

In analytical chemistry, we find particularly grateful examples for balancing equations, e.g.:

$$Na_2S_2O_3 + 4\,Br_2 + 5\,H_2O \rightarrow 2\,NaHSO_4 + 8\,HBr$$
$$Cr_2O_7^{2-} + 6\,J^- + 14\,H^+ \rightarrow 2\,Cr^{3+} + 3\,J_2 + 7\,H_2O \qquad (13.23)$$
$$2\,MnO_4 + 5\,(C_2O_4)^{2-} + 16\,H^+ \rightarrow 2\,Mn^{2+} + 10\,CO_2 + 8\,H_2O.$$

To balance, we replace the charges by $e^{\pm}$ and perform the same formal procedure as with the atoms.

Question 13.3. Could purely stoichiometrically metallic manganese develop with the last reaction? Further, could the reaction remain stopped with carbon monoxide?

13.5.2 Fermentation

We consider the fermentation of carbohydrates. It is well known that glucose converts into carbon dioxide and ethyl alcohol. We treat the case more general. We set up the chemical equation as

$$(C(H_2O)_m \rightarrow \beta CO_2 + \gamma C_n H_{2n+2}O. \qquad (13.24)$$

From Eq. (13.24), the mathematical equation system arises

$$
\begin{aligned}
C: &\quad m - \beta - n\gamma = 0 \\
H: &\quad 2m - 2\gamma(n+1) = 0 \\
O: &\quad m - 2\beta - \gamma = 0
\end{aligned}
$$

We find the solution $n = 2$, $\beta = \gamma$, and $m = 3\gamma$. Thus, from carbohydrates only ethanol may be formed according to Eq. (13.24) and the minimum possible carbohydrate is a triose, since $\gamma = 1, 2, 3, \ldots$. We extend now Eq. (13.24) and introduce water as reactant:

$$(C(H_2O)_m + \alpha H_2O \rightarrow \beta CO_2 + \gamma C_n H_{2n+2}O.$$

Proceeding in the same way as before, we obtain the following set of equations:

$$
\begin{aligned}
\alpha &= m(2-n)/3n \\
\beta &= m/3 \\
\gamma &= 2m/3n
\end{aligned}
$$

Again, here the formation of ethanol is possible; however, also the formation of methanol is possible, but higher alcohols are not possible because for $n > 2$, α turns negative.

13.5.3 Oxidation Numbers

In chemical compounds, numbers are assigned to the atoms, which designate the charge of the atom based on the assumption of an ionic bond. In the broader sense, the oxidation number means the valency of the particular atom. Obviously, elements can have different oxidation numbers in different compounds.

13.5.3.1 Equivalent Oxidation Numbers

First, we assume that the oxidation number of a certain element is alike in all compounds. We replace the element symbol S by the generating function $S = \exp(sx)$ and the index by the exponent. The lower case letter s means the oxidation number. For example, $H_2O \rightarrow \exp(2hx)\exp(ox) = \exp((2h + o)x)$. By differentiating with respect to x at $x = 0$, we obtain $2h + o$. This formalism simplifies the application in symbolic computational programs.

The sum of the oxidation numbers in neutral molecules must complement to zero because of electrical neutrality. In the case of ions, which can consist of several atoms, the sum must be the total electrical charge of the ion. In this kind, we obtain a homogeneous set of equations for a set of chemical compounds. This is shown as follows.

Example 13.1. In Table 13.3, the oxidation numbers are denoted in lower case letters corresponding to the respective element symbol. The solution of this set of equations does not result in absolute values for the oxidation numbers, but in parametric solutions:

$$c = \lambda; \quad h = \lambda/4; \quad k = \lambda/4; \quad o = -\lambda/2; \quad n = -5\lambda/4.$$

We search for integral solutions and select the parameter λ in such a way that the smallest oxidation number in the set becomes 1. This is the term where the divisor

Table 13.3 System of equations for the determination of oxidation numbers in a set of chemical compounds

Compound	Equation
KOH	$k + o + h = 0$
KHCO$_3$	$k + h + c + 3o = 0$
H$_2$O	$2h + o = 0$
K$_2$CO$_3$	$2k + c + 3o = 0$
CO$_2$	$c + 2o = 0$
H$_2$CO$_3$	$2h + c + 3o = 0$
HCN	$h + c + n = 0$

has the largest value. Apart from positive parameters also a negative parameter is permitted. Thus, two solution sets with respect to the charge, which are mirror images, emerge. Thus

$$c = +4; \quad h = +1; \quad k = +1; \quad o = -2; \quad n = -5$$

and

$$c = -4; \quad h = -1; \quad k = -1; \quad o = +2; \quad n = +5.$$

arise.

In the mathematical sense, this means that we can select freely one and only one oxidation number in the set of equations and also the sign. That applies also to the physical interpretation, after which it is unimportant whether we take charges positive or negative. However, from this is taken off the question, whether two mirror worlds can exist, with respect to the charges of the various particles built up from. Mathematics permits both worlds. ☐

13.5.3.2 Different Oxidation Numbers of the Same Element

If we add to the set of equations in Table 13.3 further equations, which are contrary to the previous added, then it can happen that no more a nonzero solution can be obtained, e.g., the oxidation numbers for CO_2 in Table 13.3 are contradictory to CO. Such a solution behavior points straight on the occurrence different oxidation numbers. Prussian blue $Fe_4[Fe(CN)_6]_3$ is a famous example of a compound, with which an element, the iron, occurs in two different oxidation numbers. A resonance structure could be excluded by Mößbauer spectroscopy [23, 24].

13.5.3.3 Principle of Minimum Differences of Oxidation Numbers

If we add to the set of equations in Table 13.3, the equation for the oxidation numbers of $Fe_4[Fe(CN)_6]_3$, i.e., $4fe + 3[fe + 6(c + n)]$, then we obtain a solution, to which we would have to assign suddenly unusually high integral oxidation numbers $c = 28, fe = 18, h = 7, k = 7, n = -35, o = -14$. We can treat the matter differently and associate to the two iron atoms different oxidation numbers: $4fex + 3[fe + 6(c + n)]$. Then we obtain a solution with a free parameter while maintaining the earlier obtained oxidation numbers:

$$c = 4; \quad fe = \lambda; \quad fex = 3(6\lambda)/4; \quad h = 1; \quad k = 1; \quad n = -5; \quad o = -2.$$

The integer solutions for the oxidation numbers of iron in Prussian blue are shown in Table 13.4. Indeed, we observe in nature the oxidation numbers, whose difference is a minimum. Iron arises by the way mainly in the oxidation numbers 0, 2.3, 6. So far known, oxidation numbers are from -2 to $+6$, e.g., $FeH_2(CO)_4$.

Table 13.4 Integer solutions for the oxidation numbers of iron in Prussian blue

Fe	FeX
6	0
2	3
−2	6
−6	9
...	...

13.5.4 Top-down and Bottom-Up Techniques

The terms top-down and bottom-up arise from software technology. Bottom-up programing means that a program is developed starting with the basic subroutines and developing the main program at the end. In the top-down programming technique, the main program is filled with subroutines that may initially not do their correct job; however, they are filled successively with further subroutines as long as the whole job works as intended.

Here we start with a process in that stoichiometry is involved either with considerations related to top-down and bottom-up techniques. The particular task is relevant in steel industries: How to make iron from iron oxide? For simplicity, we are engaged with FeO. Thus

$$FeO \rightarrow Fe + \frac{1}{2}O_2. \tag{13.25}$$

Actually, we can write down a complete equation thinking of using carbon for the reduction

$$FeO + \frac{1}{2}C \rightarrow Fe + \frac{1}{2}CO_2, \tag{13.26}$$

or

$$FeO + C \rightarrow Fe + CO. \tag{13.27}$$

Equations (13.26) and (13.27) correspond to a bottom-up technique of finding a viable process for the reduction of iron oxide. However, there is a difference in Eq. (13.25), as we have introduced carbon to get a viable process. Moreover, the final product is no longer oxygen.

In order to illustrate a top-down process of finding a solution, we rewrite and split Eq. (13.25) into

$$\begin{aligned} FeO &\rightarrow \ldots \\ \ldots &\rightarrow Fe \\ \ldots &\rightarrow \frac{1}{2}O_2 \end{aligned} \tag{13.28}$$

Next we replace the ... accordingly.

$$FeO + H_2SO_4 \rightarrow FeSO_4 + H_2O$$
$$FeSO_4 + 2e^- \rightarrow Fe + SO_4^{2-}$$
$$H_2O \rightarrow \frac{1}{2}O_2 + 2H^+ + 2e^- \qquad (13.29)$$
$$2H^+ + SO_4^{2-} \rightarrow H_2SO_4$$

The last equation in the set of Eq. (13.29) we had to add. Actually, the set of Eq. (13.29) adds up to Eq. (13.25), and we have invented the electrochemical cleavage of iron oxide into iron and oxygen. Clearly, the sulfuric acid that we have introduced acts in a catalytic way, unlike the introduction of carbon in the previous processes.

References

1. Richter, J.B.: Anfangsgründe der Stöchyometrie oder Meßkunst chymischer Elemente. Reprint: Olms, Hildesheim (1968). First published 1792, Hirschberg near Breslau
2. Radhakrishnamurty, P.: Stoichiometry and chemical reactions. J. Chem. Educ. **72**(7), 668 (1995)
3. Riley, J., Richmond, T.G.: Redox equations with an infinite number of balanced solutions. J. Chem. Educ. **69**(2), 114–115 (1992)
4. Subramaniam, R., Goh, N.K., Chia, L.S.: The relationship between the number of elements and the number of independent equations of elemental balance in inorganic chemical equations. J. Chem. Educ. **72**(10), 894 (1995)
5. West, G.B., Woodruff, W.H., Brown, J.H.: Allometric scaling of metabolic rate from molecules and mitochondria to cells and mammals. Proc. Natl. Acad. Sci. USA **99** (Suppl. 1), 2473–2478 (2002)
6. Wink, D.J.: The use of matrix inversion in spreadsheet programs to obtain chemical equations. J. Chem. Educ. **71**, 490–494 (1994)
7. Herndon, W.C.: On balancing chemical equations: Past and present. J. Chem. Educ. **74**(11), 1359–1362 (1997)
8. Bottomly, J.: Method of indetermined coefficients. J. Chem. News **37**, 110 (1878)
9. Guo, C.: A new and general method for balancing chemical equations by inspections. J. Chem. Educ. **74**(11), 1365–1366 (1997)
10. Stout, R.: Redox challenges – good times for puzzle fanatics. J. Chem. Educ. **72**(12), 1125 (1995)
11. Freeman, W.A., Goh, N.K., Chia, L.S., Hart, D.M., Lucas, E.A., Perry, D.J., Subramaniam, R., ten Hoor, M.J., Toby, S., Treptow, R.S.: Letter to the editor about "how do i balance thee? ... let me count the ways!" by Lawrence A. Ferguson (J. Chem. Educ. 1996, 73, 1129). J. Chem. Educ. **74**(11), 1271 (1997)
12. Kern, G.: Und noch einmal incredibly challenging (Anmerkungen und Ergänzungen zu "incredibly challenging !?" von Reibnegger/Estelberger). Chem. Sch. (Salzburg/Austria) **1**, 15–16 (1997)
13. Ludwig, O.G.: Response to letter to the editor about *On Balancing Redox Challenges* by Oliver G. Ludwig (J. Chem. Educ. 1996, 73, 507). J. Chem. Educ. **74**(11), 1270 (1997)
14. Reibnegger, G., Estelberger, W.: Incredibly challenging !?. Eine einfache Methode zur Richtigstelluing auch sehr komplizierter Redoxgleichungen. Chem. Sch. (Salzburg/Austria) **2/96**, 14–17 (1996)

15. Stout, R.: Redox challenges – reply. J. Chem. Educ. **73**(10), A 227 (1996)
16. Tóth, Z.: Balancing chemical equations by inspection. J. Chem. Educ. **74**(11), 1363–1364 (1997)
17. Tóth, Z.: Letter to the editor about *On Balancing Redox Challenges* by Oliver G. Ludwig (J. Chem. Educ. 1996, 73, 507). J. Chem. Educ. **74**(11), 1270 (1997)
18. Woolf, A.A.: Letter to the editor about *On Balancing Redox Challenges* by Oliver G. Ludwig (J. Chem. Educ. 1996, 73, 507). J. Chem. Educ. **74**(11), 1270 (1997)
19. Moore, J.W.: Balancing the forest and the trees. J. Chem. Educ. **74**, 1253 (1997)
20. Smith, W.R., Missen, R.W.: Using mathematica and maple to obtain chemical equations. J. Chem. Educ. **74**(11), 1369–1371 (1997)
21. Smith, W.R., Missen, R.W.: Chemical Reaction Equilibrium Analysis: Theory and Algorithms. Krieger, Malabar, FL (1991)
22. Glaister, P.: A challenging balance. J. Chem. Educ. **74**(11), 1368–1368 (1997)
23. Fluck, E., Kerler, W., Neuwirth, W.: The Mößbauer effect and its significance in chemistry. Angew. Chem. **2**, 461–472 (1963)
24. Fluck, E., Kerler, W., Neuwirth, W.: The structures of complex iron compounds. Z. Anorg. Allg. Chem. **333**, 235–247 (1964)

Chapter 14
Nomenclature

Nomenclature is an essential part of chemoinformatics. The terms chemoinformatics, cheminformatics, or chemical informatics refer to the use of computers and informational techniques in chemical problems. The discipline is a young one and is assumed revolutionary in the field of chemistry [1]. We will confine ourselves here rather to nomenclature and methods for notation of chemical structures. Actually, the topic of chemoinformatics covers much more issues [2, 3].

14.1 Nomenclature Systems

Nomenclature refers to a set of rules for establishing chemical names in a unique manner. In early alchemy and modern chemistry, compounds were named often deliberately or according to the origin of the compound. First attempts to a systematic nomenclature emerged in 1597, when the *Alchemia* of *Libavius*[1] was published [4]. The material presented there is to some extent still valid.

Several systems enable searching a chemical compound [5]. Nomenclature systems are summarized in Table 14.1. Undoubtedly, the most well known and famous system is the IUPAC nomenclature which was fixed for the first time during a conference in Geneva in 1892.

14.1.1 Hantzsch – Widman System

The Hantzsch – Widman system serves for the nomenclature of heterocyclic compounds. The affixes for the Hantzsch – Widman system are shown in Table 14.2.

The system tries to name heterocyclic compounds so that several well-known heterocyclic compounds are named in the same way or at least very similar as their corresponding trivial name. For example, ethylene oxide is oxiran. Other examples are dioxan and dioxin. The systematic name for sulfolane would be thiolan.

---

[1] Andreas Libau, born 1555 in Halle, Germany, died Jul. 25, 1616, in Coburg, Germany.

J.K. Fink, *Physical Chemistry in Depth*,
DOI 10.1007/978-3-642-01014-9_14, © Springer-Verlag Berlin Heidelberg 2009

Table 14.1 Nomenclature systems [6]

Nomenclature system	Remark/reference
IUPAC[a] nomenclature	General recommendations [7]
Hantzsch – Widman system	Heterocyclic compounds
Chemical abstracts number	A running number
Beilstein registration number	Beilstein data bank
Wiswesser line notation	Machine readable
Dyson IUPAC notation	
GREMAS	[b]
ROSDAL	[c] [8]
Markush generic formulas	Retrieval in patents
SMILES	Machine readable[d]
IUPAC International Chemical Identifier	Machine readable
PDB Code	Protein data bank
CML[TM]	Chemical markup language [9]

[a] International Union of Pure and Applied Chemistry
[b] Genealogisches Recherchieren durch Magnetbandspeicherung
[c] Representation of Organic Structure Descriptions Arranged Linearly
[d] Simplified Molecular Input Line Entry Specification

Table 14.2 Hantzsch – Widman system

Members	Unsaturated	Saturated
3	iren, irin (N)	iran, iridin (N)
4	et	etan, etidin (N)
5	ol	olan, olidin (N)
6	in	an
7	epin	epan
8	ocin	ocan
9	onin	onan
10	ecin	ecan

Heteroatom	Acronym
Oxygen	ox
Nitrogen	az
Sulfur	thi

(N) Rings with nitrogen

14.2 A Nomenclature for Simple Chemicals

Going into chemical labs, it can be readily observed that basic organic chemicals are abbreviated in a different way. The acronyms used vary from department to department, even within a single university or company site. Obviously, nobody takes care that the experimenters label their bottles with the chemicals in a unique manner. For example, phthalic anhydride is abbreviated as PA, PSA, or PAN, the latter acronym being readily confused with poly(acrylonitrile).

Basic organic compounds can be addressed using a simple nomenclature system (SNS). In particular, for basic industrial chemicals standardized abbreviations can be established. In certain cases, these abbreviations are still shorter than the molecular formula.

The system presented here is just good for labeling bottles containing basic chemicals in the laboratory. The advantage is that the names that are generated are pronounceable, in contrast to those generated as machine-readable notations. Actually, this system was developed and used by the author during his temporary research time in industry.

From Table 14.3, we can easily generate some examples for this nomenclature system. In Table 14.4 we give a few examples for this type of nomenclature. Of course, this nomenclature is applicable only to basic organic chemicals with a comparatively simple structure.

Table 14.3 Elements of nomenclature (SNS)

Acronym	partial structure
un	$-CH_2-$, CH_3-
dos	$-CH_2CH_2-$, CH_3CH_2-
ol	$-CH_2-OH$, etc.
al	$-CH=O$
am	$-CH_2-NH_2-$, $-C-NH-$, etc.
car	$-COO-$, $COOH$
en	$-CH=CH_2$, $-CH=CH-$, etc.
fi	Phenyl, phenyliden, etc.
o	$-O-$
r	Ring

Table 14.4 Examples for the simple nomenclature system (SNS)

Standard name	SNS name
Methanol	ol
Ethanol	unol
1,2-Propanediol	unolol
Aminoethanol	olam
Oxiran	rodos
Diethyl ether	dosodos
Phenol	ofi
Benzyl alcohol	olfi
Formic acid	car
Acetic acid	uncar
Oxalic acid	carcar
Vinyl acetate	uncaren
Acrylic acid ethyl ester	encardos
Benzaldehyde	alfi
Benzoic acid	carfi
Salicylic acid	carfiol

14.3 Machine-Readable Representations

With the advent of chemical drawing programs, many file formats have been developed. The majority of these file formats show how to plot a chemical structure and its bonds, by storing certain internal coordinates needed for plotting. On the other hand, there are also formats that store a chemical structure rather only in the way in that the molecule is built up. These formats do not take care neither in which size the bonds would be plotted nor in which size the atomic labels would appear. These formats are based on prescriptions resulting from graph theory. Computer searchable structures and substructures are playing a key role in chemical informatics.

These notation systems are loosing importance for the ordinary user, as computer programs can read now structures that are drawn on the screen. However, in contrast, structures can be drawn via such notation systems conveniently on the screen and as output in chemical drawing programs. These forms of representation are the base of molecular modeling.

14.3.1 Wiswesser Line Notation

The Wiswesser line notation was already invented in 1938, i.e., before the advent of computers that could effectively use it [10–14]. In 1960, the Institute for Scientific Information (ISI) developed the Index Chemicus, based on the Wiswesser line notation. The Wiswesser line notation is based on sequences of letters, numbers, and other characters as shown in Table 14.5. Numbers refer to the number of carbon atoms. For example, $QV1$ refers to acetic acid, and $Q2$ refers to ethyl alcohol.

Table 14.5 Wiswesser line notation

Acronym	partial structure
Q	($-$OH)
V	($-$CO$-$)
M	($-$NH$-$)
Z	($-$NH$_2$)
R	($-$C$_6$H$_5$)
O	($-$O$-$)
E	($-$Br)
F	($-$F)
G	($-$Cl)

14.3.2 SMILES and SMARTS Notation

SMILES is an acronym for *Simplified Molecular Input Line Entry Specification*. The algorithm is a universal nomenclature. It was developed in 1988 by Weininger [15]. Several computer programs use now the SMILES algorithm. We briefly show a few most important rules to represent chemical structures by SMILES. We definitely refer to a more detailed version that is given in the Internet [16].

Atoms are represented by their usual symbols. Hydrogens are not written. So C stands for methane and CC stands for ethane. Special symbols can be represented in square brackets.

Single, double, triple, and aromatic bonds are represented by the symbols —, =, #, and :. However, the bond type may be omitted when suggestive, and sp2-hybridized bonded atoms can be written in lower case letters. So C=C—C=C means 1,3-butadiene, which is equivalent to cccc. Branches are indicated by parentheses, e.g., CC(C)C(=O)O is isobutyric acid. To specify rings, the atom that closes the ring is numbered and specified, as C1CCCCC1 for cyclohexane and c1cc2ccccc2cc1 for naphthalene. The SMILES notation can be used in chemical structure drawing programs for quick input of a structure by the skilled user.

The SMILES notation is not unique. For example, for naphthalene we could write

c1cc2ccccc2cc1,

or else

c2cc1ccccc1cc2,

etc. However, there are algorithms to generate unique representations for a certain structure [17].

SMARTS is a language that allows specifying substructures using rules that are straightforward extensions of the SMILES language. In addition, there are several other extensions of the algorithm, e.g., searching for sequences of amino acids, and for chemical reactions, both for reactants and educts.

14.3.3 IUPAC International Chemical Identifier

The IUPAC International Chemical Identifier (InChI) [18] was invented in 2001. There are programs (World Wide Molecular Matrix) that can create an InChI from the molecular structure, e.g., via a MOL file [19]. For example, for benzene, the InChI is

$$InChI = 1.12Beta/C6H6/c1 - 2 - 4 - 6 - 5 - 3 - 1/h1 - 6H$$

14.4 File Formats

14.4.1 MOL File

The MOL file belongs to the group of chemical table files. Molecular Design Limited did pioneering work in the development of such file formats [20]. Nowadays, many programs related to chemoinformatics are able to read this format.

A MOL file contains information about a single connected structure. For example, the representation of ethane in a MOL file is shown in Table 14.6.

Table 14.6 Representation of ethane in a MOL file

```
2  1  0  0  0  0  0  0  0  0  1 V2000
5.6725  -5.6725  0.0 C 0 0 0 0 0 0 0 0 0 0 0 0
7.1948  -5.6725  0.0 C 0 0 0 0 0 0 0 0 0 0 0 0
2  1  1  0  0  0  0
M    END
```

A MOL file consists of a header block and a connection table. The connection table starts with a count line, followed by an atom block and a bound block. The file closes with a properties block. In the atom block, the x-,y-,z- coordinates of the atom are given, followed by the atom symbol. The next 12 numbers refer to mass difference, charge, atom stereo-parity, hydrogen count + 1, stereo- care box, valence, H0 designator, not used, not used, atom – atom mapping number, inversion/retention flag, and exact change flag. The bond block contains first atom number, second atom number, bond type, bond stereo, not used, bond topology, and reacting center status.

14.4.2 Z-Matrix

An example for the representation of ethane as a Z-matrix is shown in Table 14.7. The representation in the z-matrix is as follows:

1. For the first atom, note only the atomic symbol.
2. For the second atom, note the atomic symbol, the number 1, and a variable for the distance between atoms 1 and 2.
3. For the third and all further atoms, note the atomic symbol, the number N, a variable to describe the distance between the current atom and NA, the atom number NB, and a variable to describe angle between the current atom, and two others.
4. Then add a blank line.
5. Finally resolve each variable (distances and angles) with its corresponding value.

Table 14.7 Representation of ethane in a Z-Matrix

```
c
c 1 11
h 2 12 1 a1
h 2 12 1 a1 3    120.0
h 2 12 1 a1 3   -120.0
h 1 12 2 a1 3    180.0
h 1 12 2 a1 6    120.0
h 1 12 2 a1 6   -120.0

11 1.54
12 1.09
a1 110.0
```

14.5 Chemical Markup Language

The chemical markup language (CML) is based on the extensible markup language (XML) in order to code chemical structures [21–24]. Besides chemical structures and reactions, CML is capable of drawing spectra, etc.

14.6 Molecules as Graphs

In graph theory, a graph is an ensemble of points, some of them connected with lines. We call the points vertices and the lines edges. A graph is suitable to represent a molecule. Graph theory plays an important role in searching of structures and substructures.

The representation of molecules according to graph theoretical rules plays a role in the development of structure property relationships for organic molecules.

References

1. Gasteiger, J. (ed.): Handbook of Chemoinformatics. From Data to Knowledge, vol. 1–4. Wiley-VCH, Weinheim (2003)
2. Chen, W.L.: Chemoinformatics: Past, present, and future. J. Chem. Inf. Model. **46**(6), 2230–2255 (2006)
3. Engel, T.: Basic overview of chemoinformatics. J. Chem. Inf. Model. **46**(6), 2267–2277 (2006)
4. Libavius, A.: Die Gerätschaft der chymischen Kunst. No. 34 in Boethius. Franz Steiner Verlag, Stuttgart (1995). Edited by Bettina Meitzner, reprint from 1606
5. Kochev, N., Monev, V., Bangov, I.: Searching chemical structures. In: J. Gasteiger, T. Engel (eds.) Chemoinformatics: A Textbook, Chap. 6, pp. 291–318.Wiley-VCH, Weinheim (2003). DOI 10.1002/3527601643.ch6
6. Gasteiger, J., Engel, T. (eds.): Chemoinformatics: A Textbook. Wiley-VCH, Weinheim (2003). DOI 10.1002/3527601643.ch6
7. Panico, R., Powell, W.H., Richer, J.C. (eds.): A Guide to IUPAC Nomenclature of Organic Compounds: Recommendations 1993. Blackwell Scientific Publications, Oxford (1993)
8. Barnard, J.M., Jochum, C.J., Welford, S.M.: ROSDAL: A universal structure/substructure representation for PC-host communication. In: W.A. Warr (ed.) Chemical Structure Information Systems, *ACS Symposium Series*, vol. 400. American Chemical Society, Washington, DC (1989)
9. Murray-Rust, P.: Chemical markup language. World Wide Web J. **2**(4), 135–147 (1997)
10. Hellwinkel, D.: Systematic Nomenclature in Organic Chemistry: A Directory to Comprehension and Application on its Basic Principles. Springer, Berlin (2001)
11. Lynch, M.F., Willett, P.: The production of machine-readable descriptions of chemical reactions using Wiswesser line notations. J. Chem. Inf. Comput. Sci. **18**(3), 149–154 (1978)
12. Smith, E.G., Baker, P.A. (eds.): The Wiswesser Line-Formula Chemical Notation (WLN), 3rd edn. Chemical Information Management, Inc., Cherry Hill, NJ (1976)
13. Vollmer, J.J.: Wiswesser line notation: An introduction. J. Chem. Educ. **60**, 192–196 (1983)
14. Wiswesser, W.J.: 107 years of line-formula notations (1861–1968). J. Chem. Doc. **8**(3), 146–150 (1968)
15. Weininger, D.: SMILES 1. Introduction to methodology and encoding rules. J. Chem. Inf. Comput. Sci. **28**, 31–36 (1988)

16. A full SMILES language tutorial. [electronic] http://www.daylight.com/smiles/ (2005)
17. Weininger, D., Weininger, A., Weininger, J.L.: SMILES 2. Algorithm for generation of unique SMILES notation. J. Chem. Inf. Comput. Sci. **29**, 97–101 (1989)
18. Stein, S., Heller, S., Tchekhovskoi, D.: The IUPAC chemical identifier. In: CAS/IUPAC Conference on Chemical Identifiers and XML for Chemistry, Columbus, OH. National Institute of Standards and Technology, Gaithersburg, MD (2002)
19. Murray-Rust, P., Rzepa, H.S., Tyrrell, S.M., Zhang, Y.: Representation and use of chemistry in the global electronic age. Org. Biomol. Chem. **2**, 1484–1491 (2004) [electronic] http://www.dspace.cam.ac.uk/bitstream/1810/741/6/article.pdf
20. Dalby, A., Nourse, J.G., Hounshell, W.D., Gushurst, A.K.I., Grier, D.L., Leland, B.A., Laufer, J.: Description of several chemical structure file formats used by computer programs developed at molecular design limited. J. Chem. Inf. Comput. Sci. **32**(3), 244–255 (1992)
21. Gkoutos, G.V., Murray-Rust, P., Rzepa, H.S., Wright, M.: Chemical markup, XML, and the world-wide web. 3. Toward a signed semantic chemical web of trust. J. Chem. Inf. Comput. Sci. **41**(5), 1124–1130 (2001)
22. Murray-Rust, P., Rzepa, H.S.: Chemical markup, XML, and the worldwide web. 1. Basic principles. J. Chem. Inf. Comput. Sci. **39**(6), 928–942 (1999)
23. Murray-Rust, P., Rzepa, H.S.: Chemical markup, XML and the world-wide web. 2. Information objects and the CMLDOM. J. Chem. Inf. Comput. Sci. **41**(5), 1113–1123 (2001)
24. Murray-Rust, P., Rzepa, H.S.: Chemical markup, XML, and the world wide web. 4. CML schema. J. Chem. Inf. Comput. Sci. **43**(3), 757–772 (2003)

Chapter 15
Symmetry

In daily life, we associate appealing shapes and proportions, periodicity and harmonics with symmetry. The property of the symmetry plays an important role in various fields of science [1, 2], and therefore naturally also in chemistry [3–6]. Hermann *Weyl*[1] may be considered as a pioneer in connecting the common arts to natural philosophy.

We discuss some historical and more recent problems in chemistry and physics, where symmetry plays a role in the solution to the problem.

15.1 Noether's Symmetry Theorem

This theorem [7, 8] was published in 1918 and is named after *Noether*,[2] The theorem states that a symmetry in a system leads to a conserved quantity.

Applications of Noether's theorem deal with the invariance of physical laws against translations of space, rotations in space, and transformation of time. The consequences of invariance or, in other words, symmetry are summarized in Table 15.1.

Table 15.1 Consequences of invariance

Symmetry	Invariance
Translation	Momentum
Rotation	Angular momentum
Time	Energy

Thus, if a physical law is not changing if the time is transformed, it is always the same at any instant of time. From this property, the conservation of energy can be concluded.

---

[1] Hermann Klaus Hugo Weyl, born Nov. 9, 1885, in Elmshorn, Germany, died Dec. 8, 1955, in Zurich, Switzerland.

[2] Amalie Emmy Noether, born Mar. 23, 1882, in Erlangen, Germany, died Apr. 14, 1935, in Bryn Mawr, PA, USA.

J.K. Fink, *Physical Chemistry in Depth*,
DOI 10.1007/978-3-642-01014-9_15, © Springer-Verlag Berlin Heidelberg 2009

A simple illustration of *Noether*'s theorem has been presented by *Baez* [9]. For a single particle, its position should be represented by a generalized coordinate q. The generalized velocity of the particle is $\dot{q}$. In terms of Lagrangian mechanics, the generalized momentum p and the generalized force F are as follows:

$$p = \frac{\partial L(q, \dot{q})}{\partial \dot{q}} \quad \text{and} \quad F = \frac{\partial L(q, \dot{q})}{\partial q}. \tag{15.1}$$

As in ordinary mechanics, the Euler – Lagrange equations demand $\dot{p} = F$, which arises from a variational problem. Now we want to transform the position q into another new position. We may express this as the position q is a function of some parameter s, i.e., $q = q(s)$. In the same way the velocity transforms when the position is changed, $\dot{q} = \dot{q}(s)$. When the Lagrangian is invariant for such a transformation, we address this as a kind of *symmetry*, here the translational homogeneity of space. This means, changing the parameter s by a small value, the Lagrangian should remain the same:

$$\frac{\mathrm{d}L(q(s), \dot{q}(s))}{\mathrm{d}s} = 0.$$

This implies that the expression

$$p\frac{\mathrm{d}q(s)}{\mathrm{d}s} = \mathrm{C}. \tag{15.2}$$

Here, C is a constant. This can be seen by forming the differential of Eq. (15.2) with respect to time, interchanging the order of differentials, and inserting the expressions of Eq. (15.1).

15.2 Group Theory

Examples where group theory can be applied successfully are molecular orbital theory, molecular spectroscopy, multipole expansions, and ligand field theory. In general, group theoretical results can be used as classification scheme and/or to simplify numerical calculations.

15.2.1 Transformation Groups

We start with the view of groups from the mathematical view. The symmetries of the regular triangle include the three reflections and the rotations by 120°, 240°, and 360° around the center, as shown in Fig. 15.1. Note that the reflection can be performed also as a rotation. We lift the triangle out of the plane, then turn it bottom side up and place it again in the plane. We may think that in the case of a reflection the lines that are connecting the points must be broken up temporarily. However, on doing a rotation, we leave the mutual distances of the points completely intact during the rotation process. On the other hand, by doing a rotation out of the plane that results in the reflection, we must leave the two-dimensional space for a moment. The rotations of angles of 120°, 240°, and 360° can be done, by resting in the two-dimensional space.

Fig. 15.1 Symmetries of the regular triangle

We address the transformations, i.e., rotation, reflection, etc., also as symmetry operations. We summarize the properties of these transformations:

The rotation by 360° gives back the original triangle. This operation is an identity transformation. If any two of these transformations are performed consecutively, then the result is again one of these six transformations. We realize that there is also an inverse transformation. This is again one of the set of transformations. For example, after doing one rotation and one reflection we can come back to the original triangle by one reflection along another axis.

We call the possible transformations a transformation group G. The transformation acts on the set of structures, i.e., the six triangles. The individual transformations have the following properties:

1. The identity transformation belongs to G
2. The result of any two transformations in G is again in G
3. The inverse of any transformation in G is again in G

A subset S of the structures is invariant if transformations in G result in structures belonging to S.

15.2.2 Wallpaper Groups

Space groups in two dimensions are called wallpaper groups. Symmetry in two dimensions emerges in tiling and tesselation artwork [10, 11]. Many of the pictures of *Escher*[3] demonstrate symmetry relationships [12].

---

[3] Maurits Cornelis Escher, born Jun. 17, 1898, in Leeuwarden, The Netherlands, died Mar. 27, 1972, in Laren.

15.2.2.1 Symmetry Operations

If we inspect a certain repeating pattern, we can slide it along a certain direction a certain distance and the result will be a pattern that matches exactly the pattern where it was moved to. This symmetry is called a translation. In the plane, there may be more than one direction in which we could move to get a congruency. Likewise, we could rotate the pattern to cover another, which is nearby. A rotation fixes one point in the plane and turns the rest of it some angle around that point. We can even perform the rotation several times. Further, we could do a reflection. A reflection fixes one line in the plane, and exchanges points from one side of the axis to the other. Note the difference in between mirroring and exchanging.

Finally we discuss the glide reflection. This is a process composed of two more basic processes. First, the pattern is reflected along an axis and then it is moved in the direction of the axis.

International Union of Crystallography Notation

There is an international notation (IUC, International Union of Crystallography) which describes the pattern properties such as m (mirror), g (gliding mirror), p1 (translation), p2 (180° rotation), p3 (120° rotation), ...

The full international symbol is read from left to right [12] and is interpreted as follows:

1. a letter p denotes the primitive cell;
2. a letter c denotes the centered cell;
3. an integer n denotes highest order of rotation, e.g., 6 ($= 2\pi/6$) means a rotation of 60°, thus including divisors as 120° and 180°;
4. an m (mirror) indicates a reflection axis;
5. a g indicates no reflection, but a glide reflection axis;
6. no symbols in the third and fourth positions indicate that the group contains no reflections or glide reflections.

Orbifold Notation

The orbifold notation consists of symbols that represent the generators of the group. An integer n indicates an n-fold rotation. An asterisk * indicates reflections. A cross x indicates glide reflections. If the numbers indicating the rotation are behind the asterisk, then the rotation centers are the intersection of mirror lines. For example, the group *2222 has all its rotations centered on mirror lines. The group 4*2 has only the two-fold rotation on mirror lines. The orbifold notation uses topology to explain symmetry and results in a more geometric understanding.

15.2.2.2 Two-Dimensional Point Lattices

We can abstract the pattern into simple points. We construct the point lattice simply by translations in two directions. By inspection of the symmetries of these translations of the points we classify the lattice. To tell in advance, there are five kinds of lattices. These are collected in Table 15.2 and are shown in Fig. 15.2. We refer to a translation in x-direction as X and we denote translation in y-direction as Y. Therefore, every point of the lattice can be found by the repeated operation $X^n Y^m$. In the parallelogram lattice, there are translations and 180° rotations. However, there are no reflections and glide reflections. In the rectangular lattice there are still more symmetry elements than in the parallelogram lattice, i.e., besides the translation there are 180° rotations and reflections. The square lattice has translations, 180° rotations, 90° rotations, reflections, and glide reflections. The rhombic lattice shows translations, 180° rotations, reflections, and glide reflections. Finally the hexagonal lattice shows translations, 60° rotations, 120° rotations, and 180° rotations, reflections, and glide reflections.

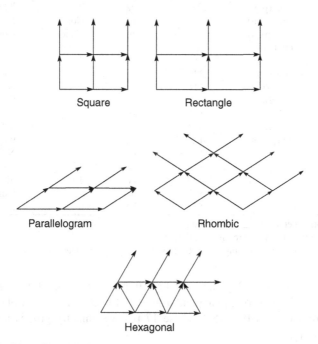

Fig. 15.2 The five wallpaper lattices

Table 15.2 The five types of two-dimensional lattices

Lattice type	Symmetry group [a]	Rotation orders	Reflection axes
Parallelogram	2 (p2)	2	None
Rectangular	6 (pmm)	2	90°
Rhombic	9 (cmm)	2	90°
Square	11 (p4m)	4	45°
Hexagonal	17 (p6m)	6	30°

[a] IUC Classification (International Union of Crystallography, 1952)

Two-dimensional tiling of a surface can be achieved with 17 different plane symmetry groups. This was pointed out by G. Pólya [13]. The 17 plane symmetry groups are listed in Table 15.3. The explanation why there are just 17 plane symmetry groups is subtle, however. The reason is that only certain symmetry operations are allowed that can create just this number of different groups [14]. The plane symmetry groups can be visually represented by using symbols in the position where the symmetry operation is valid.

Table 15.3 The 17 plane symmetry groups

No.	IUC class	Orbifold class	Lattice type	Rotation orders	Reflection axes
1	p1	o	pa	None	None
2	p2 (p2ll)	2222	pa	2	None
3	pm (p1ml)	**	re	None	Parallel
4	pg (p1gl)	xx	re	None	None
5	cm (c1ml)	x*	rh	None	Parallel
6	pmm (p2mm)	*2222	re	2	90°
7	pmg (p2mg)	22*	re	2	Parallel
8	pgg (p2gg)	22x	re	2	None
9	cmm (c2mm)	2*22	rh	2	90°
10	p4	442	sq	4	None
11	p4m (p4mm)	*442	sq	4 [a]	45*
12	p4g (p4gm)	4*2	sq	4 [b]	90*
13	p3	333	he	3	None
14	p31m	3*3	he	3 [b]	60*
15	p3m1	*333	he	3 [a]	30*
16	p6	632	he	6	None
17	p6m (p6mm)	*632	he	6	30°

[a] All rotation centers are on reflection axes
[b] Not all rotation centers are on reflection axes
pa: Parallelogram; re: Rectangle; rh: Rhombus; sq: Square; he: Hexagon

These symbols can be 2 for a twofold rotation, a 3 for a threefold rotation, etc., further a === for a mirror plane and a − − − for glide reflection plane. A few examples are shown in Fig. 15.3. These 17 plane symmetry groups result in 81 isohedral tiles [15].

15.2.3 Point Groups and Symmetry Operations

Modern methods in order to illustrate the nature of point groups and symmetry operations are available [16]. Molecules are abstracted as a set of points in space. We can associate with these points coordinates.

A symmetry element is a geometrical entity, i.e., a line, plane, or point, with respect to which some symmetry operations may be carried out. A symmetry operation is a prescription, in which way the coordinates of a molecule have to be changed.

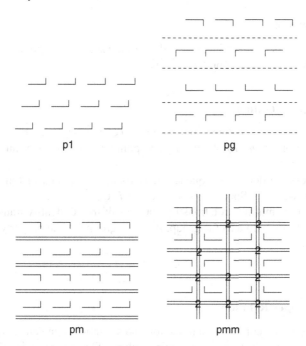

Fig. 15.3 Plane symmetries

After performing a change of the coordinates of a molecule often, the result is another molecule that can be distinguished from the original in position of the constituting atoms. We address a symmetry operation as such a change in coordinates where the resultant molecule is the same.

Symmetry operations are summarized in Table 15.4. A point group is a collection or group of symmetry operations, which leave at least one point where it was before the symmetry operation was done. We exemplify now the types of symmetry operations. We denote the symmetry operation like a function that acts on the coordinates of the points under consideration.

Table 15.4 Symmetry elements and symmetry operations [3]

Symmetry element	Symmetry operations
Plane	Reflection through plane
Center of inversion	Inversion of all atoms through the center
Proper axis	One or more rotations about the axis
Improper axis	Rotation followed by a reflection in a plane to the rotation axis

15.2.3.1 Identity

For completeness, we have a symmetry operation that does nothing in fact. This is the identity. We express the identity by $I(x, y, z) \to (x, y, z)$.

15.2.3.2 Inversion

The inversion refers to I. As the name indicates, the signs of the points are changed. We express this for a single point as $I(x, y, z) \rightarrow (-x, -y, -z)$.

15.2.3.3 Proper Rotation

The proper rotation is a rotation about an axis at a certain angle. C_n means a rotation about an angle of $2\pi/n$. To obtain the new points by rotation we must know the position of the axis.

For briefness we denote a sequence of m rotations C_n, i.e., m times about an angle of $2\pi/n$ with C_n^m. Special cases are $C_n^n = I$, $C_n^{2n} = I$, ...

In other words, multiple rotations by $360°$ result in the identity transformation. The set of rotations form a cyclic group. A cyclic group has the property

$$C_n^n = I.$$

15.2.3.4 Improper Rotation

An improper rotation consists of a reflection and a rotation along an axis perpendicular to the plane of reflection, i.e., the mirror plane. This operation is referred to as S.

15.2.3.5 Reflection

We abbreviate reflections with σ. In particular, $\sigma_{xy}(x, y, z)$ means a reflection through a plane that includes the x-axis and the y-axis. We easily see that $\sigma_{xy}(x, y, z) \rightarrow (x, y, -z)$. If we apply two times, or more general $2n$ times, the reflection by the same plane, then we arrive at the identity, e.g., $\sigma_{xy}^{2n} = I$.

15.2.4 Crystallographic Point Groups

Crystallographic point groups are point groups in which translational periodicity is required. The so-called *crystallographic restriction* states that rotations of a symmetry of 2,3,4, and 6 may occur.

It is necessary that the sum of the interior angles divided by the number of sides is a divisor of $360°$. Therefore,

$$\frac{\pi(n-2)}{n} = \frac{2\pi}{m}; \quad m = 1, 2, \ldots. \tag{15.3}$$

A solution is possible only for $n = 2, n = 3, n = 4, n = 6$. In the case of $n = 2, m \rightarrow \infty$.

There are 32 crystallographic point groups [17]. These are summarized in Table 15.5. The notation refers according to *Schönflies*. *Schönflies*[4] was a mathematician and crystallographer. In 1891 he showed that there are 320 point groups in three-dimensional space [18, 19]. Closely at the same time, independently, other important contributions from *Fedorov*[5] and *Barlow*[6] were published. The Schönflies point group symbols are used mainly by chemists. There are also other notations in use.

Table 15.5 The 32 crystallographic point groups

Point group type	Groups
Nonaxial	C_i, C_s
Cyclic	C_1, C_2, C_3, C_4, C_6
Cyclic with horizontal planes	$C_{2h}, C_{3h}, C_{4h}, C_{6h}$
Cyclic with vertical planes	$C_{2v}, C_{3v}, C_{4v}, C_{6v}$
Dihedral with horizontal planes	$D_{2h}, D_{3h}, D_{4h}, D_{6h}$
Dihedral with vertical planes	$D_{2v}, D_{3v}, D_{4v}, D_{6v}$
Dihedral with planes between axes	D_{2d}, D_{3d}
Improper rotation	S_4, S_6
Cubic groups	T, T_h, T_d, O, O_h

The dihedral group is the symmetry group for an n-sided regular polygon. The cubic group includes the octahedral group that represents the symmetries of the octahedron and the tetrahedral group that represents the symmetries of the tetrahedron.

The problem of higher dimensional crystallographic groups has been formulated as problem 18 of the famous Hilbert problems. *Hilbert*.[7] in 1900, proposed 23 fairly general problems to stimulate mathematical research. In 1910, this problem was solved by *Bieberbach*,[8] He proved that in any dimension that there were only finitely many groups [20]. Around 60 years later it was found that there are 4783 groups in the four-dimensional space.

15.3 Stereochemistry

Symmetry played a fundamental role in problems of stereochemistry. The most celebrated example is the establishment of the ring structure of benzene.

---

[4] Arthur Moritz Schönflies, born Apr. 17, 1853, in Landsberg an der Warthe (today Gorzów, Poland), died May 27, 1928, in Frankfurt am Main.

[5] Евграф Степанович Фёдоров, born Dec. 10, 1853, in Оренбург, Russia, died May 21, 1919, in Петроград, Russia.

[6] William Barlow, born Aug. 8, 1845, in Islington, London, died Feb. 28, 1934, in Great Stanmore, England.

[7] David Hilbert, born Jan. 23, 1862, in Königsberg (now Kaliningrad), Prussia, died Feb. 14, 1943, in Göttingen, Germany.

[8] Ludwig Georg Elias Moses Bieberbach, born Dec. 4, 1886, in Goddelau, Germany, died Sep. 1, 1982, in Oberaudorf, Germany.

15.3.1 Structure of Benzene

The structure of benzene was postulated from the number of substituted isomers that could be identified. *Kekulé*[9] is considered still today as the inventor of the structure of benzene (1865), although some people consider *Loschmidt*[10] as the originator. Assuming a certain structure of the benzene, only a certain number of dichloroben-benzenes, trichlorobenbenzenes, etc., can be formed.

By a careful compilation of the number of isomers of the known isomers of various aromatic compounds at that time, Kekulé postulated the benzene ring. Several other structures of benzene have been proposed, e.g., the Dewar benzene structure, the *Ladenburg*[11] benzene structure.

15.3.2 Tetrahedral Structure of Methane

The theory of stereochemistry by *van't Hoff*[12] and *Le Bel*[13] (1874) is based on similar considerations to those of Kekulé, which led to the assumption that in methane the bonds from carbon to the hydrogen atoms form the tetrahedron angle.

We illustrate the considerations by a well-known simple example. If the structure of the dichloromethane would be as that the chlorine atoms are placed in the corners of a square and the carbon atom is placed in the center of the square, then two isomers of dichloromethane would be possible. The situation is shown in Fig. 15.4.

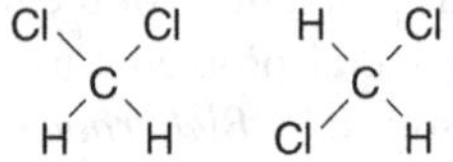

Fig. 15.4 Isomers of dichloromethane in quadratic structure

$$
\begin{array}{cc}
Cl\diagdown\quad\diagup Cl & H\diagdown\quad\diagup Cl \\
C & C \\
H\diagup\quad\diagdown H & Cl\diagup\quad\diagdown H
\end{array}
$$

On the other hand, the fact that there exists only one isomer can be explained with the assumption of a tetrahedral structure of the methane structure.

15.3.3 Optical Activity

Optical activity arises from a dissymmetric structure of matter. Before we discuss the origin of optical activity in detail, we recall the equations of nature of light.

---

[9] Friedrich August Kekulé von Stradonitz, born Sep. 7, 1829, in Darmstadt, died Jul. 13, 1896, in Bonn.

[10] Johann Joseph Loschmidt, born Mar. 15, 1821, in Putschirn (Počerny), Czech Republic, died Jul. 8, 1895, in Vienna.

[11] Albert Ladenburg, born Jul. 2, 1842, in Mannheim, Germany, died Aug. 15, 1911, in Breslau, now Poland.

[12] Jacobus Henricus van't Hoff, born Aug. 30, 1852, in Rotterdam, died Mar. 1, 1911, in Berlin-Steglitz.

[13] Joseph Achille Le Bel, born Jan. 21, 1847, in Pechelbronn, Alsace, died Aug. 6, 1930, in Paris.

15.3.3.1 The Nature of Light

Maxwell's equations read as

$$\nabla \times \mathbf{E} = -\dot{\mathbf{B}}; \qquad \nabla \cdot B = 0;$$
$$\nabla \times \mathbf{H} = \mathbf{J} + \dot{\mathbf{D}}; \qquad \nabla \cdot D = \rho_{el}. \tag{15.4}$$

Here

E Electric field,
D Electric displacement field,
H Magnetic field,
B Magnetic flux density,
J Free current density,
ρ_{el} Free electric charge density.

The electric displacement field is related to electric field by $\mathbf{D} = \varepsilon\mathbf{E}$, and the magnetic flux density is related to the magnetic field by $\mathbf{B} = \mu\mathbf{H}$. ε is the electrical permittivity, which is related to the dielectric constant and μ is the magnetic permeability.

Forming the differential with respect to time of the third equation of Eq. (15.4) with $\mathbf{J} = 0$, we obtain $\nabla \times \dot{\mathbf{H}} = -\ddot{\mathbf{D}} = -\varepsilon\ddot{\mathbf{E}}$ and from the first equation of Eq. (15.4) we have $\dot{\mathbf{H}} = \nabla \times \mathbf{E}/\mu$. In combining these two equations we obtain immediately $\nabla \times \nabla \times \mathbf{E} + \mu\varepsilon\ddot{\mathbf{E}} = 0$, which is equivalent to

$$\ddot{\mathbf{E}} + \frac{1}{\mu\varepsilon}\Delta\mathbf{E} = 0. \tag{15.5}$$

Equation (15.5) is a vector equation in a wave with a propagation velocity of $1/\sqrt{\mu\varepsilon}$. Since this expression was found to be close to the velocity of light, *Maxwell* concluded that light should be an electromagnetic wave in the sense of Eq. (15.5).

Equation (15.5) deals with the most simple case of the behavior of light either in vacuum or in a nonconductive, nonmagnetic, isotropic medium.

15.3.3.2 Dissymmetric Media

In anisotropic media, the dielectric permittivity is no longer a constant, but we have to treat the dielectric displacement a function of multiple terms [21, pp. 503], e.g., for the x-component of **D**

$$D_x = \varepsilon E_x +$$
$$a_{1,1}\frac{\partial E_x}{\partial x} + a_{1,2}\frac{\partial E_x}{\partial y} + a_{1,3}\frac{\partial E_x}{\partial z} +$$
$$a_{1,4}\frac{\partial E_y}{\partial x} + a_{1,5}\frac{\partial E_y}{\partial y} + a_{1,6}\frac{\partial E_y}{\partial z} + \tag{15.6}$$
$$a_{1,7}\frac{\partial E_z}{\partial x} + a_{1,8}\frac{\partial E_z}{\partial y} + a_{1,9}\frac{\partial E_z}{\partial z}$$

Equation (15.6) may be considered as the start of Taylor expansion. Similar equations hold for the other components of **D** and **B**. The expansion must be invariant for a transformation of the coordinates, e.g., a rotation.

In Fig. 15.5, the rotation of a right-handed coordinate system around the z-axis is shown. In this particular rotation, the coordinates will be transformed as follows:

$$x \rightarrow y; \quad y \rightarrow -x; \quad z \rightarrow z.$$

In order to maintain invariance on this rotation, we have the relations

$$D_x = D_y \qquad D_y = -D_x \qquad D_z = D_z$$
$$\frac{\partial E_x}{\partial x} = \frac{\partial E_y}{\partial y} \qquad \frac{\partial E_y}{\partial y} = -\frac{\partial E_y}{\partial x} \qquad \dots = \dots$$

Inspection of other transformation reveals several other relations. It turns out that a lot of the coefficients in Eq. (15.6) are zero, namely Eq. (15.6) for the electric displacement field and the corresponding equation for the magnetic flux density must be of the form

$$\mathbf{D} = \varepsilon \mathbf{E} + g_1 \nabla \times \mathbf{E}$$
$$\mathbf{B} = \mu \mathbf{H} + g_2 \nabla \times \mathbf{H} \qquad (15.7)$$

By means of *Poynting's*[14] theorem, it can be shown that $g_1 = g_2 = g$. Equation (15.7) holds for a dissymmetric isotropic medium. With Maxwell's equations (15.4), Eq. (15.7), neglecting the rotational term, become

$$\mathbf{D} = \varepsilon \mathbf{E} - g \dot{\mathbf{B}}$$
$$\mathbf{B} = \mu \mathbf{H} + g \dot{\mathbf{D}} \qquad (15.8)$$

Using the same procedure as in the derivation of Eq. (15.5), a coupled system of two differential equations emerges [21, p. 505] with, e.g., the solution

$$E_x = \frac{A}{2} \cos \omega (t - \frac{z}{v_1}) + \frac{A}{2} \cos \omega (t - \frac{z}{v_2})$$
$$E_y = \frac{A}{2} \cos \omega (t - \frac{z}{v_1}) + \frac{A}{2} \cos \omega (t - \frac{z}{v_2}) \qquad (15.9)$$

This solution consists in the superposition of two harmonic waves with the same amplitude. The position vector moves along the z-direction in a circular screw. Thus, the plane of light is rotating. In a crystal, as the light beam is reflected back as it wants to leave the crystal, the rotation of the plane is turned back.

---

[14] John Henry Poynting, born Sep. 9, 1852, in Monton, England, died Mar. 30, 1914, in Birmingham, England.

Fig. 15.5 Rotation of a right-handed coordinate system around the z-axis

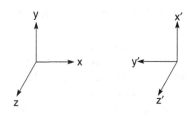

The velocities v_1 and v_2 are approximately

$$v_1 \approx v_0 - \frac{g\omega}{n_0^2}; \quad v_2 \approx v_0 + \frac{g\omega}{n_0^2},$$

or with Taylor expansion

$$\frac{v_0}{v_1} \approx 1 + \frac{g\omega}{n_0^2}; \quad \frac{v_0}{v_2} \approx 1 - \frac{g\omega}{n_0^2}.$$

n_0 is the refractive index. With the addition theorem of the trigonometric functions, and the relation $n_0 v_0 = c_0$ the trigonometric terms in Eq. (15.9) can be rearranged into

$$E_x = A \cos \frac{\omega^2 g z}{c_0^2} \cos \omega \left(t - \frac{z}{v_0} \right)$$

$$E_y = -A \sin \frac{\omega^2 g z}{c_0^2} \cos \omega \left(t - \frac{z}{v_0} \right)$$

Obviously, the plain of propagation is inclined by

$$\tan \frac{E_y}{E_x} = - \tan \alpha_z = - \tan \frac{\omega^2 g z}{c_0^2}.$$

Thus, for $\alpha_0 = 0$, $z = 0$,

$$\alpha_d = - \frac{\omega^2 g d}{c_0^2} = - \frac{4\pi^2 g d}{\lambda^2}.$$

The angle of rotation is proportional to the length of the path d and proportional to the reciprocal of the wavelength. The latter proportionality is addressed as the law of *Biot*,[15]

---

[15] Jean Baptiste Biot, born Apr. 21, 1774, in Paris, France, died Feb. 3, 1862, in Paris, France.

15.3.4 Separation of Racemic Compounds

Pasteur[16] could separate the optical isomers due to the crystalline forms of sodium ammonium tartrate [22]. Natural tartaric acid rotates the plane of polarized light. However, tartaric acid from laboratory synthesis does not show any rotation of the plane of polarized light.

Pasteur observed that the crystals of sodium ammonium tartrate are a mixture of two asymmetric forms that are mirror images of one another. He could separate the two forms of crystals manually and he arrived at two forms of sodium ammonium tartrate. One form rotated the plane of polarized light clockwise and the other form rotated the plane of polarized light anticlockwise. The separation in this way is possible for tartaric acid only below 27°C, because above this temperature the isomers of tartaric acid do not form separate crystals.

15.4 Woodward – Hoffmann Rules

The *Woodward – Hoffmann* rules explain which products will be obtained in the course of certain concerted organic reactions. In particular, they are applicable for pericyclic reactions, in which the reaction consists of a reorganization of pairs of electrons. The rules were set up around 1965 by *Woodward*[17] and *Hoffmann*,[18] Originally, the rules were used as guideline in the total synthesis of the vitamin B_{12}.

Precursors of these rules are the *Wigner*[19] – *Wittmer* rules [23], concerning the symmetry of chemical reactions. The *Wigner – Wittmer* rules deal with the conservation of spin and orbital angular momentum in the course of the reaction of diatomic molecules.

15.5 Chemical Kinetics

Physicochemical systems with coupled processes having different length scales can exhibit stationary spatially periodic structures. These arise from symmetry-breaking instabilities.

---

[16] Louis Pasteur, born Dec. 27, 1822, in Dole, France, died Sep. 28, 1895, in Saint-Cloud, near Paris.

[17] Robert Burns Woodward, born Apr. 10, 1917, in Boston, died Jul. 8, 1979, in Cambridge, MA, USA.

[18] Roald Hoffmann, born Jul. 18, 1937, in Złoczów, Poland.

[19] Jenő Pál Wigner, born Nov. 17, 1902, in Budapest, Hungary, died Jan. 1, 1995, in Princeton, NJ, USA.

A common example is the *Belousov*[20] – *Zhabotinsky*[21] reaction [24]. Beautiful patterns of chemical wave propagation can be created in a chemical reaction – diffusion system with a spatiotemporal feedback. The wave behavior can be controlled by feedback-regulated excitability gradients that guide propagation in the specified directions [25, 26].

Lotka[22] came to the United States in 1902 and wrote a number of theoretical articles on chemical oscillations. In his paper dated 1910 [27] he concludes

Eine Reaktion, welche diesem Gesetze folgte, scheint zurzeit nicht bekannt zu sein. Der hier behandelte Fall ergab sich auch ursprünglich aus Betrachtungen, welche ausserhalb des Gebietes der physikalischen Chemie liegen.

The equations became later well known as Lotka – Volterra equations and used rather in biological sciences. *Volterra*[23] established his equations in 1926. It seems that he did not know the work of *Lotka*.

The *Lotka – Volterra* equations deal with the response of a biological population that is oscillating. *Volterra* observed that in some years the adriatic fishermen came back with a lot of fishes in their nets, and in some years there was a lack of fishes. *Volterra* attributed this that the fishes are predominately not the prey of the fishermen, but of other fishes. He concluded that the variations in the amount of prey fishes arise from the fact that the big (predator) fishes eat the small fishes. Therefore, the population of the big fishes is growing, but simultaneously the amount of small fishes is decreasing.

This effects that the big fishes are left hungry and weak and decrease in population, after which population of the small fishes is again growing. Therefore, a periodic fluctuation in the population of both types of fishes occurs.

We can set up the *Lotka – Volterra* equations for a simple case,

$$
\begin{aligned}
\frac{dP}{dt} &= +k_P P - k_{P,H} P H \\
\frac{dH}{dt} &= -k_H H - k_{H,P} P H
\end{aligned}
\qquad (15.10)
$$

H Number of hunting (predator) animals
P Number of prey animals
$k_{X,Y}$ Kinetic constants

The first right-hand side term of the first equation in Eq. (15.10) states that the prey will increase in population, because there is always enough vegetarian food available, but the second term of the first equation states that the population of the

---

[20] Борис Павлович Белоусов (1893–1970).

[21] Анатол Маркович Жаботинскы born 1938 in Moscow.

[22] Alfred James Lotka, born Mar. 2, 1880, in Lemberg, now Ukraine, died Dec. 5, 1949, in New York, USA.

[23] Vito Volterra, born May 3, 1860, in Ancona, Papal States [Italy], died Oct. 11, 1940, in Rome.

prey decreases because of the presence of the predators. The second equation in Eq. (15.10) states that the predators will die out naturally. But if sufficient prey is available, then the predators can grow in population.

15.6 Dipole Moments

Permanent dipole moments may not occur in molecules with a certain symmetry. For example, *p*-dichlorobenzene should not have a dipole moment, whereas *m*-dichlorobenzene should have a dipole moment. This plays a role in spectroscopy. The intensity of light absorption, when a system moves from the state *A* to the state *B* with wave functions Ψ_A and Ψ_B, can be characterized by the transition integral

$$\overline{I} = \int \Psi_A \mu \Psi_B \, dV. \tag{15.11}$$

If the integral is zero, then no light absorption can occur. This relation is called *selection rule*. The dipole moment μ may be a permanent dipole moment or an induced dipole moment, which is caused by oscillation. The asymmetrical stretch vibration and the bending vibration of CO_2 are examples for oscillations that induce a dipole moment. Molecules like N_2 and O_2 have a zero dipole moment. These molecules are called infrared inactive.

15.7 Mechanics

15.7.1 Mechanical Systems

In mechanical systems a symmetry often arises by the absence of a certain constraint. During the derivation of the law of *Hagen*[24] and *Poiseuille*[25] we use cylindrical coordinates, in order to simplify the description of the mechanical system. We make assumption the flow is radially symmetrical. Moreover, the angle dependence of the flow does not arise. If we place a capillary diagonally in the gravitational field, then we probably expect for a suspension with different densities of the individual particles an angle dependence of the flow. Therefore, the radial symmetry of the flow arises, because the flow does not have a reason to start to turn with respect to the angular coordinate.

In the course of the derivation of the law of *Stokes*, the situation is very similar. However, the flow can be explained purely mechanically by the principle of the

---

[24] Gotthilf Heinrich Ludwig Hagen, born Mar. 3, 1797, in Königsberg, Prussia, died Feb. 3, 1884, in Berlin.

[25] Jean Louis Marie Poiseuille, born Apr. 22, 1797, in Paris, France, died Dec. 26, 1869.

least action. The principle of the least action means that a mechanical system moves along that coordinates that the integral becomes stationary over the action.

Using such an extremal principle, we can obtain also the equation of motion. *Lagrange* could show further that the use of the laws of *Newton* and the use of the principle of the least action are two equivalent formulations of the same problem. The expansion of these concepts on quantum mechanics is naturally possible. We become aware due to these considerations that symmetry obviously is related to extremal principles in the mechanical sense. On the other hand, obviously a system tries to achieve a symmetrical condition, when disturbances are missing, which tend to produce asymmetry.

We emphasize still another example from physical chemistry. In the derivation of the Kelvin equation, i.e., the vapor pressure of small droplets, we immediately and naturally accept that the shape of the droplets is spherical. Whoever spills mercury knows this. It is obviously not at all necessary to justify spherical droplets. Spherical droplets are in our common sense. But an extremal principle is behind the spherical droplets. A body with a given volume seeks to reach a minimum boundary surface, if no disturbance is aside. Particles that are bigger than a certain size exhibit an obvious deviation from the spherical shape under the effect of gravity.

15.7.2 Quantum Mechanics

The evaluation of integrals in quantum mechanics is greatly facilitated by the symmetry properties of the integrals. Integrating from $-\infty$ to $+\infty$, in the case of an odd function the integral becomes zero. An odd function is a function with the property $f(x) = -f(-x)$. An even function is a function with the property $f(x) = f(-x)$.

Examples for odd and even functions are $\sin(x)$ and $\cos(x)$. We can resolve each arbitrary function into an even part and an odd part.

15.8 Thermodynamics

15.8.1 Entropy

It is well known that the entropy is a measure for the disorder of a system. It is clear that the entropy can serve also as a measure for the order of a system. However, the fact that we can understand entropy as a measure of symmetry is rather emphasized rarely.

15.8.2 Statistical Thermodynamics

In statistical thermodynamics, it is established that we can compute from the partition function several properties. In tables of textbooks, the entropy of gases measured by caloric methods is compared with calculations from statistical thermo-dynamics. The agreement of the two methods is impressing. The substantial goal of these tables is to exemplify the third law of thermodynamics.

If we direct our interest toward the symmetry problem, then we can state that an injury of third law of thermodynamics is due to frozen nonequilibrium states, because the molecules cannot readily align themselves in the course of the cooling. Therefore, a residual energy remains.

Going down to absolute zero, the matter condenses and the energy barrier is too large than that the molecule could achieve the equilibrium. Commonly cited examples are nitric oxide and water. If we inspect the entropies of different gases at ambient temperature, then we must be aware that the partition function is composed of a translational portion and of a rotational portion.

We can still neglect the vibrational portion of the partition function and the portion for the electronically excited states. In the rotation portion of the partition function a symmetry number enters. This emerges because certain symmetries in transitions are not permitted. The entropy for a symmetrical molecule is thus as smaller, as more symmetrical such a molecule is, with otherwise same characteris-tics. We experience here a strange contradiction: If the elementary particles would be freely mobile in a molecule, then we would expect that they distribute equally. That means an asymmetrical molecule should want itself to convert into a more symmetrical molecule. On the other hand, the law of symmetry in entropy tells to us that an asymmetrical molecule has larger entropy.

15.9 Crystallography

Symmetry plays also an outstanding role to establish and to justify the possible crys-tal classes. Due to the principle of the space fulfillment, for a long time a pentagonal symmetry in crystallography was considered to be impossible.

In 1982, in metallic glasses such structures were discovered by Dan *Shechtman* [28–30]. These occur in such spatial arrangements, in that the long-range order translation is waived. Such structures are addressed as quasicrystals.

15.9.1 Snowflakes

In 1611, *Kepler*[26] investigated the hexagonal symmetry of snow crystals [31–33]. *Bentley*[27] photographed more than 5,000 snowflakes during his lifetime, not finding any two alike. He published many articles for magazines and journals. In 1931, his

---

[26] Johannes Kepler (Ioannes Keplerus), born Dec. 27, 1571, in Weil der Stadt, died Nov. 15, 1630, in Regensburg.

[27] Wilson Alwyn Bentley, born Feb. 9, 1865, in Jericho, Vermont, died Dec. 23, 1939, in Jericho, Vermont.

book *Snow Crystals* was published, containing more than 2,400 snow crystal images [34–36].

15.10 Biosciences

A lot of animals show symmetry on the macroscopic scale. Birds would have problems to fly if they had only one wing. In fact, farmers used to clip one wing of those chickens that were skilled in escaping the run, inducing thus an asymmetry.

In addition, subtle asymmetries in otherwise symmetric organisms occur. Already Aristotle noted that in some crabs the right claw is larger and stronger than the left claw. The most familiar of subtle asymmetries is the fluctuating asymmetry. This term was coined by the German biologist Wilhelm Ludwig [37]. Everybody knows that there are left-handers and right-handers. Less common is the easily verifiable fact that the breath flows only through one of the two nasal orifices. From time to time, the flow of breath changes from one nostril to the other as the reader can check by himself.

There are snails with a distinct direction of coiling, i.e., sinistral and dextral as pointed out by *Thompson* [38, 39].

Slightly more than 1/4 of the dancing bees displayed statistically significant individual handedness in initiating their bouts of dancing with either clockwise or counterclockwise turns [40]. How the symmetry is broken, in the course of the development of asymmetrical structures in organisms that are otherwise bilaterally symmetrical, is still unsolved in most cases [41, 42].

15.10.1 Chirality

On the microscopic scale, chirality is an important property in life science. All amino acids in proteins are left-handed. On the other hand, all the sugars in the nucleic acids are right-handed. This is in contrast to reagent glass organic chemistry, where usually racemates are produced. However, nature synthesizes completely pure chiral compounds. How chiral substances arrived in living materials is still an open question, it is addressed as chirality problem. Several environmental conditions during synthesis have been considered as the source of chirality, such as [43]

- circularly polarized ultraviolet light;
- decay;
- optically active catalysts;
- self-selection;
- fluke seeding in crystallization; and
- magnetic fields.

Many of the environmental conditions have only a little effect on the degree of asymmetry produced. On the other hand, it may be argued that nature had a long time to develop asymmetry.

15.10.1.1 Circularly Polarized Ultraviolet Light

Circularly polarized ultraviolet light interacts differently with left and right enantiomers. The selectivity is not unique, however. The light would not produce 100% homochirality required for life.

15.10.1.2 β-Decay

The β- decay is governed by the weak force. This force has a slight handedness, called parity violation. Therefore, theories have proposed that the β- decay could be responsible for the chirality in living organisms. These theories were opposed by the argument that the energy difference of the enantiomeric compounds is too small to achieve chirality in the required amount [44].

15.10.1.3 Optically Active Catalysts

Quartz and other minerals, e.g., clays, have chiral sites on the surface. It was attempted to separate chiral compounds selectively. A very small chiral selection effect by clay minerals was reported. It was argued that the chiral sites of these minerals would arise from the adsorption of optically active biomolecules. This means that prebiotic clays would then have had no chiral sites.

15.10.1.4 Fluke Seeding in Crystallization

It is reasonable that a fluke seeding of a supersaturated solution with a chiral crystal would cause the crystallization of the same enantiomer. However, for some researchers it is difficult to imagine that such concentrated regions existed in the early global environment.

15.10.1.5 Magnetic Fields

It was reported that a very strong magnetic field produced chiral products from achiral reagents. However, nobody could reproduce these experiments.

Summarizing, the chirality problem is still under discussion.

References

1. Lax, M.J.: Symmetry Principles in Solid State and Molecular Physics. Dover Publications, Mineola, NY (2001)
2. Weyl, H.: Symmetry, Princeton Science Library, 2nd edn. Princeton, NJ (1989)

3. Cotton, F.A.: Chemical Applications of Group Theory, 3rd edn. John Wiley, New York (1990)
4. Fackler, J.P. (ed.): Symmetry in Chemical Theory; the Application of Group Theoretical Techniques to the Solution of Chemical Problems, *Benchmark Papers in Inorganic Chemistry*, vol. 4. Dowden, Hutchinson & Ross, Stroudsburg, PA. (1973)
5. Hargittai, I., Hargittai, M.: Symmetry Through the Eyes of a Chemist, 2nd edn. Plenum Press, New York (1995)
6. Jaffé, H.H., Orchin, M.: Symmetry in Chemistry. Dover Publications, Mineola, NY (2002)
7. Byers, N.: The heritage of Emmy Noether in algebra, geometry, and physics. In: M. Teicher (ed.) The Heritage of Emmy Noether, *Israel Mathematical Conference Proceedings*, vol. 12. Gelbart Research Institute for Mathematical Sciences & Emmy Noether Research Institute of Mathematics, Ramat-Gan, Israel, Research Institute of Mathematics, Bar-Ilan University, Israel (1999). [electronic] http://www.physics.ucla.edu/~cwp/articles/noether.asg/noether.html
8. Teicher, M. (ed.): The Heritage of Emmy Noether, *Israel Mathematical Conference Proceedings*, vol. 12. Gelbart Research Institute for Mathematical Sciences & Emmy Noether Research Institute of Mathematics, Ramat-Gan, Israel, Research Institute of Mathematics, Bar-Ilan University, Israel (1999)
9. Baez, J.: Noether's theorem in a nutshell. [electronic] http://math.ucr.edu/home/baez/noether.html (2002)
10. Bigalke, H.G., Wippermann, H.: Reguläre Parkettierungen – Mit Anwendungen in Kristallograhie, Industrie, Baugewerbe, Design und Kunst. BI Wissenschaftsverlag, Mannheim-Leipzig-Wien-Zürich (1994)
11. Grünbaum, B., Shepard, G.C.: Tilings and Patterns: An Introduction. W. H. Freeman & Co., New York (1989)
12. Schattschneider, D.: Visions of Symmetry: Notebooks, Periodic Drawings, and Related Work of M.C. Escher. W.H. Freeman and Co., New York (1990)
13. Pólya, G.: Über die Analogie der Kristallsymmetrie in der Ebene. Z. Kristallogr. **60**(278–282) (1924)
14. Bix, R.: Topics in Geometry. Academic Press, Boston, MA (1994)
15. Grünbaum, B., Shephard, G.C.: Tilings and Patterns. W. H. Freeman, New York (1987)
16. De Graef, M.: A novel way to represent the 32 crystallographic point groups. J. Mater. Educ. **20**, 31–42 (1998)
17. Zwillinger, D.: Crystallographic groups. In: CRC Standard Mathematical Tables and Formulae, pp. 259–264. CRC Press, Boca Raton, FL (1995)
18. Burckhardt, J.J.: Zur Entdeckung der 230 Raumgruppen. Arch. Hist. Exact. Sci. **4**, 235–246 (1967)
19. Schönflies, A.M.: Kristallsysteme und Kristallstruktur. Teubner Verlag, Leipzig (1891)
20. Bieberbach, L.: Über die Bewegungsgruppen des n-dimensionalen euklidischen Raumes mit einem endlichen Fundamentalbereich. Gött. Nachr. pp. 75–84 (1910)
21. Schaefer, C.: Elektrodynamik und Optik, *Einführung in die theoretische Physik*, vol. 3–1, 1st edn. De Gruyter, Berlin (1932)
22. Pasteur, L.: Recherches sur les relations qui peuvent exister entre la forme cristalline et la composition chimique, et le sens de la polarisation rotatoire. Ann. Chim. Phys. **24**(3), 442–459 (1848)
23. Wigner, E., Wittmer, E.W.: Über die Struktur der zweiatomigen Molekelspektren nach der Quantenmechanik. Z. Phys. **51**, 859–886 (1928)
24. Hildebrand, M., Skødt, H., Showalter, K.: Spatial symmetry breaking in the Belousov-Zhabotinsky reaction with light-induced remote communication. Phys. Rev. Lett. **87**(8), 883031–883034 (2001)
25. Kapral, R., Showalter, K. (eds.): ChemicalWaves and Patterns. Kluwer Academic Publishers, Dordrecht, Holland (1995)
26. Sakurai, T., Mihaliuk, E., Chirila, F., Showalter, K.: Design and control patterns of wave propagation patterns in excitable media. Science **296**, 2009–2012 (296)

27. Lotka, A.: Zur Theorie der periodischen Reaktionen. Z. Phys. Chem. **72**, 508 (1910)
28. Jaric, M.V., Gratias, D. (eds.): Aperiodicity and Order: Extended Icosahedral Structures. Academic Press, New York (1989)
29. Senechal, M.: Quasicrystals and Geometry. Cambridge University Press, Cambridge (1995)
30. Shechtman, D., Blech, I., Gratias, D., Cahn, J.W.: Metallic phase with long-range orientational order and no translational symmetry. Phys. Rev. Lett. **53**, 1951–1953 (1984)
31. Kepler, J.: Strena seu de Nive sexangula. Godfey Tampach, Frankfurt/M. (1611)
32. Kepler, J.: The Six-Cornered Snowflake. Clarendon Press, Oxford (1966). Edited by Colin Hardie
33. Kepler, J.: Vom sechseckigen Schnee. Geest & Portig, Leipzig (1987). Edited by Dorothea Götz
34. Bahr, M.: My Brother Loved Snowflakes: The Story of Wilson A. Bentley, the Snowflake Man. Boyds Mills Press, Honesdale, PA (2002)
35. Bentley, W.A.: Snow Crystals: Snowflakes in Photographs. Dover Publications, Mineola, NY (2002)
36. Blanchard, D.C.: The Snowflake Man: A Biography of Wilson A. Bentley. McDonald & Woodward Pub., Blacksburg, VA (1998)
37. Ludwig, W.: Das Rechts-Links-Problem im Tierreich und beim Menschen. Springer, Berlin (1970). Reprint from 1932
38. Thompson, D.W.: On Growth and Form. Cambridge University Press, Cambridge (2000). Edited by John Tyler Bonner, first published in 1917
39. Thompson, D.W.: Über Wachstum und Form. Eichborn, Frankfurt (2006). Edited by Anita Albus and John Tyler Bonner, first published in 1917
40. Fergusson-Kolmes, L., Kolmes, S.A., Winston, M.L.: Handedness and asymmetry in the waggle dance of worker honey bees (Hymenoptera: Apidae). J. Kans. Entomol. Soc. **65**(1), 85–86 (1992)
41. Hallgrimsson, B., Hall, B.K. (eds.): Variation: A Central Concept in Biology. Elsevier, Amsterdam (2005)
42. Palmer, A.R.: From symmetry to asymmetry: Phylogenetic patterns of asymmetry variation in animals and their evolutionary significance. Proc. Natl. Acad. Sci. USA **93**, 14279–14286 (1996)
43. Sarfati, J.: Origin of life: The chirality problem. Creation Ex Nihilo Technical Journal **12**(3), 263–266 (1998)
44. Bada, J.L.: Origins of homochirality. Nature **374**, 594–595 (1995)

Chapter 16
Harmony

In common sense, harmony refers to agreement and well fitting together of parts of any set that are forming the whole construction. In particular, harmony means

- well-sounding music,
- appealing proportions, e.g., in architecture, or
- compatibility in opinion and action.

The ancient Greek philosophers searched harmony in the universe as well as in the human arts. Certain geometric constructions are addressed as platonic bodies. The school named after *Pythagoras* tried to measure harmony in numbers. As simpler a ration of numbers that are characterizing a system, more harmonic the system is itself. The starting point was problems of music, because here harmony is very easy to demonstrate.

Alberti[1] was an excellent mathematician and geometrician. He wrote 10 books on architecture [1]. He experimented with proportions in space and he stated

> We shall therefore borrow all our rules for the finishing our proportions, from the musicians, who are the greatest masters of this sort of numbers, and from those things wherein nature shows herself most excellent and compleat.

Thompson[2] stated in his famous book *On Growth and Form* in 1917 [2]:

> For the harmony of the world is made manifest in Form and Number, and the heart and soul and all poetry of Natural Philosophy are embodied in the concept of mathematical beauty.

In this chapter, we will merely focus on topics that are accessible by harmonic analysis.

---

[1] Leon Battista Alberti, born Feb. 14, 1404, in Genoa, died Apr. 25, 1472, in Rome.

[2] D'Arcy Wentworth Thompson, born May 2, 1860, in Edinburgh, Scotland, died Jun. 21, 1948, in St. Andrews, Scotland.

J.K. Fink, *Physical Chemistry in Depth*,
DOI 10.1007/978-3-642-01014-9_16, © Springer-Verlag Berlin Heidelberg 2009

16.1 Tools of Harmony

16.1.1 Means

There are three classical means

1. arithmetic mean $\bar{x} = \frac{1}{2}(x_1 + x_2)$,
2. geometric mean $\bar{x} = \sqrt{x_1 x_2}$, and
3. harmonic mean $\bar{x} = 2x_1 x_2/(x_1 + x_2)$.

These means were extensively discussed by *Palladio*,[3] an Italian architect [3]. Further examples for proportion in traditional architecture have been reviewed by *Langhein* [4].

The means have some interesting properties that were known already to the ancient mathematicians. The arithmetic mean has the property of $x_1 - \bar{x} = \bar{x} - x_2$, the geometric mean has the property of $x_1/\bar{x} = \bar{x}/x_2$, and the harmonic mean has the property of $x_1/x_2 = (x_1 - \bar{x})/(\bar{x} - x_2)$. The reciprocal of the harmonic mean is

$$\frac{1}{\bar{x}} = \frac{1}{2}\left(\frac{1}{x_1} + \frac{1}{x_2}\right).$$

Thus, if the reciprocals are formed, the harmonic mean turns into the arithmetic mean. Such a formula appears in mechanics, e.g., in the vibration of molecules.

16.1.2 Harmonic Analysis

Harmonic analysis, or Fourier analysis, is the decomposition of a periodic function into a sum of simple periodic components. In particular, Fourier series are expansions of periodic functions $f(x)$ in terms of an infinite sum of sines and cosines of the form

$$f(x) = \sum_{n=0}^{n=\infty} a'_n \cos(nx) + b'_n \sin(nx) \quad = \frac{1}{2}a_0 + \sum_{n=1}^{n=\infty} a_n \cos(nx) + b_n \sin(nx).$$

$$(16.1)$$

The coefficients a_n and b_n can be obtained by the orthogonal property of the sine and cosine functions

---

[3] Andrea Palladio, born Nov. 30, 1508, in Padua, died Aug. 1580 in Vicenza.

$$\int_{-\pi}^{+\pi} \sin(mx)\sin(nx)\mathrm{d}x = \pi\delta_{n,m}, n, m \neq 0;$$

$$\int_{-\pi}^{+\pi} \cos(mx)\cos(nx)\mathrm{d}x = \pi\delta_{n,m}, n, m \neq 0;$$

$$\int_{-\pi}^{+\pi} \sin(mx)\cos(nx)\mathrm{d}x = 0; \qquad\qquad , \qquad (16.2)$$

$$\int_{-\pi}^{+\pi} \sin(mx)\mathrm{d}x = 0;$$

$$\int_{-\pi}^{+\pi} \cos(nx)\mathrm{d}x = 0$$

for integers m and n. Further, $\delta_{m,n}$ is the *Kronecker*[4] delta. If we multiply Eq. (16.1) by $\sin(nx)$, integrate and use the orthogonality relations (16.2), we obtain, for example

$$\int_{-\pi}^{+\pi} f(x)\sin(mx) = \pi b_m. \qquad (16.3)$$

Similarly, we obtain the coefficient a_n by multiplying by $\cos(nx)$ and integrating. The Fourier series of a linear combination of two functions is the linear combination of the corresponding two series. If $f(x)$ periodic with period $2L$, with an appropriate transformation again the formulas can be used. If the period goes to the limit $L \rightarrow \infty$, then the Fourier integral emerges.

Fourier[5] did his important mathematical work on the theory of heat from 1804 to 1807. The reason why he was engaged on the theory of heat was purely personal, namely he suffered from rheumatic pains. Contemporaries report that he kept his dwelling so hot since his stay in Egypt that visitors could not bear the hotness, while he wrapped himself additionally still into thick coats. His memoir *On the Propagation of Heat in Solid Bodies* was at that time accepted with controversy, but is nowadays highly commended. In 1808, Lagrange and Laplace opposed the idea of trigonometric expansion.

The heat transfer equation is a parabolic partial differential equation. We rewrite it in modern notation in the absence of a heat source and convective heat transfer:

$$\rho C \frac{\partial T}{\partial t} = \nabla(k\Delta T) \qquad (16.4)$$

---

[4] Leopold Kronecker, born Dec. 7, 1823, in Liegnitz (now Legnica, Poland), died Dec. 29, 1891, in Berlin, Germany.

[5] Jean Baptiste Joseph, Baron Fourier, born Mar. 21, 1768, in Auxerre, France, died May 16, 1830, in Paris.

ρ Density
C Heat capacity
k Coefficient of heat conduction

The heat transfer equation can be solved by means of Fourier series. Soon it turned out that Fourier series can find application in inundated topics.

16.2 Vibrating String

The differential equation of the vibrating string is the one-dimensional wave equation, which is a partial differential equation, second order in each of t and x:

$$\frac{\partial^2 y(x,t)}{\partial t^2} = c^2 \frac{\partial^2 y(x,t)}{\partial x^2},$$

where $y(x,t)$ is the lateral elongation from rest, x is the axial direction of the string, c is the velocity of propagation of sound in the string, and t is the time.

We have certain boundary conditions. If the string is fixed at the ends ($x = 0, x = L$), then the motion ($y'(0,t) = 0$, $y'(L,t) = 0$) and the elongation ($y(0,t) = 0$, $y(L,t) = 0$) is zero there at any time. The differential equation can be solved by the method of separation of variables.

In general, the elongation of the vibrating string is given by

$$y(x,t) = \sum_1^\infty \sin\left(\frac{n\pi x}{L}\right)\left(C_n \sin\left(\frac{n\pi c}{L}t\right) + D_n \cos\left(\frac{n\pi c}{L}t\right)\right).$$

The coefficients C_n and D_n can be found via the boundary conditions by the orthogonality properties of the triangular functions.

16.3 Particle in a Box

We recall the quantum mechanical treatment of the particle in the box with infinite high potential barriers. The Hamilton operator is

$$\hat{H} = \frac{\hbar^2 d^2}{2m dx^2} + U.$$

U is the potential energy with $U = 0, 0 \le x \le L$, otherwise $U = \infty$. The differential equation for the stationary state is

$$\hat{H}\Psi = E\Psi.$$

We transform $\xi = x/L$ and obtain the solution

$$\Psi_n(\xi) = \sqrt{2}\sin(n\pi\xi).$$

The solutions correspond to the vibrating string.

16.4 Fourier Transform Techniques

Fourier transform techniques in spectroscopy [5–7] that are useful in chemistry include all kinds of spectroscopy, in particular, infrared spectroscopy [8] and *Raman*[6] spectroscopy [9], mass spectrometry, nuclear magnetic resonance spectroscopy [10], and X-ray crystallography [11].

Even when for all these techniques the Fourier transform is essential, the experimental techniques and thus the apparatus are completely different. A famous general monograph on Fourier transform techniques has been compiled by *Bracewell,*[7] one of the pioneers [12].

16.4.1 Fourier Transform Infrared Spectroscopy

In Fourier transform infrared spectroscopy the instrument consists of a *Michelson*[8] interferometer, cf. Fig. 16.1. The light entering into the interferometer is split by a semi-silvered mirror into two beams. Each beam goes a different path, finally the beams being recombined again. The recombined beam is directed to a detector. In the case of monochromatic and coherent light, the intensity of the recombined beam is dependent entirely on the difference of the path length, because of interference.

Fig. 16.1 Michelson interferometer

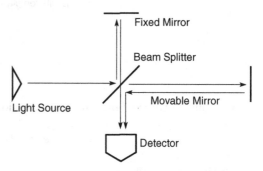

---

[6] Chandrasekhara Venkata Raman, born Nov. 7, 1888, in Tiruchirappalli, India, died Nov. 21, 1970, in Bangalore, India.

[7] Ronald Newbold Bracewell, born Jul. 22, 1921, in Sydney, Australia, died Aug. 13, 2007, in Stanford, CA, USA.

[8] Albert Abraham Michelson, born Dec. 19, 1852, in Strelno (now Strzelno, Poland), died May 9, 1931, in Pasadena, CA, USA.

The path is artificially altered, by a movable mirror for one beam and the intensity of the light signal is recorded in relation to the position of the mirror. In the case of a monochromatic light, a sinusoidal signal is obtained. The Fourier transform is just a single peak, ideally the δ function. In the case of nonmonochromatic light, still interference occurs. We can imagine the signal as a superposition of monochromatic light sources.

The Fourier transform of the interferogram results from the spectrum of the light introduced. In this way, the spectrum or interferogram, respectively, without sample (background) and with sample can be obtained, and thus the spectrum of the pure sample by subtraction.

16.4.2 Fourier Transform Mass Spectroscopy

The apparatus for cyclotron resonance spectrometry is shown in Fig. 16.2. A magnetic field is applied perpendicular to the plane. There are two pairs of electrodes, one pair to excite the ions inside by a pulse or by an alternating current and one pair to measure the electric response. The whole cell is kept in high vacuum, in order to allow the ions without collisions.

Fig. 16.2 Cyclotron resonance spectrometry. The magnetic field is perpendicular to the plane

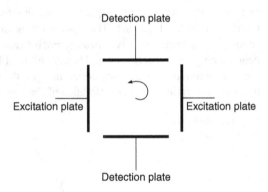

According to *Lorentz*,[9] the force exerted on ions in the cell is

$$F = m\dot{\mathbf{v}} = e\,(\mathbf{v} \times \mathbf{H} + \mathbf{E}). \qquad (16.5)$$

m Mass of the particle
e Charge of the particle
$\mathbf{v}$ Velocity of the particle
$\mathbf{H}$ Magnetic field strength
$\mathbf{E}$ Electric field strength

[9] Hendrik Antoon Lorentz, born Jul. 18, 1853, in Arnhem, The Netherlands, died Feb. 4, 1928, in Haarlem.

Therefore, if the magnetic field acts only in z-direction, then we obtain from Eq. (16.5) the set of equations

$$m\frac{dv_x(t)}{dt} = e\left(E_x(t) + H_z v_y(t)\right)$$

$$m\frac{dv_y(t)}{dt} = e\left(E_y(t) - H_z v_x(t)\right) \qquad (16.6)$$

$$m\frac{dv_z(t)}{dt} = e\left(E_z(t)\right)$$

In the absence of an electrical field, a circular motion is possible. We show this by forming again the derivative with time with the second and third equations of the set of (16.6),

$$m\frac{dv_x(t)}{dt} = +e H_z v_y(t)$$

$$m\frac{d^2 v_y(t)^2}{dt} = -e H_z \frac{dv_x(t)}{dt} \qquad (16.7)$$

$$m\frac{d^2 v_z(t)^2}{dt} = 0$$

Substituting for $dv_x(t)/dt$ in Eq. (16.7) we arrive at

$$\frac{d^2 v_y(t)}{dt^2} = -\left(\frac{e H_z}{m)}\right)^2 v_y(t),$$

which is the equation of a harmonic oscillation. If an appropriate electric field with the respective frequency is available, an undamped oscillation will be induced.

16.4.3 Diffraction

The process of diffraction is shown for a one-dimensional crystal in Fig. 16.3. The rays behind the pattern emerge with a different phase, as before. Therefore, for positive interference the well-known Bragg condition holds. This law is named after *Bragg*[10] and his son,[11] who did pioneering work on X-rays and crystal structure around 1915. In fact, the intensity of the light will not completely concentrated in spots where is positive interference, but rather more or less distributed over the whole region in peaks that appear more and more sharp, as the number of scattering

[10] Sir William Henry Bragg, born Jul. 2, 1862, in Wigton, Cumberland, England, died Mar. 12, 1942, in London.

[11] William Lawrence Bragg, born Mar. 31, 1890, in Adelaide, South Australia, died Jul. 1, 1971, in Waldringford, England.

Fig. 16.3 Diffraction

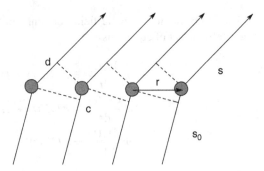

points increase. For example, the scattering on a single point will give rise to a rather broad peak of intensity distribution.

The diffraction pattern becomes more and more complicated, as the crystal becomes complicated in structure.

We calculate the intensity via the path difference. Suppose, we have two centers of scattering in a distance of $\mathbf{r}$. The unit vector of the incident path is $\mathbf{s_0}$ and the direction of the scattered beam is $\mathbf{s_0}$. Then, with the phase vector $\mathbf{S} = (\mathbf{s_0} - \mathbf{s})/\lambda$, we obtain for the phase difference

$$\alpha = 2\pi\,\mathbf{r}\cdot\mathbf{S}.$$

In general, for i atoms in a grid as shown in Fig. 16.3, with a scattering factor f_i, we obtain for the scattering function

$$F(\mathbf{S}) = \sum_i f_i \exp(2\pi i\mathbf{r_i}\cdot\mathbf{S}).$$

The intensity is the absolute value of the scattering function, i.e., $FF^\star$, the product of the function and its complex conjugate.

16.4.4 Fourier Transform Gas Chromatography

With the advent of computers, Fourier transform techniques in data acquisition have become quite common. Thus, naturally the question may arise, why there is no Fourier transform gas chromatography? To anticipate the answer ordinary gas chromatography can be considered as a degenerate Fourier transform technique. We will illustrate this issue after becoming familiar with basic dynamic system analysis.

16.4.4.1 Dynamic System Analysis

Dynamic system analysis originates from and is widely used in the field of electrotechnics [13]. The technique is useful in other fields. We explain the basic issues in terms of electrotechnics and eventually apply the method to gas chromatography.

In electrotechnics, in the simplest case, a system is a black box with an input for an electrical signal and an output for an electrical signal. In order to obtain information of the system's behavior, the system is probed with an electrical signal and the electrical output signal is measured. Without going into details, the system must exhibit certain properties such as linearity, causality, as well as it must behave deterministic and invariant with respect to a time shift.

A central tool in the technique is the convolution integral, as briefly explained also in Sect. 20.4.2.

For a linear system the input signal, e.g., a voltage, may be a function of time, $f_i(t)$. The same is true for the output signal $f_o(t)$. The input signal and the output signal are related by the convolution integral as

$$f_o(t) = \int_{\tau=-\infty}^{\tau=+\infty} f_i(\tau)g(t-\tau)\,d\tau. \tag{16.8}$$

In short notation, the convolution is written as

$$f_o(t) = f_i(t) * g(t).$$

The integrand in Eq. (16.8), $f_i(\tau)g(t-\tau)$ is the contribution of the input signal at time τ to the output signal at time t. To get all the contributions for the output of time t, the contributions are summed up for the full time. The function $g(t-\tau)$ reflects the behavior of the black box. Actually, the function $g(t-\tau)$ can be considered as the characteristics of the system.

Of course if $\tau > t$, there should be no contribution, otherwise a signal that appears in the future will act on the presence. This would break the principle of causality. For example, if the black box is simply a short wire, then the output signal is expected to be nearly identical to the output signal, i.e., $f_i(t) = f_o(t)$. In this case, $g(t-\tau)$ must be adjusted to contribute only at $t = \tau$. This means, $g(x) = \delta(x)$. Here $\delta(x)$ is the Dirac delta function.

The convolution theorem states that the Fourier transform of the convolution is simply the product of the functions. We write the respective Fourier transform in capitals:

$$F_o(\iota\omega) = F_i(\iota\omega)G(\iota\omega). \tag{16.9}$$

For convenience we present some rules how to apply the Fourier transform in Tables 16.1 and 16.2.

In electrotechnics, it is preferred to test the response of a system by a needle-like pulse of voltage or current, which is mathematically represented by the Dirac delta function. As can be seen from Eq. (16.9), since the Fourier transform of the Dirac delta function is 1, so $F_i(\iota\omega) = 1$, the Fourier transform of the system characteristics, $G(\iota\omega)$ is directly the Fourier transform of the output response, $F_o(\iota\omega)$. Of course this implies that also the corresponding functions in the time domains are equal.

Table 16.1 Rules of the Fourier transform [13]

Definition		
$f(t) = \frac{1}{2\pi}\int_{-\infty}^{+\infty} F(\iota\omega)\exp(\iota\omega t)\,d\omega$		$F(\iota\omega) = \int_{-\infty}^{+\infty} f(t)\exp(-\iota\omega t)\,dt$
Rules		
$f(t)$	$\rightleftharpoons$	$F(\iota\omega)$
Linearity		
$\sum a_k f_k(t)$	$\rightleftharpoons$	$\sum a_k F_k(\iota\omega)$
Derivative		
$\frac{d^n f(t)}{dt^n}$	$\rightleftharpoons$	$(\iota\omega)^n F(\iota\omega)$
$(-\iota t)^n f(t)$	$\rightleftharpoons$	$\frac{d^n F(\iota\omega)}{d\omega^n}$
Integral		
$\int_{-\infty}^{t} f(\tau)\,d\tau$	$\rightleftharpoons$	$\frac{1}{\iota\omega}F(\iota\omega)$
Scaling		
$f(at)$	$\rightleftharpoons$	$\frac{1}{|a|}F(\frac{\iota\omega}{a\iota\omega})$
Shift		
$f(t - t_0)$	$\rightleftharpoons$	$\exp(-\iota\omega t_0)F(\iota\omega)$
$\exp(\iota\omega_0)f(t)$	$\rightleftharpoons$	$F(\iota(\omega - \omega_0))$
Convolution		
$f_1(t) * f_2(t) = \int_{-\infty}^{+\infty} f_1(\tau)f_2(t - \tau)\,d\tau$	$\rightleftharpoons$	$F_1(\iota\omega)F_2(\iota\omega)$
$f_1(t)f_2(t)$	$\rightleftharpoons$	$\frac{1}{2\pi}F_1(\iota\omega) * F_2(\iota\omega)$

Table 16.2 Correspondences of the Fourier transform [13]

$f(t)$	$\rightleftharpoons$	$F(\iota\omega)$
$\delta(t)$	$\rightleftharpoons$	1
$s(t) = \int \delta(t)\,dt$	$\rightleftharpoons$	$\pi\delta(\omega) + \frac{1}{\iota\omega}$
1	$\rightleftharpoons$	$2\pi\delta(\omega)$
$\sin(\omega_0 t)$	$\rightleftharpoons$	$\iota\pi(\delta(\omega - \omega_0) + \delta(\omega + \omega_0))$
$\cos(\omega_0 t)$	$\rightleftharpoons$	$\pi(\delta(\omega - \omega_0) + \delta(\omega + \omega_0))$

16.4.4.2 System Analysis of Chromatography

We are ready now to transform the knowledge obtained from electrotechnics into chromatographic systems. Initially, we want to assume that the column exhibits an idealized behavior. No dispersion should be there and no losses by irreversible absorption in the column. This means that the peak eluted by the column retains its shape. In electrotechnics, the corresponding situation is of primary importance, since it guarantees undistorted signal transmission.

If the injection of a single occurs instantly, which corresponds to an initial profile of a Dirac delta function the peak should leave the column as a Dirac delta function delayed by the retention time t_R. We denote the concentration profile now with $c(t)$, thus the injected concentration profile as $c_i(t)$ and the eluted, outputted concentration profile as $c_o(t)$. The corresponding Fourier transforms we denote with capital letters, as before. Due to the needle-like shape of the injected amount we have $c_i(t) = \delta(t)$. Equation (16.9) now will be read as

$$C_o(\iota\omega) = C_i(\iota\omega)G(\iota\omega). \tag{16.10}$$

Due to the assumption that the injection concentration profile is a Dirac delta function and the output profile is simply shifted by the retention time, we find, using the shift theorem from Table 16.1 and the transform of the Dirac delta function in Table 16.2, by inserting in Eq. (16.10)

$$\exp(-\iota \omega t_R) = G(\iota \omega). \tag{16.11}$$

If the injected sample consists of multiple components $(1, 2, \ldots)$ with retention times $t_{R,1}, t_{R,1}, \ldots$, then the injected sample will split at the output into multiple peaks, with similar shape. Then by the addition theorem we have

$$\begin{aligned} C_{o,1}(\iota \omega) + C_{o,2}(\iota \omega) + \ldots = \\ \exp(-\iota \omega t_{R,1})C_{i,1}(\iota \omega) + \exp(-\iota \omega t_{R,2})C_{i,2}(\iota \omega) + \ldots = \\ [C_{i,1}(\iota \omega) + C_{i,2}(\iota \omega) + \ldots]G(\iota \omega). \end{aligned} \tag{16.12}$$

However, all the $C_{i,k}$ have the same form, but differing only by a constant which is proportional to the concentration of the species k. In the case of a Dirac delta function, they are actually the concentration of the species k.

Obviously, $G(\iota \omega)$ in Eq. (16.12) represents the chromatogram due to the performance of the column. However, if we rely on the assumption that the shape of the concentration profile is not distorted, but simply split up in similar looking peaks that are delayed by different retention times, the concentration profile of injection is immaterial. Instead of injecting with a needle-like concentration profile, we could likewise adjust the dosage in a continuous, say sinus-like varying profile or something else. We would always get the same function for $G(\iota \omega)$.

If the peaks are distorted when moving through the column, we could still get the chromatogram in $G(\iota \omega)$, when forming F_o/F_i, which does not cancel longer to 1.

It should now be clear that the common method of doing chromatography is a special and limiting case of dynamic system analysis. A chromatographic analysis is equivalent to electrotechnics in probing the system by a needle-like activation. In fact this is the quickest way to get the chromatogram and there is no need to make use of the full apparatus of Fourier technique. However, it has been shown that basically the injection profile is not limited to a needle-like form. The chromatogram could be obtained from arbitrary concentration profiles. However, such a technique would not prove to be economic.

16.4.5 Time-of-Flight Mass Spectroscopy

Eventually, we emphasize that the situation in gas chromatography is very similar to that in time-of-flight mass spectroscopy. There, basically a technique addressing a phase focussing [14, p. 131] can be used.

In Fourier transform time-of-flight mass spectroscopy, the flow of ions from the ion source through the drift tube region to the detector is modulated at a certain modulation frequency which is swept and the output is detected as a function of the modulation frequency to obtain an ion mass interferogram output. The signal is then Fourier transformed into the time domain to obtain a time-of-flight mass spectrum of the substance under analysis. An improved signal-to-noise ratio is obtained [15, 16].

References

1. Grafton, A.: Leon Battista Alberti: Master Builder of the Italian Renaissance, 1st edn. Hill and Wang, New York (2000)
2. Thompson, D.W.: On Growth and Form. Cambridge University Press, Cambridge (2000). Edited by John Tyler Bonner, first published in 1917
3. Rybczynski, W.: The Perfect House: A Journey with the Renaissance Architect Andrea Palladio. Scribner, New York (2002)
4. Langhein, J.: Proportion and traditional architecture. INTBAU: International Network for Traditional Building, Architecture and Urbanism 1(10) (2001). [electronic] http://www.intbau.org/essay10.htm
5. Kauppinen, J., Partanen, J.: Fourier Transforms in Spectroscopy. Wiley, New York (2001)
6. Marshall, A.G. (ed.): Fourier, Hadamard, and Hilbert Transforms in Chemistry. Plenum Press, New York (1982)
7. Wilson, R.G.: Fourier Series and Optical Transform Techniques in Contemporary Optics: An Introduction. Wiley, New York (1995)
8. Griffiths, P.R., DeHaseth, J.A.: Fourier Transform Infrared Spectrometry, *Chemical Analysis*, vol. 83. Wiley, New York (1986)
9. Hendra, P., Jones, C., Warnes, G.: Fourier Transform Raman Spectroscopy: Instrumental and Chemical Applications. Ellis Horwood, New York (1991)
10. Roberts, J.D.: ABCs of FT-NMR. University Science Books, Sausalito, CA (2000)
11. McPherson, A.: Introduction to Macromolecular Crystallography. Wiley-Liss, Hoboken, NJ (2003)
12. Bracewell, R.N.: The Fourier Transform and its Application, *McGraw-Hill Series in Electrical Engineering*, 3rd edn. McGraw-Hill, Boston, MA (2000)
13. Fritzsche, G.: Systeme Felder Wellen, 1st edn. VEB Verlag, Berlin (1975)
14. Simonyi, K.: Physikalische Elektronik. B.G. Teubner Verlag, Stuttgart (1972)
15. Brock, A., Rodriguez, N., Zare, R.N.: Time-of-flight mass spectrometer and ion analysis. US Patent 6 300 626, assigned to Board of Trustees of the Leland Stanford Junior University (Stanford, CA), Oct. 9 2001
16. Knorr, F.J. Fourier transform time of flight mass spectrometer. US Patent 4 707 602, assigned to Surface Science Laboratories, Inc. (Mountain View, CA), Nov. 17 1987

Chapter 17
Generating Functions in Polymer Science

17.1 Polymerization

The concept of generating functions is useful to obtain the distribution and averages of the degree of polymerization in the course of polymerization reactions and polycondensation reactions. Most of the material described in the literature deals with branched systems. In these systems, classical statistical methods often run against walls and only advanced methods like the application of generating functions will be successful.

Even in less complicated cases, the use of generating functions will provide advantages over the classical methods; papers on this topic are missing, however, in introductory texts. In this chapter, we will show a series of simple examples illustrating the application of the method of generating functions to illustrate the power and usefulness of the machinery.

Generating functions provide a powerful tool in general in probability theory, number theory, and graph theory and in particular in such fields of science that can use the results of those disciplines. Generating functions have also found entrance into polymer science since about some 50 years. Pioneers who used generating functions in the field of polymers are *Burchard* [1], *Dušek* [2, 3], and *Gordon* [4]. *Gordon*[1] introduced the cascade theory [5] to derive finally the degree of polymerization in various systems. *Burchard* pointed out the importance of the concept of directionality [6]. This concept was also used by *Peppas* [7]. Multivariate distributions were derived in [8–10]. Integral distributions were obtained by maximum entropy methods [11]. Numerical integration was used by introducing generating functions into the kinetic equations [12, 13]. The mathematics of generating functions has already been developed, however. There is a monograph by *Wilf* on general aspects of this subject [14].

---

[1] Manfred Gordon, born May 10, 1917, in Leipzig, Germany, died Mar. 3, 2000, in Cambridge, England.

J.K. Fink, *Physical Chemistry in Depth*,   443
DOI 10.1007/978-3-642-01014-9_17, © Springer-Verlag Berlin Heidelberg 2009

The method of generating functions is in concurrence with combinatorial methods and recursive techniques. Combinatorial methods have been used most early in particular by *Flory*[2] and *Stockmayer*[3] to derive critical branching [15].

The recursive technique has been introduced by *Macosko* and *Miller* [16–18]. The technique has been extended to a variety of different systems such as living polymers [19]. On the other hand, generating functions were used by *Yan* in the field of living polymerization of divinyl compounds to resolve certain sums [20].

Generating functions have been used in the past mainly to derive averages of molecular weight and distributions for branched and cross-linked polymers. These systems are difficult or even impossible to handle with combinatorial methods. Other fields of applications like the degradation of polymers and application for thermodynamic equilibria for polymers seem to have completely dismissed until now. In the case of cross-linked structures, the procedure as such becomes complicated and difficult to perceive to the ground. As a matter of fact, this seems to be the reason why generating functions are not popular in textbooks. But the elegance and beauty of the application of generating functions can be felt, even if more simple examples are treated. These have not been fully worked out in the literature hitherto. Therefore, this topic should be the starting point to get common with generating functions.

Most of the literature on generating functions concerning branched polymers is written in a rather advanced style. This makes the material difficult to read and to understand. For this reason, the presentation here has been purposely kept tutorial by including an introductory example.

17.1.1 An Introductory Example

We illustrate now the machinery is shown with a very simple introductory example.

17.1.1.1 Mechanism of Radical Polymerization

Consider the mechanism of well-known radical polymerization at low conversions.

$$I \rightarrow 2S\cdot$$
$$S\cdot + M \rightarrow R_1\cdot$$
$$R_1\cdot + M \rightarrow R_2\cdot$$
$$\cdots\cdots\cdots\cdots\cdots$$
$$R_n\cdot + M \rightarrow R_{n+1}\cdot$$
$$\cdots\cdots\cdots\cdots\cdots$$
$$R_n\cdot + R_m\cdot \rightarrow \text{Products}$$

---

[2] Paul John Flory, born Jun. 19, 1910, in Sterling, Illinois, died Sep. 9, 1985, in Big Sur, CA, USA.

[3] Walter Hugo Stockmayer, born Apr. 7, 1914, in Rutherford, New Jersey, died May 9, 2004, in Norwich, Vermont.

We illustrate the start of the polymerization with azoisobutyronitril and the first two steps of growth of the chain with styrene in Fig. 17.1. The notation follows the usual in literature. The initiator decomposes into two initiator radicals that are adding successively to monomers resulting in polymer radicals. At the end of the kinetic chain, the radicals deactivate mutually. The rate constants are assumed to be independent of the degree of polymerization.

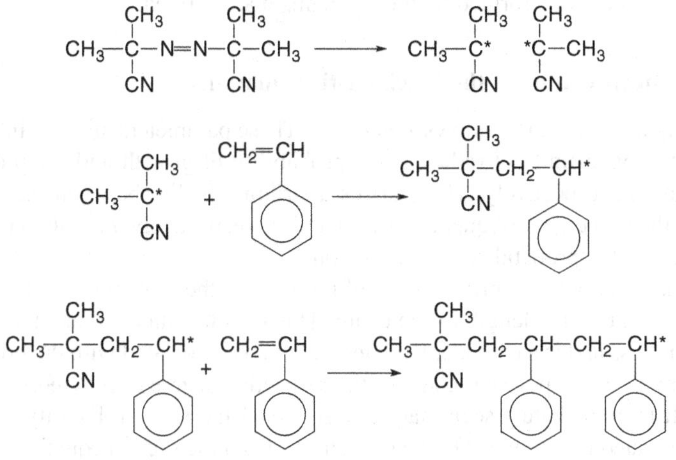

Fig. 17.1 Start of the polymerization with azoisobutyronitril; growth with styrene

17.1.1.2 Mechanistic Equations

The reaction mechanism above can be boiled down to a set of symbolic equations, referred to as mechanistic equations because they reflect the mechanism,

$$\begin{aligned} s &= sr \\ r &= e + mr \end{aligned} \qquad (17.1)$$

The symbolic equations (17.1) should be read as

1. A starter radical becomes in the next step a polymer radical, and
2. A polymer radical becomes in the next step either an end group or a polymer radical with one more monomer incorporated.

Recursive substitution of r gives

$$\begin{aligned} \text{Step 1}: & \quad se + smr \\ \text{Step 2}: & \quad se + sme + smmr \\ \text{Step 3}: & \quad se + sme + smme + smmmr \\ & \quad \text{etc.} \quad \ldots \end{aligned}$$

The mathematical structure is built up just in the same way as a polymeric chain would grow in the reagent glass. The specific properties of multiplication of an expression $s \ldots m$ with the binomial $mr + e$ effect that the chain is copied and put aside with the end group, whereas the other copy with r at the end is subject to further growth. The *plus* in the binomial stands for a logic *or* as will become more clear later.

Further, it is easy to see that by substitution of r *ad infinitum* all the possible polymers that could be formed by the mechanism are formed.

17.1.1.3 Interpretation of the Mechanistic Equations

Let r be equal to 1, then it follows $1 = m + e$. These parameters are now interpreted not as a monomer unit but rather as the probability of growth and the probability of termination, respectively. There arises a striking duality between the structure itself and the associated frequency or probability, respectively. m is the probability of growth, e is the probability of termination.

Note that here both the probabilities of growth and the probability of termination are independent on the length of the chain. This is in fact an alternate notation of a *Markov* process and is a cascade. Each term $s \ldots e$ is associated with the probability of the formation of a polymer chain of the respective degree of polymerization that is $p(P)$. It is immediately seen that because $r = 1$ the total probability for all the chains generated becomes 1. On the other hand, from the second equation in the set (17.1), r can be calculated to be $r = e/(1 - m)$. This can be inserted into the first equation in the set (17.1).

17.1.1.4 Generating Function

Now substitute $m \to m \exp(x)$; $e \to e \exp(x)$ and obtain

$$F(x) = \frac{se \exp(x)}{1 - m \exp(x)}.$$

Expansion results in

$$F(x) = \frac{se \exp(x)}{1 - m \exp(x)} = p(1) \exp(x) + p(2) \exp(2x) + \ldots = \sum_{P \geq 1} p(P) \exp(Px).$$

In general, to form the generating function, every unit that counts as a monomer will be replaced by $\exp(x)$. In the example, the end group is treated as a monomer unit, which is approximately true for disproportionation.

For some purposes it is of advantage to use instead of $\exp(x)$ or $\exp(-x)$ or $\exp(ix)$ or simply x or $x/P!$ Possible types of generating functions are summarized in Table 17.1. The coefficient of the respective term is the probability that a degree of polymerization of P is obtained. For example, by using the characteristic function it can be seen that the orthogonality of the sine and cosine functions can be used to obtain the desired coefficients. Further, approximating the sum of the generating function by an integral it turns out that it is some well-known integral transformation.

Table 17.1 Types of generating functions

Term	Name
$\exp(Px)$	Moment generating function
$\exp(\iota Px)$	Characteristic function
x^P	Probability generating function or ordinary power generating function
$x^P/P!$	Exponential power generating function

17.1.1.5 Properties of the Moment Generating Function $F(x)$

Subsequently we show some important properties of the moment generating function. This include the number and weight average of the degree of polymerization and other averages, as well as the distribution function. There is a potential confusing situation in nomenclature. In polymer chemistry, the term *cumulative distribution* is used for what is called in mathematics the *cumulative distribution function* or simply *distribution function*. Further, polymer chemistry uses the term *differential distribution* for the mathematical term *probability density function*.

Number and Weight Average

Forming the derivatives, we obtain

$$F'(x) = \sum_{P \geq 1} P p(P) \exp(Px)$$
$$F''(x) = \sum_{P \geq 1} P^2 p(P) \exp(Px)$$

So the number average P_n and the weight average P_w of the degree of polymerization are

$$P_n = \frac{F'(0)}{F(0)}; \quad P_w = \frac{F''(0)}{F'(0)}.$$

The notation $F'(0)$ above means that first the derivative at some arbitrary x should be formed and afterward the derivative should be evaluated at $x = 0$. For the introductory example the number and weight averages are

$$P_n = \frac{1}{1-m}; \quad P_w = \frac{1+m}{1-m} \approx \frac{2}{1-m}.$$

Viscosity Average

In order to obtain the viscosity average, it is necessary to form a non-integral derivative [21, 22], e.g., $d^{1.7} [\bullet]/dx^{1.7}$, so non-integer moments are obtained. The procedure will be illustrated below in detail. The viscosity average is defined as

$$P_\eta = \left(\frac{\sum P^{1+\alpha} p(P)}{\sum P p(P)} \right)^{1/\alpha} = \left(\frac{F^{\{1+\alpha\}'}(0)}{F'(0)} \right)^{1/\alpha}.$$

To obtain an expression for the non-integer derivative, observe that for long chains, $m \to 1$ and further for analysis at $x \to 0$ that is finally required the approximation below is in order

$$\frac{1}{1 - m \exp(x)} \approx \frac{1}{-\ln m - x}$$

and thus

$$\frac{d^l}{dx^l} \left(\frac{1}{-\ln m - x} \right) \approx \frac{l!}{(1 - m)^{l+1}} \tag{17.2}$$

for $x \to 0$. Here the end group e is neglected as monomer unit in the polymer. The formula in Eq. (17.2) can be used for non-integer l, thus being a heuristic approach to fractional calculus from the view of mathematics. Further, $\sum p(P)P = 1/(1 - m)^2$. This results in

$$\left(\frac{\sum p(P)P^{1+\alpha}}{\sum p(P)P} \right)^{1/\alpha} = \left(\frac{(1+\alpha)!}{(1 - m)^\alpha} \right)^{1/\alpha}.$$

Finally with $((1 + \alpha)!)^{1/\alpha} \approx (3 + \alpha)/2$,

$$\left(\frac{\sum p(P)P^{1+\alpha}}{\sum p(P)P} \right)^{1/\alpha} = \frac{3 + \alpha}{2(1 - m)}.$$

This result is given in the book of *Henrici-Olive* [23] derived by integrating to obtain the viscosity average. The approximation

$$((1 + \alpha)!)^{1/\alpha} \approx \frac{3 + \alpha}{2}$$

can be found by *Taylor* expansion of the integral definition of the factorial function for non-integers

$$l! = \int_0^\infty t^l \exp(-t) \, dt \tag{17.3}$$

around $n = 1$ and numerical integration:

$$[(1+n)!]^{\frac{1}{n}} = 2 + 2(1-n)\ln 2 - (1-n)\left.\frac{dn!}{dn}\right|_{n=2} + \dots$$

$$n! = \int_0^\infty \exp(-t)t^n\,dt; \qquad \frac{dn!}{dn} = \int_0^\infty \exp(-t)t^n \ln(t)\,dt.$$

Cumulative Distribution

The cumulative generating function that is actually a formal series can be obtained from the moment generating function by division through $1 - \exp(x)$. It is easily seen that the partial sums are obtained when the divisor is expanded in a geometric series and resolved term by term, namely

$$\frac{\left[a_1 \exp(x) + a_2 \exp(2x) + \dots\right]}{1 - \exp(x)} =$$
$$\left[a_1 \exp(x) + a_2 \exp(2x) + \dots\right]\left[1 + \exp(x) + \exp(2x) + \dots\right].$$

Counting the Number of Items in an Expression

Assume an expression $abb + aabbb$. There are clearly totally 3 A and 5 B inside. In general, the A's and the B's can be counted by forming the generating function

$$F(\alpha, \beta) = ab^2 \exp(\alpha + 2\beta) + a^2b^3 \exp(2\alpha + 3\beta) \quad .$$

This function is obtained simply by replacing a by $a \exp(\alpha)$ and b by $b \exp(\beta)$. Now

$$\left.\frac{\partial}{\partial \alpha} F(\alpha, \beta)\right|_{\alpha=0; \beta=0} = ab^2 + 2a^2b^3$$

$$\left.\frac{\partial}{\partial \beta} F(\alpha, \beta)\right|_{\alpha=0; \beta=0} = 2ab^2 + 3a^2b^3$$

The numbers coming down in this way are just the numbers required. To obtain the pure sum, A and B are put as one ($a = b = 1$). On the other hand, A and B may reflect some statistical weight so that the sum has directly a physical meaning.

17.1.2 Other Examples

17.1.2.1 Bifunctional Initiator

The mechanistic equations become

$$s = rsr = sr^2; \quad r = e + mr.$$

So two chains are growing from the initiator radical. To obtain the respective averages, the same procedure as explained thoroughly in this chapter can be applied.

17.1.2.2 Termination by Recombination

When a polymeric chain is not terminated by recombination, the end group e is no longer a monomer unit but a whole primary chain. The mechanistic equation for the first chain remains unchanged. But the mechanistic equation for the second chain that has actually the same form ends in a head-to-head structure (hh),

$$s = sr; \quad r = h + mr.$$

Inserting $shh/(1-m)$ for e, the generating function for recombination is obtained

$$G(x) = \left[\frac{sh \exp(x)}{1 - m \exp(x)} \right]^2.$$

Note that the probability of growth m is the same for both primary chains. Thus, $e = h$ and $G(x) = F(x)^2$. The number average for recombination is then twice than those for disproportionation.

17.1.2.3 Chain Scission

The equation $G(x) = F(x)^2$ can also be read in this way that the function $G(x)$ is known and $F(x)$ is unknown. Solving $F(x)$ means that the chains will be split in just the reverse way as they would be formed according to the multiplication rule for series according to *Cauchy*. This is a very special type of chain scission.

17.1.2.4 Higher Degree of Coupling

Recombination is a coupling of degree 2. According to the multiplication rule it is straightforward, without using combinatorics, to handle higher degrees of coupling. A degree of coupling of n means simply to raise the generating function of the primary chain to the power of n

$$K(x) = F(x)^n.$$

The term $\exp(Px)$ in the coefficients collects just the probabilities for a degree of polymerization of P. The coefficients reflect the possibilities, in which ways a degree of polymerization of P can be obtained from n primary chains.

17.1.2.5 Composite Distributions

Generating functions for composite distributions can be summed, e.g., if disproportionation is together with recombination.

$$G(x) = F(x)^2; \quad H(x) = aF(x) + bG(x).$$

17.1.2.6 Modification of the Monomer Units

Suppose a polystyrene polymer with a certain generating function is sulfonated. In this process, a sulfonic acid group is attached randomly to any monomer moiety. Only one group can be attached to a monomer unit. To obtain the distribution of the unsulfonated (u) and sulfonated groups, i.e., vulcanized (v), replace simply m by $u + v$. So v/m is the probability that a monomer is sulfonated. For the introductory example the mechanistic equations become

$$s = sr; \quad r = e + (u + v)r.$$

This expression expands into

$$s = se + sue + sve + suue + suve + svue + svve + \ldots$$

To cut away the chains with sulfur groups from the distribution simply set $v = 0$.

17.1.2.7 Graft Polymers

First, a main chain will be generated as shown above. Next, m is substituted by $u+v$. It is assumed that every unit v bears a side chain. This side chain is then clicked into the main chain. Then the generating function looks like

$$F(x) = \frac{se \exp(x)}{1 - (u + vS(x)) \exp(x)}.$$

$S(x)$ is the generating function for the side chain. This function can have a similar form like that for the main chain, i.e., a geometric distribution with different numerical values for the parameters.

If u approaches zero, a comb-like structure is obtained. The generating function of such a structure is

$$F(x) = \frac{se \exp(x)}{1 - vS(x) \exp(x)} =$$
$$se \exp(x)[1 + vS(x) \exp(x) + v^2 S^2(x) \exp(2x) + \ldots].$$

The expansion above shows the similarity of the generating function to a mixture with degrees of couplings with respect to the side chain varying from 1 to infinity.

17.1.2.8 Chain Transfer

If chain transfer to the solvent occurs, an additional mechanism of termination takes place. The mechanism of chain transfer can be basically written as

$$M\cdot + Cl - C\,(Cl)_3 \to M - Cl + C\,(Cl)_3.$$

$$\quad\quad r \quad\quad u \quad\quad\quad\quad\quad\quad v$$

A growing radical now may still grow or terminate, or transfer. Therefore, the mechanistic equations must be changed into

$$s = sr; \quad r = mr + e + u; \quad v = vr.$$

Every time when a transfer occurs a new polymer chain with a v unit at the begin starts. So there are now two reactions that start a chain, one is the ordinary start with (s) and the other is the start with a residual radical fragment from the chain transfer agent (v). (u) contains a monomer unit and will count as such.

The mechanistic equation and the generating function become

$$\frac{s + v}{1 - m}(e + u) \quad \text{and} \quad F(x) = \frac{s + v}{1 - m\exp(x)}(e\exp(x) + u\exp(x)).$$

Again, the number average becomes

$$\frac{1}{P_n} = \frac{F\,(0)}{F'\,(0)} = 1 - m.$$

From the view of kinetics, the probability of growth is

$$m = \frac{v_p}{v_p + v_t + v_{A''}} \approx 1 - \frac{v_t}{v_p} - \frac{v_{A''}}{v_p},$$

v_p: Rate of propagation
v_t: Rate of termination
$v_{A''}$: Rate of transfer

from which immediately the *Mayo* equation emerges. In contrast to the common derivation of the *Mayo* equation, with the concept of generating functions also other averages than the number average are readily accessible.

Since the chains starting with (s) and with (v) have the same distribution, mixing of these chains does not affect the distribution at all. Therefore, for the averages it is not necessary to get explicit values for s and v. But these data can be obtained by balancing the number of the units occurring in the structure. For every (s) an (e) must be in the structure expandable from the mechanistic equation; and for every (v) there must be a (u). The number of (s) will be designated as $N(s)$, the number of (e) as $N(e)$, etc. These numbers and others can be counted by the method given in Sect. 17.1.1.5. Here, the following set of equations is obtained:

$$N(s) = \frac{s(e+u)}{1-m} = s; \quad N(e) = \frac{e(s+v)}{1-m};$$

$$N(u) = \frac{u(s+v)}{1-m}; \quad N(v) = \frac{v(e+u)}{1-m} = v;$$

$$\frac{N(s)N(u)}{N(e)N(v)} = \frac{su}{ev} = 1; \quad \frac{s}{v} = \frac{e}{u} = \frac{v_t}{v_{A''}}.$$

So the ratio of the number of chains beginning with (s) to the number of chains beginning with (v) is the ratio of the rate of termination to the rate of transfer. We would expect intuitively a ratio of the rate of generation of initiator radicals to the rate of generation transferred radicals $s/v = v_s/v_{A''}$, because this reflects the relative rate of generation of the respective species (s) and (v). But there is still some other aspect behind, namely the rate of generation of initiator radicals must be equal to the rate of termination $vs = vt$. So in using the balance of starting groups (s) and end groups (e) in terms of generating functions suddenly *Bodenstein*'s[4] principle emerges.

The generating function is now easily split into the portions of chains that are initiated by the initiator radical (s) and the portion that is initiated by the transfer agent radical (v)

$$F(x) = S(x) + V(x) = \frac{s(e\exp(x) + m\exp(x)u)}{1 - m\exp(x)} + \frac{v(e\exp(x) + m\exp(x)u)}{1 - m\exp(x)}.$$

17.1.2.9 Primary Chain Termination

The mechanistic equations for this case are

$$s = sr_1; \quad r_1 = m_1 r + e_1; \quad r = mr + e.$$

This means that the first polymer radical r_1 has a different growth probability than the radicals r with a higher degree of polymerization. The generating function in the presence of primary chain termination is

$$F(x) = s\left[m_1 \exp(x) \left(\frac{e\exp(x)}{1 - m\exp(x)} \right) + e_1 \exp(x) \right]$$

and the number average and the weight average of the degree of polymerization are

$$P_n = \frac{1 - (m_1 - m)}{1 - m}; \quad P_w = \frac{m^2 - m(m_1 + 2) + 3m_1 + 1}{m^2 - m(m_1 + 2) + m_1 + 1}.$$

---

[4] Max Bodenstein, born Jul. 15, 1871, in Magdeburg, died Sep. 3, 1942, in Berlin.

17.1.2.10 Generalization of Primary Chain Termination

In the introductory example, all reactivities are independent of the chain length. In the primary chain termination example in Sect. 17.1.2.9, only the first radical with a degree of polymerization of 1 differs in reactivity from the others. If every radical has a different reactivity then the mechanistic equation will look like

$$s = sr_1$$
$$r_1 = m_1r_2 + e_1$$
$$r_2 = m_2r_3 + e_2$$
$$\dots\dots\dots\dots\dots$$
$$r_n = m_nr_{n+1} + e_n$$

In favorite cases, an analytic formula for the generating function can be obtained.

17.1.2.11 End-to-End Length

In a three-dimensional lattice, a growing radical being at a specific point may move one step right (r), left (l), up (u), down (d), front (f), back (b). So the mechanistic equation, now with r' for the growing radical, looks like

$$s = sr'; \quad r' = mr'(r + l + u + d + f + b) + e.$$

The second mechanistic equation expresses that a growing radical r' in the next step will

1. take up a monomer *and* will now turn a step to the right or to the left, or upward, or downward, or in front, or backward, or
2. the radical will terminate.

To obtain the generating function monomer, both a monomer counting variable t and dummy position variables x, y, z are introduced. We substitute

$$m \rightarrow m \exp(t); \quad e \rightarrow e \exp(t);$$
$$r \rightarrow r \exp(b_0x); \quad l \rightarrow l \exp(-b_0x);$$
$$u \rightarrow u \exp(b_0y); \quad d \rightarrow d \exp(-b_0y);$$
$$f \rightarrow f \exp(b_0z); \quad b \rightarrow b \exp(-b_0z)$$

b_0 is the bond length of the monomer unit. Moreover, if the space is isotropic then all probabilities concerning the direction of movement are equal to $r = l = u = d = f = b = 1/6$. The generating function now turns out as

$$F(t, x, y, z) = \frac{se\exp(t)}{1 - m\exp(t)D(x, y, z, b_0)}$$

$$D(x, y, z, b_0) = \frac{1}{6}(\exp(b_0 x) + \exp(-b_0 x) +$$

$$\exp(b_0 y) + \exp(-b_0 y) + \quad \cdot$$

$$\exp(b_0 z) + \exp(-b_0 z))$$

The function $D(x, y, z, b_0)$ we have introduced just for brevity. If only a polymer with a degree of polymerization of P is of interest, then the coefficient attached to the term with $\exp(Pt)$ must be extracted. This is for geometric series simply

$$G(P, x, y, z) = D(x, y, z, b_0)^P,$$

where $G(P, x, y, z)$ is the generating function for a chain with length P. The second moment is obtained by forming the second derivative

$$\frac{\partial^2 G(P, x, y, z)}{\partial x^2} + \frac{\partial^2 G(P, x, y, z)}{\partial y^2} + \frac{\partial^2 G(P, x, y, z)}{\partial z^2}$$

at $x, y, z = 0$. Thus, with $\overline{r^2} = \overline{x^2} + \overline{y^2} + \overline{z^2}$ it turns out that

$$\overline{r^2} = b_0^2 P. \tag{17.4}$$

The average end-to-end distance of the mean is even more simply obtained by forming the derivative of the generating function, i.e., $\Delta F(t, x, y, z)$ itself without extracting the particular term for P. In our example the number average will appear for P.

Several refinements are possible, e.g.,

- unequal probabilities for each direction;
- a penultimate effect, meaning that a radical that has grown to right just before may not grow to the left in the next step;
- copolymerization introducing two types of growing radicals and monomer bond lengths.

The treatment is similar to the primary chain termination effect. Configurational statistics of copolymers has been described by *Gordon* and *Malcolm* [24].

17.1.2.12 Binary Copolymerization

The mechanism of the binary copolymerization is given in the following scheme:

Start	Propagation	Termination
$I \rightarrow 2S\cdot$	$R_{a,n} \cdot + A \rightarrow R_{a,n+1}\cdot$	$R_{a,n} \cdot + R_{a,n}\cdot \rightarrow$ Products
$S \cdot + A \rightarrow R_{a,1}\cdot$	$R_{a,n} \cdot + B \rightarrow R_{b,n+1}\cdot$	$R_{a,n} \cdot + R_{b,n}\cdot \rightarrow$ Products
$S \cdot + B \rightarrow R_{b,1}\cdot$	$R_{b,n} \cdot + A \rightarrow R_{a,n+1}\cdot$	$R_{b,n} \cdot + R_{b,n}\cdot \rightarrow$ Products
	$R_{b,n} \cdot + B \rightarrow R_{b,n+1}\cdot$	

The mechanistic equations are

$$s = s\,(r_a + r_b)/2$$
$$r_a = a\,(f r_a + (1 - f)r_b) + a_e$$
$$r_b = b\,(g r_b + (1 - g)r_a) + b_e$$

This means that an initiator radical $S\cdot$ may add either monomer A or monomer B with equal probability. A growing radical resulting from a monomer at the end (referred to as growing radical $A\cdot$) will be converted in a chain unit of type A on and it may add on growth a monomer A or a monomer B. The relative probability if a growing radical will add, if it adds at all, a monomer A is f, and the relative probability that it will add a monomer B is $1 - f$.

On the other hand, the growing radical can terminate to give a final polymer with a chain end resulting from a monomer A. The same is true, mutatis mutandis for a growing radical $B\cdot$. After elimination of r_a and r_b, the following mechanistic equation is obtained:

$$s\frac{r_a + r_b}{2} = -s\frac{2 a b_e f + 2 a_e b g - a b_e - a_e b - a_e - b_e}{2 a b f + 2 a b g - 2 a b - 2 a f - 2 b g + 2}.$$

The probabilities have a kinetic interpretation, i.e.,

$$a = \frac{v_{aa} + v_{ab}}{v_{aa} + v_{ab} + v_{t,a}}; \qquad a_e = \frac{v_{t,a}}{v_{aa} + v_{ab} + v_{t,a}};$$

$$b = \frac{v_{bb} + v_{ba}}{v_{bb} + v_{ba} + v_{t,b}}; \qquad b_e = \frac{v_{t,b}}{v_{bb} + v_{ba} + v_{t,b}};$$

$$f = \frac{v_{aa}}{v_{aa} + v_{ab}}; \qquad g = \frac{v_{bb}}{v_{bb} + v_{ba}}.$$

$$m = \frac{[A]}{[B]}; \quad f = \frac{1}{1 + \frac{1}{r_1 m}}; \quad g = \frac{1}{1 + \frac{m}{r_2}}$$

$v_{aa}, v_{ab}, v_{ba}, v_{bb}$ are the growth rates and $v_{t,a}, v_{t,b}$ are termination rates. From the growth rates the monomer reactivities r_1 and r_2 are obtained in the usual way.

Generating Function

The generating function for binary copolymerization is obtained by substituting $a \to a \exp(t)$, $a_e \to a_e \exp(t)$, $b \to b \exp(t)$, $b_e \to b_e \exp(t)$. The averages of the degree of polymerization are obtained in the usual way by forming the derivatives.

Molar Masses

In the case of homopolymers, the end groups can be neglected and the molar masses can be safely obtained multiplying the degree of polymerization with the mole mass of the monomer unit. In copolymerization, this is in general not allowed.

The average mole masses are obtained if another type of generating function is formed. Substitute

$$a \to a \exp(M_a t); \qquad\qquad a_e \to a e \exp(M_{ae} t);$$
$$b \to b \exp(M_b t); \qquad\qquad b_e \to b e \exp(M_{be} t).$$

If such a type of generating function is converted into the derivative with respect to t then the total mole mass of the aggregate is coming down from the exponent, namely by expansion of the generating function, terms like

$$p(\alpha, \beta) \exp(\alpha M_a t) \exp(\beta, M_b t)$$

are occurring that will give in the first derivative $p(\alpha, \beta)(\alpha M_a + \beta M_b)$.

Copolymerization Equation

To derive the copolymerization equation the following substitutions are appropriate:

$$a \to a \exp(x); \qquad\qquad a_e \to a_e \exp(x);$$
$$b \to b \exp(y); \qquad\qquad b_e \to b_e \exp(y).$$

Then the generating function will contain terms like

$$p(\alpha, \beta) \exp(\alpha x) \exp(\beta y).$$

The derivative with respect to x yields the total number of monomer units A in the polymer (in fact a number proportional to that) and the derivative with respect to y yields the total number of monomer units B in the polymer. The ratio of the number of monomer units A to the number of monomer units B in the copolymer is obtained from

$$\frac{[A]}{[B]}\Bigg|_{Polymer} = \frac{\frac{\partial F(x,y)}{\partial x}}{\frac{\partial F(x,y)}{\partial y}}\Bigg|_{x=y=0} .$$

When the limit for infinite long chains, i.e., $ae = be \rightarrow 0$ is formed, the traditional copolymerization equation is obtained. The derivation given here is more general.

Distribution of the A and B Units in the Polymer

If the ordinary power generating function is formed from the mechanistic equation (i.e., $a \rightarrow at$, $b \rightarrow bt$) and this is expanded into powers of t then for the coefficient at t^P basically the following is obtained:

$$\left[t^P\right] F(t) = c_{P,0} a^P b^0 + c_{P-1,1} a^{P-1} b^1 + \ldots + c_{0,P} a^0 b^P.$$

The ordinary power representation has been chosen because the expansion of the generating function works with the method of partial fractions. The coefficient at t^P gathers all the probabilities for a total degree of polymerization of P. Here again the end groups are neglected.

The homogeneous polynomial in A and B reflects the distribution of the units A and B in a copolymer of length P. The expressions become particularly simple in the case of ideal copolymerization. This was already commented by *Stockmayer* [25] more than 50 years back.

Copolymer Sequence Length

To obtain an average sequence length, the mechanistic equations must be changed as follows:

$$r_a = a\,(f r_a + u(1 - f)r_b) + a_e$$
$$r_b = b\,(g r_b + v(1 - g)r_a) + b_e$$

The parameters u and v are pure dummy. But as can be seen readily from the mechanistic equations, every time when the growing chain will switch from an A-unit to a B-unit a u-symbol will be built in as separator. Similarly, a v-symbol is built in when the chain is switching from a B-unit to an A-unit.

So at the beginning of a B sequence a u-symbol is preceding and at the end of a B sequence a v is added behind. In order to keep this true even at the start of the chain, we will also modify the first equation into

$$s = \frac{s}{2}(v r_a + u r_b).$$

From the modified generating function, the average sequence length L_a of the A- and B-units built in can be obtained. It is simply the ratio of the number of A units to the number of u-units.

Substitute in the mechanistic equation $a \rightarrow ax$ and $b \rightarrow by$, then

$$L_a = \frac{\frac{\partial F}{\partial x}}{\frac{\partial F}{\partial u}}\bigg|_{x=y=u=v=1} \quad ; \quad L_b = \frac{\frac{\partial F}{\partial y}}{\frac{dF}{dv}}\bigg|_{x=y=u=v=1} .$$

For long chains in this way the following result is obtained:

$$L_a = 1 + r_1 m; \quad L_b = 1 + r_2/m.$$

If both monomer reactivity ratios r_1 and r_2 will become in the limit 0, then alternating copolymerization is indicated. Clearly, the ratio of both sequence lengths will give the expression for the copolymerization equation.

17.1.2.13 Living Polymers

The formation of living polymer obeys a completely different mechanism than faced in radial polymerization. With living polymers, highly uniform materials can be synthesized. Living polymerization includes not only anionic polymerization but also atom transfer radical polymerization, living controlled free radical polymerization.

Mechanistic Equations

In contrast to the classical radical polymerization, living polymerization has no termination reaction. The mechanistic equations are

$$s = sl; \quad l = (m + w)l.$$

The system is watched in time intervals Δt. After a time interval, it could have happened that a living species (l) has added a monomer (m) or it was waiting for a polymerization step. m and w are the respective probabilities for reaction or waiting in the time interval. Recursive substitution shows that after n time intervals the expression $s(m + w)^n l$ is obtained. Since only m counts for the degree of polymerization, the generating function will become

$$F(n, x) = s(m \exp(x) + w)^n.$$

Definitely, the probabilities of growth and waiting are some functions of the time interval of observation and in fact, a *Taylor* expansion for small times shows that it is proportional to the time interval and must approach zero for a zero time interval.

$$m(\Delta t) = k_0 + k_1 \Delta t + \dots$$
$$k_0 = 0.$$

The probabilities of reaction and waiting add up to 1 and so

$$F(n, x) = s(k_1 \Delta t \exp(x) + 1 - k_1 \Delta t)^n =$$

$$= s \left[1 - k_1 \Delta t (1 - \exp(x)) \right]^{\frac{-k_1 t(1-\exp(x))}{-k_1 \Delta t(1-\exp(x))}} = s \left[1 + h \right]^{\frac{-k_1 t(1-\exp(x))}{h}} \approx$$

$$\approx s \exp[-k[M]t(1 - \exp(x))]. \quad (17.5)$$

Here we take the limit of $h \to 0$ as $\Delta t \to 0$ and substitute $n\Delta t = t; k_1 = k[M]$. Equation (17.5) is the moment generating function of the *Poisson* distribution.

Poisson Process

Living polymers follow a kinetic reaction scheme looking like

$$\frac{dc_1}{dt} = -kc_1[M]$$

$$\frac{dc_2}{dt} = -kc_2[M] + kc_1[M]$$

$$\frac{dc_3}{dt} = -kc_3[M] + kc_2[M] \qquad \cdot$$

$$\cdots \cdots \cdots \cdots \cdots \cdots \cdots \cdots$$

$$\frac{dc_n}{dt} = -kc_n[M] + kc_{n-1}[M]$$

This system of equations can be solved in one stroke if the equations are successively multiplied by $x, x^2, x^3, \ldots, x^n$. A generating function of the following type is defined now:

$$C(x, t) = c_1(t) x + c_2(t) x^2 + \ldots.$$

When the set of equations, each equation being multiplied by the appropriate power of x, is added together and the generating function is introduced then the following simple differential equation is obtained:

$$\frac{\partial C(x, t)}{\partial t} = -k[M](1 - x)C(x, t).$$

The differential equations can be easily solved and if x is substituted by $\exp(x)$, the moment generating function for the *Poisson* distribution is obtained.

17.2 Polycondensation

The use of generating functions in the field of condensation polymers is more established than in the field of polymerization. Even *Flory* used generating functions when he modified power series by forming the derivative, but he did not address them as such [15]. In some sense what is called partition function in science is a generating function in the mathematical notation.

The field of generating functions was actually opened to polymer science with the work of Burchard [1], Dušek [2], and Gordon [4]. Gordon introduced the cascade theory [5] to derive finally the degree of polymerization in various systems. In contrast to radical polymerization the concept of directionality [6] is not of importance for equilibrium condensation reactions. Extensive work on the subject has been done by Tang [26, 27].

Here we point out the subtle difference with equilibrium and nonequilibrium polycondensation and we point out the equivalence between radical polymerization and polycondensation.

17.2.1 An Introductory Example

The machinery how to obtain the generating function from a kinetic scheme is shown in a simple introductory example.

17.2.1.1 Scheme

Consider a monomer of type $HO - R - COOH$. This monomer forms polycondensates of the type

$$HO - R - COO - R - (COO - R)_{n-1} - COOH.$$

Subsequently condensation reactions and hydrolysis reactions are examined.

17.2.1.2 Condensation Reactions

The formation and the disappearance of the various species with a certain degree of polymerization can be balanced and clarified in the following scheme. The scheme looks like a dyadic product

$$
\begin{array}{llll}
c_1 c_1 & c_1 c_2 & c_1 c_3 & \dots \\
c_2 c_1 & c_2 c_2 & c_2 c_3 & \dots \\
c_3 c_1 & c_3 c_2 & c_3 c_3 & \dots \\
\cdot\,\cdot & \dots\dots\dots\dots
\end{array}
$$

Here the first species in $c_i c_j$ is reacting with the hydroxyl group and the second species is reacting with the carboxyl group. The total concentration is abbreviated as

$$c = c_1 + c_2 + \ldots = \sum_{i=1}^{i=\infty} c_i.$$

Polymers with a degree of condensation i disappear exclusively from the i-th row and from the i-th column. At the diagonal element $c_i c_i$, two monomers are disappearing in an elementary step. It is sufficient to consider the respective row and column crossing at the diagonal element.

A polymer with a certain degree of condensation is formed from the elements in the parallels to the minor diagonal, for example, c_4 from $c_1 c_3$, $c_2 c_2$, $c_3 c_1$. Thus, for the condensation reaction the kinetic scheme is set up as

$$\frac{dc_1}{dt} = -k_p(c_1 c + c c_1)$$

$$\frac{dc_2}{dt} = -k_p(c_2 c + c c_2) + k_p(c_1 c_1)$$

$$\frac{dc_3}{dt} = -k_p(c_3 c + c c_3) + k_p(c_1 c_2 + c_2 c_1) \qquad .$$

$$\ldots\ldots\ldots\ldots\ldots\ldots\ldots\ldots\ldots\ldots\ldots \ldots$$

$$\frac{dc_i}{dt} = -k_p(c_i c + c c_i) + k_p \sum_{j=1}^{j=i-1} c_j c_{i-j}$$

Note that the sum in the second right-hand term is just the definition of the *Cauchy* formula for series multiplication.

17.2.1.3 Hydrolysis Reactions

On hydrolysis, a piece of length i from a polymer of length $k > i$ can be formed in two ways, for example, from a polymer of length 4 the following fragments can be formed by the scission of the chain:

$$c_4 \to c_1 + c_3$$
$$c_4 \to c_2 + c_2 \qquad .$$
$$c_4 \to c_3 + c_1$$

The hydrolysis reaction can occur at each ester linkage with equal probability. A molecule with a degree of condensation i has $i - 1$ ester linkages.

Monomers will be formed from dimers, tetramers, ... and dimers will be formed from trimers, tetramers, ..., etc. But there are only two ways of formation, namely formation from either ends of the chain. But the polymers that disappear in the course of hydrolysis are attacked according to their number of ester linkages in the molecule. So a higher condensed polymer will hydrolyze more readily. These considerations result in the following kinetic equations for the formation of the hydrolysis products:

$$\frac{dc_1}{dt} = 2k_d c_{H_2O}(c_2 + c_3 + c_4 + \ldots)$$

$$\frac{dc_2}{dt} = 2k_d c_{H_2O}(c_3 + c_4 + \ldots) - k_d c_{H_2O} c_2$$

$$\frac{dc_3}{dt} = 2k_d c_{H_2O}(c_4 + \ldots) - 2k_d c_{H_2O} c_3$$

$$\ldots = 2k_d c_{H_2O}(+ \ldots) - \ldots$$

and in general,

$$\frac{dc_i}{dt} = 2k_d c_{H_2O}(c_{i+1} + c_{i+2} + \ldots) - (i-1)k_d c_{H_2O} c_i.$$

The first term on the right-hand side describes the rate of formation of hydrolyzed products, whereas the second term describes the rate of disappearance due to hydrolysis.

17.2.1.4 Combined Equations

The combined kinetic equation for both condensation and hydrolysis can be obtained adding the two rate equations,

$$\frac{dc_1}{dt} = -2k_p c_1 c + 2k_d c_{H_2O}(c_2 + c_3 + c_4 + \ldots)$$

$$\frac{dc_2}{dt} = -2k_p c_2 c + k_p c_1 c_1 + 2k_d c_{H_2O}(+c_3 + c_4 + \ldots) - k_d c_{H_2O} c_2$$

$$\ldots = \ldots \qquad \qquad (17.6)$$

$$\frac{dc_i}{dt} = -2k_p c c_i + k_p \sum_{j=1}^{j=i-1} c_j c_{i-j} + 2k_d c_{H_2O}(+c_{i+1} + c_{i+2} + \ldots)$$

$$\qquad - (i-1)k_d c_{H_2O} c_i$$

17.2.1.5 Solution with Generating Functions

A suitable generating function for the set of equations in Eq. (17.6) is defined in the following way:

$$c = c_1 + c_2 + \ldots = \sum_{i=1}^{i=\infty} c_i, \qquad (17.7)$$

$$F(s, t) = c_1(t)s + c_2(t)s^2 + \ldots = \sum_{i=1}^{i=\infty} c_i(t)s^i,$$

$$F(1, t) = c(t).$$

The series of equations Eq. (17.7) are multiplied with the dummy s, s^2, s^3 and are added. Note that

$$\sum_{i=1}^{i=\infty} \sum_{j=1}^{j=i-1} c_j c_{i-j} s^i = \sum_{i=1}^{i=\infty} \sum_{j=1}^{j=i-1} c_j s^j c_{i-j} s^{i-j} = F^2(s, t)$$

and

$$c_i \left(s + s^2 + \ldots + s^{i-1} \right) = c_i \frac{s - s^i}{1 - s}.$$

This results in a differential equation for the generating function

$$\frac{dF(s, t)}{dt} = -2k_p c F(s, t) + k_p F^2(s, t)$$

$$+ 2k_d c_{H_2O} \left[\frac{(c - c_1)s}{1 - s} - \frac{F(s, t) - c_1 s}{1 - s} \right] - k_d c_{H_2O} s^2 \frac{d \frac{F(s,t)}{s}}{ds}$$

or

$$\frac{dF(s, t)}{dt} = -2k_p c F(s, t) + k_p F^2(s, t) +$$

$$+ 2k_d c_{H_2O} \left[\frac{cs}{1 - s} - \frac{F(s, t)}{1 - s} \right] - k_d c_{H_2O} s^2 \frac{d \frac{F(s,t)}{s}}{ds}$$

17.2.1.6 Stationary State

There are many special cases of interest. The most simple case is the stationary state. In the stationary state, the rate of hydrolysis reactions and the rate of condensation reactions are the same. Therefore, $dF(s, t)/dt = 0$. The time dependency is suppressed in the arguments in this section.

The following solution for $F(s)$ is obtained:

$$F(s) = \frac{cs}{1 + ck - cks}; \quad k = \frac{k_p c}{k_p c_{H_2O}}.$$

The generating function indicates a distribution of the *Schulz*[5] – Flory type. In the equation are four terms. These terms have a very special meaning. The first two terms are associated with the condensation mechanism (a formation mechanism: Φ) and the second two terms are associated with the hydrolysis mechanism (a degradation mechanism: Δ). Both mechanisms remove and give some species to the pot. These mechanisms are further denoted with $(+)$ and $(-)$, respectively. We can split the terms as

$$\Phi(-) = -2k_p c F(s)$$
$$\Phi(+) = +k_p F^2(s)$$
$$\Delta(+) = +2k_d c_{H_2O} \left[\frac{cs}{1-s} - \frac{F(s)}{1-s} \right] \tag{17.8}$$
$$\Delta(-) = -k_d c_{H_2O} s^2 \frac{d\frac{F(s)}{s}}{ds}$$

In the stationary state the rate at which one mechanism is giving species to the system must be equal to the rate that the other mechanism is consuming the species. Therefore, in the stationary state we obtain

$$0 = \Phi(-) + \Delta(+) = -2k_p c F(s) + 2k_d c_{H_2O} \left[\frac{cs}{1-s} - \frac{F(s)}{1-s} \right]$$
$$0 = \Phi(+) + \Delta(-) = +k_p F^2(s) - k_d c_{H_2O} s^2 \frac{d\frac{F(s)}{s}}{ds} .$$

This means that in equilibrium the equation will split up into two subsidiary equations, each of them being zero. For this reason, for equilibrium treatment there is no need to set up the whole generating function, but only the expressions for the two corresponding opposing (sub-) mechanisms of reaction.

Note that in the example presented here, one subsidiary equation is an ordinary equation, whereas the other subsidiary equation is a differential equation.

17.2.1.7 Instationary State

A lot of papers in this field are dealing with probabilities to find a bond reacted, and so on. In fact, these studies often describe a state of equilibrium. Next, the instationary state is analyzed in the absence of hydrolysis reactions. The equation for the generating function reduces now to

---

[5] Günter Victor Schulz, born Oct. 4, 1905, in Łodź, (now) Russia, died 1999.

$$\frac{\mathrm{d}F(s,t)}{\mathrm{d}t} = -2k_p F(1,t) F(s,t) + k_p F^2(s,t)$$

and has the solution

$$F(s,t) = \frac{c_{1,0}(s-1)}{1 - c_{1,0}k_p t(s-1)} + \frac{c_{1,0}}{1 + c_{1,0}k_p t}.$$

The solution meets the conditions given in Eq. (17.9):

$$\left.\frac{\mathrm{d}F}{\mathrm{d}s}\right|_{s=1} = c_{1,0}$$

$$F(1,t) = \frac{c_{1,0}}{1 + k_p c_{1,0} t} \qquad\qquad (17.9)$$

$$F(s,0) = c_{1,0}s$$

In words, these conditions mean

1. the number of monomer units are constant, because $\mathrm{d}F/\mathrm{d}s$ at $s = 1$ counts the monomer units;
2. the reaction follows a second-order law; and
3. initially the number of monomer units is $c_{1,0}$.

$F(1,t)$ is the total concentration in the course of the time. Forming the ratio $F(s,t)/F(1,t)$, it is seen that the solution is a time-dependent geometric distribution, the *Schulz – Flory* distribution

$$\frac{F(s,t)}{F(1,t)} = \frac{s}{1 + c_{1,0}kt(1-s)}.$$

Averages of Degree of Polymerization

Going for a moment from the ordinary power generating function to the moment generating function

$$G(x,t) = F(\exp(x),t),$$

we obtain readily the number average, the weight average, and the polydispersity

$$\frac{G_x(0,t)}{G(0,t)} = P_n \qquad = 1 + c_{1,0}kt$$

$$\frac{G_{xx}(0,t)}{G_x(0,t)} = P_w \qquad = 1 + 2c_{1,0}kt \qquad .$$

$$D = \frac{P_w}{P_n} \qquad = \frac{1 + 2c_{1,0}kt}{1 + c_{1,0}kt}$$

We use subscripts to indicate the order of differentiation. It is seen that the averages are increasing continuously with time starting from 1. The polydispersity D starts with 1 and reaches a limit of 2.

17.2.2 Direct Substitution into the Kinetic Equations

From the definition of the averages of the degree of polymerization it is obvious that in certain cases it is possible to calculate some of them by making use of the kinetic laws directly. There is no need to have the generating function solved as such for equilibrium.

For example, the second subsidiary equation given already in Eq. (17.8)

$$0 = +k_p F^2 (s, t) - k_d c_{H_2O} s^2 \frac{d \frac{F(s,t)}{s}}{ds}$$

provides a relation in between derivative and original function. Therefore, this equation can be resolved to obtain the number average. After forming the derivative, the second moment and thus the weight average is obtained. Note that for the ordinary power series the $x \frac{d}{dx}$ operator must be used instead of the simple derivative in the exponential case.

But by changing from s to $\exp(x)$ and rewriting the differential, the following elegant relation follows:

$$F' = K F^2 + F.$$

The averages are immediately obtained

$$P_n = \frac{F'(0)}{F(0)} = 1 + K F(0) = \frac{1}{1 - p}$$

$$P_w = \frac{F''(0)}{F'(0)} = 1 + 2K F(0) = \frac{1 + p}{1 - p}$$

$$P_w \approx 2P_n; \qquad p \to 1$$

The equilibrium constant is $K = k_p/k_d c_{H_2O}$ and p is the conversion. Here we have made use of the fact that in equilibrium the rate of formation equals the rate of hydrolysis: $v_p = k_p c^2 = v_d = k_p c_{H_2O} c_E$.

17.2.2.1 Ester Hydrolysis

As a further example, consider the hydrolysis of a polyester. The loss of ester groups with time is assumed as

$$-\frac{d[E]}{dt} = k[E]. \tag{17.10}$$

The ester groups can be expressed in terms of the generating function as $[E] = s F' - F$ at $s = 1$. The number average is $P_n(t) = \frac{F'(1,t)}{F(1,t)}$. Inserting into Eq. (17.10) results in

$$-\frac{d\left(s F' - F\right)}{dt} = k\left(s F' - F\right) - \dot{F}' + \dot{F} = +\dot{F} = k\left(F' - F\right)$$

and therefore,

$$\frac{d P_n}{dt} = \frac{d \frac{F'}{F}}{dt} = \frac{\dot{F}'}{F} - \frac{F'\dot{F}}{F^2} = -\frac{F'\dot{F}}{F^2} = -\frac{F'k\left(F' - F\right)}{F^2} = -k P_n\left(P_n - 1\right).$$

Thus, a kinetic relation for the number average degree of polymerization is established. There is no assumption upon the structure of the polycondensate.

There arises the question when such substitutions and such operations are allowed. Consider, for example, the condensation reaction for hydroxyacetic acid. Here every molecule contains exactly one hydroxyl group. The generating function for the hydroxyl groups is

$$\frac{d F(s, t)}{dt} = -2k_p F(1, t) F(s, t) + k_p F^2(s, t). \tag{17.11}$$

Therefore, the concentration of the hydroxyl groups is $[OH] = F(1, t)$. If the generating function is equated at $s = 1$, then Eq. (17.11) reduces to

$$\frac{d F(1, t)}{dt} = -k_p F^2(1, t).$$

Of course, the expression gives exactly the disappearance of the hydroxyl groups with time, but the expression may not be manipulated in certain aspects.

On the other hand, if the generating function is known at any argument then it is in order to insert the generating function for special arguments such as 1 into the kinetic equations. The reverse way, say to find a solution for $\frac{d F(1,s)}{dt} = -k_p F^2(1, s)$ and treat this as the general generating function, may not be correct in general.

17.2.2.2 Branched Polyethers

We discuss now the case of branched polyethers such as resulting from trimethylol-propane with a functionality of 3, but in general f. There is an easy way to verify the relation between

1. the number of ether groups (e) and the degree of polymerization (P) and
2. the hydroxyl groups (h) and the degree of polymerization (P).

We assume that no cyclic macromolecules are formed

$$e(P) = P - 1$$
$$h(P) = fP - 2e(P) = (f - 2)P + 2.$$

We cross now to the moment generating function,

$$F(x) = c_1 \exp(x) + c_2 \exp(2x) + \ldots .$$

If any c_P should be replaced by Pc_P then make the derivative of the generating function. This is the rule that is used now. The generating functions for the ether groups and for the hydroxyl groups are

$$E(x) = F'(x) - F(x)$$
$$H(x) = (f - 2) F'(x) + 2F(x) \quad .$$

The conversion p is based on the maximal number of hydroxyl groups available

$$p = 1 - \frac{H(0)}{H_0(0)}.$$

$H_0(0)$ is the total number of hydroxyl groups at the start of the reaction. A P-*mer* would have fP hydroxyl groups, if completely hydrolyzed into the monomers,

$$h_0 = fP; \quad H_0(x) = fF'(x).$$

From relation of H and F, the number average is obtained as

$$P_n = \frac{F'(0)}{F(0)} = \frac{1}{1 - \frac{f}{2}p}.$$

The kinetic equation reads as

$$-\frac{d[OH]}{dt} = k_p[OH][OH] - k_d[H_2O][-O-] \tag{17.12}$$

and for the stationary state, the kinetic equation becomes

$$0 = k_p((f - 2)F' + 2F)^2 - k_d[H_2O](F' - F).$$

The equilibrium constant becomes

$$K = \frac{k_p}{k_d[H_2O]}.$$

Essentially, the equation above originates from a consideration at $x = 0$ and is plausible for this special case. But we ignore this fact, manipulate the equation at arbitrary x and later, again we are coming back to the case for $x = 0$.

The weight average can be found by forming the derivative of the kinetic equation. Then, substituting into this equation one by one the following terms

$$F'' = P_w F'; \quad K = \frac{F' - F}{((f - 2)F' + 2F)^2}; \quad F' = \frac{F}{1 - \frac{f}{2}p}$$

and resolving to the weight average P_w, finally we arrive at

$$P_w = \frac{1 + p}{1 - (f - 1)p}.$$

This is exactly the equation given already by *Flory* and *Stockmayer* [15, p. 373].

17.2.3 Cascade Theory and the Equivalency to Dynamic Equilibrium

There is an equivalency between the approach of cascade theory and the corresponding kinetic equation.

17.2.3.1 Dendrimers

First, recall that for building up linear chains in radical polymerization the recursive substitution has been used (cf. Eq. (17.1)). Dendrimers can be built up in a very similar way using

$$\begin{aligned} G(x) &= s \exp(x) r^f \\ r &= m \exp(x) r^{f-1} + e \end{aligned} \qquad . \qquad (17.13)$$

Here already the dummy x is inserted. Combination of the two equations (17.13) will result in

$$r^2 - e \left(r + \frac{m}{es} G(x) \right) = 0. \qquad (17.14)$$

Successive substitution of r yields a tree, as shown subsequently in Eq. (17.15).

$$
s \left\{ \begin{array}{c} r \\ r \\ \cdots \\ r \end{array} \right\} (: \times f)
$$

$$
s \left\{ \begin{array}{c} e \\ e \\ \cdots \\ e \end{array} \right\} (: \times f) + s \left\{ \begin{array}{c} m \left\{ \begin{array}{c} r \\ r \\ \cdots \\ r \end{array} \right\} (: \times f - 1) \\ e \\ \cdots \\ e \end{array} \right\} (: \times f) + \cdots \tag{17.15}
$$

$$\cdots\cdots\cdots\cdots\cdots\cdots\cdots\cdots\cdots$$

In Eq. (17.15) (: $\times f$) means that the column consists of f elements. The structure formed here in mathematics is addressed as a graph-like tree. In Eq. (17.14) we assume that the end group does not contribute to a monomeric unit. We will illustrate Eq. (17.15) in a more chemical way. Figure 17.2 shows the beginning etherification of glycerol. In glycerol there are two different hydroxyl groups, i.e., a primary hydroxyl group and a secondary hydroxyl group. We do not distinguish here between these two groups. Instead we are treating them as equal. In fact this is allowed, if the reactivity of primary hydroxyl groups is equal to the reactivity of secondary hydroxyl groups. On the other hand, from the view of the detailed structure the links of a primary hydroxyl group and a secondary hydroxyl group can be distinguished. If the compound is fully symmetrical, then there is no difference in the structure. This is true for trimethylolpropane, which is a symmetrical trifunctional alcohol, cf. Fig. 17.3.

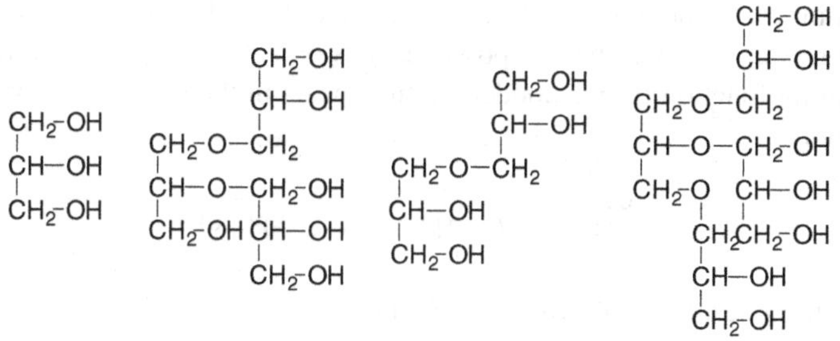

Fig. 17.2 Etherification of glycerol

In fact, the polycondensation reaction does not proceed like a polymerization reaction. There is no active species, like the growing radical or the growing ion, or a

Fig. 17.3 Trimethylolpropane

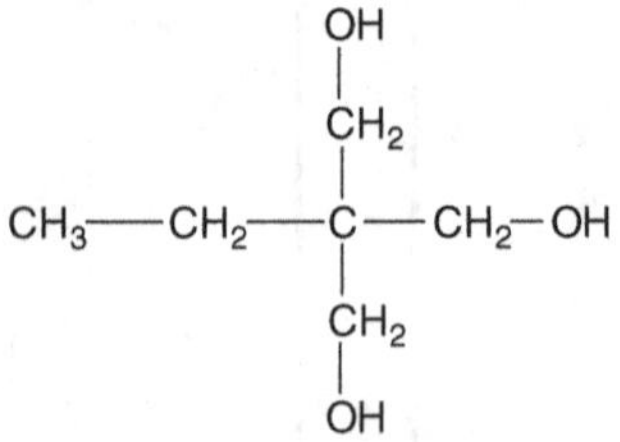

$$
\begin{array}{c}
\text{OH} \\
| \\
\text{CH}_2 \\
| \\
\text{CH}_3\!-\!\!-\text{CH}_2\!-\!\!-\text{C}\!-\!\text{CH}_2\!-\text{OH} \\
| \\
\text{CH}_2 \\
| \\
\text{OH}
\end{array}
$$

Ziegler[6] – *Natta*[7] complex as active site. The formation of the polycondensate does not proceed in one stroke. Rather each potential functionality can start to be active at a certain time and react with another monomer or likewise with an already formed polymer to form a polymer with a still higher degree of polymerization. After such a reaction step, the functionalities may be dormant for a certain time and after the dormant period again a reaction step will occur. For this reason, polycondensation is also addressed as step growth polymerization.

17.2.3.2 Kinetic Equation

If Eq. (17.12) is slightly modified using

$$
u = -\frac{2F - F'(2 - f)}{f},
$$

the following equation is obtained:

$$
u^2 - \frac{1}{2Kf}\left(u + F'\right) = 0.
$$

Note the striking similarity with Eq. (17.14). It turns out that $F'(x)$ corresponds to $G(x)$. Thus, in polymerization a definite root (s) is established, while in polycondensation every monomer unit can act as root to build up the tree. Now a tree with n monomer units will have n possible roots. For this reason, the derivative of the generating function of polycondensation appears in the expression for a single rooted chain

$$
G(x) = \frac{dF(x)}{dx} = c_1 \exp(x) + 2c_2 \exp(2x) + \ldots .
$$

Resuming these results it follows that there is a

---

[6] Karl Ziegler, born Nov. 16, 1898, in Helsa, Germany, died Aug. 11, 1973, in Mülheim an der Ruhr, Germany.

[7] Giulio Natta, born Feb. 26, 1903, in Imperia, Italy, died May 2, 1979, in Bergamo, Italy.

1. certain calculus to build up a structure,
2. the structure itself that can be visualized, and
3. an associated differential equation to this particular structure.

17.2.3.3 Solving the Nonlinear Differential Equations

The second equation in Eq. (17.13)

$$r = m \exp(x) r^{f-1} + e$$

is a functional equation that can be solved term by term with the *Lagrange* inversion procedure [14]. So there is a method emerging to solve nonlinear differential equations of a certain type. A special example above has been worked out in the literature [4].

Lagrange Inversion

We show here briefly the Lagrange inversion procedure. Given is a functional equation

$$u = t\Phi(u).$$

Here u is a function of t. To get the power series in t of u and more general the power series in t of a function of u, namely $F(u)$, the *Lagrange* inversion formula may be applied:

$$[t^n]\{F(u)\} = \frac{1}{n}[u^{n-1}]\{F'(u)\Phi(u)^n\}. \tag{17.16}$$

The symbol $[t^n]$ refers to the coefficient at the nth power of t in the series expansion of $\{\bullet\}$. In words, Eq. (17.16) may be formulated as the coefficient of t^n of $F(u)$ is the same as the coefficient of u^{n-1} of $\frac{F'(u)\Phi(u)^n}{n}$. By substituting $\exp(x) \to t; r - e \to u$, we can bring the second equation in Eq. (17.13) in the form

$$u = mt(u + e)^{f-1}. \tag{17.17}$$

Most desirable, we want to calculate r^f. Inspection of Eq. (17.17) and comparing with (17.16) results in

$$F(u) = (u + e)^f$$
$$\Phi(u) = m(u + e)^{f-1}$$

Thus, we arrive at

$$[t^n]\{F(u)\} = \frac{1}{n}[u^{n-1}]\{fm^n(u + e)^{(f-1)(n+1)}\}.$$

Substituting for a moment $(f - 1)(n + 1) \rightarrow k$ we have

$$(u + e)^k = \sum_j \binom{k}{j} u^j e^{(k-j)}.$$

Therefore, the term of u with the power of $n - 1$ is obtained to be

$$\binom{(f - 1)(n + 1)}{n - 1} u^{n-1} e^{n(l-2)+l}.$$

From this follows

$$[t^n]\{r^f\} = \binom{(f - 1)(n + 1)}{n - 1} \frac{e^{n(l-2)+l}}{n}.$$

17.2.3.4 Derivatives with Respect to the Functional Group

A simple example may be in order to explain how the algorithm works. A set of molecules may consist of a certain portion of trimethylolpropane and hydroxyacetic acid, as given in Fig. 17.4. Forming the derivative with respect to the hydroxyl group (F_{OH}), the four possibilities where a hydroxyl group in the whole set can be removed are obtained. The same is true mutatis mutandis for the derivative with respect to the carboxyl group. The generating function of the initial composition can be defined as

$$F(r_1, r_2, oh, cooh, t) = c_{1,0,3,0}(t) r_1 (oh)^3 + c_{0,1,1,1}(t) r_2 (oh)(cooh)$$
$$c(t) = c_{r1,r2,oh,cooh}(t); t = 0$$

r_1, r_2 are the monomer units apart from the functional groups. Subscripts oh and $cooh$ are single symbols. The indexes of the coefficient c that means the concentration are the numbers of $R_1, R_2, OH, COOH$ groups in the compound that are also attached at the exponent of the dummies $r_1, r_2, oh, cooh$. The coefficient c is further dependent on the time of reaction, $c = c(t)$.

It can be seen that the derivative with respect to the hydroxyl groups removes just one hydroxyl group from the compound and takes care that every possibility to remove just one hydroxyl group in the molecule is taken into account, cf. Fig. 17.4. Thus, because trimethylolpropane contains three hydroxyl groups on forming the derivative, the exponent 3 will come down. In other words, the derivative generates the reactive species in the correct portion.

Condensation means that every reactive species reacts with the appropriate reactant, so the hydroxyl group reacts with the carboxyl group. Within a sufficiently small time step, only a single reaction step will occur at each species. This means mathematically that the two derivatives are to be multiplied to obtain the product that will be formed. Therefore, the products formed by condensation (and only due to condensation) are

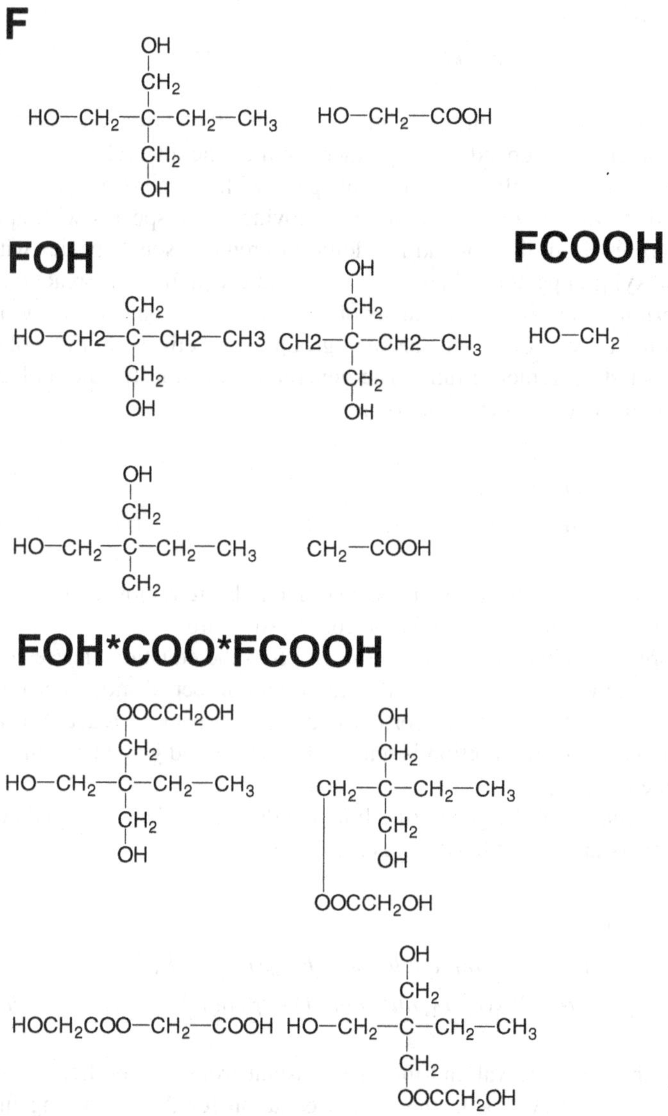

F

$$OH$$
$$|$$
$$CH_2$$
$$|$$
$$HO-CH_2-C-CH_2-CH_3 \quad HO-CH_2-COOH$$
$$|$$
$$CH_2$$
$$|$$
$$OH$$

FOH   **FCOOH**

$$OH$$
$$|$$
$$CH_2 \qquad\qquad CH_2$$
$$|\qquad\qquad\qquad |$$
$$HO-CH2-C-CH2-CH_3 \quad CH2-C-CH2-CH_3 \qquad HO-CH_2$$
$$|\qquad\qquad\qquad |$$
$$CH_2 \qquad\qquad CH_2$$
$$|\qquad\qquad\qquad |$$
$$OH \qquad\qquad\quad OH$$

$$OH$$
$$|$$
$$CH_2$$
$$|$$
$$HO-CH_2-C-CH_2-CH_3 \qquad CH_2-COOH$$
$$|$$
$$CH_2$$

FOH*COO*FCOOH

$$OOCCH_2OH \qquad\qquad OH$$
$$|\qquad\qquad\qquad\qquad |$$
$$CH_2 \qquad\qquad\qquad CH_2$$
$$|\qquad\qquad\qquad\qquad |$$
$$HO-CH_2-C-CH_2-CH_3 \quad CH_2-C-CH_2-CH_3$$
$$|\qquad\qquad\qquad\qquad |$$
$$CH_2 \qquad\qquad\qquad CH_2$$
$$|\qquad\qquad\qquad\qquad |$$
$$OH \qquad\qquad\qquad OH$$
$$|$$
$$OOCCH_2OH$$

$$OH$$
$$|$$
$$CH_2$$
$$|$$
$$HOCH_2COO-CH_2-COOH \quad HO-CH_2-C-CH_2-CH_3$$
$$|$$
$$CH_2$$
$$|$$
$$OOCCH_2OH$$

Fig. 17.4 A set of molecules consisting of trimethylolpropane and hydroxyacetic acid (F). Derivatives with respect to the hydroxyl group (F_{OH}) and carboxyl group (F_{COOH}) and the product of these derivatives together with (COO). The derivative is indicated as subscript as usual for partial derivatives

$$F(r_1, r_2, oh, cooh, t + \Delta t) =$$
$$\gamma F_{oh}(r_1, r_2, oh, cooh, t)\, coo\, F_{cooh}(r_1, r_2, oh, cooh, t).$$

Here the function is multiplied with coo to retain the correct chemical structure. γ is a coefficient that depends among others on the time interval.

But this is not the only event what happens with the generating function. The whole condensation reaction consists in removing some species with appropriate hydroxyl groups from the pot and in addition to remove some species with appropriate carboxyl groups, then allowing to react and giving back the reaction product.

Appropriate means that the numbers of compounds are removed according to the number of hydroxyl groups and carboxyl groups, respectively, in the molecule and in the correct stoichiometric ratio. The generating function for the molecules that disappear in the course of the reaction is

$$F(r_1, r_2, oh, cooh, t) =$$
$$- \alpha oh\, F_{oh}(r_1, r_2, oh, cooh, t) - \beta cooh\, F_{cooh}(r_1, r_2, oh, cooh, t).$$

The minus sign now indicates that some compounds are removed and the multiplication with oh and $cooh$ is to retain the correct structure.

There are two unknown coefficients α and β in the equation. These coefficients must be determined in this way so that the total number of monomer units in the portion that is removed and the portion that is given back in reacted form should not change. Note that the portion is removed at time t and given back in condensed form at time $t + \Delta t$.

The total change of the generating function due to condensation and conversion of the reactants due to condensation is therefore

$$\Delta F(r_1, r_2, oh, cooh, t) =$$
$$\gamma F_{oh}(r_1, r_2, oh, cooh, t) coo\, F_{cooh}(r_1, r_2, oh, cooh, t)$$
$$- \alpha oh\, F_{oh}(r_1, r_2, oh, cooh, t) - \beta cooh\, F_{cooh}(r_1, r_2, oh, cooh, t).$$

Extracting the time interval and the rate constant from the coefficients and going to the time limit of $\Delta t \to 0$, the kinetic equation for the generating function is obtained

$$\dot{F}(r_1, r_2, oh, cooh, t) = \frac{dF(r_1, r_2, oh, cooh, t)}{dt} =$$
$$k\gamma\prime F_{oh}(r_1, r_2, oh, cooh, t) coo\, F_{cooh}(r_1, r_2, oh, cooh, t)$$
$$- k\alpha\prime k oh\, F_{oh}(r_1, r_2, oh, cooh, t) - k\beta\prime cooh\, F_{cooh}(r_1, r_2, oh, cooh, t).$$

The remaining stoichiometric coefficients are determined using the fact that the monomer units must be constant or should not change with time. Therefore,

$$\dot{F}_{r_1}(1, 1, 1, 1, t) = 0$$
$$\dot{F}_{r_2}(1, 1, 1, 1, t) = 0$$

Since the equation is homogeneous in the coefficients, we can arbitrarily set one coefficient to 1, say $\gamma' = 1$. In this way, we obtain

$$F_{oh,r_1}(\mathbf{1}, t)\, F_{cooh}(\mathbf{1}, t) +$$
$$F_{oh}(\mathbf{1}, t)\, F_{cooh,r_1}(\mathbf{1}, t) - \alpha'\, F_{oh,r_1}(\mathbf{1}, t) - \beta'\, F_{cooh,r_1}(\mathbf{1}, t) = 0.$$

Here, $\mathbf{1} = 1, 1, 1, 1$. The relations may become simpler for special cases.

17.3 Degradation

17.3.1 Olefins in Thermodynamic Degradation Equilibrium

Consider a homologous series of α-olefins

$$CH_2 = CH_2$$
$$CH_3 - CH = CH_2$$
$$CH_3 - CH_2 - CH = CH_2$$
$$CH_3 - CH_2 - CH_2 - CH = CH_2$$
$$\cdots \cdots \cdots \cdots \quad \cdots \cdots$$

There should be a formal equilibrium by disproportionation which reads as

$$-CH_2 - CH_2 - CH_2 - CH_2- \rightleftharpoons$$
$$-CH_2 = CH_2 + CH_3 - CH_2-$$
$$\cdots \cdots \cdots \rightleftharpoons \cdots \cdots \cdots$$
$$CH_2 = CH_2 + CH_2 = CH_2 \rightleftharpoons$$
$$CH_3 - CH_2 - CH = CH_2$$
$$CH_2 = CH_2 + CH_3 - CH = CH_2 \rightleftharpoons$$
$$CH_3 - CH_2 - CH_2 - CH = CH_2 \qquad . \qquad (17.18)$$
$$CH_2 = CH_2 + CH_3 - CH_2 - CH = CH_2 \rightleftharpoons$$
$$CH_3 - CH_2 - CH_2 - CH_2 - CH = CH_2$$
$$\cdots \cdots \cdots \rightleftharpoons \cdots \cdots \cdots$$
$$CH_3 - CH = CH_2 + CH_3 - CH = CH_2 \rightleftharpoons$$
$$CH_3 - CH_2 - CH_2 - CH = CH_2$$
$$\cdots \cdots \cdots \rightleftharpoons \cdots \cdots \cdots$$

17.3.1.1 Formal Equations

Equations (17.18) can be rewritten in short symbolic form as

$$O(2) + O(2) \rightleftharpoons O(4)$$
$$O(2) + O(3) \rightleftharpoons O(5)$$
$$O(2) + O(4) \rightleftharpoons O(6)$$
$$\ldots \rightleftharpoons \ldots$$

etc., where $O(n)$ stands for an α-olefin with n carbon atoms.

17.3.2 Kinetic Equation

It is assumed that each functionality is equally reactive. The reaction can be split into a degradation term and a recombination term. Subsequently, $[V]$ denotes the overall concentration of vinyl groups, one in each chain and $[DL]$ denotes the concentration of degradable links. For example, in a chain of four carbon atoms, there is one degradable link, in a chain of length of five there are two degradable links. The recombination reaction, which is forward reaction in the scheme, will occur by the reaction of a methyl end group M with a vinyl group of another chain.

Notably, $[M] = [V]$, because in each chain is one methyl group and one vinyl group. By each recombination step, one vinyl group and one methyl group will disappear. The rate of disappearance of vinyl groups is

$$-\frac{d[V]}{dt} = k_r[M][V] = k_r[V]^2.$$

The degradation reaction (backward reaction) reads as

$$+\frac{d[V]}{dt} = k_d[DL].$$

Here by an elementary step, one vinyl group will disappear and two vinyl groups will be formed. Combination of both half reactions will result in the overall reaction

$$-\frac{d[V]}{dt} = k_r[V]^2 - k_d[DL].$$

In equilibrium, the forward reaction and the backward reaction will become equally fast. Thus, $\frac{d[V]}{dt} = 0$. For equilibrium we obtain

$$K = \frac{k_r}{k_d} = \frac{[DL]}{[V]^2}. \tag{17.19}$$

17.3.3 Generating Function

A generating function for the respective concentrations or likewise the partial pressures of the homologous olefin series can be established,

$$F(x) = c_2 x^2 + c_3 x^3 + c_4 x^4 + \dots.$$

The total concentration of vinyl groups and methyl groups is simply $F(1)$. The generating function for the degradable links DL is

$$G(x) = c_4 x^4 + 2 c_5 x^5 + 3 c_6 x^6 + \dots$$

and the generating function can be related to F(x) by the expression

$$G(x) = x^4 \frac{d}{dx} \left(\frac{F(x) - c_2 x^2 - c_3 x^3}{x^3} \right).$$

17.3.4 Differential Equation for the Generating Function

The differential equation for the generating function in equilibrium, Eq. (17.19), is obtained by inserting the generating functions for V and DL and is

$$x^4 \frac{d}{dx} \left(\frac{F(x) - c_2 x^2 - c_3 x^3}{x^3} \right) - K F(x)^2 = 0.$$

The equation is strictly a priori valid only at $x = 1$. However, we insert for general x and perform the mathematical manipulations needed. We think that if the process is valid for general x, then it should hold also for a special value of x:

$$F(x) = \frac{c_2 x^2}{1 - \sqrt{c_2 K x}}.$$

The equation can be normalized, because $F(1)$ will refer to the sum of the partial pressures. The equilibrium constant is dependent on the temperature.

17.3.5 Thermodynamic Approach

We denote $\Delta G(i)$ as the free enthalpy of formation of a species of a degree of polymerization of i. Equations (17.18) can be rewritten in the language of thermodynamics as

$$\Delta G(2) + \Delta G(2) = \Delta G(4)$$
$$\Delta G(2) + \Delta G(3) = \Delta G(5)$$
$$\ldots\ldots\ldots\ldots\ldots\ldots = \ldots\ldots$$
$$\Delta G(i) + \Delta G(j) = \Delta G(i + j)$$

17.3.6 Simple Approach

It is immediately seen that a solution of the set of equations is $\Delta G(i) = Q * i$. Further,

$$\Delta G(i) = \Delta G_0(i) + RT \ln p_i. \tag{17.20}$$

From Eq. (17.20) and from $\Delta G(i) = Q * i$, the ratios of partial pressures of various homologues can be calculated.

17.3.7 Generating Function for Free Enthalpy

The general equation for the free enthalpy of the homologue compounds is

$$G(i) + G(j) = G(i + j); \quad i, j : 1, 2, 3, \ldots.$$

Multiplying by x_i and x_j and summing over the set, we obtain

$$\sum_{j>1} \sum_{i>1} [\Delta G(i) + \Delta G(j) - \Delta G(i + j)] x^i x^j =$$

$$\sum_{j>1} x^j \sum_{i>1} [\Delta G(i)] x^i + \sum_{j>1} [\Delta G(j)] x^j \sum_{i>1} x^j - \sum_{j>1} \sum_{i>1} [\Delta G(i + j)] x^i x^j =$$

$$\frac{2x^2}{1 - x} \sum_{k>1} [\Delta G(k)] x^k - [\Delta G(4)x^4 + 2\Delta G(5)x^5 + 3\Delta G(6)x^6 + \ldots]$$

by using

$$F(x) = \Delta G(2) x^2 + \Delta G(3) x^3 + \Delta G(4) x^4 + \ldots.$$

Finally the differential equation

$$\frac{2x^2}{1 - x} F(x) - x^4 \frac{d}{dx} \left[\frac{F(x) - \Delta G(2)x^2 - \Delta G(3)x^3}{x^3} \right] =$$

$$\frac{2}{1 - x} F(x) - x^2 \frac{d}{dx} \left[\frac{F(x) - \Delta G(2)x^2 - \Delta G(3)x^3}{x^3} \right] = 0.$$

The solution of this equation is

$$F(x) = \frac{x^2 \exp(-2x)[2xC' - \exp(2x)(x - 2)\Delta G(2)]}{2(x - 1)^2}.$$

The condition $\Delta G(m) + \Delta G(n) = \Delta G(m+n)$ suggests $C_1 = 0$. From the coefficient expansion $\Delta G(3) = C' + 3\Delta G(2)/2$ means that a propene will be thermodynamically explained as formed from 1.5 ethene units in some way, which does not need to be specified in more detail. In general the partial pressure of the respective oligomer will be found from the coefficients

$$\Delta G(n) = \frac{n}{2} \Delta G(2)$$

$$\Delta G_0(n) + RT \ln \frac{p_n}{p_0} = \frac{n}{2}[\Delta G_0(2) + RT \ln \frac{p_2}{p_0}]$$

p_0 is the standard pressure and p_2, p_3, ... are the partial pressures of the α-olefins in the ideal assumed mixture. The result reflects the fact common in thermodynamics that the energy required to form a certain compound is independent on the path of reaction. Note that the form of the equation leaves freedom to the state and to irregularities of free energy of formation of an n-mer.

17.3.7.1 Interpretation of the Terms

Basically, the free enthalpy of polymerization from ethene up to n-olefin is obtained:

$$\frac{n}{2} CH_2 = CH_2 \rightarrow CH_2 = CH - (CH_2)_{n-3} - CH_3,$$

$$\Delta G_{0,P}(n) = \Delta G_0(n) - \frac{n}{2}\Delta G_0(2) = +\frac{n}{2}RT \ln \frac{p_2}{p_0} - RT \ln \frac{p_n}{p_0}.$$

If no side reactions would occur then the partial pressures can be summed up to yield the total pressure p_{tot} which will be under standard conditions p_0. This procedure may serve as normalization. The plot of the free enthalpy against n should give a straight line. From the slope the equilibrium temperature can be obtained. The total pressure must be known.

17.3.7.2 Equilibrium Temperature

From the distribution of olefins the equilibrium temperature can be equated, the free enthalpy of formation being known and other assumptions, i.e., ideal mixture, ideal gas behavior, independence of standard reaction enthalpy, standard reaction entropy of temperature, etc., being valid. The absolute pressure must be known. This is restrictive to methods of measurement of composition. Note that the application for the *Gibbs* free enthalpy G requires a free exchange of entropy into and from environment and free exchange of volume work during the process. Thus, thermal equilibrium and manal equilibrium must be established (T, p). Otherwise, for example, the usage of the enthalpy H will be suitable for an isobaric adiabatic reaction, and so on. The heats of formations of a homologous series of α-olefins are shown in Table 17.2.

Table 17.2 Heat of formation of the homologous series of α-olefins

n	$\Delta H_f(g)$ [kJ mol^{-1}][a]	$\Delta H_f(g)$ [kJ mol^{-1}][b]
2	46	52.3
3	15	20.4
4	−7	1.17
5	−29	−20.9
6	−51	−41.7
7	−73	−62.3
8	−95	−82.9
9	−117	
10	−139	−124
11	−161	−145
12	−183	−165
13	−205	
14	−227	
15	−245	

[a] From increments after van Krevelen [28]
[b] From D'Ans-Lax [29]

17.3.8 Plot of Amounts α-Olefins

From GC-MS data, the amounts of the series of olefins are obtained from high-temperature pyrolysis from polyethylene, by throwing PE on the surface of liquid steel. The gas chromatogram of α-olefins is shown in Fig. 17.5. Table 17.3 lists the relative heights of peaks, which correspond roughly to the weights of the homologues. In Fig. 17.6, the logarithm of the amount of olefin is plotted against the number of carbons. A straight line is expected, if $\Delta G(n) \propto n$. If chemical equilibrium is not established the plot will be expected to be distorted. Here, we did not prove that without equilibrium established, a similar curve could be obtained. So we will assume equilibrium.

In order to use these relationships as a chemical thermometer, a normalization must be established to obtain p_2. In principle, the temperature can be obtained from the equilibrium distribution of the olefins – if the system is in equilibrium at all.

17.3.8.1 OAD Equilibrium

The OAD Equilibrium means equilibrium with respect to olefin, alkane, and diene of each homologue with a number of carbon atoms n. Consider the reaction

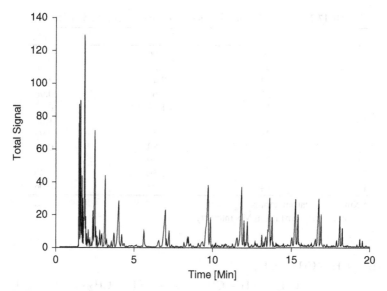

Fig. 17.5 Gas chromatogram of α-olefins

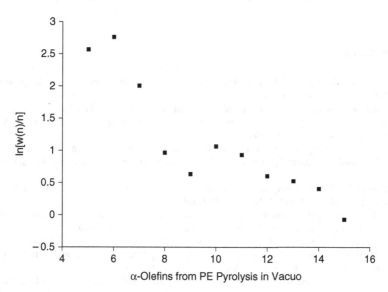

Fig. 17.6 Plot of carbon atoms, viz., relative amount of α-olefins in mol

Table 17.3 Peak heights of α-olefins related to number of carbon atoms

n^a	h^b	$h(n)/n$	$\ln(h/n)$
5	65	13.00	2.56
6	95	15.83	2.76
7	52	7.42	2.00
8	21	2.62	0.96
9	17	1.88	0.63
10	29	2.90	1.06
11	28	2.54	0.93
12	22	1.83	0.60
13	22	1.69	0.52
14	21	1.50	0.40
15	14	0.93	−0.06

[a] Number of carbon atoms
[b] Peak height [mm] From MS intensity

$$2\,CH_2 = CH - CH_2 - CH_3 \rightarrow$$
$$CH_2 = CH - CH = CH_2 + CH_3 - CH_2 - CH_2 - CH_3 \quad .$$

This means that 1-butene $O(4)$ is in equilibrium to 1,3-butadiene $D(4)$ and n-butane $A(4)$. Here O stands for olefin, D for diene, and A for alkane. Since the number of molecules does not change, the equilibrium is not dependent on total pressure at all.

References

1. Burchard, W.: Static and dynamic light scattering from branched polymers and biopolymers. Adv. Polym. Sci. **48**, 1–124 (1983)
2. Dušek, K.: Epoxy resins and composite. Adv. Polym. Sci. **78**, 1–58 (1986)
3. Dušek, K., Šomvársy, J.: Modelling of ring-free crosslinking chain (co)polymerization. Polym. Int. **44**(3), 225–236 (1997)
4. Gordon, M., Temple, W.B.: The graph-like state of matter and polymer science. In: A.T. Balaban (ed.) Chemical Applications of Graph Theory, pp. 299–332. Academic Press, London (1976)
5. Gordon, M.: Good's theory of cascade processes applied to the statistics of polymer distributions. Proc. R. Soc. Lond. A **268**, 240–259 (1962)
6. Wolf, C., Burchard, W.: Branching in free radical polymerization due to chain transfer. application to poly(vinyl acetate). Makromol. Chem. **177**, 2519–2538 (1976)
7. Scranton, A.B., Klier, J., Peppas, N.A.: Statistical analysis of free-radical copolymerization/crosslinking reactions using probability generating functions: Reaction directionality and general termination. Macromolecules **24**, 1412–1415 (1991)
8. Asteasuain, M., Sarmoria, C., Brandolin, A.: Peroxide modification of polyethylene. Prediction of molecular weight distributions by probability generating functions. Polymer **43**(8), 2363–2373 (2002)
9. Tobita, H.: Bivariate distribution of chain length and composition in multicomponent polymerization. Polymer **39**(11), 2367–2372 (1998)
10. Whiteley, K.S., Garriga, A.: Derivation of continuous molecular weight distributions from the generating function. Comput. Theor. Polym. Sci. **11**(4), 319–324 (2001)

11. Poland, D.: Time evolution of polymer distribution functions from moment equations and maximum-entropy methods. J. Chem. Phys. **111**(17), 8214–8224 (1999)
12. Cavin, L., Rouge, A., Meyer, T., Renken, A.: Kinetic modeling of free radical polymerization of styrene initiated by the bifunctional initiator 2,5-dimethyl-2,5-bis(2-ethyl hexanoyl peroxy)hexane. Polymer **41**(11), 3925–3935 (2000)
13. Krajnc, M., Poljanšek, I., Golob, J.: Kinetic modeling of methyl methacrylate free-radical polymerization initiated by tetraphenyl biphosphine. Polymer **42**(9), 4153–4162 (2001)
14. Wilf, H.S.: Generating Functionology, 2nd edn. Academic Press, New York (1994)
15. Flory, P.J.: Principles of Polymer Chemistry. Cornell University Press, Ithaca, NY (1953)
16. Macosko, C.W., Miller, D.R.: A new derivation of average molecular weights of nonlinear polymers. Macromolecules **9**, 198–206 (1976)
17. Miller, D.R., Macosko, C.W.: Average property relations for nonlinear polymerization with unequal reactivity. Macromolecules **11**, 656–662 (1976)
18. Miller, D.R., Macosko, C.W.: A new derivation of postgel properties of network polymers. Macromolecules **9**, 207–211 (1976)
19. Macosko, C.W., Miller, D.R.: Calculation of average molecular properties during nonlinear, living copolymerization. Makromol. Chem. **192**, 377–404 (1991)
20. Yan, D.Y., Zhou, Z.P.: Theoretical study on the linking process of divinyl compounds with living precursors. Macromol. Theory Simul. **6**, 1211–1235 (1997)
21. Oldham, K.B., Spanier, J.: The Fractional Calculus, Mathematics in Science and Engineering, vol. 3. Academic Press, New York (1974)
22. Prudnikov, A.P., Brychov, Y.A., Marichev, O.I.: Integrals and Series, vol. 1. Gordon and Breach Science Publishers, New York (1986)
23. Henrici-Olivé, G., Olivé, S.: Polymerization. Katalyse-Kinetik-Mechanismen. Chemische Taschenbücher. Verlag Chemie, Weinheim (1969)
24. Gordon, M., Malcolm, G.W.: Configurational statistics of polymer chains. Proc. R. Soc. Lond. A **295** (1966)
25. Stockmayer, W.H.: Distribution of chain length and compositions in copolymers. J. Chem. Phys. **13**, 199–207 (1945)
26. Tang, A., Tang, X., Sun, C., Li, L.: Polycondensation reaction of RAa + RBb type involving intramolecular cyclization. Macromolecules **22**, 3424–3428 (1989)
27. Tang, A.C., Li, Z.S., Sun, C.C., Tang, X.Y.: Crosslinking reaction of type AA involving intramolecular cyclization. J. Macromol. Sci. Chem. **A26**, 693–714 (1989)
28. van Krevelen, D.W.: Properties of Polymers. Their Correlation with Chemical Structure. Their Numerical Estimation and Prediction from Additive Group Contributions, 3rd edn., pp. 627–639. Elsevier, Amsterdam (1990)
29. D'Ans, J., Lax, E.: Taschenbuch für Chemiker und Physiker, 2. Organische Verbindungen. Springer, Berlin (1983)

Chapter 18
Chemical Kinetics

Chemical kinetics describes the change of species with time under the conditions of nonequilibrium. For qualitative statements, we require a reaction mechanism and the associated kinetic constants of the individual reactions. For special areas, like pyrolysis and flames, there are already sophisticated computer programs, which develop automatically kinetic models [1] and perform a data reduction [2]. Also in the field of geochemistry, such programs are available [3].

A large number of kinetic data are well known and compiled in individual tables, like *Landolt Börnstein* [4], and further in the review literature [5]. Also machine-readable databases were arranged, including the NIST database for chemical kinetics [6]. Version 7.0 contains over 38,000 records on over 11,700 reactant pairs. It also includes reactions of excited states of atoms and molecules [7]. Kinetic calculations are substantially more complex in comparison to the field of equilibrium thermodynamics.

18.1 Formal Kinetics

Often, the laws of chemical kinetics, expressed in terms of differential equations, can be considerably simplified by the introduction of dimensionless variables. We show examples of dimensionless variables with regard to chemical kinetics in Sect. 11.3.8. Nowadays systems of differential equations can be solved by the use of computer programs, even by a simple spreadsheet calculator.

Here, we exemplify a simple consecutive reaction of first order. We modify the scheme of Eq. (11.5) in a way that we force an addition of the component A to the reactor in periodic time intervals t_0

$$-\frac{d[A]}{dt} = k_A[A] - [A_\delta] \sum_{l=0}^{l=\infty} \delta(t - lt_0)$$

$$-\frac{d[B]}{dt} = k_B[B] - \frac{d[A]}{dt}.$$

(18.1)

J.K. Fink, *Physical Chemistry in Depth*,
DOI 10.1007/978-3-642-01014-9_18, © Springer-Verlag Berlin Heidelberg 2009

Here $\delta(t)$ is the Dirac delta function. The first equation of Eq. (18.1) has on the right-hand side two terms. If $k_A = 0$, then we have for the first term in the sum of the Dirac delta functions

$$\frac{d[A]}{dt} = [A_\delta]\delta(t - t_0).$$

The Dirac delta function integrates to the unit step function at $t = t_0$. Thus, the concentration of A jumps by $[A_\delta]$ at $t = t_0$. The same holds in general for the further terms in the sum for $t = lt_0; \ l = 2, 3, \dots$.

The system of Eq. (18.1) can still be solved by analytic methods, but initially we demonstrate the use of a numerical method. To implement a numerical method, recall the Taylor expansion

$$f(t + \Delta t) = f(t) + \frac{d f(t)}{dt}\Delta t + \dots .$$

When we at first omit the sum of the Dirac delta functions, we can convert the differential equations in Eq. (18.1) in a system of difference equations, as

$$-\frac{[A](t + \Delta t) - [A](t)}{\Delta t} = k_A[A](t)$$

$$\frac{[B](t + \Delta t) - [B](t)}{\Delta t} = k_B[B](t) + k_A[A](t). \tag{18.2}$$

From this system, $[A](t + \Delta t)$ and $[B](t + \Delta t)$ can be readily calculated from terms that contain only terms at time t. We set discrete time steps in sufficient small fraction of t_0, e.g., $t_0/24$. The time step depends on the rate constants k_A and k_B. The relative change of the concentration should be in the range of a few percent in the course of a single time step. To simulate the periodic addition of A, we increase $[A]$ by $[A_\delta]$, whenever the time reaches an integer multiple of t_0. The numeric calculation can be performed with a spreadsheet calculator. An example for time dependence of the concentration is shown in Fig. 18.1.

The intermediate B starts from zero concentration and stabilizes after some time, hereby still fluctuating. If we change the dosage from 1 concentration units every 24 h to 0.5 concentration units every 12 h, the same amount is added to the system, however, in comparison the half amount twice in the original period. The simulation is shown in Fig. 18.2. As expected, the stationary value of the intermediate B remains the same; however, the fluctuation becomes less in comparison to the simulation in Fig. 18.1.

We proceed now with some attempt for an analytical solution in the stationary state. The stationary state is characterized not by the common conditions $d[A](t)/dt = 0$ and $d[B](t)/dt = 0$ but by a periodic fluctuation of the concentrations.

Assuming that the concentration $[A](t)$ is periodic in time with a period of t_0 in the stationary state, we can calculate the fluctuations in a simple way. Immediately

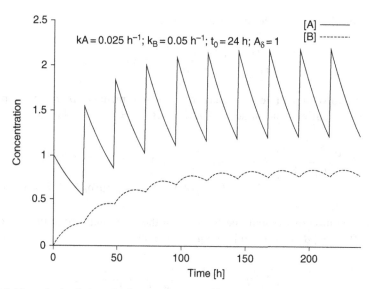

Fig. 18.1 Numeric simulation of the time dependence of Eq. (18.1), dosage 1 unit every 24 h

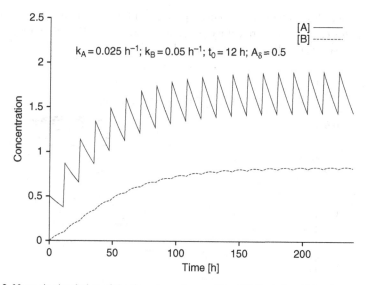

Fig. 18.2 Numeric simulation of the time dependence of Eq. (18.1), dosage 0.5 units every 12 h

after feed the concentration should be $[A](t)$. After the period t_0, there is a decrease to the value $[A](t + t_0) = [A](t) \exp(-k_A t_0)$, just before the next portion $[A]_\delta$ is added. After the addition, the concentration returns to that of $[A](t)$. We inspect now a single period in the stationary state. For convenience, we set the time at the beginning of the period as zero, i.e., $t \to 0$. In this notation, we get the ratio of lower to upper limiting value in the stationary state as

$$\frac{[A](0+t_0)}{[A](t)} = \frac{[A](0) - [A_\delta]}{[A](0)} = \exp(-k_A t_0).$$

Further, we solve the differential equation for the intermediate $[B](t)$ for one period with nonzero starting concentration of B, i.e., $[B](0) \neq 0$. The solution is

$$[B](t) = \frac{e^{-(k_A-k_B)t-k_Bt}}{k_A - k_B} \big(-[A](0)k_A + [A](0)e^{(k_A-k_B)t}k_A +$$
$$[B](0)e^{(k_A-k_B)t}k_A - [B](0)e^{(k_A-k_B)t}k_B \big). \quad (18.3)$$

The concentration of B should be the same at the beginning and at the end of the period $[B](0) = [B](t)$. From this condition, the relation

$$[B](0) = \frac{k_A[A](0)(e^{(-k_A+k_B)t_0} - 1)}{(e^{k_Bt_0} - 1)(-k_A + k_B)}$$

arises.

Finally, we show a parametric plot of $[A](t)$ and $[B](t)$ from the data plotted in Fig. 18.2 in Fig. 18.3. The trajectory starts at $[A] = 0.5$; $[B] = 0$ and moves up, until it approaches to a limit cycle, or periodic orbit.

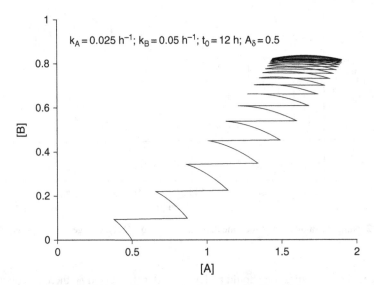

$k_A = 0.025$ h^{-1}; $k_B = 0.05$ h^{-1}; $t_0 = 12$ h; $A_\delta = 0.5$

Fig. 18.3 Parametric plot of $[B]$ and $[A]$ from the data plotted in Fig. 18.2

The example with periodic feed is relevant in chemical reactors. Further, it is the starting point for the kinetics of drug administration. However, in real live the situation is much more complicated. For example, in the drug administration for insulin the intake of carbohydrates interferes the kinetics [8].

18.2 Stochastic Nature of Chemical Kinetics

If we deal with an individual particle, then we cannot predict whether this particle will react in the next instant of time or not. In contrast, we can make statements for a large number of particles, e.g., how the concentration of a certain species in a chemical reaction changes in time.

18.2.1 Simple Reaction with First Order

A simple reaction with first order is the decomposition of a particle, e.g., azoisobutyronitril. We inspect now a certain time interval, which we divide into small time steps Δt. We observe now the molecule at the times $t, t+\Delta t, t+2\Delta t, \ldots, t+n\Delta t$. It could happen that in between the observations the molecule decomposes according to the reaction in Fig. 18.4. The probability for this decomposition is p.

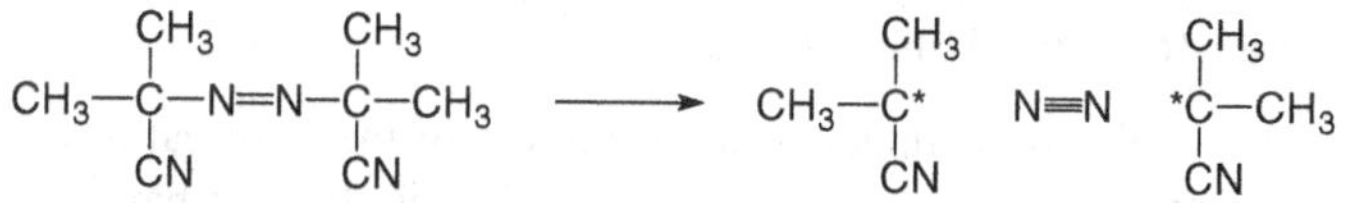

(1)   (2)

Fig. 18.4 Decomposition of azoisobutyronitril

18.2.2 Markov Chains

We can formalize the above process in a certain way.

18.2.2.1 States

We define the possible states:

1. The molecule is intact.
2. The molecule has decomposed already.

18.2.2.2 Transition Matrix

During a small time interval Δt in between the observations the state of the system can change. The molecule changes from state (1) into state (2). We define now the

Table 18.1 Transition probabilities

Transition probability	Description: molecule changes in the time interval Δt
p_{11}	Remains intact
p_{12}	Decomposes
p_{21}	Backward reaction
p_{22}	Remains decomposed

transition probabilities in Table 18.1. The transition probabilities can be summarized in a scheme, as shown in Eq. (18.4).

$$A_1 = \begin{pmatrix} p_{11} & p_{12} \\ p_{21} & p_{22} \end{pmatrix} = \begin{pmatrix} 1-p & p \\ 0 & 1. \end{pmatrix} \tag{18.4}$$

According to the lower row a backward reaction is impossible, because of $p_{21} = 0$. If the molecule has decomposed, then it remains decomposed. Further, the probabilities are not dependent on the time. No matter, whether the molecule is in the condition (1) or (2), to a further transition the same probabilities apply. This is a so-called homogeneous *Markov*[1] chain.

18.2.2.3 State After Two Time Intervals

We presume the molecule is intact at the beginning of the observations. If we find the molecule after two time intervals decomposed, then this can have happened in two kinds:

1. The molecule is already decomposed within the first time interval.
2. The molecule decomposed only within the second time interval.

We can inspect also the possibilities to find the molecule after two intact time intervals:

1. the molecule remained intact the whole time;
2. or the molecule decomposed in the meantime into state (2), therefore it has decomposed and then again regenerated.

However, for the backward reaction the probability (p_{21}) in the special example is equal to zero. In general, it could also be that the molecule changes more than twice the state within the two time intervals. Later we will make the time interval so small that no transition can escape the eyes of the observer. Therefore, we can exclude this possibility. The mathematical formulation of statements above is given

---

[1] Андреы Андреыевич Марков, born Jun. 14, 1856, in Ryazan, Russia, died Jul. 20, 1922, in Petrograd (now St. Petersburg).

in Eq. (18.5).

$$
\begin{aligned}
p_{12}(\Delta t) &= p \\
p_{11}(\Delta t) &= 1 - p \\
p_{12}(2\Delta t) &= p_{11}(\Delta t)p_{12}(\Delta t) + p_{12}(\Delta t)p_{22}(\Delta t) \\
p_{11}(2\Delta t) &= p_{11}(\Delta t)p_{11}(\Delta t) + p_{12}(\Delta t)p_{21}(\Delta t).
\end{aligned}
\tag{18.5}
$$

Here the terms are not explicit functions of the time steps n, at which we are check-ing the state of the system. However, the probabilities are a function of the length of the time interval Δt. In a similar way, we obtain the expressions for the other transition probabilities. In fact, the procedure turns out as a matrix multiplication. After two time units, we find in the general case

$$
A_2 = A_1 \times A_1 = A_1^2
$$

and for the special example

$$
A_2 = \begin{pmatrix} (1-p)^2 & 2p - p^2 \\ 0 & 1 \end{pmatrix}.
$$

18.2.2.4 The State After Multiple Time Intervals

The general result after multiple time intervals arises simply as a result of repetition of the process

$$
A_n = A_{n-1} \times A_1 = A_1^n.
$$

For the special example discussed in this section we obtain

$$
A_n = \begin{pmatrix} (1-p)^n & 1 - (1-p)^n \\ 0 & 1 \end{pmatrix}.
$$

18.2.2.5 Dependence of the Transition Probability on the Time Interval

The transition probability p depends on the length of the selected time interval. The expansion into a Taylor series results in

$$
p = p(\Delta t) = p(0) + k\Delta t + \dots
$$

If the time interval limits to zero $\Delta t \to 0$, then the transition probability should also tend to zero $p \to 0$. Thus, we presume that the chemical process is not infinitely fast. Therefore, we postulate $p(0) = 0$.

18.2.2.6 Transition Probability After Multiple Time Steps

The probability that the molecule did not decompose after n time steps yet is $p_{11}(n)\Delta t$. We set

$$p = k\Delta t; \quad t = n\Delta t \tag{18.6}$$

and get

$$p_{11}(t) = (1 + k\Delta t)^n = (1 + k\Delta t)^{-\frac{nk\Delta t}{k\Delta t}} = (1 + x)^{-\frac{kt}{x}} \tag{18.7}$$

If the limit $x \rightarrow 0$ in Eq. (18.7), then we obtain

$$p_{11}(t) = \exp(-kt).$$

$p_{11}(t)$ is the probability that a particle does not react and remains intact at least until t. This probability decreases exponentially with time. If many particles react in this way independently, then we can identify this probability with the frequency, or the fraction of molecules that are still there as such. The well-known law of a first-order reaction emerges as

$$\frac{c(t)}{c(0)} = \exp(-kt).$$

The derivation given above shows clearly that the kinetic laws are actually no deterministic laws, but have rather a probabilistic character. In the common sense, it is strongly engraved that by the development of quantum mechanics the interpretation of states as probabilities is forced. In fact, the interpretation of states emerging from nondeterministic laws is sound even in the classical statistical mechanics, in addition, like here, in chemical kinetics.

By the way, if we identify in the above derivation the time as the path of a particle, light quantum, etc., through an absorbing medium, and the rate constant as the absorption coefficient, then we obtain the law of *Lambert*[2] and *Beer*.[3]

18.2.3 Opposing Reactions of First Order

The transition probabilities of opposing reactions of first order can be summarized in a scheme given as

$$A_1 = \begin{pmatrix} p_{11} & p_{12} \\ p_{21} & p_{22} \end{pmatrix} = \begin{pmatrix} 1 - p & p \\ q & 1 - q \end{pmatrix}, \tag{18.8}$$

in that p is the probability for the forward reaction and q is the probability for the backward reaction in a certain time interval Δt. The probabilities should be

---

[2] Johann Heinrich Lambert, born Aug. 26, 1728, in Mühlhausen, Switzerland, died Sep. 25, 1777, in Berlin.

[3] August Beer, born Jul. 31, 1825, in Trier, Germany, died Nov. 18, 1863, in Bonn.

independent on the entire number of the steps n, but dependent on the time interval Δt. We substitute now $z = p + q$ and obtain

$$A_1 = \begin{pmatrix} 1 - p & p \\ z - p & 1 - (z - p) \end{pmatrix}.$$

By repeated application of the matrix multiplication we obtain for $p_{11}(n)\Delta t$

$$p_{11}(n\Delta t) = \left[(1 - z)^n - 1\right]\frac{p}{z} + 1$$

and for $p_{12}(n)\Delta t$

$$p_{12}(n\Delta t) = -\frac{p}{z}\left[(1 - z)^n - 1\right] \tag{18.9}$$

is obtained. Further, we substitute

$$(k_1 + k_2)\Delta t = -x; \quad p = -k_1\Delta t; \quad t = n\Delta t.$$

Inserting these shorthands into Eq. (18.9) results in

$$p_{11}(t) = \left[(1 + x)^{-\frac{(k_1+k_2)t}{x}} - 1\right]\frac{k_1}{k_1 + k_2} + 1.$$

Next, we go to the limit $x \to 0$ and obtain

$$p_{11}(t) = \{\exp\left[-(k_1 + k_2)t\right] - 1\}\frac{k_1}{k_1 + k_2} + 1. \tag{18.10}$$

Equation (18.10) can be transformed [9, p. 789] into

$$\frac{[A]}{[A_0]} = \frac{k_2 + k_1 \exp\left[-(k_1 + k_2)t\right]}{k_1 + k_2}. \tag{18.11}$$

In Eq. (18.11), $[A]$ refers to the concentration of the reacting species and $[A_0]$ to the initial concentration of the species at the beginning of the reaction.

18.2.3.1   Generalization of the Scheme

After the preceding comments, the generalization of the scheme should not cause problems. Consider a reaction scheme as given in Eq. (18.12).

$$(1) \rightleftharpoons (2) \rightleftharpoons (3) \rightleftharpoons (\ldots). \tag{18.12}$$

We have designated the individual compounds in place of A, B, C, ... with (1), (2), (3),.... Further, the individual components are numbered in such a way that only compounds with successive numbers can be converted into one another. The transition matrix reads thereby

$$P = \begin{pmatrix} p_{11} & p_{12} & 0 & \cdots & 0 \\ p_{21} & p_{22} & p_{23} & \cdots & 0 \\ & & \cdots\cdots\cdots & & \\ 0 & 0 & 0 & p_{n,n-1} & p_{n,n} \end{pmatrix}.$$

Into this transition matrix now the individual transition probabilities are inserted. There are a lot of zeros in the transition matrix. For example, $p_{13} = 0$. This arises from the kinetic scheme. It is not allowed that a species is directly transformed from (1) into (3). Rather the reaction runs via the intermediate compound (2). Therefore, the matrix is a tridiagonal one.

Usually the required transition probabilities are not very well known. For this reason, as a first approximation, we set them all equal to a certain number, which will be later multiplied by an appropriate factor. With progressive reaction, the transition matrix rises to a certain power. We find the coefficients then directly in the first row of the matrix. These coefficients refer to the probability that the reaction has been taken place, e.g., from compound (1) to compound (n) and these coefficients are to be weighted. If we raise this matrix to a sufficiently large power, then we approximate the equilibrium state. All elements in the first line become then equivalently large. The weighting factors for the probabilities are chosen now in this way that in equilibrium the correct concentrations will turn out.

If we know that a certain reaction step with a clearly smaller reaction rate takes place, then we can correct the scheme in this way for this particular p_{ij}, of course. However, in this case it is compulsory that the individual terms in the transition matrix are again properly standardized. Each element of the transition matrix for a certain epoch out of the equilibrium is then divided by the corresponding elements for the equilibrium, so that the weighting factors for the equilibrium tend to approach again to 1.

18.2.3.2 Infinite Number of Compounds

In the limit of an infinite number of compounds the transition probabilities $p_{1,k}(t)$, the reaction scheme approaches a mathematical expression similar to the diffusion equation. Starting with Eq. (18.12), we get the conventional scheme as follows:

$$\frac{dc_1}{dt} = -k_{1,2}c_1 + k_{2,1}c_2$$

$$\frac{dc_2}{dt} = -k_{2,3}c_2 + k_{3,2}c_3 - k_{2,1}c_2 + k_{1,2}c_1$$

(18.13)

$$\cdots\cdots\cdots\cdots\cdots\cdots\cdots\cdots$$

$$\frac{dc_i}{dt} = -k_{i,i+1}c_i + k_{i+1,i}c_{i+1} - k_{i,i-1}c_i + k_{i-1,i}c_{i-1}$$

If all the rate constants are equal, i.e., $k_{i,j} = k$, Eq. (18.13) reduces to

$$\frac{dc_1}{dt} = +kc_2 - kc_1$$

$$\frac{dc_2}{dt} = +kc_3 - 2kc_2 + kc_1 \tag{18.14}$$

$$\cdots\cdots\cdots\cdots$$

$$\frac{dc_i}{dt} = +kc_{i+1} - 2kc_i + kc_{i-1}$$

We presume that a certain function exists for that $f(i) = c_i$. We treat now i as a continuous variable. Taylor expansion yields

$$f(i + \Delta i) = \ f(i) + f'(i)\Delta i + \frac{1}{2}f''(i)(\Delta i)^2 + \dots$$

$$f(i - \Delta i) = \ f(i) - f'(i)\Delta i + \frac{1}{2}f''(i)(\Delta i)^2 - \dots$$

and consequently

$$f(i + \Delta i) - 2f(i) + f(i - \Delta i) = f''(i)(\Delta i)^2 = \frac{d^2 f(i)}{di^2}(\Delta i)^2. \tag{18.15}$$

Inserting (18.15) with $\Delta k = 1$ in the last equation of Eq. (18.14) yields

$$\frac{\partial c_i}{\partial t} = k\frac{\partial^2 c_i}{\partial i^2}$$

as an approximation for large i. Here we crossed over to the ∂ notation.

On the other hand, set (18.14) can be consecutively multiplied by x, x^2, x^3 to result in

$$\frac{dc_1}{dt}x = +kc_2 x - kc_1 x$$

$$\frac{dc_2}{dt}x^2 = +kc_3 x^2 - 2kc_2 x^2 + kc_1 x^2 \tag{18.16}$$

$$\cdots\cdots\cdots\cdots$$

$$\frac{dc_i}{dt}x^i = +kc_{i+1}x^i - kc_i 2x^i + kc_{i-1}x^i$$

Summing Eq. (18.16) and defining a generating function

$$F(t, x) = c_1(t)x + c_2(t)x^2 + c_3(t)x^3 + \ldots$$

finally yields

$$\frac{\partial F(t, x)}{\partial t} = F(t, x)(-2 + 1/x + x) + c_1(x - 1).$$

18.3 Mechanical – Chemical Analogy

The mechanical – chemical analogy can predict mechanisms of degradation and pathways of degradation. The thermal degradation of polymers in the course of heavy thermal stress will result usually in the formation of small molecules that can be analyzed by means of conventional methods of organic chemistry, namely infrared spectroscopy or gas chromatography coupled with mass spectrometry. The observation of these products will serve to the chemist to establish a reaction mechanism concerning the chemical process of degradation.

The reaction mechanism is proposed on the basis of certain common rules that are undoubtedly sound and founded by the general accepted up-to-date opinion common in organic chemistry. The experienced expert is able to sketch from scratch several different mechanisms of degradation, including a series of postulated unstable molecules or radicals that are not accessible by chemical analysis. It is often just a challenge to postulate certain intermediates in order to describe the experimental findings most elegantly. One of the favorite intermediates is, for example, the six-membered ring.

Because of the complexity of thermal degradation it happens in fact that it is difficult to decide according to which proposed mechanism the degradation process will follow in reality, or which pathway would be the preferred degradation path. Setting up the complete set of elementary reaction steps may become a task beyond the difficulty of playing chess. Computer programs [10–14] that can do the job for special fields have been developed, e.g., CHEMKIN and EXGAS. The authors suggest combining the results with thermodynamic and kinetic data. For example, in the oxidation of cyclohexane, 513 intermediates and 2446 reactions have been identified [15].

18.3.1 Extremal Principles in Mechanics

Here we will present a most simple tool that will offer more certainty to decide in between possible reaction mechanisms of degradation, apart from the pure intuition. The starting point is the mechanics of a system. It is well established that the mechanical equilibrium is associated with a minimum of energy. There is a complete analogue to chemical equilibrium.

Crossing over from mechanical and chemical equilibrium to kinematics in mechanical sciences and to kinetics in chemistry the analogy is not established as well. Descriptive chemical kinetics is a very different concept than the concept of kinematics. Kinematics is governed by the Hamiltonian equations or other extremal principles.

A simple procedure to calculate the energies of the intermediates in a complex chemical degradation along the reaction coordinate, using the method of increments by *van Krevelen*, will be developed. From the energies along the reaction coordinate we will discriminate among various proposed reaction mechanisms.

The equilibrium of mechanical systems and chemical systems is described just in the same way. The concept of virtual work can be applied in both cases successfully to obtain the equilibrium state.

The path of mechanical systems has been described by extremal principles. We emphasize the principles of *Fermat*,[4] *Hamilton*.[5] The principle of least action is named after *Maupertuis*[6] but this concept is also associated with *Leibnitz*,[7] *Euler*, and *Jacobi*.[8] For details, cf. any textbook of theoretical physics, e.g., the book of Lindsay [16, p. 129]. Further, it is interesting to note that the importance of minimal principles has been pointed out in the field of molecular evolution by Davis [17]. So, in his words

> a developing system traces a path of minimal cost in terms of the kinetic action. A least action principle applies also to the thermodynamic branch during molecular evolution, ...

18.3.2 The Principle of Least Action in Chemical Kinetics

The principle of least action has also been applied to systems where the motion is not readily seen, e.g., for acoustic and electric phenomena.

Therefore, it would be intriguing to establish a mechanical theory of chemical kinetics. In a heuristic way we could relate the kinetic energy of an ensemble of molecules to the rate at which the move along a given path of reaction. If the ensemble has a constant total energy, the path with a maximum kinetic energy would be a path with lowest average potential energy. Pictorially speaking, the preferred

---

[4] Pierre de Fermat, born at end 1607 or at begin 1608 in Beaumont-de-Lomagne, France, died Jan. 12, 1665, in Castres, France.

[5] William Rowan Hamilton, born Aug. 4, 1805, in Dublin, Ireland, died Sep. 2, 1865, in Dunsink, Ireland.

[6] Pierre Louis Moreau de Maupertuis, born Jul. 7, 1698, in Saint-Malo, France, died Jul. 27, 1759, in Basel, Switzerland.

[7] Gottfried Wilhelm Leibnitz (or Leibniz), born Jul. 1, 1646, in Leipzig, Germany, died Nov. 14, 1716, in Hannover, Germany.

[8] Carl Gustav Jacob Jacobi, born Dec. 10, 1804, in Potsdam, Germany, died Feb. 18, 1851, in Berlin, Germany.

reaction path would be characterized with a minimum of elementary reaction steps and with minimum potential energy of the respective intermediates.

Even when such a principle seems to be suggestive, we will not been work out it rigorously. Indeed, there may appear conceptual difficulties. Actually, chemical kinetics approaches the problem in a different way [18, pp. 372–412]. The use of potential energy surfaces is established in general. However, potential energy surfaces have been calculated for the reaction paths for rather simple systems. In the case of composite reactions, the theory of consecutive reactions is used to identify the rate-controlling steps. The rate-controlling step is that elementary reaction step with the highest energy of activation in the whole reaction path. In this way, it can be concluded from a variety of proposed reaction mechanisms whether a certain reaction path has sufficient *throughput* to be regarded as a reliable reaction mechanism.

The theory of consecutive reactions treats the elementary steps in an obviously isolated manner. In a sequence of elementary reactions

$$X \to Y \to Z,$$

namely the kinetic equations for a first-order sequence are

$$-\frac{d[X]}{dt} = k_1[X]$$

$$-\frac{d[Y]}{dt} = -k_1[X] + k_2[Y] \quad .$$

$$\frac{d[Z]}{dt} = k_2[Y]$$

We focus on the third equation of the set above. The form of the equation suggests that we could assess the kinetic constant k_2 by measuring the decay of the species Y, no matter whether this species is poured from a reagent bottle into the reagent glass and brought to temperature; or whether the species Y is formed as an intermediate in the course of the reaction sequence given above. If the species Y is obtained in the latter mode, then it may enter additional energy from the previous elementary step into the reaction, in contrary to the first experimental setup. Therefore, the kinetic constant k_2 may depend on the history of the formation of the intermediate. This aspect is not pointed out readily in the literature. In polymer chemistry there is at least the concept of hot radicals that deal in some way with the origin and history of reactive intermediates [19].

18.3.3 Calculation of the Energies Along the Reaction Coordinates

We will use the energy profile along the reaction coordinates to decide which mechanism would be the most favorable one. For complicated systems, the calculation of the energy of a molecule or radical in terms of the mutual distances of the atoms is a tedious procedure and therefore it seems that this approach is often dismissed, when a reaction mechanism is set up. An example that illustrates the expenditure is the detailed discussion of the thermal degradation of polyvinylchloride by *Bacaloglu* and *Fisch* [20, 21].

There is a simplified procedure available, in particular for degradation reactions. We will focus instead on unstable intermediates of the reaction instead on the activated complexes themselves. The energy of the intermediates can be obtained by the method of the increments given by van Krevelen [22]. van Krevelen summarizes in tables increments of the Gibbs energy of formation for certain groups that would constitute an organic molecule. From these increments, the Gibbs energy of a certain molecule, fragment, or radical can be calculated. We illustrate the procedure to calculate the Gibbs energy of formation of a polystyrene unit in Table 18.2.

Table 18.2 Calculation of the Gibbs energy of formation of a polystyrene unit from increments via enthalpy and entropy increments of formation

Group	ΔH_{298} [J mol^{-1}]	ΔS_{298} [J mol^{-1}K^{-1}]	ΔG_{298} [J mol^{-1}]
—CH$_2$—	$-22,000$	-102	
> CH—	$-2,700$	-120	
Phenyl	$87,000$	-167	
Sum = polystyrene unit	$62,300$	-389	$178,000$

The calculated Gibbs energy of formation is valid strictly for gas phase species, but van Krevelen gives for some cases correction for other states.

18.3.3.1 Degradation of a Polystyrene Chain

A simple example should illustrate the concept. Consider the degradation of a polystyrene chain. We have chosen a segment with head-to-tail links including one head-to-head link. We consider now three mechanisms of degradation that can result in the formation of styrene. These mechanisms are shown in Figs. 18.5, 18.6, and 18.7.

We are starting from the same polystyrene chain in each mechanism. We are also ending essentially with the same products. In between, however, there are different intermediates.

Mechanism 1 starts with a chain scission at the head-to-head link (Intermediate 1–2). After this an unzipping reaction starts, resulting in the formation of one styrene unit (1–3). Finally, a recombination is taking place, leaving a polystyrene chain, now one unit shorter behind (1–4). Mechanism 2 is very similar in comparison to mechanism 1. The only difference is that the chain scission is now taking place in a regular head-to-tail sequence. Mechanism 3 suggests the formation of a tertiary polymer radical and a hydrogen radical (3–2). An intermolecular reaction like disproportionation results in the formation of intermediate (3–3). From this structure, a depolymerization step occurs to result in (3–4). The polymer chain is then restored (3–5) and the hydrogen radical finds its way back to the main chain (3–6). Mechanism 3 has two more steps than the other mechanisms discussed.

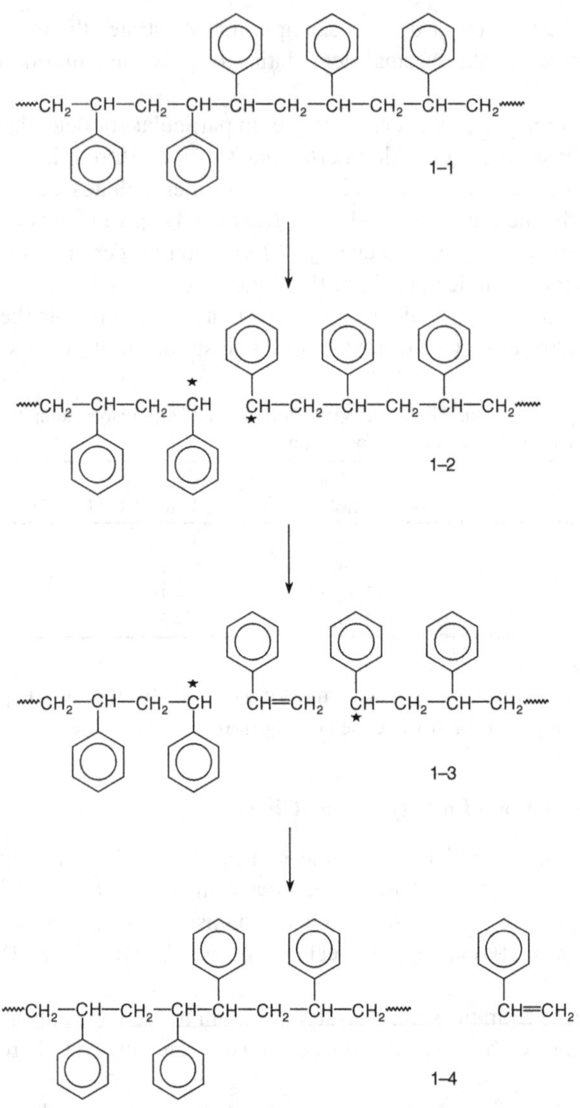

Fig. 18.5 Degradation of polystyrene to yield styrene. Mechanism 1: chain scission at the head-to-head link

Of course, the expert will know for this simple example what is the more probable mechanism of degradation without doing any calculation. However, our goal rather is to illustrate the general concept, with a simple example. There may be cases that are more complicated, where the result will not be as obvious as here.

The *Gibbs* energy of formation of the intermediates has been calculated and is presented in Table 18.3. For the calculation of the *Gibbs* energy of formation at

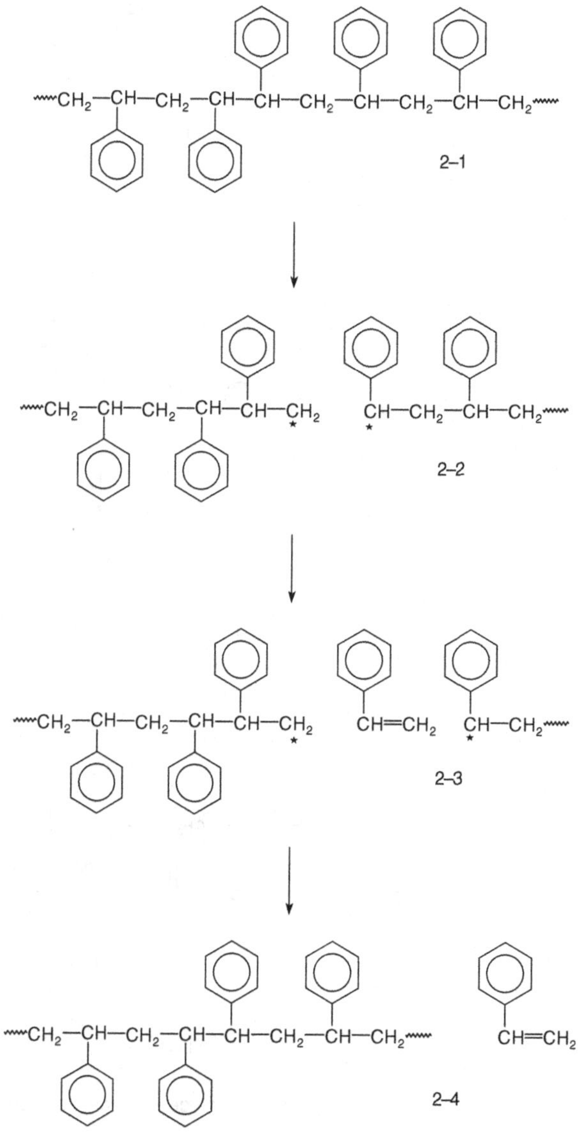

Fig. 18.6 Degradation of polystyrene to yield styrene. Mechanism 2: chain scission at the head-to-tail link

600 K the first approximation of *Ulich* has been used, i.e., the standard enthalpy of formation and the standard entropy of formation have been used.

Actually, it is not necessary to calculate the Gibbs energy of the whole molecule, only the reacting parts are needed. This is a consequence of the increment method. On the other hand, it is not allowed to omit or to add some parts of the structure

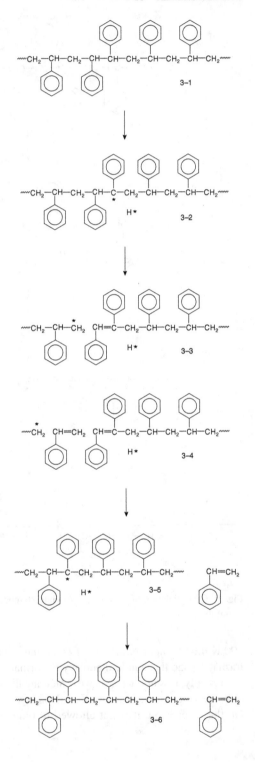

Fig. 18.7 Degradation of polystyrene to yield styrene. Mechanism 3: hydrogen abstraction followed by chain scission

in the course of the mechanism. This is often done in the abbreviated notation of reaction mechanisms, e.g., from somewhere hydrogen radicals are coming and introduced in the scheme. Further, we can subtract the Gibbs energy of the initial structure from all others. In this case, the table would start with zero. It is also possible to form the difference of the Gibbs energies between two subsequent species occurring in the mechanism. This will give a picture of the energetic situation of a single elementary reaction. In Fig. 18.8, a plot of the energy profiles of the three reaction mechanisms is shown.

Table 18.3 Energies along the reaction coordinate for the three mechanisms of the degradation of polystyrene

Species	ΔH_{298} J mol^{-1}	ΔS_{298} J mol^{-1}K^{-1}	ΔG_{298} J mol^{-1}	ΔG_{600} J mol^{-1}
Mechanism (1–..)				
1–1	3.11E5	−1945	8.906E5	1.478E6
1–2	6.16E5	−1781	1.147E6	1.685E6
1–3	7.02E5	−1627	1.187E6	1.678E6
1–4	3.97E5	−1791	9.307E5	1.472E6
Mechanism (2–..)				
2–1	3.11E5	−1945	8.906E5	1.478E6
2–2	6.28E5	−1757	1.152E6	1.682E6
2–3	7.13E5	−1603	1.191E6	1.675E6
2–4	3.97E5	−1791	9.307E5	1.472E6
Mechanism (3–..)				
3–1	3.11E5	−1945	8.906E5	1.478E6
3–2	6.91E5	−1839	1.239E6	1.794E6
3–3	7.86E5	−1638	1.274E6	1.769E6
3–4	8.72E5	−1484	1.314E6	1.762E6
3–5	7.76E5	−1685	1.278E6	1.787E6
3–6	3.97E5	−1791	9.307E5	1.472E6

Inspection of Fig. 18.8 shows that, as expected, the mechanism involving free hydrogen radicals has higher Gibbs energy values than the other two proposed mechanisms. These are running nearly at the same energy profile. Mechanism (1) is preferred over mechanism (2), because it is elementary knowledge in organic chemistry that tertiary radicals will be more stable than secondary radicals. Therefore, the head-to-head link is regarded as a weak link.

Inspection of Table 18.3 shows that the structures 1–2 and 1–3 of mechanism (1) have slightly lower Gibbs energy than the corresponding structures 2–2 and 2–3 of mechanism (2) at 298 K but the situation is reversed at 600 K. This reflects the temperature dependence of the Gibbs energy. The effect is small, however, and the data are not as precise that this effect should be discussed seriously. In other situations, such effects may be more pronounced.

Summarizing, it may be stated that the calculation of the energy profile for different reaction mechanisms is helpful to discriminate between the proposed mechanisms. Of course, it is obvious that the method is suitable in general for the discussion of the reliability of proposed reaction mechanism. One big advantage

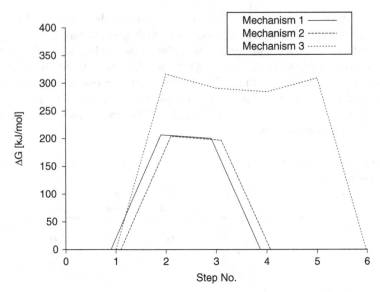

Fig. 18.8 Free Gibbs enthalpy of the species involved in the reaction mechanisms. The plot is reduced by the free Gibbs enthalpy of the reactant. Therefore, the plot starts at zero

is that the method is easy to apply. On the other hand, the tables of increments, which can be regarded as a *partial structure library*, are containing structural elements of stable groups and some free radicals. For an extensive application of the method, however, other entries, like partial structures with delocalized electrons, ionic groups, would be helpful.

References

1. Battin Leclerc, F., Glaude, P.A., Warth, V., Fournet, R., Scacchi, G., Côme, G.M.: Computer tools for modelling the chemical phenomena related to combustion. Chem. Eng. Sci. **55**(15), 2883–2893 (2000)
2. Handle, D.J., Broadbelt, L.: Mechanism reduction during computer generation of compact reaction models. AIChE J. **43**(7), 1828–1837 (1997)
3. Baverman, C., Strömberg, B., Moreno, L., Neretnieks, I.: Chemfronts: A coupled geochemical and transport simulation tool. J. Contam. Hydrol. **36**(3–4), 333–351 (1999)
4. Eucken, A., Landolt, H., Börnstein, R. (eds.): Zahlenwerte und Funktionen aus Physik, Chemie, Astronomie, Geophysik und Technik, **13**, (1985)
5. Nichipor, H., Dashouk, E., Yermakov: A computer simulation of the kinetics of high temperature radiation induced reduction of NO. Radiat. Phys. Chem. **54**(3), 307–315 (1999)
6. Chase, M.W., Sauerwein, J.C.: NIST Standard Reference DATA Products Catalog 1994, NIST Chemical Kinetics Database: Version 2Q98 Standard Reference DATA Program. No. 782 in NIST Special Publication. National Institute of Standards and Technology, Gaithersburg, MD (1994)
7. N.N.I. Technology of Standards: Chemical kinetics database on the web, standard reference database 17, version 7.0 (web version), release 1.3 a compilation of kinetics data on gasphase reactions. [electronic] http://kinetics.nist.gov/kinetics/index.jsp (2000)

8. Makroglou, A., Li, J., Kuang, Y.: Mathematical models and software tools for the glu-coseinsulin regulatory system and diabetes: An overview. Appl. Numer. Math. **56**(3), 559–573 (2006)

9. Atkins, P.W.: Physical Chemistry, 4th edn. Oxford University Press, Oxford (1990)

10. Dahm, K.D., Virk, P.S., Bounaceur, R., Battin-Leclerc, F., Marquaire, P.M., Fournet, R., Daniau, E., Bouchez, M.: Experimental and modelling investigation of the thermal decom-position of n-dodecane. J. Anal. Appl. Pyrolysis **71**(2), 865–881 (2004)

11. Edwards, K., Edgar, T.F., Manousiouthakis, V.I.: Kinetic model reduction using genetic algo-rithms. Comput. Chem. Eng. **22**(1–2), 239–246 (1998)

12. Kee, R.J., Rupley, F.M., Miller, J.A.: Chemkin-II: A Fortran Chemical Kinetics Package for the Analysis of Gas-Phase Chemical Kinetics. Report SAND-89-8009, Sandia Natl. Lab., Livermore, CA, USA (1989)

13. Klinke, D.J., Broadbelt, L.J.: Mechanism reduction during computer generation of compact reaction models. AIChE J. **43**(7), 1828–1837 (1997)

14. Warth, V., Battin-Leclerc, F., Fournet, R., Glaude, P.A., Côme, G.M., Scacchi, G.: Computer based generation of reaction mechanisms for gas-phase oxidation. Comput. Chem. (Oxford) **24**(5), 541–560 (2000)

15. Buda, F., Heyberger, B., Fournet, R., Glaude, P.A., Warth, V., Battin-Leclerc, F.: Modeling of the gas-phase oxidation of cyclohexane. Energy Fuels **20**, 1450–1459 (2006)

16. Lindsay, R.B.: Concepts and Methods of Theoretical Physics. Dover Publications, Inc., New York (1969)

17. Brian, K.D.: The forces driving molecular evolution. Prog. Biophys. Mol. Biol. **69**, 83–150 (1998)

18. Laidler, K.J., Meiser, J.H.: Physical Chemistry. The Benjamin Cummings Publishing Com-pany, Menlo Park, CA (1982)

19. Tüdős, F.: Kinetics of radical polymerization. X. The assumption of hot radicals in inhibition processes. Makromol. Chem. **79**(1), 8–24 (1964)

20. Bacaloglu, R., Fisch, M.: Degradation and stabilization of poly(vinyl chloride). IV. Molecu-lar orbital calculations of activation enthalpies for dehydrochlorination of chloroalkanes and chloroalkenes. Polym. Degrad. Stabil. **47**(1), 9–32 (1995)

21. Bacaloglu, R., Fisch, M.: Degradation and stabilization of poly(vinyl chloride). V. Reaction mechanism of poly(vinyl chloride) degradation. Polym. Degrad. Stabil. **47**(1), 33–57 (1995)

22. van Krevelen, D.W.: Properties of Polymers. Their Correlation with Chemical Structure. Their Numerical Estimation and Prediction from Additive Group Contributions, 3rd edn., pp. 627–639. Elsevier, Amsterdam (1990)

Chapter 19
Transport Phenomena

Transport phenomena are associated with the flow or the movement of particles and more generally the general change of thermodynamic variables in time. Transport phenomena include mainly

- momentum transfer,
- heat transfer, and
- mass transfer.

There is a close analogy between these phenomena. The balance equations can be derived in a quite similar manner. Actually, the basic principle is simple. In a certain fixed volume the accumulation of a certain quantity is described by the difference of inflow, outflow, and by the rate of generation or loss in the particular volume.

19.1 The Equations of Continuity

The equation of continuity is of fundamental importance in balancing systems. There are several equations of continuity, dealing with the continuity of matter, momentum, energy, and other physical quantities.

Consider the scheme shown in Fig. 19.1. We are setting the coordinates at the left lower foreground corner to be x, y, z. The length of the edges is Δx, Δy, Δz. Therefore, the center of the cube is at the position $x + \Delta x/2$, $y + \Delta y/2$, $z + \Delta z/2$. There is still another method of derivation which centers the cube at x, y, z. However, in this case, the derivatives must be defined as $\partial \bullet / \partial x = (\bullet(x + \Delta x/2) - \bullet(x - \Delta x/2)/\Delta x$, etc., which arrives essentially at the same result.

Matter is flowing into a cube across the face $\Delta y \Delta z$ at the position $x, y + \Delta y/2, z + \Delta z/2$ and flowing out at position $x + \Delta x, y + \Delta y/2, z + \Delta z/2$. The mass flow m_x in x-direction is a function of the position x

$$m(x, y + \Delta y/2, z + \Delta z/2) =$$
$$\rho(x, y + \Delta y/2, z + \Delta z/2)v_x(x, y + \Delta y/2, z + \Delta z/2)\Delta y \Delta z.$$

J.K. Fink, *Physical Chemistry in Depth*,
DOI 10.1007/978-3-642-01014-9_19, © Springer-Verlag Berlin Heidelberg 2009

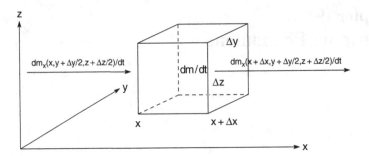

Fig. 19.1 Balance of matter flowing in and out of a volume element

v_x is the linear flow rate. Then the difference of inflow and outflow is the accumulation of matter with time in the volume element which is

$$\frac{dm(x, y + \Delta y/2, z + \Delta z/2)}{dt} =$$

$$\frac{d\rho(x + \Delta/2x, y + \Delta y/2, z + \Delta z/2)}{dt} \Delta x \Delta y \Delta z =$$

$$\rho(x, y + \Delta y/2, z + \Delta z/2)v_x(x, y + \Delta y/2, z + \Delta z/2)\Delta y \Delta z$$

$$- \rho(x + \Delta x, y + \Delta y/2, z + \Delta z/2)v_x(x + \Delta x, y + \Delta y/2, z + \Delta z/2)\Delta y \Delta z.$$

Division by the volume element $\Delta x \Delta y \Delta z$ and expansion of the last term in a Taylor series with respect to Δx and going to the limit of zero volume results in

$$\frac{d\rho(x, y, z)}{dt} = -\frac{\partial \rho(x, y, z)v_x(x, y, z)}{\partial x}.$$

Repeating the consideration for flow in y- and z-directions, in terms of the elegant notation of analysis, results in the equation of continuity. The equation of continuity is a fundamental equation for balancing operations

$$\dot{\rho} = -(\nabla \cdot \rho \mathbf{v}). \tag{19.1}$$

The equation of continuity goes back to *Euler*.[1] According to the same principle, several other equations based on balances can be formulated. These include the following:

- The equation of motion, which is a momentum balance instead of a balance of matter, the equation of motion plays a role for flow problems.
- The equation of energy.
- The equation of angular momentum.

---

[1] Leonhard Euler, born Apr. 15, 1707, in Basel, Switzerland, died Sep. 18, 1783, in St. Petersburg, Russia.

If inside the cube shown in Fig. 19.1 is a source or a sink of matter, Eq. (19.1) changes into

$$\dot{\rho} = -(\nabla \cdot \rho \mathbf{v}) + \dot{R}.$$

Here $\dot{R}$ is the rate of production of mass in the respective volume element. Such a situation arises in petroleum technology, when an injection or production well is ending in the volume element. More interesting for the physical chemist is the flow of a multicomponent mixture with chemical reaction.

19.1.1 Energy Equation

The energy equation arises from a balance of energy inflow and outflow by convective transport, energy inflow and outflow by heat conduction, rate of work done by pressure forces, rate of work done by viscous forces, and rate of work done by external forces [1, p. 335]. The terms on the right-hand side of Eq. (19.2) describe the flows mentioned just before in this order:

$$\frac{\partial}{\partial t}(1/2\rho v^2 + \rho \hat{U}) =$$
$$- \left(\nabla \cdot (1/2\rho v^2 + \rho \hat{U})\right) \mathbf{v} - \nabla \cdot \mathbf{q} - \nabla \cdot p\mathbf{v} - \nabla \cdot [\tau \cdot \mathbf{v}] + \rho(\mathbf{v} \cdot \mathbf{g}).$$
$$(19.2)$$

$\hat{U}$ is the specific internal energy, thus $1/2\rho v^2 + \rho \hat{U}$ is the total energy per unit volume as the sum of internal and kinetic energies. Further, $\mathbf{q}$ is the heat flux relative to the motion, τ is the momentum flux tensor. The enthalpy can be introduced by the relation $\hat{U} = \hat{H} - p\hat{V} = \hat{H} - p/\rho$.

There are several important special cases of the energy equation. If there is no kinetic energy, $\mathbf{v} = 0$, the velocity becomes the null vector. Equation (19.2) then reduces at constant pressure and density to

$$\frac{\partial}{\partial t}(\rho \hat{H}) = \rho \hat{C}_p \frac{\partial T}{\partial t} = -\nabla \cdot \mathbf{q} = k\nabla^2 T,$$

which dismantles as Fourier's law.

19.1.2 Entropy Equation

The equation of change of entropy can be obtained by a balance [1, p. 372]. In a volume element, the accumulation of entropy with time arises from convective transport of entropy attached to matter, from inflow and outflow by molecular transport, and from local entropy production in the volume element itself. Thus, the equation of change of entropy reads as

$$\frac{\partial \rho \hat{S}}{\partial t} = -\nabla \cdot (\rho \hat{S} \mathbf{v}) - \nabla \cdot \mathbf{s} + g_S. \tag{19.3}$$

The first term on the right-hand side of Eq. (19.3) refers to convective transport, the second term with the entropy flow vector $\mathbf{s}$ refers to the entropy flow by molecular transport, and term g_S is the rate of local entropy production. Equation (19.3) goes back to *Jaumann*.[2]

The heat flow is related to the entropy flow by $\mathbf{s} = \mathbf{q}/T$. Equation (19.3) is one of the starting points for irreversible thermodynamics.

19.1.3 Diffusion Equation

The diffusion equation describes the flow of matter by molecular transport. At constant diffusion coefficient D it reads for the one-dimensional case as

$$\frac{\partial c(x,t)}{\partial t} = D \frac{\partial^2 c(x,t)}{\partial x^2}. \tag{19.4}$$

$c = c(x,t)$ is the concentration as function of position and time. Equation (19.4) is also known as Fick's second law:

$$J = -D \frac{\partial c(x)}{\partial x}. \tag{19.5}$$

Fick's first law is Eq. (19.5). It is told that *Fick*[3] did not find out the laws of diffusion by experiment, but rather by analogy with Fourier's law of heat conduction.

19.1.3.1 Stationary State

In the stationary state we have

$$0 = \frac{\partial c(x,t)}{\partial t} = D \frac{\partial^2 c(x,t)}{\partial x^2}; \quad J = -D \frac{\partial c(x)}{\partial x}; \quad \frac{J}{D} x + C_2 = c(x).$$

The total amount of diffusing substance through a plane sheet [2] is $Q(t)$,

$$Q(t) = \frac{D C_1}{l} \left(t - \frac{l^2}{6D} \right).$$

---

[2] Gustav Andreas Johannes Jaumann, born Apr. 18, 1863, in Karánsebes, Hungary, died Jul. 21, 1924, in Ötztaler Alpen, Austria. Both Jaumann and Einstein were candidates for a chair at the university of Prague.

[3] Adolf Fick, born Sep. 3, 1829, in Kassel, Germany, died Aug. 21, 1901, in Blankenberghe, Flanders.

19.2 Capillary Flow

Capillary flow is detailed in [1, p. 50]. From the momentum balance follows

$$\frac{d}{dr}(r\tau_{rz}) = \left(\frac{\Delta P}{L}\right) r. \tag{19.6}$$

τ_{rz} is the viscous stress tensor, L is the length of the capillary, and r is the radius. Newton's law for the viscous stress tensor yields $\tau_{rz} = -\eta \partial v_z / \partial r$.

Integration of Eq. (19.6) gives

$$\tau_{rz} = \left(\frac{\Delta P}{2L}\right) r + \frac{C_0}{r} = -\eta \frac{dv_z}{dr}. \tag{19.7}$$

The right-hand side in Eq. (19.7) is Newton's law. η is the viscosity. This is the general form of the dependence of the flow velocity only on z-direction. Since the momentum flux may not become infinite, $C_0 = 0$.

19.2.1 Capillary Flow of Nonmixing Fluids

Nonmixing fluids are not necessarily immiscible fluids. Actually, the term nonmixing should indicate that the two fluids do not mix in the course of the flow through the capillary. In the case of miscible fluids this means that the residence time is as short that in laminar flow the mixing by diffusion is negligible. Of course, the most frequent situation is the flow of immiscible fluids. The situation for a two-phase flow in a capillary is shown in Fig. 19.2. At the entrance of the capillary, there must be a device that ensures no premixing in the case of basically miscible fluids, e.g., a Daniell tap as commonly used in a blowpipe for burners. Two fluids (1) and (2) are flowing through the capillary. The flow is directed in increasing z-direction.

For the outer layer (fluid 1) we integrate Eq. (19.7) with the boundary condition $r = R$; $v_z = 0$ and obtain

$$v_{z,1} = \frac{\Delta P R^2}{4\eta_1 L} \left[1 - \left(\frac{r}{R}\right)^2 \right]. \tag{19.8}$$

Here we have introduced subscript (1) to indicate that we are dealing with fluid (1). For fluid (2) we integrate Eq. (19.7) with the boundary condition $r = r_b$; $v_{z,1} = v_{z,2}$. We know already $v_{z,1}$ as a function of r. Integration gives

$$v_{z,2} = \frac{\Delta P R^2}{4\eta_1 L} \left[1 - \left(\frac{r_b}{R}\right)^2 + \frac{\eta_1}{\eta_2} \left(\left(\frac{r_b}{R}\right)^2 - \left(\frac{r}{R}\right)^2 \right) \right]. \tag{19.9}$$

If the viscosities of both fluids are equal ($\eta_1 = \eta_2$), then Eq. (19.9) reduces to Eq. (19.8). In order to obtain the total volume flow $\dot{V}$ through the capillary, we

Fig. 19.2 Two-phase flow in
a capillary

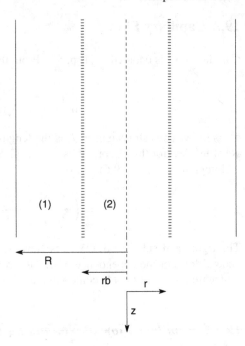

integrate v_z as shown in Eq. (19.10),

$$\dot{V} = \int_{r_b}^{R} 2\pi r v_{z,1} \mathrm{d}r + \int_{0}^{r_b} 2\pi r v_{z,2} \mathrm{d}r. \qquad (19.10)$$

The result is

$$\dot{V} = \frac{\Delta P R^4 \pi}{8\eta_1 L}\left[1 + \left(\frac{\eta_1}{\eta_2} - 1\right)\left(\frac{r_b}{R}\right)^4\right]. \qquad (19.11)$$

There are two limiting cases for Eq. (19.11). If $\eta_1 = \eta_2$ then

$$\dot{V} = \frac{\Delta P R^4 \pi}{8\eta_1 L}.$$

Further, if $\eta_2 \gg \eta_1$, then

$$\dot{V} = \frac{\Delta P R^4 \pi}{8\eta_1 L}\left[1 - \left(\frac{r_b}{R}\right)^4\right].$$

This is a situation, similar, as a plunger would move in the center of the capillary. Likewise, we could rearrange Eq. (19.11) to obtain

$$\dot{V} = \frac{\Delta P R^4 \pi}{8 \eta_2 L} \left[\left(\frac{r_b}{R} \right)^4 + \frac{\eta_2}{\eta_1} \left(1 - \left(\frac{r_b}{R} \right)^4 \right) \right].$$

19.3 Irreversible Thermodynamics

Transport phenomena can be explained by the theory of irreversible thermodynamics [3]. First of all we recall the nature of energy. Energy has the physical dimension of $kg\,m^2\,s^{-2}$. Consequently, the derivative of energy with respect to a length x, $\partial U / \partial x$, has the physical dimension of a force $\mathbf{F}$, namely $kg\,m\,s^{-2}$.

This differential is nonzero, if the energy is changing in space. Pictorially, a force is acting on some properties or variables that make up the energy, such as entropy, volume, matter, in order to change these variables to boil down the force to zero.

We exemplify the situation with matter. At constant temperature and pressure, as explained elsewhere, instead of ordinary energy we must use the free enthalpy $G(T, p, n)$ instead of energy $U(S, V, n)$. Moreover, we use molar quantities, $\tilde{G}(T, p)$.

Now we subdivide the system into small subsystems and we assume that the molar free enthalpy is varying with position. Thus, in addition, the molar free enthalpy will become a function of position $(x, y, z) = \mathbf{r}$ and in addition, if we allow a variation in time, also a function of time t.

We form the derivative to a single direction in space,

$$\frac{\partial \tilde{G}(T, p, x, y, z, t)}{\partial x} = \frac{\partial \mu(T, p, x, y, z, t)}{\partial x} = \mu_x = F_x. \tag{19.12}$$

More generally, in three dimensions we can express Eq. (19.12) in terms of the gradient, $\nabla \mu = \mathbf{F}$.

In a multicomponent system, additionally, the molar fractions must be introduced as functional parameters, and these are the most important parameters if changes in concentration are under consideration. More precisely, the force acting on a particle of type i is then $F_{i,x} = \mu_{i,x}$.

The force appearing in Eq. (19.12) is the force acting on 1 mol of particles and it can be likewise related to a single particle, simply by dividing Avogadro's number. In a viscous medium, assuming spherical particles the force can be related to Stokes' law as

$$\frac{F_{i,x}}{N} = 6\pi \eta r_i v_{i,x}.$$

Here η is the viscosity and r_i is the radius of the particle of type i, and N is Avogadro's number.

Now, in a multicomponent system, the variation of the chemical potential with space can be expressed in terms of the molar fractions, or concentrations as function of space. Further the velocity of the particles can be expressed in terms of a material flux across an imaginary perpendicular surface to the respective axis. In this way, the equation of diffusion can be derived from thermodynamic arguments. We emphasize that we have now silently crossed over from equilibrium thermodynamics to irreversible thermodynamics.

The diffusion coefficient thus defined is derived in the *Stokes – Einstein* relation. Similar considerations are used in the derivation of the *Einstein – Smoluchowski*[4] equation.

In the introduction of this section, we showed that the spacial derivative of the energy is a force and we illustrated associated changes, which we addressed as a flux. In irreversible thermodynamics the concept of force and flux is generalized. Actually, the development of the thermodynamics of irreversible processes was launched in the 1930s by *Onsager*[5] [4, 5].

19.3.1 Thermokinetic Potential

Actually prior to Onsager, *Natanson*[6] published a series of papers on the problems of thermodynamically irreversible processes, thus becoming a pioneer in the modern thermodynamics of irreversible processes [6]. However, Natanson's thermokinetic principle was, because of its generality, not understood by his contemporaries [7].

The thermokinetic potential was introduced by splitting the internal energy in an extra term containing the heat energy and the work and inserting the respective terms into the Lagrange equations [8–10].

We follow now the ideas of *Natanson*. For the description of a system generalized variables q_i and the derivatives with respect to time P_i are sufficient. The total energy is composed of the kinetic energy K which is a second-degree homogeneous function of the variables q_i and P_i, as usual. The potential energy U is a function of the variables q_i alone. The system can take up and release energies of work W and heat Q. Note that most literature will distinguish between heat and work; however, both are energy forms. Both energy forms can be expressed in differential form as the sum of functions the generalized variables q_i

$$\delta W = \sum_i A_i \delta q_i; \quad \delta Q = \sum_i B_i \delta q_i \tag{19.13}$$

---

[4] Marjan Smoluchowski, born May 28, 1872, in Vorderbrühl, Austria, died September 5, 1917, in Cracow, Poland.

[5] Lars Onsager, born November 27, 1903, in Oslo, Norway, died October 5, 1976, in Coral Gables (Florida).

[6] Władysław Natanson (1864–1937), Professor of Theoretical Physics in Cracow.

It is assumed that the coefficients can be split into reversible parts and irreversible parts, i.e.,

$$\delta A_i = A_i^0 + A_i'; \quad \delta B_i = B_i^0 + B_i'.$$

Inserting in Eq. (19.13) gives the reversible and irreversible parts of work and heat

$$\begin{aligned} \delta W^0 &= \sum_i A_i^0 \delta q_i \; ; \; \delta W' = \sum_i A_i' \delta q_i \\ \delta Q^0 &= \sum_i B_i^0 \delta q_i \; ; \; \delta Q' = \sum_i B_i' \delta q_i. \end{aligned} \tag{19.14}$$

The sign of the reversible energies will change when all the differentials of the variables δq_i change its sign, but the sign of the irreversible energies will not change.

Inserting the thermodynamic expressions in the Lagrange equation, the most general expression of the thermokinetic principle is

$$\int_{t_0}^{t_1} (\delta K - \delta U + \delta Q + \delta W) = 0.$$

For example, with this principle Natanson gave the equation of change in osmotic systems [11].

After 1960 two new approaches to the thermodynamics of irreversible processes emerged, rational thermodynamics and extended irreversible thermodynamics [12]. The latter formulation was based on similar assumptions to that of Natanson.

References

1. Bird, R.B., Stuart, W.E., Lightfoot, E.N.: Transport Phenomena, 2nd edn. John Wiley & Sons Inc., New York (2002)
2. Crank, J.: The Mathematics of Diffusion, 2nd edn. Oxford University Press, Oxford (2001 [reprinted])
3. de Groot, S.R., Mazur, P.: Non-Equilibrium Thermodynamics. Dover Books on Physics and Chemistry. Dover, New York (1984). URL http://books.google.at/books?id= HFAIv43rlGkC&pg=PP11&hl=en&source=gbs_selected_pages&cad=0_1&sig= Ri9B6lvI ce2t3hc9ybUs6-RUGXo
4. Onsager, L.: Reciprocal relations in irreversible processes. I. Phys. Rev. **37**, 405–426 (1931)
5. Onsager, L.: Reciprocal relations in irreversible processes. II. Phys. Rev. **38**, 2265–2279 (1931)
6. N. N.: Polish chemistry. Chem. Int. **20**(5), 131–138 (1998). [electronic] www.iupac.org/ publications/ci/1998/september/poland.pdf
7. Średniawa, B.: Władysław Natanson (1864–1937), the physicist who was ahead of his epoch. [electronic] http://www.wiw.pl/wielcy/kwartalnik/42.97s.3_22_english.asp (2005)
8. Natanson, L.: On the laws of irreversible phenomena. Philos. Mag. **41**, 385 (1896)
9. Natanson, L.: Ü [on the laws of irreversible phenomena]. Z. Phys. Chem. **21**, 193–217 (1896)
10. Natanson, L.: Ü [On the thermokinetic properties of thermokinetic potentials]. Z. Phys. Chem. **24**, 302–314 (1897)
11. Natanson, L.: Ü [On the thermokinetic properties of thermokinetic properties of solutions]. Z. Phys. Chem. **30**, 681–704 (1899)
12. Sieniutycz, S., Farkas, H. (eds.): Variational and Extremum Principles in Macroscopic Systems. Elsevier, Amsterdam (2005)

Chapter 20
Separation Science

In chemistry, in the original sense, separation science deals with processes suitable for the separation of distinguishable substances that are initially mingled together. Inundated sophisticated techniques have been invented and described. Most well-known separation techniques include

- crystallization;
- distillation;
- sublimation;
- extraction; and
- chromatography.

Often combined techniques are used. In classical analytical chemistry, separation processes play an essential role to achieve its target. However, modern tools used in analytical chemistry do not really require a separation and an isolation step of the constituting compounds. For example, in infrared or nuclear resonance techniques, in favorite cases the individual compounds can be analyzed without performing a separation before throwing the sample into the apparatus. It is sufficient that the individual compounds exhibit a different and distinguishable response to the method applied. We will regard these techniques as some extended field of separation science.

Subsequently we will focus rather to some general aspects of separation science. In particular, we will discuss the base of the development of hitherto unknown separation methods rather from a theoretical view and we will find out why the methods could not work properly.

20.1 Particle Motion in a Homogeneous Field

The motion of a particle in a homogeneous field is shown in Fig. 20.1. There is an alternating electric field on the plates of the condenser. The equation of motion due to the field is given as follows:

$$M \ddot{y}(t) + f \dot{y}(t) + A \cos(\omega t + \phi) = 0. \tag{20.1}$$

J.K. Fink, *Physical Chemistry in Depth*,
DOI 10.1007/978-3-642-01014-9_20, © Springer-Verlag Berlin Heidelberg 2009

Fig. 20.1 Motion of a
particle in a plate condenser

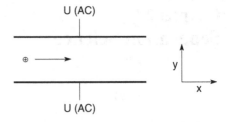

There may also be a motion in x-direction, as indicated in Fig. 20.1. The first
term describes the acceleration of the particle in y-direction, as the inert force.
The second term contains the velocity of the particle and a friction term, i.e., the
friction force rises with the velocity, counteracting the acceleration. The third term
describes the electric force due to the voltage across the plates of the condenser.
The friction law is similar to the law of *Stokes*, that is, the friction force is proportional to the velocity of the particle. f is a friction factor, in terms of the law of
Stokes

$$f = 6\pi \eta r.$$

η is the viscosity of the environment of the particle, ϕ is the phase angle of the
electric field.

We will later state that at the initial time the particle will enter when the field
has the phase ϕ. If there is a continuous flow of particles in the condenser, then
the phase will range from $\phi = 0$ to $\phi = 2\pi$. It remains to explain the constant A.
This is the maximum electric force acting on the particle. Since the condenser has
homogeneous field, this means that the field strength is present in every position,
apart from the border of the plate, i.e., in the interior of the condenser $A = eU/2L$.
Here $2L$ is the distance of the plates, e is the charge that the particle is bearing, and
U is the maximum AC voltage across the plates. We introduce now dimensionless
variables by substituting

$$\tau = \omega t; \quad y(t) = \Upsilon(\tau); \quad \alpha = \frac{A}{ML\omega^2}; \quad \beta = \frac{f}{M\omega}.$$

τ is a dimensionless time, ω is the frequency of the alternating current, L is a characteristic length of the system. For example, we could adjust the distance of the
condenser plate as $2L$. This means further that we can place the coordinate system
so that the origin is in the middle in between the plates. If the particle enters the
condenser region in the middle, then the initial dimensionless $\Upsilon = 0$. The particle
touches the plates at $\Upsilon = 1$. Eventually we arrive at the dimensionless form of
Eq. (20.1) as follows:

$$\ddot{\Upsilon}(\tau) + \beta \dot{\Upsilon}(\tau) + \alpha \cos(\tau + \phi) = 0. \tag{20.2}$$

Before we start to solve Eq. (20.2) we explain the physical meaning of the coefficients α and β. If we insert for A we arrive at

$$\alpha = \frac{eU}{2mL^2\omega^2}.$$ (20.3)

Inspection of Eq. (20.3) shows that α is the ratio of electric energy to kinetic energy, i.e., the energy of the harmonic motion. Similarly, Eq. (20.4) states that β is the ratio of the viscous momentum to the momentum of the harmonic motion

$$\beta = \frac{f}{mL\omega}.$$ (20.4)

At a first glance it looks that both α and β are proportional only and exclusively reciprocal with the mass; however, this is not true.

We solve Eq. (20.2) first or the frictionless case, i.e., $\beta = 0$. We use the initial conditions

$$\Upsilon(0) = 0 \quad \text{and} \quad \dot{\Upsilon}(0) = 0.$$ (20.5)

The solution of Eq. (20.2) is

$$\Upsilon(\tau) = \alpha \cos(\phi + \tau) + \alpha t \sin(\phi) - \alpha \cos(\phi).$$ (20.6)

The first term in Eq. (20.6) is an oscillating one, the second term is linearly increasing with the phase that acts on the particle, when it enters the condenser. We presume that the particle has an x-component of velocity. Thus, the particle is spreading around an oscillation in a direction depending on the initial phase.

We investigate now the damped oscillation, i.e., $\beta \neq 0$ with the same initial conditions. Here the solution is

$$\Upsilon(\tau) = \frac{\alpha}{\beta(1 + \beta^2)} \times$$
$$[-\beta\exp(-\beta\tau)\cos\phi - \exp(-\beta\tau)\sin\phi$$
$$+ \beta\cos(\phi + \tau) + \sin\phi + \beta^2 \sin\phi$$
$$- \beta^2 \sin(\tau + \phi)]$$

We can calculate an example with the data

η:	1002×10^{-6}	Pa s	e:	1.6×10^{-19}	A s
r:	1×10^{-10}	m	M:	1.66×10^{-27}	kg
U:	200	V	ω:	$2\pi \times 10^2$	s^{-1}.
L:	5×10^{-2}	m			
α:	9.8×10^6		β:	1.8×10^{12}	

We can see that the friction factor is much higher than the factor of the oscillatory driving force in the example. We could increase the voltage or reduce that oscillation frequency. The example is calculated for small particles on a molecular scale. The situation is different with larger particles, because of the dependence of the radius with the mass. Nevertheless, the method is not satisfactory for particle separation by forced dilution. This means, if we inject a particle via a small tube into the condenser, we find a controlled dilution by the oscillations at the other end of the condenser that can tell us about the mobility of the ions.

20.2 Charged Particle in a Quadrupole Field

The basic mathematical equation of the quadrupole mass spectrometer is Mathieu's equation.

20.2.1 Mathieu's Equation

Mathieu's equation is a linear second-order differential equation with periodic coefficients. It belongs to the family of *Hill*'s[1] equations. The one-dimensional Mathieu's equation is written in standard form as

$$\frac{\mathrm{d}^2 x(t)}{\mathrm{d}t^2} + x(t)(a - 2q\cos(2t)) = 0. \tag{20.7}$$

The French mathematician *Mathieu*[2] investigated this equation in 1868 to describe the vibrations of an elliptical membrane. Mathieu functions are applicable to a wide variety of physical phenomena, e.g., problems involving waveguides, diffraction, amplitude distortion, and vibrations in a medium with modulated density. Hill was interested in the motion of planets and was thus engaged in differential equations with periodic integrals [1].

The solutions to *Mathieu*'s equation comprise an orthogonal set and possess the curious property that the coefficients of their Fourier series expansions are identical in magnitude, with alternating signs, to corresponding coefficients of their *Bessel*[3] series expansions [2, 3]. *Floquet*'s[4] theorem asserts that any solution of equation (Eq. 20.7) is of the form

---

[1] George William Hill, born Mar. 3, 1838, in New York, USA, died Apr. 16, 1914, in West Nyack, New York, USA.

[2] Claude Louis Mathieu, born Nov. 25, 1783, in Macon, France, died Mar. 5, 1875, in Paris.

[3] Friedrich Wilhelm Bessel, born Jul. 22, 1784, in Minden, Germany, died Mar. 17, 1846, in Königsberg, now Russia.

[4] Achille Marie Gaston Floquet, born Dec. 15, 1847, in Épinal, France, died Oct. 7, 1920, in Nancy, France.

$$x(t) = C_1 \text{MathieuC}(a, q, t) + C_2 \text{MathieuS}(a, q, t). \tag{20.8}$$

MathieuC and MathieuS are the even and odd Mathieu functions. These functions may be bounded or not bounded, depending on the parameters a and q. All bounded solutions to the Mathieu equation are infinite series of harmonic oscillations whose amplitudes decrease with increasing frequency.

20.2.1.1 Stability Regions

The Mathieu equation has stable, i.e., bounded trajectories and unstable, i.e., unbounded, trajectories. Stability regions for the one-dimensional Mathieu equation correspond to regions of the a, q parameter space with bounded solutions for $x(t)$. In order to understand the a, q values which comprise the Mathieu stability regions, we must consider the integral-order solutions to the Mathieu equation.

To each value of q, a countable infinite set of characteristic values of a is associated for which $x(t)$ is an odd or even function that is $n\pi$-periodic in time, n being an integer. Series approximations for the characteristic values are obtained by expressing the integral-order Mathieu function as a series of harmonic oscillations, plugging the resultant expression into Eq. (20.8), and equating coefficients of each (orthogonal) frequency component to zero. These laborious calculations yield infinite series in q where each coefficient of q can be expressed as a continued fraction [4].

20.2.2 Mathieu's Equation with a Friction Force

Mass spectrometry is done in high vacuum. High vacuum guarantees the absence of molecular collision and thus the motion occurs without friction. We introduce now a frictional term into Eq. (20.7).

$$\frac{d^2x(t)}{dt^2} + b\frac{dx(t)}{dt} + x(t)(a - 2q\cos(2t)) = 0. \tag{20.9}$$

For instance, in an electrolyte the electrophoretic motion in a quadrupole field would be described by Eq. (20.9). The solution of Eq. (20.9) is now very similar to Eq. (20.8)

$$x(t) = \exp(-bt/2)C_1\text{MathieuC}(a - b^2/4, q, t) +$$
$$\exp(-bt/2)C_2\text{MathieuS}(a - b^2/4, q, t).$$

We observe an exponential decay in the equation. This means that if the Mathieu functions are bounded, the particle will come eventually at rest at the axis $x = 0$. However, if the Mathieu functions are unbounded, also $x(t)$ is unbounded, as can be verified by inserting appropriate parameters.

If the exponential term is governing the solution, the oscillation will approach to zero after sometime. For this reason, the quadrupole principle is useless in separation of charged particles in solution environment.

20.3 The Filter as an Instrument in Analytical Chemistry

These are, in alphabetical order, ash-free filters, bandpass filters, black band filters, blue band filters, broadband filters, cigarette filter, fluted filters, high-pass filters, low-pass filters, qualitative filters, and quantitative filters. The filter is catholic. In German, in chemistry the filter has neutral gender. In communication theory, the filter has male gender, however.

20.3.1 The Ideal Filter

Filtering is a method for the separation of materials. Materials with a particle diameter larger than the pores of the filter are collected as a filter cake, particles, which are smaller than the pores, can pass through the filter. With a fine filter one expects a transmission curve in the shape of a step function. For particles, which are smaller than the pore size, the transmission is 1; for particles, which are larger than the pore size, the transmission is zero, cf. Fig. 20.2. We will refer the diameter of particles that are in the range of the pore size of the filter the critical particle diameter, or consequently also as the critical pore size in the reverse view.

We inspect once more Fig. 20.2. If we replace the abscissa label with *frequency*, then in electrotechnics, this particular transmission behavior mimics just an ideal low-pass filter.

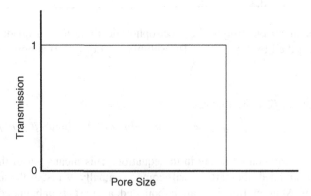

Fig. 20.2 Transmission of an ideal filter vs particle size

20.3.2 The Leaching Filter

In practically realizable cases, however, the transmission curve is not a step function, but the step drops finally steeply within a more or less narrow range of the critical particle size. Subsequently, the transmission, which we did not define initially very clearly will be examined more into depth.

20.3.3 Transmission

In practice, a filtration process runs in a way that a slurry, a suspension, or a dispersion, etc., of solid particles in a liquid will be pressed through a filter. The overwhelming amount of solids of this suspension is expected to be retained by the filter. Ideally, a filter should be completely impermeable for the suspended particles.

The suspension should be distributed in the liquid completely homogeneous at a concentration of say $c_1[\text{kg/m}^3]$ in a volume of $V_1[\text{m}^3]$. After filtering off the liquid with the suspension, there is still a concentration of the solid particles of $c_2[\text{kg/m}^3]$. The filtrate has only a volume of $V_2[\text{m}^3]$. From the total amount of $c_1 V_1[\text{kg}]$, only $c_2 V_2[\text{kg}]$ have passed the filter. Thus, the transmission we will define as

$$T = \frac{c_2 V_2}{c_1 V_1} \approx \frac{c_2}{c_1}. \tag{20.10}$$

The last term in Eq. (20.10) is valid only at small concentrations of the suspended material, in particular, the volume will be changed only negligibly by the filtration process.

20.3.4 Selectivity of the Filter

20.3.4.1 Definitions and Problem Specifications

In a mixture, there should be three substances with different particle sizes. We address these kinds of particles with (1), (2), and (3). The mixture is suspended in a liquid. The substances (1), (2), and (3) are present in the amounts w_1, w_2, w_3. They have the transmissions for the filter of T_1, T_2, T_3. We presume that we know the numerical values of these transmissions. But we do not know the mass amounts of the respective particles.

Calculation of the Accumulation in the Filter

Using the transmissions we can calculate the amount that is accumulating as filter cake, if we know the initial feed, w_1, w_2, w_3. After filtration the filter cake will be found:

$$W_K = (1 - T_1)w_{1,K} + (1 - T_2)w_{2,K} + (1 - T_3)w_{3,K}.$$

In the filtrate the following will be found:

$$W_F = T_1 w_{1,F} + T_2 w_{2,F} + T_3 w_{3,F}.$$

If we would evaporate the liquid, from the suspension before filtering, then the total amount of the initial feed present would be recovered:

$$W_T = w_1 + w_2, +w_3.$$

The index F represents the filtrate, the index K represents the filter cake, and the index T represents the total amount of the initial feed of the suspended particles to be filtered, without the liquid.

Linearity

Obviously, also with a mixture of more than three substances with known transmission such a calculation can be performed. However, we must assume that the individual substances do not interact somehow. Thus,

$$W_K = \sum_i w_i (1 - T_i)$$

$$W_F = \sum_i w_i T_i$$

$$W_T = \sum_i w_i$$

Calculation of the Feed from the Accumulation

In the preceding example, the accumulation in the filter cake and in the filtrate of the individual substances with different particle sizes was predicted from the transmission factors and from the amounts initially present.

Practically another situation is of more interest, i.e., the calculation of the amounts initially present from the measured accumulation in the filter cake. In the previous example the known quantities were the amounts w_1, w_2, w_3 and the unknowns were W_f, W_K, W_G. We can easily obtain these amounts as

$$\begin{aligned}
W_K &= (1 - T_1)w_{1,K} + (1 - T_2)w_{2,K} + (1 - T_3)w_{3,K} \\
W_F &= T_1 w_{1,F} + T_2 w_{2,F} + T_3 w_{3,F} \\
W_T &= w_1 + w_2 + w_3
\end{aligned} \qquad (20.11)$$

Inverse Transformation

The exact calculation of w_1, w_2, w_3 from the quantities W_f, W_K, W_G is not possible, because these equations are not linearly independent. This means that one equation in the system of the three equations can be obtained by combining two other equations. We elucidate this in detail. We inspect once more the Eq. (20.11).

By direct addition of the first equation and the second equation, the third equation in Eq. (20.11) is obtained. In the sense of pure mathematics, this means that the addition of a new equation will not produce more information.

20.3.5 Measurement of the Particle Size Distribution

If two filters with decreasing pore sizes in the direction of flow are coupled together, then in between these two filters particles with a particle size smaller then the pore size of the first filter and larger than the pore size of the second filter can be collected.

If a series of filters with decreasing pore size are collected, then the particle size distribution can be obtained by such a device. If the amount of particles collected in each cascade are plotted against the pore sizes then a histogram is obtained. An example for such a plot is shown in Fig. 20.3. The histogram in this example is equidistant. This means that the pore sizes of the filter decreased linearly here.

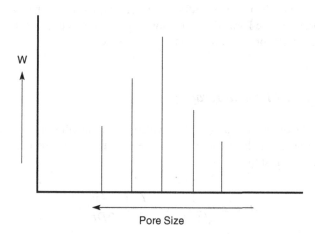

Fig. 20.3 Histogram of the weight (W) collected particles against the pore size of a filter

20.3.6 Continuous Filters – Separation Columns

A chromatographic column is somehow similar to a filter cascade. However, in the case of a chromatographic column a substance does not stop with the motion in a certain part of the column, as it occurs in a classical filter. Through the column a fluid, either liquid or gas, flows with a constant velocity v. Using the mass balance and the molecular material transport as a driving force, Eq. (20.12) is obtained

$$\frac{\partial c}{\partial t} + v\frac{\partial c}{\partial x} = D\frac{\partial^2 c}{\partial x^2}. \tag{20.12}$$

$c = c(x, t)$ is the concentration of the substance, D is the diffusion coefficient. If the mobility is zero then the second law of diffusion is obtained. By the substitution $y = x - vt$, i.e., a coordinate system that is moving with the flow, Eq. (20.13) is obtained, namely

$$\frac{\partial c}{\partial t} = D\frac{\partial^2 c}{\partial y^2}. \tag{20.13}$$

A possible solution of Eq. (20.13) is

$$c(y, t) = c_0 \frac{1}{\sqrt{2\pi Dt}} \exp\left(\frac{-y^2}{2Dt}\right). \tag{20.14}$$

Without going into details, it is sufficient to emphasize that for small times the function approaches the Dirac delta function. This conforms to the usual conditions under those a separation column is operated in practice.

20.3.7 Elution Chromatography

The method of elution chromatography involves observing the variation of the concentration at the end of the column, i.e., at position $x = L$. Introducing this condition in Eq. (20.14) results in

$$c(t) = c_0 \frac{1}{\sqrt{2\pi Dt}} \exp\left(-\frac{(L - vt)^2}{2Dt}\right). \tag{20.15}$$

It is well known that the separation of substances arises because each substance is moving with a different velocity v. In the deeper sense, the velocity is characteristic for the nature of the substance. Usually, not the velocity is recorded, but the time is given at which the peak maximum appears. This time is addressed as retention time t_R. With this adaption, Eq. (20.15) becomes

$$t_R = \frac{L}{v}; \quad c(t) = \frac{c_0}{\sqrt{2\pi Dt}} \exp\left(-\frac{(Lt_R)^2(t_R - t)^2}{2Dt}\right).$$

20.3.8 Concentration Signal in the Case of Mixtures

Usually mixtures are regarded as separated, if the individual components in the chromatogram are obviously separated. Theoreticians have worked out separation factors, etc. [5].

If the signals of the individual components are not separated completely, then we must add to the signal also the contributions of substances that are super-positioned. This is valid for each individual component that should be resolved. In general, a chromatogram is obtained by summing the individual signals of concentration,

$$S(t) = \sum_i \frac{c_{0,i}}{\sqrt{2\pi\, Dt}} \exp\left(\frac{-(Lt_{R,i})^2(t_{R,i} - t)^2}{2Dt}\right).$$

Here for each component a retention time was associated with an index i.

20.3.9 Infrared Spectroscopy

For certain samples, in particular fossil fuels, highly overlapped infrared spectra are obtained. The overlap can be reduced by special methods that artificially reduce the bandwidth of the bands [6].

Infrared bands show a bandwidth intensity according to a Lorentz curve

$$A_i(v) = A_{0,i} \frac{\gamma_i^2}{\gamma_i^2 + 4(v - v_i^0)^2}. \tag{20.16}$$

γ_i refers to the spectral width, v_i^0 refers to the frequency at maximum absorption. The function in Eq. (20.16) can be regarded as a convolution of

$$\frac{\gamma_i^2}{\gamma_i^2 + 4v^2}$$

with the δ-function. According to the rules of the convolution, we must form the Fourier transform of $A_i(v)$, which is equal to the product of the Fourier transform, the functions in the convolution integral

$$FT[A(v)_i] = G(x) = \frac{1}{2} A_{0,i} \gamma_i \cos(2\pi\, v_i^0) \exp(-\pi \gamma_i x). \tag{20.17}$$

The argument of the function G is a length because the wave number has the dimension cm^{-1}. If we divide the function $G(x)$ by the exponential term at the right-hand side of Eq. (20.17), then the contribution of the bandwidth will be filtered off. The approximate bandwidth should be known initially. The back transformation of a so-modified function should be a δ-function, or a series of δ-functions, in the case of a sum of individual bands. The procedure described just now is a so-called self-de-convolution.

20.4 Alternating Regression in GC/MS

In gas chromatography coupled with mass spectrometry a procedure has been given, how to separate the inadequate resolved peaks [7, 8]. It is remarkable that the mass spectra of the individual peaks that are overlapping need not to be known. Such a procedure is only possible, because in a mass spectrum of an individual compound there is so much information inside. The mass spectrum in a gas chromatographic peak is, for each mass number, the sum of the concentration of each individual compound, weighted by the relative intensity of the respective mass number and the relative amount of the respective compound. In the procedure, a normalized mass spectrum is constructed as a vector with the relative intensities of the mass numbers. Here it is necessary that the number of the mass numbers constituting a compound is equal for each compound, i.e., all the vectors should have the same dimension. Because of the formalism of the matrix multiplication, the overlapped spectrum is the product of the matrix of the spectra and the concentration vector. The overlapped spectrum is also addressed as matrix of the observable intensities

$$S = M * C.$$

To solve the matrix equation the method of the alternating regression is used. Initially the mass spectra are filled with random numbers. The peak shape curve (concentration curve) is calculated from the overlapped spectrum. In the next step, the peak shape curve is treated as given and from this the mass spectra are calculated. In a not resolved region, the mass spectra should not change at all. In the iteration step from the newly obtained mass spectra, the peak shapes are calculated.

The data must be filtered in some way, in order that the procedure remains stable. The matrix of the observable intensities contains also a certain content of noise. In particular, the high-frequency part of the noise could make the procedure instable. It is possible to filter off the high-frequency noise, as the spectra are subjected to a Fourier transform and the high-frequency part is removed in the transformed spectra. This method is referred to as digital filtering. If too much of the high-frequency part is filtered off, then significant information will be lost.

20.4.1 Tung's Equation

In size exclusion chromatography, the contributions of the polymers with different molar masses are highly overlapped, or smeared, as it is usual to say in polymer chemistry. The signal $S(t)$ as a function of time t of a chromatogram will be described by Tung's equation

$$S(t) = \int\limits_{\tau-\infty}^{\tau=\infty} g(t, \tau)u(\tau)\mathrm{d}\tau, \tag{20.18}$$

where $g(t, \tau)$ is a broadening function $u(\tau)$. The central task [9] in size exclusion chromatography is to solve the integral equation (20.18).

20.4.2 Convolution

Convolution is an integral that expresses the interference of a function with another function that is shifted by some varying value. The convolution of two functions $f * g$ is mathematically expressed as the integral

$$s(x) = \int f(x)g(x - y)\mathrm{d}y = f * g.$$

Depending on the nature of the integral transformation, the limits of integration will change. According to the convolution theorem, the transformed function of s is the product of the transformed functions of f and g. In probability theory the convolution is sometimes defined as

$$f * g = \int f(x - y)\mathrm{d}g(y),$$

which is *Stieltjes*[5] integral. In this form, sometimes the convolution appears in size exclusion chromatography.

The convolution theorem states that the Fourier transforms, as well as other integral transforms, follow a multiplication law [10]. If $FT[f]$ denotes the Fourier transform of some function f, then

$$FT[f * g] = FT[f]FT[g].$$

It is possible to prove this theorem for numerous integral transformations.

20.4.3 Uncertainty Relation in Analytical Chemistry

It has been shown that the application of the de-convolution method will result in a significant information gain in problems of analytical chemistry.

Therefore, we raise the question how much the selectivity can be improved by using such methods. We may also formulate the problem, as we ask how much more

---

[5] Thomas Joannes Stieltjes, born Dec. 29, 1856, in Zwolle, The Netherlands, died Dec. 31, 1894, in Toulouse, France.

compounds we can analyze or identify by consequent application of data transformations.

We illustrate the basic restriction of such methods with an example. The Fourier transform of a sine wave gives the delta function. To obtain the delta function, in fact, we must integrate over a time interval from $-\infty$ to $+\infty$. If over finite time interval a wave train is integrated, then in the Fourier transform contributions of other frequencies creep in that are not there originally in fact. This is known as the classical uncertainty relation.

The same is true for problems related to analytical chemistry. In order to compress the response of a chemical compound to an infinite small range in the spectral density function, in the spectrum the data in the infinite range of wave numbers must be known. Or in the case of a chromatogram, it must be sampled over an infinite range of time. In this case, basically an infinite number of compounds could be separated. If only the measurement in a finite domain is possible, then basically only a finite amount of compounds can be separated. However, the consideration is still limited by the fact that the measurement of the peaks, etc., is not possible to a limited accuracy. In practice, the latter objection seems to be even more restrictive.

References

1. Hill, G.W.: On differential equations with periodic integrals. Ann. Math. **3**(5), 145–153 (1887)
2. Morse, P.M., Feschbach, H.: Methods of Theoretical Physics. *International Series in Pure and Applied Physics*. McGraw-Hill, New York (1953). Reprint from 1999
3. National Bureau of Standards, Computation Laboratory (ed.): Tables Relating to Mathieu Functions; Characteristic Values, Coefficients, and Joining Factors. US Department of Commerce, National Bureau of Standards, Washington, DC (1967)
4. Fischer, S.L.: Ions in electromagnetic traps. Ph.D. thesis, Davidson College, Davidson, North Carolina (1995). [electronic] http://webphysics.davidson.edu/Projects/SuFischer/thesis.html
5. Giddings, J.C.: Unified Separation Science. John Wiley & Sons, New York (1991)
6. Griffith Peter, R., Pierce John, A., Hongjin, G.: Curve fitting and Fourier self-deconvolution for the quantitative representation of complex spectra. In: H.L.C. Meuzelaar, T.L. Isenhour (eds.) Computer Enhanced Analytical Spectroscopy, vol. 2, pp. 29–33. Plenum Press, New York (1990)
7. Karjalainen, E.J.: Isolation of pure spectra in GC/MS by mathematical chromatography: Entropy considerations. In: H.L.C. Meuzelaar (ed.) Computer Enhanced Analytical Spectroscopy, vol. 2, pp. 49–71. Plenum Press, New York (1990)
8. Karjalainen, E.J., Karjalainen, U.P.: Simultaneous analysis of multiple chromatographic runs and samples with alternating regression. Chemometr. Intell. Lab. Syst. **14**(1–3), 423–427 (1992)
9. Gugliotta L.M., Alba, D., Meira, G.R.: Correction for instrumental broadening in size exclusion chromatography using a stochastic matrix approach based on the Wiener filtering theory. In: T. Provder (ed.) Detection and Data Analysis in Size Exclusion Chromatography, *ACS Symposium Series*, p. 283. American Chemical Society, Washington, DC (1987)
10. Bracewell, R.N.: The Fourier Transform and its Application, *McGraw-Hill Series in Electrical Engineering*, 3rd edn. McGraw-Hill, Boston, MA (2000)

Chapter 21
Stochastic Processes

While *Laplace*[1] is considered as the father of celestial mechanics, which forecasts the movements of the planets from eternity to eternity, his compatriot *Poincaré*[2] one century later pointed out that the forecast is restricted, because by tiny small disturbances the movements can add up within larger periods in such a way that they have nothing more in common with the forecasts. This realization was accepted, but dismissed as a theoretical strange thing.

In recent times, Poincaré in this regard, in connection with chaotic processes, becomes again modern. Wiener mentioned an example, in which two worlds hit one on the other in its book cybernetics [1, 2]. It is the text in the child song by *Hey*[3] [3]:

> Weißt Du, wieviel Sternlein stehen an dem blauen Himmelszelt? Weißt Du, wieviel Wolken gehen weithin über alle Welt? Gott der Herr hat sie gezählet, daß ihm auch nicht eines fehlet an der ganzen großen Zahl, an der ganzen großen Zahl.
>
> Weißt Du, wieviel Mücklein spielen in der heißen Sonnenglut? Wieviel Fischlein sich auch kühlen in der hellen Wasserflut? Gott der Herr rief sie mit Namen, daß sie all ins Leben kamen, daß sie nun so fröhlich sind, daß sie nun so fröhlich sind.
>
> Weißt Du, wieviel Kinder frühe stehn aus ihren Bettlein auf, daß sie ohne Sorg und Mühe fröhlich sind im Tageslauf? Gott im Himmel hat an allen seine Lust, sein Wohlgefallen, kennt auch dich und hat dich lieb, kennt auch dich und hat dich lieb.

Do you know how many stars .... Directly in the second sentence it continues with the text *do you know, how many clouds* .... *Wiener*[4] uses this song as an opportunity to discuss deterministic and stochastic processes.

Stochastic processes arise more frequently, than perhaps one may think at first sight. We remind the actual movement of a projectile with appropriate dispersion on the target, the formation of crystallization germs, the formation of blisters in

---

[1] Pierre Simon, Marquis de Laplace, born Mar. 23, 1749, in Beaumount-en-Auge, Normandy, France, died Mar. 5, 1827, in Paris.

[2] Henri Poincaré, born Apr. 29, 1854, in Nancy, France, died Jul. 17, 1912, in Paris.

[3] Johann Wilhelm Hey, born Mar. 16, 1789, in Leina, Thuringia, Germany, died May 19, 1854, in Ichtershausen, Thuringia, Germany.

[4] Norbert Wiener, born Nov. 26, 1894, in Columbia, MO, USA, died Mar. 18, 1964, in Stockholm, Sweden.

J.K. Fink, *Physical Chemistry in Depth*,
DOI 10.1007/978-3-642-01014-9_21, © Springer-Verlag Berlin Heidelberg 2009

meteorological events, gas bubble formation during simmering, etc. In particular, wave mechanics is a special kind of stochastic mechanics.

21.1 Ludo

In the game ludo we describe a stochastic movement to exemplify the movement of a playing unity on a game board. We regard first the straight line unaccelerated movement of a particle. To the speed of the particle applies as the well-known relation

$$\frac{dx(t)}{dt} = k \quad \text{or} \quad x(t + \Delta t) = x(t) + k\Delta t.$$

We focus now to the movement of a gaming piece on consecutively numbered fields. The gaming piece may shift itself in one time unit a field, thus moving from the field n into the field $n + 1$. We can force this behavior by continuing to shift the gaming piece in each time unit a field. We know completely exactly, in which field the gaming piece is after a certain time. The gaming piece moves with a constant speed forward, at least, if we are generous and observe the game board in sufficiently large time intervals.

In contrast, we could throw a dice. Now the gaming piece has to be shifted the number of pips, which the dice shows upside. The situation does not make a difference to the above described, if each surface of the cube is occupied with only one pip in the same way. The result of a throw is then always 1. The situation changes, if we dice with a common dice that has different pips on each face, running from 1 to 6. If we ask now for the position of the gaming piece on the playing board, then we can answer this question no longer clearly.

If we begin the journey on the field with the number zero, then after one time unit we can find the gaming piece on the field number 1, or on the field 2..., or on the field 6. Thus, the concept of the position on the playing board loses its original meaning. We cannot make definitive statements on whatever field the gaming piece will be after the first throw, i.e., after the first time unit.

On the other hand, we can predict with a certain probability on which field the stone after a certain number of throws could be.

Thus, a small change in the initial situation causes a strong change of the method, how the forecast of the position can be handled. We must leave thereby the concept of the position of the particle and the following becomes insignificant:

$$x(t + \Delta t) = x(t) + k\Delta t. \tag{21.1}$$

As already explained, in place of the absolute position, the probability of the position appears as function of the time. Only in this way, we can treat the process of the movement reasonably.

We describe now the transition from the deterministic process to a stochastic process in more detail. The gaming piece, or more generally, the particle, is at the beginning with certainty in the position zero. Then after time t it is in position 1, or 2..., or 6 in the special case with the probability of 1/6.

After the second epoch, thus after $2\Delta t$ the particle, which was before at the position 1, is now at position 2, or 3..., or 7. Further, the particle which was before at position 2 is now in place 3, or 4..., or 8. This continues in such a way up to the particle, which was at position 6 before. This particle is now in position 7, or 8..., or 12. We emphasize the situation in Table 21.1.

Table 21.1 Possible positions of a particle after one and two intervals of time in the course of a stochastic motion

Δt	1	2	3	4	5	6	7	8	9	10	11	12
1	x	x	x	x	x	x						
2		x	x	x	x	x	x					
		x	x	x	x	x	x					
			x	x	x	x	x	x				
			x	x	x	x	x	x				
				x	x	x	x	x	x			
				x	x	x	x	x	x	x		
					x	x	x	x	x	x		x

x in the table: possible position after the time step concerned

If we go in the table along the columns, then we see the number of possibilities, with which a certain position can be reached. For example, position 7 in the second time interval can be reached in six different kinds and this position is the most probable, after two time steps.

In the special case, the probabilities are all the same so that the number of x in the columns is directly proportional to the probability of reaching this particular position. We know this problem also from the theory of the probability, where the question is answered, what is the probability or throwing a certain number of pips with two dices in one stroke. Here we have dressed the problem into another garb.

We could continue the consideration arbitrarily for a long time and the table became super-proportionally more extensive than the time steps. In the third time step we would have to add six possibilities for each of the six possibilities again, etc. The table would be simplified substantially, if there would be only a single increment in the course of a single time step. In the scheme there is only one diagonal of the x-es. This would be the deterministic case.

As already pointed out, now the probability of a position emerges in place of the definitive position. We deal now with the problem how to find this probability conveniently and how to modify the deterministic equation $x(t + \Delta t) = kn\Delta t + x(t)$ for the stochastic case. It turns out that x and k are now stochastic variables and we denote them simply in capitals.

Further, we characterize somehow the kind of the stochastic increment, in that casually K can vary from 1 to 6, for example,

$$X(t + \Delta t) = K \Delta t + X(t),$$

which means that the probability of the position after a time step Δt increases by some value $K \Delta t$. K is not necessarily a constant. We must elaborate more clearly the meaning of the stochastic variable. We define now a generating function for the probability,

$$F(s, t) = a_0(t)s^0 + a_1(t)s^1 + a_2(t)s^2 + \dots.$$

The exponents of s refer to the position and the coefficients of s^k refer to the probability that the position is k at a certain time t. We notice that the position can take only integral values. Further, observe that $F(s, 0) = 1 \times s^0 = 1$ holds.

After the first time step, the function changes F to $F(s, 1) = K$, with $K = (s + s^2 + \dots + s^6)/6$. After the second time step, $F(s, 2) = K^2$, and after the nth time step $F(s, n) = K^n$. We write the second time step in detail, for closer illustration:

$$F(s, 2) = 0.027s^{13} + 0.055s^{12} + 0.083s^{11} + 0.11s^{10} + 0.13s^9 +$$
$$+ 0.16s^8 + 0.13s^7 + 0.11s^6 + 0.083s^5 + 0.055s^4 + 0.027s^3. \quad (21.2)$$

We can interpret the quantity K for the dice with its six possibilities as the beginning of a geometrical series. For the term $K(6)$ we can write

$$K(6) = \frac{s}{6}\left[\frac{1 - s^6}{1 - s}\right].$$

More generally, we replace the number of 6 by j, then

$$K(j) = \frac{s}{j}\left[\frac{1 - s^j}{1 - s}\right].$$

With $j = 1$ the relation simplifies to $K(1) = s$. Here, we turn to the deterministic case, i.e., that with each throw the gaming piece advances around a unit.

Summarizing, we place a probability function for the position steps the position with a stochastic process. While in the deterministic case a fixed increment is added, in the stochastic case the generating function is multiplied by an increment. The assigned equation for the probability generating function to the stochastic equation $X(t + \Delta t) = K \Delta t + X(t)$ is therefore

$$F(s, t + \Delta t) = F(s, t)K.$$

21.2 Wiener or Brown Process

We can and must extend the train of thought consequently to a continuous variable. In addition, we deduce a difference equation for a stochastic process and form the limit to the corresponding differential equation.

21.2.1 Random Walk

We derive the probability for the position of a particle for the random walk [4]. We follow the position of a particle that moves in a linear spatial lattice with a mesh size of Δx. We observe the particle after uniform time steps. The particle may move in a time step with the probability $p(x)$ at one grid point forward and move with the probability $q(x)$ one grid point backward. Therefore,

$$p(x) + q(x) = 1. \tag{21.3}$$

The particle will not rest at the same position during a time step. It will move either forward or backward.

21.2.2 Generating Function

We set up the moment generating function of the distribution for n independent steps,

$$F(s) = \left[p(x) \exp(\Delta x s) + q(x) \exp(-\Delta x s) \right]^n. \tag{21.4}$$

21.2.2.1 Expectance

We get the expectance, as we form the differential with respect to s at $s = 1$,

$$E = F'(1) = n\Delta x \left[p(x) - q(x) \right] = t\frac{\Delta x}{\Delta t} \left[p(x) - q(x) \right]. \tag{21.5}$$

If one time step lasts Δt then n time steps will last $t = n\Delta t$. This is what we have made use of in Eq. (21.5). We set now

$$E = \mu(x)t. \tag{21.6}$$

21.2.2.2 Variance

From the generating function, the variance can be obtained as

$$Var = F''(1) - [F'(1)]^2 = 4p(x)q(x)t\frac{\Delta x^2}{\Delta t} = t\frac{\Delta x^2}{\Delta t} - \frac{\Delta t\mu^2}{t}. \tag{21.7}$$

We emphasize the special case for $p = q \to \frac{1}{2}$

$$Var = \frac{\Delta x^2}{\Delta t}t; \quad \mu(x) = 0. \tag{21.8}$$

Otherwise, the variance will be in the limit for $\Delta t \to 0$ with the condition

$$\frac{\Delta x^2}{\Delta t} = D$$

equal to $Var = Dt$. We will use these relations to form the limit.

21.2.3 Difference Equation

If we select a fixed place x, then a particle moves within a time step from a rear place to forward, with the probability p. Exactly in the same way a particle moves from $x + \Delta x$ with the probability q backward, thus to the grid point at x. This phenomenon is described by the difference equation

$$P(x, t + \Delta t) = p(x - \Delta x)P(x - \Delta x, t) + q(x + \Delta x)P(x + \Delta x, t). \tag{21.9}$$

$P(x, t)$ is the probability to find the particle at time t in position x.

21.2.4 Differential Equation

We expand now Eq. (21.9) into a Taylor series in space and time and obtain

$$\frac{\partial P(x, t)}{\partial t} + \ldots = -\frac{\Delta x}{\Delta t}\frac{\partial}{\partial x}[(p(x) - q(x))P(x, t)] + \frac{(\Delta x)^2}{2\Delta t}\frac{\partial^2}{\partial x^2}[P(x, t)] + \ldots . \tag{21.10}$$

We substitute now the expressions $\mu(x)$, Var in Eq. (21.10) and go to the limit for $\Delta t \to 0$:

$$\frac{\partial P(x, t)}{\partial t} = -\frac{\partial}{\partial x}[\mu(x)P(x, t)] + \frac{Var}{2t}\frac{\partial^2}{\partial x^2}[P(x, t)]. \tag{21.11}$$

Equation (21.11) is known as *Fokker – Planck* equation [5]. We observe the similarity with the equation of continuity, if the variance approaches zero. Or else, we realize the similarity with the equation of diffusion. In fact, the motion is a dispersive process. Also the *Schrödinger* equation belongs to this type of equations. The Schrödinger equation deals with the energy. μ has the dimension of a velocity.

21.3 Generalization of the Formalism

21.3.1 Stochastic Differential Equations

We indicate a generalization of the equations for the random walk without exact proof. If a differential equation contains also nondeterministic terms, i.e., noise, then the solution can be described only statistically. The equations are often written down as system of equations, or as a vector equation. We are content here with the linear formulation:

$$\frac{\mathrm{d}x\,(t)}{\mathrm{d}t} = f(x) + g\,(x)n\,(t). \tag{21.12}$$

Here $n\,(t)$ is a white noise. This function obeys the following conditions:

$$E[n(t)] = 0,$$
$$E[n(t)n(t + \tau)] = \Delta(\tau). \tag{21.13}$$

Thus, the expectance is zero and the autocorrelation function is equal to the Dirac delta function. In addition, we postulate the initial condition $x(0) = x_0$.

21.3.2 Fokker – Planck Equation

For a stochastic differential equation, there exists an associated *Fokker*[5] *– Planck* equation, which describes the probability that the variable takes the value concerned. The Fokker – Planck equation is also called the forward *Kolmogorov*[6] equation. To the particular stochastic differential equation (21.13) the following *Fokker – Planck* equation is associated:

$$\frac{\partial P\,(x, t)}{\partial t} = -\frac{\partial}{\partial x}\,[f(x)P\,(x, t)] + \frac{1}{2}\frac{\partial^2}{\partial x^2}\,\big[g^2(x)P\,(x, t)\big]. \tag{21.14}$$

$P(t, x)$ is the probability to find the particle at time t at position x. The associated initial condition is

$$P\,(0, x) = \Delta\,(x - x_0).$$

A more general derivation of this type of equations can be found in the book of *Fisz* [6].

---

[5] Adriaan Daniël Fokker, born Aug. 17, 1887, in Buitenzorg (Bogor), Java, died Sep. 25, 1972, in Beekbergen, The Netherlands.

[6] Андреы Николаыевич Колмогоров, born Apr. 25, 1903 in Tambov, Russia, died Oct. 20, 1987, in Moscow, Russia.

21.4 Stochastic Motion

We want to study the movement of a particle, which is in the deterministic sense force free, using the above equations. The kinetic energy during a stochastic process may change. We formulate the equation for the entire energy, as the Hamilton function with noise,

$$\frac{m}{2}\left[\frac{dx\,(t)}{dt}\right]^2 = U + \varepsilon n\,(t).$$

(21.15)

The term at the left-hand side of Eq. (21.15) is the kinetic energy and at the same time the Hamilton function. We assumed here that the particle has on the average a constant energy, but this energy is subjected to a fluctuation, which has the character of a white noise. We can form the square root of this expression. If the noise is small in relation to the energy, then we obtain

$$\frac{dx\,(t)}{dt} = \pm\sqrt{\frac{2U}{m}}\left(1 + \frac{\varepsilon n\,(t)}{2U}\right).$$

Here we expanded the root into a Taylor series and stopped after the first-order term. We notice that in the classical, deterministic mechanics the kinetic energy, which is the entire energy, is

$$K = \frac{mv^2}{2} = U.$$

Therefore,

$$\frac{dx\,(t)}{dt} = \pm v\left(1 + \frac{\varepsilon n\,(t)}{2U}\right).$$

21.4.1 The Fokker – Planck Equation for Stochastic Motion

We set up now the *Fokker – Planck* equation for the special case that the noise is small

$$\frac{\partial P(x,t)}{\partial t} = -v\frac{\partial P(x,t)}{\partial x} + \frac{1}{2}\left(\frac{\varepsilon}{mv}\right)^2\frac{\partial^2 P(x,t)}{\partial x^2}.$$

(21.16)

We have chosen the positive root; the negative root can be introduced also by introducing the negative sign into the velocity. The equation above has the same form like the equation of diffusion for a moving system. If we are moving together with the coordinate system with the velocity v, i.e., introducing a transformation of a variable $x' = x - vt$, then

$$\frac{\partial P(x,t)}{\partial t} = \frac{\partial P(x'-vt,t)}{\partial t} = \frac{\partial P(x'-vt,t)}{\partial t} - v\frac{\partial P(x'-vt,t)}{\partial x}. \quad (21.17)$$

We have used the chain rule for functions with more than one argument. Therefore, the equation results with the same form as the well-known equation of diffusion,

$$\frac{\partial P(x',t)}{\partial t} = +\frac{1}{2}(\frac{\varepsilon}{mv})^2\frac{\partial^2 P(x',t)}{\partial x'^2}. \quad (21.18)$$

The function $P(x,t)$ is the probability to find a particle at time t at position x. The particle is moving with a constant velocity. This means, it has a kinetic energy, which is modified by a small stochastic term. The solution of Eq. (21.18) is

$$\frac{\partial P(x,t)}{\partial t} = -v\frac{\partial P(x,t)}{\partial x} + k\frac{\partial^2 P(x,t)}{\partial x^2}. \quad (21.19)$$

A particular solution of Eq. (21.19) is

$$P(x,t) = \frac{\exp(-\frac{(x-vt)^2}{4kt})}{\sqrt{kt}}. \quad (21.20)$$

If we are again moving with the velocity v, then a *Gauß*[7] distribution will arise that broadens in time. Often this behavior is addressed that the particle flows away or melts in time. Thus, we are finding in the stochastic mechanics a very similar situation to that in comparison to the quantum mechanics. The concepts of the position and the momentum must be revisited and replaced by a new concept. In stochastic mechanics, we can reasonably justify that a particle never can be observed in an isolated manner. Here an interaction with other particles in the neighborhood appears. This interaction cannot be described by a deterministic method. Therefore, we chose a statistic description of the motion. Quantum mechanics starts with certain postulates that cannot be justified in detail. The success of the method defends it, even when the vividness suffers.

We introduce now several stochastic processes. These processes are generalizations of the classic random walk. A particle should be at a certain time t in one box of a collection of boxes. After a time step, it moves into another box, or it stays in this box. We can set up Fig. 21.1. The arrows show, where a particle can move within the next time step. The boxes have a width of Δx.

The particle moves with a certain probability, which is referred to as transition probability. The transition probability can be a function of the position, i.e., of a box, or can be a function of the time, or both. In general, the transition probability can be a function of the whole history of what happened to the particle. So the transition probability can be dependent on the whole path that the particle went before.

---

[7] Johann Carl Friedrich Gauß, born Apr. 30, 1777, in Brunswick, Germany, died Feb. 23, 1855, in Göttingen, Hannover.

Fig. 21.1 Examples for
stochastic processes. *Top*:
Brown's process, *Mid*:
Poisson's process, *Bottom*:
ludo. The *arrows* show where
a particle can move within the
next time step. The *boxes*
have a width of Δx

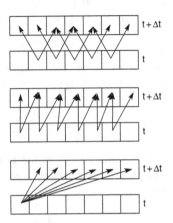

We can relate the transition probability to the probability to find the particle at a distinct position at a certain time. We focus at a certain position for the second time step and ask from where a particle can enter in this position from the previous time step. To elucidate this problem, the scheme in Fig. 21.1 is helpful.

In the case of ludo we obtain the following difference equation:

$$P(x, t + \Delta t) = \frac{P(x - 6\Delta x, t) + P(x - 5\Delta x, t) + \ldots + P(x - \Delta x, t)}{6}. \quad (21.21)$$

We address the probability to find a particle at a certain time at a distinct box as the probability of residence or probability density. We differentiate clearly between the probability density and the transition probability.

The probability density, i.e., the probability to find the particle after a time step between the positions x and $x + \Delta x$ is here the sum of the probabilities, at which the particle is at the positions $x - 6 \times \Delta x$ to $x - 1 \times \Delta x$ before the time step Δt times the transition probability. The transition probability is $1/6$ for this case.

The transition probability is the probability, with which within a time step a particle moves from one box into another box. If we label the boxes consecutively in the discrete case, then we call these probabilities $p_{i,j} = p_{i,i+k}$. For the transition into the continuous case we would write $p_{x,x+k\Delta x}$.

The transition probabilities are in general exactly the same as the probability densities as functions of the position, the time, and the history. The transition probabilities can be represented as a transition matrix. In each row of this matrix, the transition probabilities on 1 must sum up to 1.

The transition probabilities $P(x, t)$ in the individual boxes do not necessarily sum up to 1. If, for instance, a box acts as a trap for the particle, then it cannot leave this box any more. Thus, the process dies out in this case. In the next time step, the probability densities sum up to a number smaller than 1.

In addition, we emphasize that the particle may be at the beginning at position 1, thus in the first box and that for this reason, for the process after the first time step it holds $x \geq 7\Delta x$. Otherwise, we would get access to negative positions that do not exist, however. The algorithm must be cut off at the start of the process in reasonable way to the rear. We indicate the transition probabilities for the board game:

$$p_{i,i+k} = \frac{1}{6}; \quad k = 1, \ldots, 6; \qquad\qquad i = 1, 2, \ldots$$

$$p_{i,i+k} = 0; \quad k = \ldots, -1, 0, 7, 8, \ldots; \quad i = 1, 2, \ldots. \tag{21.22}$$

The transition probabilities do not depend on the time in this special case. The sum of the probability densities equals exactly 1, at each time. That means that in the course of the process no particle is caught permanently in a box.

21.4.2 Further Examples

We present the transition probability for the *Ehrenfest*[8] process. These transition probabilities are

$$p_{i,i+1} = 1 - \frac{i}{\rho}; \quad p_{i,i-1} = \frac{i}{\rho}. \tag{21.23}$$

All other transition probabilities are zero. The Ehrenfest process is a diffusion process. The probability to move to the next position is as smaller, as more far away is this respective position. There are ρ boxes present. The particle can move only in between the boxes with number zero and the box with number ρ. Therefore, the Ehrenfest process describes a motion with a central force.

In the case of the Bernoulli – Laplace model we deal with two different liquids that are mixing. From two boxes, a single molecule is removed coincidentally and placed into the other box. In each box, there are ρ molecules, initially in one box k molecules of kind 1 and $\rho - k$ molecules of kind 2. In the other box, the numbers are reverse. The transition probability is then

$$p_{i,i+1} = \left(\frac{\rho - i}{\rho}\right)^2; \quad p_{i,i-1} = \left(\frac{i}{\rho}\right)^2; \quad p_{i,i} = 2i\frac{\rho - i}{\rho^2}. \tag{21.24}$$

All other transition probabilities are zero, otherwise. It is clear that the transition probabilities are needed in order to set up the difference equation for the probabilities for the respective stochastic process.

Further, we present the difference equation for the probabilities for Brown's process:

$$P(x, t + \Delta t) = pP(x - \Delta x, t) + (1 - p)P(x + \Delta x, t). \tag{21.25}$$

---

[8] Paul Ehrenfest, born Jan. 18, 1880, in Vienna, Austria, died Sep. 25, 1933, in Leiden, The Netherlands.

For Poisson's process, Eq. (21.26) represents the difference equation for the probabilities

$$P(x, t + \Delta t) = pP(x - \Delta x, t) + (1 - p)P(x, t). \qquad (21.26)$$

21.5 Related Differential Equations

We form the operator

$$L[\bullet] = p_2(x)\frac{\mathrm{d}^2 \bullet}{\mathrm{d}x^2} + p_1(x)\frac{\mathrm{d}\bullet}{\mathrm{d}x} + p_0(x) \bullet. \qquad (21.27)$$

The function

$$\phi(x) = u(x) + \imath v(x)$$

and its complex conjugate function

$$\phi^*(x) = u(x) - \imath v(x)$$

should be a solution of the differential equation, i.e.,

$$L\left[\phi(x)\right] = p_2(x)\frac{\mathrm{d}^2 \phi}{\mathrm{d}x^2} + p_1(x)\frac{\mathrm{d}\phi}{\mathrm{d}x} + p_0(x)\phi = 0$$

and

$$L\left[\phi^*(x)\right] = p_2(x)\frac{\mathrm{d}^2 \phi^*}{\mathrm{d}x^2} + p_1(x)\frac{\mathrm{d}\phi^*}{\mathrm{d}x} + p_0(x)\phi^* = 0.$$

Then

$$L\left[\phi(x)\phi^*(x)\right] - 2\phi^*(x)L\left[\phi(x)\right] -$$
$$2\phi(x)L\left[\phi^*(x)\right] + p_0(x)\phi(x)\phi^*(x) = 2p_2(x)\left[u'^2 + v'^2\right]. \qquad (21.28)$$

In Eq. (21.28), we abbreviate $(\mathrm{d}u(x)/\mathrm{d}x)^2 = u'^2$ and $(\mathrm{d}v(x)/\mathrm{d}x)^2 = v'^2$, because $L[\phi(x)] = 0$ and $L[\phi^*(x)] = 0$, we simplify Eq. (21.28) as

$$L\left[\phi(x)\phi^*(x)\right] - p_0(x)\phi(x)\phi^*(x) = 2p_2(x)\left[u'^2 + v'^2\right]. \qquad (21.29)$$

If a certain function $\phi(x)$ is a solution of $L[\bullet]$, then the function

$$\phi(x)\phi(x)^* = u(x)^2 + v(x)^2$$

is a solution of the equation

$$p_2(x)\frac{\mathrm{d}^2\phi\phi^\star}{\mathrm{d}x^2} + p_1(x)\frac{\mathrm{d}\phi\phi^\star}{\mathrm{d}x} + 2p_0(x)\phi\phi^\star - 2p_2(x)\left[u'^2 + v'^2\right]. \qquad (21.30)$$

The operators in Eqs. (21.27) and (21.30) are very similar. They differ with respect to the linear terms only in the factor 2 at $p_0(x)$ and in an additional nonlinear term of $2p_2(x)\left[u'^2 + v'^2\right]$.

The similarity of stochastic mechanics with quantum mechanics has been pointed out and used by several authors [7–9]. The relationship goes back to Nelson [10, 11], who published in the mid-1960s a stochastic theory that eventually led to *Schrödinger*'s equation. At that time, Nelson's theory did not become very popular, because it did not explain the origin of the stochastic noise, which is essential for this theory. However, nowadays this theory seems to have been rediscovered [12].

References

1. Wiener, N.: Cybernetics, 2nd edn: Or the Control and Communication in the Animal and the Machine, 2nd edn. MIT Press, Cambridge, MA. (1965)
2. Wiener, N.: Kybernetik: Regelung und Nachrichtenübertragung im Lebewesen und in der Maschine. Econ, Düsseldorf, Wien (1992)
3. Hey, J.W.: Do you know how many stars there are. [electronic] http://www. recmusic.org/lieder/get_text.html?TextId=29357 (2007). English translation by Steve Butler (2007)
4. Feller, W.: An Introduction to Probability Theory and its Applications, *Wiley Publications in Statistics*, vol. 1, 3rd edn. pp. 354–359. John Wiley, New York (1968)
5. Zwillinger, D.: Fokker-Planck equation. In: Handbook of Differential Equations, 2nd edn., pp. 254–258. Academic Press Inc., Harcourt Brace Jovanovich, Publishers, Boston, MA (1992)
6. Fisz, M.: Wahrscheinlichkeitsrechnung und mathematische Statistik. VEB Deutscher Verlag der Wissenschaften, Berlin (1980)
7. Leschke, H., Schmutz, M.: Operator orderings and functional formulations of quantum and stochastic dynamics. Z. Phys. B: Condens. Matter **27**, 85–94 (1977). DOI 10.1007/BF01315509
8. Millonas, M.M., Reichl, L.E.: Stochastic chaos in a class of Fokker-Planck equations. Phys. Rev. Lett. **68**(21), 3125–3128 (1992). DOI 10.1103/PhysRevLett.68.3125
9. Wang, M.S.: Quantum stochastic motion in complex space. Phys. Rev. A **57**(3), 1565–1571 (1998). URL http://prola.aps.org/abstract/PRA/v57/i3/p1565_1
10. Nelson, E.: Quantum Fluctuations. *Princeton Series in Physics*. Princeton University Press, Princeton, NJ (1985). [electronic] http://www.math.princeton.edu/~nelson/books/qf.pdf
11. Nelson, E.: Dynamical Theories of Brownian Motion, 2nd edn. Princeton University Press, Princeton, NJ (2001). [electronic] http://www.math.princeton.edu/~nelson/books/bmotion.pdf
12. Durran, R., Neate, A., Truman, A.: The divine clockwork: Bohr's correspondence principle and Nelson's stochastic mechanics for the atomic elliptic state. J. Math. Phys. **49**, 1–42 (2008). [electronic] URL http://arxiv.org/abs/0711.0157 http://arxiv.org/abs/0711.0157

Chapter 22
Structure – Property Relationships

Various macroscopic properties of molecules can be derived from its quantities related to the molecules themselves. It was first *Haüy*[1] (1743–1823), the leading French mineralogist at this time [1], who proposed that since sodium chloride crystalizes as cubes, even the smallest constituting particles of the crystal must have a cubic symmetry. We can address this statement as a structure – property relationship. Haüy is also known for his research in pyroelectricity.

This chapter is somehow related to the goals of dimensional analysis. However, it does not deal with dimensional analysis in the closer sense. Structure – property relationships find use in a wide field of science, in particular in connection with molecular modeling. Most simply, structure – property relationships are qualitative thumb rules. For example, an experienced scientist can predict if a molecule would have a dipole moment or not. Even the organic chemist has an idea on the boiling point, if you present him a structural molecular formula.

Actually, most often, a property is described by a physical parameter, for example, the boiling point. The boiling point is then plotted against some other property. Relationships of this type are referred to as quantitative structure – property relationships (QSPR). Further, a relationship involving a property, such as biological activity, is called a quantitative structure – activity relationship (QSAR).

22.1 Increments

Van Krevelen discovered that both the heats and the entropies of formation of molecules in the gas phase can be obtained from group contributions. Later he extended the concept of group contributions to other properties, in particular of polymers [2]. Group increments for the heat and entropy of formation obtained from certain compounds are given in Table 22.1.

For example, from Table 22.1, for gaseous methanol, the standard heat of formation is $\Delta H_0 = -46 - 176 = -222 \, \mathrm{kJ \, mol^{-1}}$ and the entropy of formation is $\Delta S_0 = -96 - 50 = -146 \, \mathrm{J \, mol^{-1} K}$.

---

[1] René Just Haüy, born Feb. 23, 1743, in Saint-Just, France, died Jun. 3, 1822, in Paris, France.

J.K. Fink, *Physical Chemistry in Depth*,
DOI 10.1007/978-3-642-01014-9_22, © Springer-Verlag Berlin Heidelberg 2009

Table 22.1 Standard heats and entropies of formation for group increments [2]

Group	ΔH_0 [kJ mol^{-1}]	ΔS_0 [J mol^{-1} K^{-1}]
$-CH_3$	-46	-95
$-CH_2-$	-22	-102
$-CH <$	-2.7	-120
$CH_2 =$	23	-30
$-CH =$	38	-38
$-Cl$	-49	9
$-OH$	-176	-50

22.2 Graph Theoretical Numbers

Chemical graph theory is a powerful tool for the elucidation of several aspects in chemistry [3, 4]. The pioneers of chemical graph theory are Alexandru T. *Balaban*, Ivan *Gutman*, Nenad *Trinajstić*, and Haruo *Hosoya*. The history of chemical graph theory is described in the literature [4, 5].

22.2.1 Morgan Algorithm

The Morgan algorithm was initially developed for the description of chemical structures that are searchable by computers [6]. The molecule is treated as a mathematical graph. An ordered list of nodes that refers to the chemical symbols is created. This list contains entries that indicate the bond type and the list entry of atoms attached via these particular bonds.

The Morgan extended connectivity values can determine a value associated with the symmetry of a molecule [7]. The procedure is shown in Fig. 22.1. In the first step, label the molecule so that the atoms receive label numbers according to their number of neighbors. Next, replace the labels by the sum of the labels of its neighbors, and so forth until the set diminishes. This number serves as a measure for symmetry.

The Morgan extended connectivity values have been correlated with the enthalpies of formation of alkanes [8].

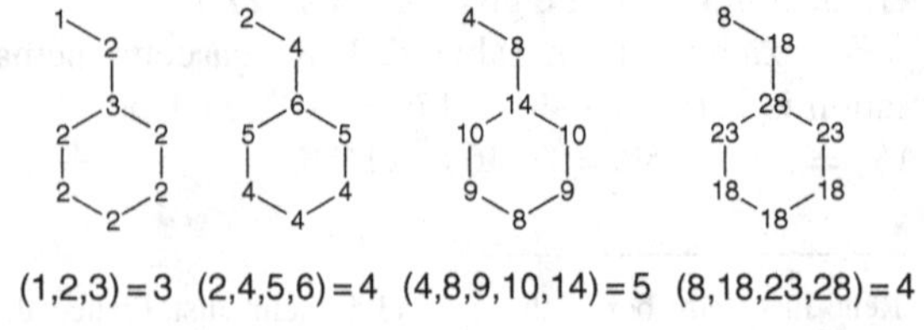

Fig. 22.1 Morgan extended connectivity values

$(1,2,3) = 3$ $(2,4,5,6) = 4$ $(4,8,9,10,14) = 5$ $(8,18,23,28) = 4$

22.2.2 Wiener Index

The Wiener index was introduced in 1947 by Harold *Wiener* [9, 10] as the path number for saturated acyclic hydrocarbons. The path number is defined as the number of the bonds between all pairs of nonhydrogen atoms in the molecule,

$$W = \frac{1}{2} \sum_{i,j} d_{i,j}.$$

Thus, the Wiener index is equal to the half-sum of the distances of the distance matrix between all pairs of the vertices in the structure [11].

22.2.2.1 Distance Matrix

For example, the distance matrix for isopentane for the labeling as shown in Fig. 22.2 is

$$D = \begin{pmatrix} 0 & 1 & 2 & 3 & 2 \\ 1 & 0 & 1 & 2 & 1 \\ 2 & 1 & 0 & 1 & 2 \\ 3 & 2 & 1 & 0 & 3 \\ 2 & 1 & 2 & 3 & 0 \end{pmatrix}.$$

The Wiener index for isopentane is 18. The Wiener index is used as a mathematical descriptor used in structure – property relationships. There are many other related descriptors for molecular graphs, including shape index, chirality index, Szeged index [12]. A comparison of the boiling points of alkanes against the different indices has been shown in the literature [13].

We show how to calculate the distance matrix from the positions of the nearest neighbors. Consider neohexane, in that the atoms are numbered as shown in Fig. 22.2. From the graphical representation we get the atoms closest connected as $2 \to 1, 3 \to 2, 4 \to 2, 5 \to 2, 6 \to 5$.

Therefore, the elements $a_{1,2}$, $a_{3,2}$, ... are equal to 1. For completeness, we can interchange the indices. If there is a bond from atom 1 to atom 2, then there is also a bond from atom 2 to atom 1. If we place only the nearest neighbors in the distance matrix, we get

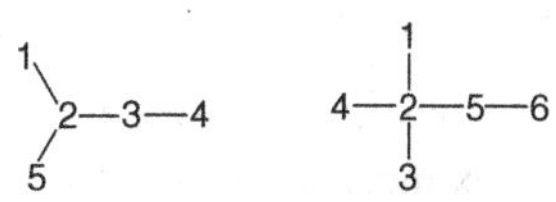

Fig. 22.2 Isopentane, neohexane

$$D_1 = \begin{pmatrix} 0 & 1 & 0 & 0 & 0 & 0 \\ 1 & 0 & 1 & 1 & 1 & 0 \\ 0 & 1 & 0 & 0 & 0 & 0 \\ 0 & 1 & 0 & 0 & 0 & 0 \\ 0 & 1 & 0 & 0 & 0 & 1 \\ 0 & 0 & 0 & 0 & 1 & 0 \end{pmatrix}.$$

We call the nearest neighbors of order 1. Next, we look which atoms are connected via a single atom. These we call atoms of order 2. For example, we observe from the graphical representation that atoms 1 and 5 are connected via atom 2. Therefore, $a_{1,2} = a_{2,5} = 1$. So we can set $a_{1,5} = 2$ in the next iteration step. The general algorithm is that when we check for a connected path from atom i to k of order 2, we must verify that $a_{i,j} = a_{j,k} = 1$ for $j = 1, \ldots n$, where n is the number of atoms constituting the molecule. In this way, we get the neighbors of order 2 that we can add to the distance matrix. So we get the distance matrix of maximum order of 2 as

$$D_2 = \begin{pmatrix} 0 & 1 & 2 & 2 & 2 & 0 \\ 1 & 0 & 1 & 1 & 1 & 2 \\ 2 & 1 & 0 & 2 & 2 & 0 \\ 2 & 1 & 2 & 0 & 2 & 0 \\ 2 & 1 & 2 & 2 & 0 & 1 \\ 0 & 2 & 0 & 0 & 1 & 0 \end{pmatrix}.$$

To get higher orders of distances we repeat the process. We look for an element that is still zero, $a_{i,k} = 0$ and inspect whether $a_{i,j} > 1$ and $a_{j,k} > 1$ for $j = 1, \ldots, n$. We insert into $a_{i,k}$ the minimal nonzero sum of $a_{i,j} + a_{j,k}$. This is justified, because in a cyclic molecule, say cyclohexane, atom 1 is connected to atom 6 either directly or via 2, 3, 4, and 5. So we are fixing the shortest path. For completeness, we give the final distance matrix that has eventually a maximum order of 3:

$$D_3 = \begin{pmatrix} 0 & 1 & 2 & 2 & 2 & 3 \\ 1 & 0 & 1 & 1 & 1 & 2 \\ 2 & 1 & 0 & 2 & 2 & 3 \\ 2 & 1 & 2 & 0 & 2 & 3 \\ 2 & 1 & 2 & 2 & 0 & 1 \\ 3 & 2 & 3 & 3 & 1 & 0 \end{pmatrix}.$$

For special cases, we can easily calculate the distance matrix and the Wiener index more conveniently. Equation (22.1) shows the distance matrix for n-hexane,

$$
L(6) = \begin{pmatrix}
0 & 1 & 2 & 3 & 4 & 5 \\
1 & 0 & 1 & 2 & 3 & 4 \\
2 & 1 & 0 & 1 & 2 & 3 \\
3 & 2 & 1 & 0 & 1 & 2 \\
4 & 3 & 2 & 1 & 0 & 1 \\
5 & 4 & 3 & 2 & 1 & 0
\end{pmatrix}. \tag{22.1}
$$

The Wiener index of n-hexane is 35. We call this particular distance matrix $L(6)$, because we are dealing with a linear chain of length 6. Inspection of the matrix in Eq. (22.1) reveals the symmetry across the zero diagonal elements and that behind and below the diagonal both rows and columns start with 1 and run consecutively up to the end of the matrix. Imagine for a moment that we are dealing with n-pentane. In this case, we have to omit the last row and the last column of the matrix $L(6)$ in order to get $L(5)$. If we again extend from n-pentane to n-hexane, we have to add $1, 2, 3, 4, 5$, both as row and as column. So the distance matrix increases with $30 = 5 \times 6$. Obviously, the Wiener index for linear chains w_n of length n runs according to the scheme

$$
w_2 = w_1 + \frac{1}{2} 1 \times 2
$$

$$
w_3 = w_2 + \frac{1}{2} 2 \times 3
$$

$$
w_4 = w_2 + \frac{1}{2} 3 \times 4 \tag{22.2}
$$

$$
\dots = \dots\dots\dots
$$

$$
w_{n+1} = w_n + \frac{1}{2} n \times (n+1).
$$

So, in general we conclude that the Wiener index for a linear chain w_n of length n increases by the recursion formula

$$
w_{n+1} = w_n + \frac{1}{2} n \times (n+1). \tag{22.3}
$$

We form a generating function for the Wiener index of linear molecules as

$$
W(x) = w_1 x^1 + w_2 x^2 + w_3 x^3 + \dots.
$$

By multiplying the rows in Eq. (22.2) consecutively by $x^2, x^3, x^4, \dots$ and using the generating function defined in Eq. (22.3), we obtain

$$
-w_1 x + W(x) = x W(x) + \frac{1}{2} \sum_{n=1}^{n=\infty} n(n+1) x^{n+1}. \tag{22.4}
$$

With $w_1 = 0$ we obtain the generating function

$$W(x) = \frac{1}{2(1-x)} \sum_{n=1}^{n=\infty} n(n+1)x^{n+1}.$$

The coefficients of the generating function at power n are the Wiener indices of the linear alkanes with a chain length of n. We inspect now the sum on the right-hand side of Eq. (22.4). Notice that $1 + x + x^2 + \ldots = 1/(1-x)$. Therefore,

$$\frac{x^2}{1-x} = \sum_{n=1}^{n=\infty} x^{n+1}$$

and

$$\frac{d^2}{dx^2}\left(\frac{x^2}{1-x}\right) = \sum_{n=1}^{n=\infty} n(n+1)x^{n-1}.$$

From this we get

$$W(x) = \frac{x^2}{2(1-x)}\frac{d^2}{dx^2}\left(\frac{x^2}{1-x}\right) = \frac{x^2}{(1-x)^4}.$$

In particular, $W(x)$ expands into

$$W(x) =$$
$$1x^2 + 4x^3 + 10x^4 + 20x^5 + 35x^6 + 56x^7 + 84x^8 + 120x^9 + 165x^{10} + \ldots,$$

the coefficients at the powers of x^n being the Wiener indices of the linear chains of length n.

In other words, we can trace back the Wiener indices for linear chains to binomial coefficients for negative powers. The Wiener index for n-hexane is calculated, omitting the negative signs, as $(4 \times 5 \times 6 \times 7)/(1 \times 2 \times 3 \times 4) = 35$.

We can obtain the Wiener index for linear chains compactly as

$$w_n = \prod_{k=0}^{k=n-3} \frac{4+k}{1+k}.$$

Finally, we show the distance matrix of cyclohexane $C(6)$ as follows:

$$C(6) = \begin{pmatrix} 0 & 1 & 2 & 3 & 2 & 1 \\ 1 & 0 & 1 & 2 & 3 & 2 \\ 2 & 1 & 0 & 1 & 2 & 3 \\ 3 & 2 & 1 & 0 & 1 & 2 \\ 2 & 3 & 2 & 1 & 0 & 1 \\ 1 & 2 & 3 & 2 & 1 & 0 \end{pmatrix}. \tag{22.5}$$

Observe the difference of $L(6)$ and $D(6)$. We can see the symmetry in a nice manner as the numbers are growing and after reaching 3 falling again. What disappears on the right end of the row comes back at the start of the row. Inspecting the diagonals, we have the main diagonal filled with 0 and the upper and lower diagonals filled with $1, 2, 3, 2, 1$. A consecutive numbering is necessary to obtain this result. In the first lower diagonal we have five times the 1 and down again one times the 1, so totally 6×1. In the second diagonal we have four times the 2 and after the next diagonal again two times the 2, so totally 6×2. Then we have three times the 3. In contrast, the numbers in the diagonal increase continuously for the distance matrix $L(6)$. From the view of the rows, each row contains the sequence of numbers $0, 1, 2, 3, 2, 1$.

A plot of the Wiener index against the boiling point of various alkanes is shown in Fig. 22.3. It can be clearly seen that for cyclic structures there are deviations. It has been shown that certain modifications in the calculus of the Wiener index will give better results [13].

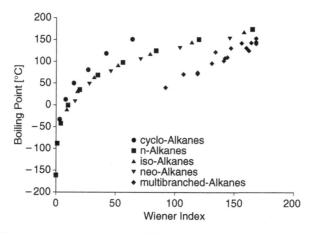

Fig. 22.3 Boiling point of alkanes against the Wiener index

References

1. Haüy, R.J.: Traité de minéralogie, vol. 1–5, 2nd edn. Bachelier, Paris (1822)
2. van Krevelen, D.W.: Properties of Polymers. Their Correlation with Chemical Structure. Their Numerical Estimation and Prediction from Additive Group Contributions, 3rd edn., pp. 627–639. Elsevier, Amsterdam (1990)
3. Bonchev, D., Rouvray, D.H. (eds.): Chemical Graph Theory: Introduction and Fundamentals, *Mathematical Chemistry*, vol. 1. Gordon and Breach Science Publishers, New York (1991)
4. Trinajstić, N.: Chemical Graph Theory, 2nd edn. CRC Press, Boca Raton, FL (1992)
5. Randić, M.: Nenad trinajstić – pioneer of chemical graph theory. Croat. Chem. Acta **77**(1–2), 1–15 (2004)
6. Morgan, H.: The generation of unique machine description for chemical structures – a technique developed at chemical abstracts services. J. Chem. Doc. **5**(2), 107–113 (1965)

7. Chen, W., Huang, J., Gilson, M.K.: Identification of symmetries in molecules and complexes. J. Chem. Inf. Comput. Sci. **44**, 1301–1313 (2004)
8. Toropov, A.A., Toropova, A.P., Nesterova, A.I., Nabiev, O.M.: Prediction of alkane enthalpies by means of correlation weighting of Morgan extended connectivity in molecular graphs. Chem. Phys. Lett. **384**(4–6), 357–363 (2004)
9. Nikolic, S., Trinajstic, N., Mihalic, Z.: The Wiener index: Development and applications. Croat. Chem. Acta **68**, 105–129 (1995)
10. Wiener, H.: Structural determination of paraffin boiling points. J. Am. Chem. Soc. **69**, 17–20 (1947)
11. Li, X.H., Li, Z.G., Hu, M.L.: A novel set of Wiener indices. J. Mol. Graphics Modell. **22**(2), 161–172 (2003)
12. Todeschini, R., Consonni, V.: Handbook of Molecular Descriptors, *Methods and Principles in Medicinal Chemistry*, vol. 11. Wiley-VCH, Weinheim (2000)
13. Randić, M.: On generalization of Wiener index for cyclic structures. Acta Chim. Slov. **49**, 483–496 (2002)

Chapter 23
Factorial Designs

In practice, frequently the problem arises to manufacture a product with certain characteristics wished by the customer. These characteristics can often be achieved by slight modifications of the recipe or the manufacturing process. In addition, some knowledge is necessary, in which way the individual components of a recipe or the process parameters could affect the product properties dependent on the production method. Usually experiments are necessary to succeed. Monographs on factorial designs are available [1–4].

Experiments, however, are time consuming and money intensive. It is desirable to achieve the goal with a minimum of experiments. In the following the foundations and methods are presented as to how to succeed with a minimum of effort. The method presented here is generally applicable to scientific problems.

Scientists becoming skilled in the method will become aware that experienced experimenters work according to such a method, even unconscientious. However, to get maximum information from the experiments, it is necessary to know a minimum of the underlying theory. The method originates from the agrarian sciences.

23.1 Simplex Method

The name of the simplex method is not deduced from simple, even when the method in its basic applications is simple to use, but the term simplex refers to the simplex in geometry, i.e., a triangle in two dimensions, the tetrahedron in three dimensions, etc. The method is attributed to *Nelder* and *Mead* in 1964 [5].

The simplex method is suitable to find optima fast. We start with an example. A resin is cured by hot injection; 2% of junk is produced. This junk appears too high and the plant manager decides to change the operating conditions. The classical procedure and the procedure using the simplex method is shown in Fig. 23.1.

With the classical method, first at constant curing temperature the curing times (abscissa) are changed and the quality of the product is measured. Then at the curing time with best properties, we change the curing temperature (ordinate) and we are of course finding an optimum. This optimum is not, however, the best optimum that could be found.

J.K. Fink, *Physical Chemistry in Depth*,
DOI 10.1007/978-3-642-01014-9_23, © Springer-Verlag Berlin Heidelberg 2009

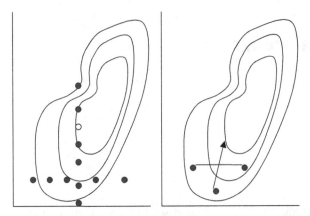

Fig. 23.1 *Left-hand side*: Finding an optimum by systematic variation of the parameters in two steps, first the curing time, x-axis, and then the temperature, y-axis; *right-hand side*: simplex method

With the simplex method, temperature and curing time are simultaneously changed. We start with the three experiments shown in Fig. 23.1, right-hand side.

Then we identify the points with the two best results, connect them by a line and mark the middle of this line. Now we mirror the point with worst product properties through the middle of the line. In this way, we obtain a further point. We do not consider the worst point longer. Instead, we perform an experiment at this point. Thus, we have the data at the new point and the data at the two best points from the previous turn. Now the procedure repeats. Again, with these three points we identify the two best points. We draw a line through them, find the middle, and mirror the worst point through the middle. The procedure explained as graphic method can be dressed easily into the appropriate formulas of analytic geometry. It is also possible to extend the procedure to more than two variables. The mathematical procedure of the generalization is not as easy as the graphical method suggests.

The advantages of the simplex method are that we arrive rapidly to an optimum. However, we have no overview of conditions outside the range under investigation.

The simplex method can be applied in more than two dimensions. There are variants of the simplex method. A big triangle can be placed over a region where the optimum is suspected.

23.1.1 Two-Level Factorial Experiments

In a boat building company, boat hulls are made from glass fiber-reinforced composites. In addition, glass fiber mats are impregnated with resin and placed in a form. The resin must be mixed before with an accelerator and a hardener. The curing begins already at this time. For optimal production on the one hand as long a gel time of the resin as possible is desired, and on the other hand, after gelling demolding should be possible after a short period, in order to have a short cycle time. At the moment 1.5% hardeners and 1.5 accelerators are used. The gel time is 16 min and the demolding time is 3.5 h.

Question 23.1. Can we vary the recipe by varying the amounts of curing agent and accelerator, in order to increase the gel time and shorten the demolding time?

We will call the amount of hardener and he amount of accelerator in general the experimental variables or variables for short. In principle, we can change these variables. The gel time and the demolding time we will address as response parameters.

It is obvious that gel time and demolding time are affected both by the concentration of the curing agent and by the concentration of the accelerator, and we will try to reach the desired goal by the variation of both variables. Response parameters are the gel time and demolding time.

The dependence of the response parameters on the variables, here quantity of hardener and quantity of accelerator is not well known in most cases. Therefore, we must find out these relationships by experiments. The crucial task is thus the adjustment of the amount of the variation of the experimental variables and the documentation of the dependence of the response parameters of the variables.

The variation of the experimental variables must take place in sufficiently small steps. We may select the increments never so largely that we would override, for instance, an extreme value as the optimum in Fig. 23.1. Otherwise, we would never find out this extremum. On the other hand, if we vary the variables in too small steps, then the response that can be measured becomes small and the number of experiments to locate optimum rises.

Usually we have an imagination, how far we could change the individual variables, without coming into areas of too big nonlinearity. Otherwise, we would have to make preliminary tests. The variation of the experimental variables is usually also limited by other constraints.

For instance, in the present case, the mechanical properties could become unacceptable within certain ranges from curing agent and accelerator additions. In the present case, the quantities of curing agent and accelerator are varied in 0.25% steps in the starting composition.

We make now experiments with different quantities of curing agent and accelerator and obtain the following response as listed in Table 23.1.

Table 23.1 Dependence of the gel time and the mold release time on the addition of curing agent and accelerator

Curing agent %	Accelerator %	Gelation time (min)	Demolding time (min)
1.25	1.25	35	330
1.25	1.75	15	30
1.75	1.25	10	300
1.75	1.75	3	210

23.1.2 Setup of the Model

The response is functionally dependent on the experimental variables. The task is to establish this functions. We set up this function as a series:

$$t_g = a + b[CA] + c[AC] + d[CA]^2 + e[CA][AC] + f[AC]^2 + \ldots$$

where

t_g: Gel time
a, b, c, d, e, f Parameters
$[CA]$: Concentration of curing agent
$[AC]$: Concentration of accelerator

A similar series we set up for the demolding time. If the series contains sufficient terms, every shape of the response can be described. Not only a power series can be set up but also logarithmic terms or exponential terms are allowed. In the present case, we have cancelled the series after the fourth term. We set up the power series in a general manner:

$$Y = \beta_0 + \beta_1 x_1 + \beta_2 x_2 + \beta_{1,2} x_1 x_2 \tag{23.1}$$

where

Y   Estimated value or estimation of the response
$\beta_{i,j}$ Coefficients
x   Experimental variable

In Eq. (23.1) there are four unknown coefficients β. We need at least four independent relations $y_i, x_{1,i}, x_{2,i}; i = 1, \ldots 4$, to solve the equation with respect to the coefficients β. This is essentially a generalization of the problem of linear regression

$$y = kx + d$$

that fits a straight line. A large coefficient β indicates a strong dependence of the response on the experimental variable.

In general case more points than needed for the solution are available and so the demand arises that the coefficients β are selected in such a way that for all experimental points i the response y (y_i) deviates from the estimated value $Y(x_{1,i}, x_{2,i})$ only minimally.

However, the data may deviate in positive and in negative directions from the estimated value. To achieve this demand of minimum absolute deviation, the sum of the squares of the deviations from estimated value and measured value of the response should become a minimum

$$S = \sum \left(y_i - Y(x_{1,i}, x_{2,i}) \right) \rightarrow \text{Min.}$$

This is equivalent with the demand that the derivatives of S with respect to the coefficients $\beta_{i,j}$ will be equal to zero,

$$\frac{\partial S}{\partial \beta_0} = 0; \quad \frac{\partial S}{\partial \beta_1} = 0; \quad \frac{\partial S}{\partial \beta_2} = 0; \quad \frac{\partial S}{\partial \beta_{1,2}} = 0. \tag{23.2}$$

From Eq. (23.2) we obtain after substitution for the estimated value and expansion of the sums into the individual terms the set of equations

$$\frac{\partial S}{\partial \beta_0} = 2 \sum \left(y_i - Y(x_{1,i}, x_{2,i}) \right) \qquad = 0$$

$$\frac{\partial S}{\partial \beta_1} = 2 \sum \left(y_i - Y(x_{1,i}, x_{2,i}) \right) x_{1,i} \qquad = 0$$

$$\frac{\partial S}{\partial \beta_2} = 2 \sum \left(y_i - Y(x_{1,i}, x_{2,i}) \right) x_{2,i} \qquad = 0$$

$$\frac{\partial S}{\partial \beta_{1,2}} = 2 \sum \left(y_i - Y(x_{1,i}, x_{2,i}) \right) x_{1,i} x_{2,i} \qquad = 0$$

The system becomes clearer, if we expand the sum term by term

$$\sum y_i - \beta_0 - \beta_1 x_{1,i} - \beta_2 x_{2,i} - \beta_{1,2} x_{1,i} x_{2,i} = 0$$

$$\sum y_i x_{1,i} - \beta_0 x_{1,i} - \beta_1 x_{1,i} x_{1,i} - \beta_2 x_{2,i} x_{1,i} - \beta_{1,2} x_{1,i} x_{2,i} x_{1,i} = 0$$

$$\sum y_i x_{2,i} - \beta_0 x_{2,i} - \beta_1 x_{1,i} x_{2,i} - \beta_2 x_{2,i} x_{2,i} - \beta_{1,2} x_{1,i} x_{2,i} x_{2,i} = 0 \qquad (23.3)$$

$$\sum y_i x_{1,i} x_{2,i} - \beta_0 x_{1,i} x_{2,i} - \beta_1 x_{1,i} x_{1,i} x_{2,i}$$
$$- \beta_2 x_{2,i} x_{1,i} x_{2,i} - \beta_{1,2} x_{1,i} x_{2,i} x_{1,i} x_{2,i} = 0$$

Each of these equations in Eq. (23.3) must be evaluated to zero.

23.1.3 Encoded Experimental Variables

If we select and encode the numerical values of the experimental variables skillfully, then the solution of the equations in Eq. (23.3) can be substantially simplified.

By the transformation $z = 4x - 6$ we obtain from Table 23.1 the result as given in Table 23.2. We observe in Table 23.2 immediately that the sums are simplified here substantially, because the experiment was so skillfully planned that half of the transformed variables is $+1$ and the other half of the transformed variables is -1. Only the terms remain with the coefficients, which appear squared. Also the coefficients β are of course subject to a transformation

Table 23.2 Dependence of the gel time and the mold release time on the addition of curing agent and accelerator, use of coded variables

Curing agent %	Accelerator %	Gelation time (min)	Demolding time (min)
−1	−1	35	330
−1	+1	15	30
+1	−1	10	300
+1	+1	3	210

$$\sum y_i - \beta_0 = 0$$

$$\sum y_i x_{1,i} - \beta_1 = 0$$

$$\sum y_i x_{2,i} - \beta_2 = 0$$

$$\sum y_i x_{1,i} x_{2,i} - \beta_{1,2} = 0$$

We note that as a result of n-times summing up of the coefficient the expression $n\beta_1$ arises. We recognize now immediately the reason, why Table 23.1 was selected in this form. Even when now computers are available to solve such type of equations; nevertheless, it is advisable to deal with coded variables, because sometimes rounding errors can be minimized.

23.1.3.1 Graphical Representation of the Model

With the data made of table the variables that influence the gel time and the demolding time can be computed. The results are shown in Fig. 23.2 as contour diagrams.

The starting point (1.5% curing agent and 1.5% accelerator) lies in the center of the diagrams. We see immediately that we arrive at more favorable ranges, if we move downward and slightly to the right in the diagram. We must thus, in order to achieve longer gel times and shorter demolding times, reduce the quantity of accelerator and increase the quantity of curing agent.

Similar to the considerations, when modeling we can develop factorial designs. Thus, we can determine even in systems that are more complicated the response. A typical setup of a complete factorial design is shown in Table 23.3. We explain now the conventions used in Table 23.3. In the head, we write always the meaning of the factors used, the absolute amounts added, and the mode of coding. Thus, in the first column we write '+' for the higher level and '−' for the lower level. Further, there is no need to use continuous variables as exemplified with the type of resin. In the table itself the order in which the experiments should be performed is given. This is predetermined. The experiments are randomized before the actual experiments are performed. For example, there is no sufficient time to finish all the experiments within 1 day. Suppose, on the next day it is raining, and the moisture is high. If all the experiments say with the factor C at the lower level are performed on the first

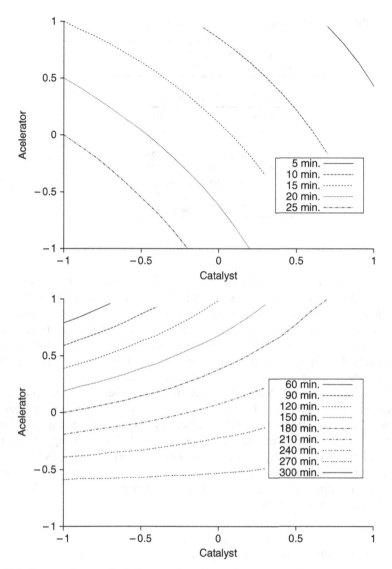

Fig. 23.2 Contour diagram for influence of catalyst and accelerator. *Top*: curing time; *bottom*: demolding time

day, and all the experiments with a factor C at a high level are performed on the second day, then a systematic influence may be overlaid to the scheme.

In the second column how the experimental variables used are set in the experiment are given in a short form, i.e., '−' for the lower level, and '+' for the upper level. The actual setting can be easily looked up for the upper part of the table.

In the next column the measured values of the response are given. The coefficients are designated with the lowercase letters of the response variables themselves.

Table 23.3 Setup of a complete factorial design, demolding time, composite

Factor	A: Resin	B: Catalyst	C: Temperature
+	Resin comp. X	1.2%	25 °C
−	Resin comp. Y	0.5%	18 °C

No.	A	B	C	Time	Var	β
3	−	−	−	47.5	1	+52.913
7	+	−	−	39.4	a	−11.175
2	−	+	−	48.2	b	+0.025
5	+	+	−	37.8	ab	+0.325
8	−	−	+	69.8	c	+19.375
1	+	−	+	54.9	ac	−1.925
4	−	+	+	68.5	bc	+0.475
6	+	+	+	57.2	abc	+1.475

1 means the average; a, b, c, ab, ac, bc, abc are the coefficients that describe the response. This means the effect, if we change the experimental variable from -1 to $+1$. a, b, c are coefficients of first order, ab, ac, bc are coefficients of interactions of second order, abc is a coefficient of interactions of third order. In practice, interactions of third order are rare. We can calculate the coefficients, say of variable a as we multiply the response the numbers in the column A sum up and divide by half of the number of experiments. This is the usual convention.

The coefficient ab is calculated by multiplying the response by the product of the numbers of column A and column B. In this way, all the coefficients can be obtained. The physical meaning of the coefficient is the estimate change in response if we switch the experimental variable from -1 to $+1$. So the change covers two units, therefore we divide by the half of the number of experiments. For modeling the situation is different. Since in the formula for estimated response we change the experimental variable of by 2, but as such, the coefficient is half of this. Therefore, the estimated response function is

$$Z = (1) + (a/2)A + (b/2)B + (c/2)C +$$
$$(ab/2)AB + (ac/2)AC + (bc/2)BC + (abc/2)ABC. \quad (23.4)$$

23.1.4 Evaluation of a Factorial Experiment

In order to evaluate the results of a factorial experiment, we inspect simply the coefficients. If a coefficient is large, then the corresponding experimental variable has a big influence on the response. If the coefficient is small, then it has only a small influence on the response.

23.1.5 Testing the Significance

Usually it is easy to say if a certain experimental variable has an influence or not. However, based on the inaccuracy of the experiment, the coefficients can have a value, different from zero, even when there is not really an effect. The problem becomes important, if the effects are in the order of the experimental noise. The following consideration can help to decide, if an effect is real or if it is simply an artifact.

Imagine, there are no real effects. In this case, the response varies only due to statistical fluctuations, caused by experimental inaccuracies. In this case, the coefficients are statistically distributed. So we check, if the coefficients are obeying a Gauß distribution. We exemplify this in a subsequent data set. In a complete factorial design with n variables, we have $2^n - 1$ coefficients. The minus one comes, because one coefficient is the average. If the coefficients obey a Gauß distribution, then in a probability plot they should show a straight line.

So we arrange the parameters in increasing size. We explain why the coefficients are plotted as a cumulative distribution. If there are seven coefficients, we divide the abscissa in seven equidistant intervals. In the range from 0 to 100% we have 100/7 for each interval. We place the coefficient in the middle of the interval. Thus, we have for the first parameter $1 * 100/(2 * 7)$, for the second parameter we have $3 * 100/(2 * 7)$, etc. The significance is shown in Table 23.4. The normal probability plot is shown in Fig. 23.3

Table 23.4 Test of significance

Variable	Coefficient	Position in plot
A	−11.175	7.1
AC	−1.925	21.4
B	+0.025	35.7
AB	+0.325	50.0
BC	+0.475	64.3
ABC	+1.475	78.6
C	+19.375	92.9

23.1.6 Incomplete Designs

A complete two-step factorial design with n experimental variables requires 2^n of experiments. With a large number of variables, the experiment becomes large. We can circumvent this drawback by preparing incomplete attempt factorial designs. In the extreme case, we lose the information from interaction effects.

Table 23.5 shows such a design with three experimental variables with a total of four experiments. The experimental variable C was switched in such a way that it confounds with the interaction of AB. In the same way, A is confounded with BC and B is confounded with AC. This is the price of information, which we

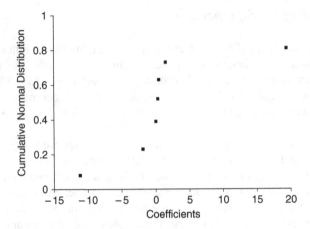

Fig. 23.3 Normal probability plot of the coefficients

Table 23.5 Incomplete factorial design with three variables and four experiments

A	B	C	A=BC	B=AC
−	−	+	−	−
+	−	−	+	−
−	+	−	−	+
+	+	+	+	+

must pay for the comfort to drive with fewer experiments. The evaluation takes place according to the same pattern as with the complete experimental design. In the literature, various factorial designs are given for different purposes. For example, we can include in a complete factorial design for three experimental variables still another experimental variable. We switch the fourth variable according to the pattern for ABC. In this way, the variable D is confounded only with the interaction ABC. As mentioned, strong interactions of three variables are rare in practice.

We show the overlapping pattern of the influence parameters that form the additive of a variable D in Table 23.6. In Table 23.7 still another experiential design with seven experiential variables with eight experiments is shown. We obtain all the seven influence parameters of first order, the so-called main effects, however, confounded with influence parameters of higher order.

23.1.7 Method of Steepest Descent

Usually after one set of experiments, the target is not reached. Therefore, we have to perform other experiments in a different rage of the experimental variables. If we have only two variables, we can see the preferred new range immediately from the contour diagram. If there are more than two variables, then the situation is

Table 23.6 Incomplete factorial design with four variables and eight experiments (2^{4-1})

A	B	C	D=ABC
−	−	−	−
+	−	−	+
−	+	−	+
+	+	−	−
−	−	+	+
+	−	+	−
−	+	+	−
+	+	+	+

Table 23.7 Incomplete factorial design with seven variables and eight experiments

A	B	C	D=AB	E=AC	F=BC	G=ABC	Shrinkage %
−	−	−	+	+	+	−	14.0
+	−	−	−	−	+	+	16.8
−	+	−	−	+	−	+	15.0
+	+	−	+	−	−	−	15.4
−	−	+	+	−	−	+	27.6
+	−	+	−	+	−	−	24.0
−	+	+	−	−	+	−	27.4
+	+	+	+	+	+	+	22.6

A: Temperature; B: Moisture content; C: Injection pressure; D: Nozzle diameter; E: Post pressure; F: Cycle time; G: Material

complicated because of the limited human imaginative power of more than three dimensions in space. In these cases, we perform a formal treatment.

We can imagine the response space in terms of the response parameters and the experimental variables as a scalar field function. The experiential variables themselves form a vector and the components of the vector are the experiential variables.

According to the rules of vector analysis, the steepest descent is the gradient of the scalar function. From factorial experiments we can obtain the response as a function of the scalar field

$$Y = \quad (1) + (a/2)\,A + (b/2)\,B + (c/2)\,C + (ab/2)\,AB + \\ + (ac/2)\,AC + (bc/2)\,BC + \dots .$$

Y Estimated response
$a/2, \dots$ Coded experimental variables

Now the gradient is the partial derivative of the response function with respect to the individual components of the vector. In the center of the interval the coded experimental variables are equal to zero. The gradient becomes then, neglecting terms of higher order,

$$
\begin{pmatrix}
\dfrac{\partial Z}{\partial A} = \dfrac{a}{2} \\[2mm]
\dfrac{\partial Z}{\partial B} = \dfrac{b}{2} \\[2mm]
\dfrac{\partial Z}{\partial C} = \dfrac{c}{2}
\end{pmatrix}.
$$

It depends now, whether the numerical value of the response function should become a minimum or a maximum. The numerical values of the experimental variable should now be changed in the direction of the gradient or against the direction of the gradient. We start from the center of the old experimental design with $A = 0$, $B = 0$, ... and chose as a new starting point $A_{new} = ka$, $B_{new} = kb$, ... Here k is a constant that should not be too large, according to a possible nonlinearity of the response function. We can estimate the nonlinearity from the interactions of higher order, i.e., ab, ...

23.1.8 Generating Functions

Here we develop a factorial design in terms of generating functions. We inspect a design in three variables, A, B, and C. First, notice that

$$(1 + A)(1 + B)(1 + C) = 1 + A + B + C + AB + AC + BC + ABC. \quad (23.5)$$

In fact, the variables representing the levels are sometimes vectors in the mathematical sense, when we are making statements valid to a whole column of a factorial design, but we will not denote them as **A**, etc. However, sometimes we refer to the very individual components of a column, belonging to a singe row. We insert now a dummy variable x. Then Eq. (23.5) changes into

$$
\begin{aligned}
(1 + Ax)(1 + Bx)(1 + Cx) = \\
1+ \\
x(A + B + C)+ \\
x^2(AB + AC + BC)+ \\
x^3(ABC)
\end{aligned}
\quad (23.6)
$$

Inspection of Eq. (23.6) reveals that the factors of first order are gathered at x, the factors of second order are gathered at x^2, etc. Next, we build another generating function from Eq. (23.5), such as

$$(1 + Ax)(1 + Bx^2)(1 + Cx^4) =$$
$$1 + Ax + Bx^2 + ABx^3 + Cx^4 + ACx^5 + BCx^6 + ABCx^7. \quad (23.7)$$

Now all the factors are separated at a single power of x. If we take the powers of x as a ranking order, then the natural position of A is 1, B is 2, AB is 3, C is 4, etc.

We explain in words how the generating functions seek to make the design. First, multiply the first and the second brackets in Eq. (23.7) to create a two-factor design of A, B, together with all the interactions AB and average (1). Then multiply by the next bracket. This multiplication says

1. keep what you have created before by the multiplication by 1 *plus* add to the new design, and
2. all what you have created before by multiplying it by the new factor C.

We can get the two levels of the individual factors of all the orders as coefficients at the powers of x when inserting for A, B, C -1 and $+1$, respectively. For example, for $A = -1$, $B = 1$, $C = 1$, Eq. (23.7) expands to

$$(1 - x)(1 + x^2)(1 + x^4) = 1 - x + x^2 - x^3 + x^4 - x^5 + x^6 - x^7. \quad (23.8)$$

Further, observe that Eq. (23.7) is closely related to the response function, given in Eq. (23.4). We have simply to replace the dummy variables $1, x, x^2, \ldots$ in Eq. (23.7) by the average (1), and the response coefficients $a/2, b/2 \ldots$, to get the response function. From this view, it becomes also more clear what the meaning of the 1 in the generating function is.

23.1.8.1 Incomplete Designs

In incomplete designs, certain factors are confounded. We elaborate the example in three variables, when C is confounded with AB, i.e., $C = AB$. To get the factors in this case we have simply to replace C by AB in Eq. (23.5) by (23.7). For example, Eq. (23.7) turns into

$$(1 + Ax)(1 + Bx^2)(1 + ABx^4) =$$
$$1 + Ax + Bx^2 + ABx^3 + ABx^4 + AABx^5 + BABx^6 + ABABx^7. \quad (23.9)$$

To find the factor levels, we can proceed like in Eq. (23.8). Now we want to find the confounded factors. Note that the multiplication of any row elements of the same column equals to 1, i.e., $AA = 1$ and $BB = 1$. Therefore, we will eliminate A^2, B^2 in Eq. (23.9) to arrive at

$$(1 + Ax)(1 + Bx^2)(1 + ABx^4) = 1 + Ax + Bx^2 + ABx^3 + ABx^4 + Bx^5 + Ax^6 + x^7.$$

Finally we have merely to gather the terms with equal variables as follows:

$$(1 + Ax)(1 + Bx^2)(1 + ABx^4) =$$
$$1(1 + x^7) + A(x + x^6) + B(x^2 + x^5) + AB(x^3 + x^4). \quad (23.10)$$

In Eq. (23.10) we can trace back the powers of x from Eq. (23.7) or (23.9). This means, to which variables these powers were originally associated. We should not forget that we have deliberately substituted $C = AB$. We substitute back this now in Eq. (23.9), or we can use Eq. (23.7). Thus, from the first term of Eq. (23.10) it follows that the average is confounded with ABC. The second term says that the variable A is confounded with BC. From the third term we learn that B is confounded with AC. Finally, the last term reflects what we have introduced, i.e., C is confounded with AB. Consider the symmetry of the powers in brackets, which is still more pronounced if we set in the first bracket on the right-hand side $1 = x^0$.

23.1.8.2 Evaluation of Generating Functions

All these relations introduced by means of generating functions are tedious to be calculated as such with pencil and paper, in particular in more complicated cases. However, a symbolic calculation program may serve as a practicable tool. In Table 23.8 a complete unabridged protocol to obtain the confounding pattern of a 2^{4-1} factorial design is shown.

Rearranging the last output in Table 23.8 and sorting gives the confounding patterns, as shown in Table 23.9. The variables of first order A, B, C, and D are confounded only with combinations of third order, BCD, ACD, ABD, and ABC.

In the last line in Table 23.8 we observe, for example, that D at x^8 is only confounded with ABC at x^7. Again, we observe that the powers of the confounded terms sum up to $2^n - 1$, with $n = 4$. However, this is not a general rule. If we would chose in this example a confounding of $D = AB$, then the sum of the powers is 12 or 19, depending on the terms.

Finally, in Table 23.10 we show the confounding patterns in a 2^{4-2} experiment. The data have been obtained just in the way as the data given in Table 23.8. Inspection of Table 23.8 shows that such a design is not reasonable, because D is confounded with the average.

23.1.9 Relationship to Binary Numbers

In this section we will denote now the factors not with -1 and $+1$ or $-$ and $+$, but instead with 0 and 1. We will address this as binary notation. A complete 2^3 two-level design transforms now from '$-+$' into '01' like

Table 23.8 Confounding pattern of a 2^{4-1} factorial design calculated by DERIVE®

```
(1+a*x)*(1+b*x^2)*(1+c*x^4)*(1+d*x^8)

a*b*c*d*x^15+a*b*c*x^7+a*b*d*x^11+a*b*x^3+a*c*d*x^13~
+a*c*x^5+a*d*x^9+a*x+b*c*d~
*x^14+b*c*x^6+b*d*x^10+b*x^2+c*d*x^12+c*x^4+d*x^8+1

d:=a*b*c

a^2*b^2*c^2*x^15+a^2*b^2*c*x^11+a^2*b*c^2*x^13~
+a^2*b*c*x^9+a*b^2*c^2*x^14+a*b^~
2*c*x^10+a*b*c^2*x^12+a*b*c*x^8+a*b*c*x^7~
+a*b*x^3+a*c*x^5+a*x+b*c*x^6+b*x^2+c*~
x^4+1

1*b^2*c^2*x^15+1*b^2*c*x^11+1*b*c^2*x^13~
+1*b*c*x^9+a*b^2*c^2*x^14+a*b^2*c*x^10~
+a*b*c^2*x^12+a*b*c*x^8+a*b*c*x^7+a*b*x^3~
+a*c*x^5+a*x+b*c*x^6+b*x^2+c*x^4+1

1*1*c^2*x^15+1*1*c*x^11+1*b*c^2*x^13~
+1*b*c*x^9+a*1*c^2*x^14+a*1*c*x^10+a*b*c^2~
*x^12+a*b*c*x^8+a*b*c*x^7+a*b*x^3+a*c*x^5~
+a*x+b*c*x^6+b*x^2+c*x^4+1

1*1*1*x^15+1*1*c*x^11+1*b*1*x^13~
+1*b*c*x^9+a*1*1*x^14+a*1*c*x^10+a*b*1*x^12+a*~
b*c*x^8+a*b*c*x^7+a*b*x^3+a*c*x^5+a*x+b*c*x^6~
+b*x^2+c*x^4+1

a*b*c*x^8+a*b*c*x^7+a*b*x^12+a*b*x^3+a*c*x^10~
+a*c*x^5+a*x^14+a*x+b*c*x^9+b*c*x~
^6+b*x^13+b*x^2+c*x^11+c*x^4+x^15+1
```

Table 23.9 Confounding patterns

$1*(x^0+x^{15})+$	$\rightarrow$	1	$\leftrightarrow$	$ABCD$
$a*(x^1+x^{14})+$	$\rightarrow$	A	$\leftrightarrow$	BCD
$b*(x^2+x^{13})+$	$\rightarrow$	B	$\leftrightarrow$	ACD
$a*b*(x^3+x^{12})+$	$\rightarrow$	AB	$\leftrightarrow$	CD
$c*(x^4+x^{11})+$	$\rightarrow$	C	$\leftrightarrow$	ABD
$a*c*(x^5+x^{10})+$	$\rightarrow$	AC	$\leftrightarrow$	BD
$b*c*(x^6+x^9)+$	$\rightarrow$	BC	$\leftrightarrow$	AD
$a*b*c*(x^7+x^8)$	$\rightarrow$	ABC	$\leftrightarrow$	D

Table 23.10 Confounding patterns in a 2^{4-2} factorial design using the variables A, B, C, D

A	$\leftrightarrow$	AD	$\leftrightarrow$	BC	$\leftrightarrow$	BCD
B	$\leftrightarrow$	AC	$\leftrightarrow$	BD	$\leftrightarrow$	ACD
C	$\leftrightarrow$	AB	$\leftrightarrow$	CD	$\leftrightarrow$	ABD
D	$\leftrightarrow$	1	$\leftrightarrow$	ABC	$\leftrightarrow$	$ABCD$

$$D = ABC: C = AB$$

$$
\begin{array}{rcl}
c\ b\ a & \to & cba \ \ \text{Hexadecimal} \\
- - - & \to & 000 \qquad 0 \\
- - + & \to & 001 \qquad 1 \\
- + - & \to & 010 \qquad 2 \\
- + + & \to & 011 \qquad 3 \\
+ - - & \to & 100 \qquad 4 \\
+ - + & \to & 101 \qquad 5 \\
+ + - & \to & 110 \qquad 6 \\
+ + + & \to & 111 \qquad 7
\end{array}
\qquad (23.11)
$$

In contrast to the usual order, we have written the factors in the reverse order. We easily realize that in the binary notation the rows are now the running numbers $0, 1, \ldots$.

Equation (23.11) suggests a specific order of the response coefficients, which is now different from the order given in Table 23.3. However, the order is the same as produced by the generating function of Eq. (23.7).

$$
\begin{array}{ll}
000 & average \\
001 & a \\
010 & b \\
011 & ab \\
100 & c \\
101 & ac \\
110 & bc \\
111 & abc
\end{array}
\qquad (23.12)
$$

In Eq. (23.12), if we add the respective response factors, then the binary number refers immediately to the order of the coefficients. In other words, the binary number refers to the coefficients with which we have to multiply the response factor to get the response function.

For example, the second binary number is 001 and is associated with the response factor $(a/2)$ and with the coefficient A in Eq. (23.4), whereas 101 is associated with the response factor $(ac/2)$ and the coefficient AC in Eq. (23.4).

The multiplication of $-1 \times -1 = +1$, whereas $0 \times +1 = 0$. However, we can redefine the multiplication rule. The multiplication rule

$$
NOT(x\ XOR\ y) = \star
$$

effects $0 \star 0 = 1$, and the other cases as well, i.e., $1 \star 0 = 0, 0 \star 1 = 0, 1 \star 1 = 1$.

23.1.9.1 Matrix Notation

We recall the formula of matrix multiplication

$$
c_{i,k} = \sum_j a_{i,j} b_{j,k}.
$$

From this formula we can identify the process of obtaining the response coefficients as a matrix product where the vector $a_{1,j}$ contains the experimental results and the matrix $b_{j,k}$ is the array of the factors in the form $[-1 + 1]$. The matrix must contain also the factors of higher orders. These are obtained by the multiplication of the appropriate columns. If we include also the average, then this matrix is a square matrix. We show such a complete matrix for three variables in Table 23.11. Contrary to the usual form, we start with the highest order terms. Also we have added a column with entirely $+1$ at the end. Inspection of Table 23.11 shows that the columns have the same pattern as the rows. In other words, the transposed matrix is equal to the original matrix. This property could be important for the implementation of a program to calculate a factorial design in detail.

Table 23.11   Factor matrix for three variables

Bit		7	6	5	4	3	2	1	0	
		ABC	BC	AC	C	AB	B	A	(1)	HEX
		↓	↓	↓	↓	↓	↓	↓	↓	↓
ABC	→	-1	$+1$	$+1$	-1	$+1$	-1	-1	$+1$	69
BC	→	$+1$	$+1$	-1	-1	-1	-1	$+1$	$+1$	C3
AC	→	$+1$	-1	$+1$	-1	-1	$+1$	-1	$+1$	A5
C	→	-1	-1	-1	-1	$+1$	$+1$	$+1$	$+1$	0F
AB	→	$+1$	-1	-1	$+1$	$+1$	-1	-1	$+1$	99
B	→	-1	-1	$+1$	$+1$	-1	-1	$+1$	$+1$	33
A	→	-1	$+1$	-1	$+1$	-1	$+1$	-1	$+1$	55
(1)	→	$+1$	$+1$	$+1$	$+1$	$+1$	$+1$	$+1$	$+1$	FF

In the last column of Table 23.11 the hexadecimal numbers of the rows are given, if the value $-1 \to 0$.

The column vectors in the matrix are mutually orthogonal. Therefore, the scalar products of different columns are zero, $\sum a_j b_j = 0$. The product of the matrix with its transposed matrix gives a diagonal matrix with $a_{jj} = n$, where n is the number of rows or columns, respectively. All other elements are zero.

23.2   Monte Carlo Methods

In some cases, the use of Monte Carlo methods is more effective than the usage of factorial experiments. However, we lose the information that we obtain from the contour diagram. There may also be the danger to arrive at a side maximum. We elucidate the principle now.

We establish in the usual way the experimental variables and the ranges of these. We encode the range from 0 to 1, however. Now we arrange experiments, say five experiments, as we associate with the coded variables random variables in the range of $0, \ldots, 1$. Now we perform these experiments and measure the response. We chose the experiment with the best response and make the range of the experimental variables smaller. Next, we associate again random values with the experimental variables and perform the experiments accordingly. We are looking for the experiment with the best response, decrease the range of the experimental variables. In this way, we repeat the procedure several times.

23.3 General Rules for Setting Up a Design

The numerical values of the experimental variables must be chosen so that they are physically reasonable. For instance, the amount of a compound to be added should be a nonnegative number. Of course, the numerical values must cover a range that is of interest for the particular problem. Within the range of interest, the variables should be interpolated but not extrapolated. The step should not be too big. Otherwise, effects in nonlinear regions could be overlooked.

If there are known relations among the experimental variables, it is possible to reduce the number of the experimental variables. For instance in a mixture of two main components, the amount of one component dictates the amount of the other component. Therefore, there is no need and no reason to include both amounts as experimental variables.

The setup of the experimental design plan is dependent on what is the target. In a system with completely unknown properties, it is not recommendable to start with a big experimental design plan that may include interactions. Rather a small design is reasonable, however, covering the main effects. If it turns out that there are experimental variables without response, it is possible to remove them in the second turn.

The simplification of the experimental variables by coding is self-explanatory. Coding effects essentially a scale transformation. Also the numerical value of the response parameter will be changed by the transformation. The response parameter reflects always a switching of the experimental variable from -1 to $+1$. This is the convention in factorial design plans, and we will not change this convention deliberately. If the response is expressed as a power series, only the half of this numerical value must be inserted.

Often there are constraints in the response function. These are often not expressed explicitly, because they are self-understanding. However, they may not be self-understanding for everybody, in particular for a mathematician who could set up the experimental design, thus supporting a chemist. On the other hand, the constraints may not be foreseeable to the experimenter himself.

Example 23.1. Suppose that one material is substituted by another cheaper material. The final properties of the article have been substantially improved. After 100 process cycles the engine that manufactures the article is off service due to corrosion. □

Often a set of different responses must be optimized together. This can be achieved, if the responses are collected in a single response function, e.g., the sum of the squares of the individual responses or even more complicated response functions.

23.3.1 Interpretation of the Results

The interpretation of the results is a substantial part of the evaluation of the results of experiments. Even the best experimental design method cannot be satisfactory, if the experimenter does not make profit from the results. Sometimes a critical inspection shows a success pointing to a topic originally not intended. You should be flexible in your targets. If you search for gold and you find diamonds, consider this as success.

It is important to think about whether the results are reasonable from the physical view. The issue is complex and therefore there are various sources of error. It starts with the fact that the newcomer is not experienced in the topic. Experimental designs shown in the textbooks are *just* not applicable to the own problem. Therefore, the newcomer should start with simple examples. Needless to say that the experiments must be performed with care.

23.3.2 The Storage Example

A certain additive for plastics has only a limited shelf-life, because it agglutinates and problems with dosage arise. Therefore, it must be ordered in regular intervals, because the product, for which it is used, is sold only in relatively small quantities to a special customer.

The contractor sells the additive in larger quantities considerably more cheaply the compounding company plans to build an air-conditioned stockroom. During visiting the stock, the foreman tells casually to the engineer that he had before-mixed this additive with another additive 2 years back erroneously, and has posed this mixture aside. When a bottleneck with this additive emerged, he remembered to this barrel again. He found that when the mixture was premixed, it could be used also after substantially longer time than expected. Thereupon the commercial leader decides to revisit the alteration of the stock again. One should do additional experiments whether premixing actually increases the stability in storage.

The technical manager thinks that a small-scale experiment should be performed on the stability in storage of the additive in the climatic chamber. In this way, the factorial design plan shown in Table 23.12 arises. The foreman stopped the experiment off after 70 days, because the additive never was longer than 70 days on stock. The commercial director asks after 3 months whether an increase of the shelf-life was achieved, because he just got an unusual inexpensive offer of the contractor when buying still larger quantities of the additive. Try to evaluate the data from Table 23.12. What was wrong at the conception of this experiment?

23.4   Time Series

We explain the method of time series in a simple story. A chemical plant is running continuously. To check the performance some data are collected every 2 h.

Table 23.12 Factorial experiment to investigate the storage stability

Premixing	Climatic chamber	Storage time [a]
–	–	> 45
+	–	> 70
–	+	> 70
+	+	> 70

[a] Agglutinates after . . . days

For instance, a sample is withdrawn every 2 h and the purity is checked by gas chromatography. The plant is running already for years satisfactorily.

However, in spring of the current year, one of the end users of the product manufactured in the plant is reclaiming that the quality is no longer satisfactorily. The plant engineer is recalling the data that have been carefully collected, and a plot is shown in Fig. 23.4. Of course, the data are scattering, but the plant engineer cannot find any abnormal behavior that could have occurred before the reclaim arrived and after the reclaim arrived. He asks also the foreman, if they have changed something in the last 3 months in plant operation and the reply is *definitely nothing*.

The plant engineer is trying out now some tricks he knows in plant operation, but the end user tells that the product is becoming still worse. Now since the quality of the product is no longer acceptable, an experienced chemist from headquarters is sent into the country to see what he can do. This chemist has a little knowledge about statistics, and with him is a pocket calculator. He analyzes the data that he received from the plant engineer in a different way and the plot he obtains is like shown in

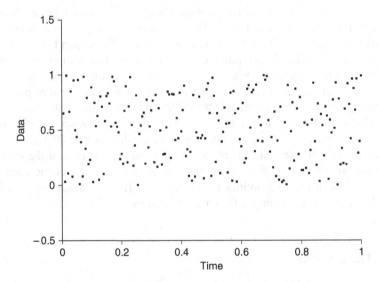

Fig. 23.4 Process parameter measured in the course of time of a chemical plant

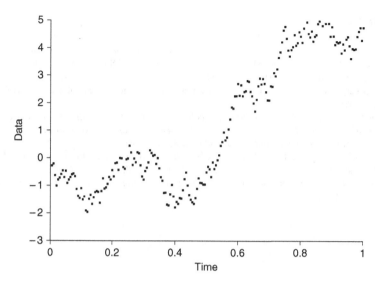

Fig. 23.5 Partial sums of process parameter measured in the course of time of a chemical plant

Fig. 23.5. He states that something happened in plant operation where the minimum in Fig. 23.5 is seen, and he can locate the date and time within 2 or 3 days. After analyzing the situation, it turned out that at this time in question the foreman has used a different zeolite quality for the water feed.

We explain now what the experienced chemist has done with the data. The data are scattering around an average. As long as the plant operation does not change, the average remains constant. If the plant operation has changed, this average may be shifted. In the direct plot of the data, this is not readily seen, since the shift of the average is small in comparison to the scattering of the data. Therefore, the change of the operation parameters remains undiscovered in a plot of the data against time. However, the chemist has plotted a partial sum of the averaged data. If the average is the same throughout all the period, then the data would still scatter around the average. However, if the average increases at a certain time, then the total average subtracted is too much, before the shift occurs. Therefore, the partial sum, i.e., the integral data, would move down. After the event what he is subtracting is too less, and the partial sum is moving up again. In this way, it is easier to see if an event occurred that is changing the average.

Finally, we explain how we have obtained the data in Figs. 23.4 and 23.5. The data are simply random numbers in the range of 0–1. At a time of 0.5 units, to each random number a constant value of 0.05 has been added. Therefore, a significant portion of only 5% of the total scattering has been introduced. In Fig. 23.4, a change is literally invisible, but in the integral plot of Fig. 23.5 the curve is moving up.

References

1. Box, G.E.P., Hunter, J.S., Hunter, W.G.: Statistics for Experimenters, An Introduction to Design, Data Analysis, and Model Building. John Wiley and Sons, New York (1978)
2. Box, G.E.P., Hunter, J.S., Hunter, W.G.: Statistics for Experimenters: Design, Innovation, and Discovery, 2nd edn. Wiley Series in Probability and Statistics. Wiley-Interscience, Hoboken, NJ (2005)
3. Kowalski, B.R. (ed.): Chemometrics, Mathematics and Statistics in Chemistry, *Nato ASI (Advanced Sciences Institutes) Series C: Mathematical and Physical Sciences*, vol. 138. D. Reidel Publishing Company, Dordrecht, Holland (1984)
4. Winer, B.J., Brown, D.R., Michels, K.M.: Statistical principles in experimental design, *McGraw-Hill Series in Psychology*, 2nd edn. McGraw-Hill, New York (1991)
5. Nelder, J.A., Mead, R.: A simplex method for function minimization. Comput. J. **7**, 308–313 (1964)

Author Index

J.K. Fink, *Physical Chemistry in Depth*,
DOI 10.1007/978-3-642-01014-9_BM2, © Springer-Verlag Berlin Heidelberg 2009

Subject Index